The Initiation, Propagation, and Arrest of Joints and Other Fractures

Geological Society Special Publications

Society Book Editors

Special Publication reviewing procedures

The Society makes every effort to ensure that the scientific and production quality of its books matches that of its journals. Since 1997, all book proposals have been refereed by specialist reviewers as well as by the Society's Books Editorial Committee. If the referees identify weaknesses in the proposal, these must be addressed before the proposal is accepted.

Once the book is accepted, the Society has a team of Book Editors (listed above) who ensure that the volume editors follow strict guidelines on refereeing and quality control. We insist that individual papers can only be accepted after satisfactory review by two independent referees. The questions on the review forms are similar to those for *Journal of the Geological Society*. The referees' forms and comments must be available to the Society's Book Editors on request.

Although many of the books result from meetings, the editors are expected to commission papers that were not presented at the meeting to ensure that the book provides a balanced coverage of the subject. Being accepted for presentation at the meeting does not guarantee inclusion in the book.

Geological Society Special Publications are included in the ISI Index of Scientific Book Contents, but they do not have an impact factor, the latter being applicable only to journals.

More information about submitting a proposal and producing a Special Publication can be found on the Society's web site: www.geolsoc.org.uk.

It is recommended that reference to all or part of this book should be made in one of the following ways:

COSGROVE, J. W. & ENGELDER, T. (eds) 2004. *The Initiation, Propagation, and Arrest of Joints and Other Fractures*. Geological Society, London, Special Publications, **231**.

BAHAT, D., BANKWITZ, P. & BANKWITZ, E. 2004. The index of hackle raggedness on joint fringes. *In*: COSGROVE, J. W. & ENGELDER, T. (eds) *The Initiation, Propagation, and Arrest of Joints and Other Fractures*. Geological Society, London, Special Publications, **231**, 103–116.

GEOLOGICAL SOCIETY SPECIAL PUBLICATION NO. 231

The Initiation, Propagation, and Arrest of Joints and Other Fractures

EDITED BY

J. W. COSGROVE

Imperial College, London, UK

and

T. ENGELDER

Pennsylvania State University, USA

2004
Published by
The Geological Society
London

THE GEOLOGICAL SOCIETY

The Geological Society of London (GSL) was founded in 1807. It is the oldest national geological society in the world and the largest in Europe. It was incorporated under Royal Charter in 1825 and is Registered Charity 210161.

The Society is the UK national learned and professional society for geology with a worldwide Fellowship (FGS) of 9000. The Society has the power to confer Chartered status on suitably qualified Fellows, and about 2000 of the Fellowship carry the title (CGeol). Chartered Geologists may also obtain the equivalent European title, European Geologist (EurGeol). One fifth of the Society's fellowship resides outside the UK. To find out more about the Society, log on to www.geolsoc.org.uk.

The Geological Society Publishing House (Bath, UK) produces the Society's international journals and books, and acts as European distributor for selected publications of the American Association of Petroleum Geologists (AAPG), the American Geological Institute (AGI), the Indonesian Petroleum Association (IPA), the Geological Society of America (GSA), the Society for Sedimentary Geology (SEPM) and the Geologists' Association (GA). Joint marketing agreements ensure that GSL Fellows may purchase these societies' publications at a discount. The Society's online bookshop (accessible from www.geolsoc.org.uk) offers secure book purchasing with your credit or debit card.

To find out about joining the Society and benefiting from substantial discounts on publications of GSL and other societies worldwide, consult www.geolsoc.org.uk, or contact the Fellowship Department at: The Geological Society, Burlington House, Piccadilly, London W1J 0BG: Tel. +44 (0)20 7434 9944; Fax +44 (0)20 7439 8975; E-mail: enquiries@geolsoc.org.uk.

For information about the Society's meetings, consult *Events* on www.geolsoc.org.uk. To find out more about the Society's Corporate Affiliates Scheme, write to enquiries@geolsoc.org.uk.

Published by The Geological Society from:
The Geological Society Publishing House
Unit 7, Brassmill Enterprise Centre
Brassmill Lane
Bath BA1 3JN, UK
(*Orders*: Tel. + 44 (0)1225 445046
          Fax + 44 (0)1225 442836)

Online bookshop: http://bookshop.geolsoc.org.uk

British Library Cataloguing in Publication Data
A catalogue record for this book is available from the British Library.

ISBN 1-86239-165-3

Typeset by Servis Filmsetting Limited, Manchester, UK

Printed by Cromwell Press, Trowbridge, UK

Distributors
USA
  AAPG Bookstore
  PO Box 979
  Tulsa
  OK 74101-0979
  USA
  Orders: Tel. + 1 918 584-2555
          Fax + 1 918 560-2652
          E-mail bookstore@aapg.org

India
  Affiliated East-West Press PVT Ltd
  G-1/16 Ansari Road, Daryaganj,
  New Delhi 110 002
  India
  Orders: Tel. + 91 11 327-9113
          Fax + 91 11 326-0538
          E-mail affiliat@nda.vsnl.net.in

Japan
  Kanda Book Trading Company
  Cityhouse Tama 204
  Tsurumaki 1-3-10
  Tama-shi, Tokyo 206-0034
  Japan
  Orders: Tel. + 81 (0)423 57-7650
          Fax + 81 (0)423 57-7651
          E-mail geokanda@ma.kcom.ne.jp

Contents

Preface

We dedicate this volume to the memory of our colleague and friend, Paul Hancock. In doing so, we second the feelings expressed by Bill Dunne, Iain Stewart and Jonathan Turner in the special issue of *Journal of Structural Geology*, the Paul Hancock Memorial Issue of 2001. Although it has been 5 years since Paul's passing, we still feel his presence as a colleague who took a special interest in brittle fracture. One of us (Terry Engelder) had the special pleasure of collaborating with Paul on developing an understanding of neotectonic joints. It was Paul who introduced both of us to the 'world-class' outcrops at Lilstock beach and we have both had the opportunity of discussing fracturing with him at numerous localities around the world ranging from Taiwan to the Appalachian Plateau.

When organizing a field workshop on joint initiation, propagation and arrest we wished to incorporate a field trip to one of the world's best outcrops for the study of joint initiation, propagation and arrest. Lilstock beach came immediately to mind. The legacy of Paul at Lilstock beach lives on in more than one paper in this Geological Society of London Special Publication titled *The Initiation, Propagation, and Arrest of Joints and Other Fractures*. Our field workshop was held in Weston super Mare, UK, less than an hour's bus drive from the outcrops. In inviting participants to our field workshop in August 2001, we announced our intention to dedicate the conference to Paul's memory. Such an invitation attracted more than two dozen participants from several countries in Europe, North America and the Middle East. The collection of chapters in this volume grew from the questions addressed during the field workshop.

We attempted to understand joint development in the crust through the following general questions. What is the mechanism by which joints are initiated? What are the mechanisms controlling the path they follow during the propagation process? What is responsible for the arrest of joints? These are the three questions implicitly found in the title to this special publication, *The Initiation, Propagation, and Arrest of Joints and Other Fractures*. Many of the answers to these questions can be inferred from the geometry of joint-surface morphology and joint patterns. Joints are a record of the orientation of stress at the time of propagation, and as such they are also useful records of ancient stress fields, regional and local. Because outcrop and subsurface views of joints are limited, statistical techniques are required to characterize joint and joint sets. Finally, joints are subject to post-propagation stresses that further localize deformation and are the focus for the development of new structures. Our special publication has one or more chapters that address in detail each of these questions or topics.

T. Engelder
J. W. Cosgrove

Opening histories of fractures in sandstone

S. E. LAUBACH[1,2], R. H. LANDER[1,3], L. M. BONNELL[1,3], J. E. OLSON[2] & R. M. REED[1]

[1] *Bureau of Economic Geology, John A. and Katherine G. Jackson School of Geosciences, University of Texas at Austin, Austin, TX 78713, USA*

[2] *Department of Petroleum and Geosystems Engineering, University of Texas at Austin, Austin, TX 78713, USA*

(e-mail: Steve.Laubach@beg.utexas.edu)

[3] *Geocosm L.L.C., 3311 San Mateo, Austin, TX 78729, USA*

Abstract: High-resolution scanning electron microscope (SEM)-based cathodoluminescence images were used to reconstruct incremental fracture opening in regional opening-mode fractures in sandstone. Opening is recorded by crack–seal texture in isolated mineral bridges that span opening-mode fractures *formed* in sandstone at moderate–great depth (*c.* 1000–6000 m). We restored opening histories of nine representative fractures with apertures of millimetres in five sandstones from five sedimentary basins. Gaps created by fracture widening in 11 bridges range from less than 1 µm to more than 1 mm, but nearly all are less than 20 µm and most are less than 5 µm. These are the opening amounts that could be spanned by cement growth in these diagenetic environments. Our observations are the first evidence of opening amounts from mostly porous, opening-mode (joint-like) fractures formed in diagenetic environments. Patterns are consistent with a new structural diagenetic model of bridge growth that can use opening patterns to indicate rate of fracture opening as a function of time.

Cement is typically present in fractures in sedimentary rocks that have been exposed to moderate–deep burial conditions (*c.* 1000–>6000 m). Thickness of these cements ranges from micron deposits that line fracture walls to crystalline masses that fill fractures with apertures of centimetres and more. Cement can fill a porous, permeable fracture to block flow (Laubach 2003). An understanding of how cement precipitates in fractures has practical value for predicting bulk permeability of a fractured rock. Cements that record fracture-opening histories (crack–seal textures) can provide insight into the timing of fractures relative to a rock's cement precipitation history (Laubach 1988). Opening amounts recorded in fracture cement constrain fracture opening, propagation and arrest history.

Fracture opening

A structural association in opening-mode fractures in moderately–deeply buried sedimentary rocks includes isolated bridges of cement deposited during fracture opening (Figs 1 and 2). Bridges are cement deposits in otherwise open, joint-like fractures. They have rod or pillar-like shapes that are typically oriented normal to, and connect opposite, opening-mode fracture walls. Bridges are composed of wall-rock fragments and cement arranged in crack–seal texture that records repeated cracking and local cementing of cracks within the bridges.

Unlike crack–seal textures described from metamorphic veins, where the entire fracture is commonly filled with cement (e.g. Ramsay 1980), these fractures in sedimentary rock can have extensive porosity and only a thin veneer of contemporaneous cement on areas of fracture wall between bridges (Laubach 1988; Lander *et al.* 2002). Although typically present only in some fractures in a given area and generally small and inconspicuous, crack–seal bridges are widespread and they provide evidence of opening amounts from mostly open, joint-like fractures formed in diagenetic environments.

Recent structural diagenetic modelling and experiments clarify how isolated crack–seal bridges form. According to Lander *et al.* (2002), such bridges arise when: (1) the increase in fracture aperture for individual fracture events (e.g. microns); (2) the rate of aperture increase integrated over geological timescales is less than the rate of precipitation on anhedral surfaces; and (3) new anhedral nucleation surfaces are periodically created by fracturing of quartz crystals. Laboratory crystal-growth experiments show that, although crystal growth is faster in the direction of the quartz *c*-axis than precipitation in the direction of the *a*-axis, slower growth rates on euhedral crystals compared with fresh fracture surfaces is a key to whether overall cement precipitation keeps up with fracture opening (Lander *et al.* unpublished results 2000). The Lander *et al.* (2002) model is based on concepts that rock surface area and temperature

From: COSGROVE, J. W. & ENGELDER, T. (eds) 2004. *The Initiation, Propagation, and Arrest of Joints and Other Fractures.* Geological Society, London, Special Publications, **231**, 1–9. 0305-8719/04/$15 © The Geological Society of London 2004.

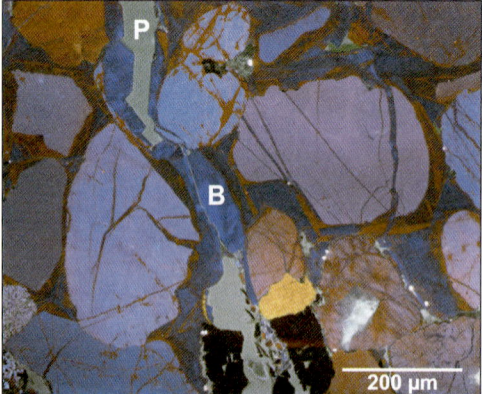

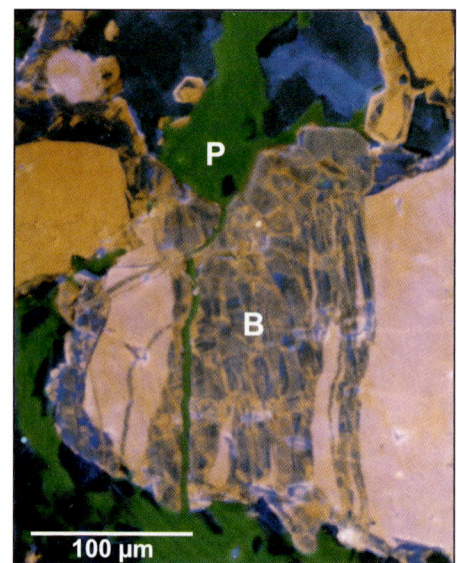

Fig. 1. Cement bridge with uniform texture marking a single opening–sealing event. Scanned-CL image: P, porosity; B, bridge. Sandstone, Venezuela, depth 4187 m. Quartz cement with red luminescence predates fracture opening; blue quartz is syn-kinematic. Note that the amount of blue quartz cement varies along fracture walls.

fundamentally control progress of quartz precipitation (Walderhaug 1996).

Bridges grow as incremental fracture opening breaks the bridge and cement precipitates in the resulting gap. To distinguish small cement-filled fractures making up crack–seal texture in bridges from the large, mostly open fracture containing the bridges, we call the former *cement-filled gaps*. The purpose of this chapter is to quantify how bridges grow by using high-resolution scanning electron microscope (SEM)-based cathodoluminescence (scanned-CL) images to identify and measure cement-filled gaps within bridges. We demonstrate that the fine detail of fracture-widening history can be recovered from typical regional fractures formed at depth. For mostly open, joint-like fractures formed in diagenetic environments, our observations document opening amounts and patterns.

Samples

In this chapter we describe 11 bridges from nine fractures. Each of these fractures contains numerous bridges, but we only examined multiple bridges in one fracture. The examples are from five sandstones from five sedimentary basins (Table 1). We selected these samples because they span a range of burial depths and they have representative patterns of microstructure. The conclusion that these features are representative is based on a survey of more than 95 sandstone formations and more than 300 oriented thin sections. Opening-mode fractures display isolated bridges containing crack–seal texture and associated fracture porosity in rocks that have expe-

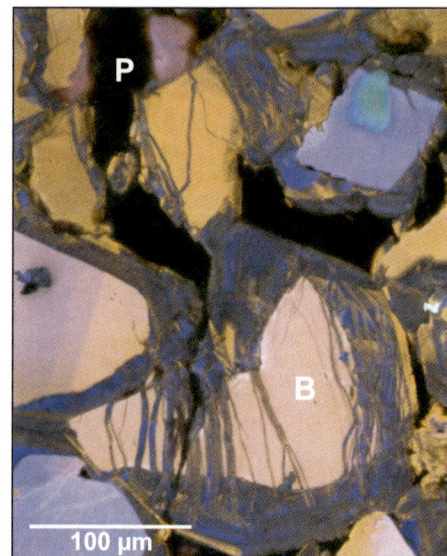

Fig. 2. Cement bridges with crack–seal textures resulting from repeated openings. Scanned-CL images: P, porosity; B, bridge. (**a**) Lower Cretaceous Fall River Formation, Wyoming, depth 3775 m. The orange and blue material is quartz cement. (**b**) Lower Cretaceous Travis Peak Formation, East Texas, depth 1865 m. The blue in the bridge is quartz cement. Note the filled fractures that extend across some cement zoning but truncate at others in (b). Crack–seal texture develops when fracture apertures are sufficiently narrow that cement can span them before any subsequent opening occurs.

Table 1. *Sample depth and location of mineral bridges*

Bridge No.	Well	Depth (m)	Unit (Age)	Basin	Aperture (μm)
1	SHCT 1	2753	Mesaverde (K)	Piceance	600
2	Gr. Valley 2 Fed.	2199	Mesaverde (K)	Piceance	450
3	Big Horn 2–3	6240	Frontier (K)	Wind River	700
4	Big Horn 3–36*	6244	Frontier (K)	Wind River	400
5	Big Horn 3–36*	6244	Frontier (K)	Wind River	400
6	Big Horn 3–36*	6244	Frontier (K)	Wind River	400
7	Big Horn 3–36	6242[†]	Frontier (K)	Wind River	312
8	Holditch SFE 2	3008[†]	Travis Peak (K)	East Texas	100
9	Holdith Howell 5	1864	Travis Peak (K)	East Texas	112
10	Linares	Outcrop[‡]	Huizachtal (J)	NE Mexico	205
11	H. L. Jenkins 1	1052	Pottsville (Penn.)	Black Warrior	35

K, Cretaceous; J, Jurassic; Penn., Pennsylvanian.
* Adjacent bridges from same fracture.
[†] Fluid-inclusion data.
[‡] Estimated maximum burial *c.* 5km.

rienced moderate–deep burial (Laubach *et al.* 2004). This association is also present in some dolomites and limestones (Gale *et al.* 2004). Typically, at least part of the cement that occurs within fractures is contemporaneous with cements infilling the rock's intergranular volume, reflecting the shared diagenetic history of fracture and host rock.

Samples used in this study are from mostly low to moderate porosity rocks that are deeply buried or that have been in the past (Table 1). Quartz is the dominant cement, generally comprising >15% whole rock volume. Although bridges occur in rocks with a spectrum of burial histories and tectonic settings, the samples described here are from foreland basins, with the exception of bridges 8 and 9 (Table 1). Samples are mostly from flat-lying rocks distant from faults or from fold limbs. We interpret the opening-mode fractures to be part of regional sets that formed in response to some combination of burial and tectonic loading and pore-pressure changes. Fluid-inclusion data sets from intermediate- (3000 m) and deep-burial (6400 m) sandstones (Table 1) suggest temperatures of bridge formation ranging between 110 and 240 °C, showing that bridges form under a range of burial conditions.

Methods

High-resolution microstructure imaging using SEM-based cathodoluminescence (scanned CL) allows effective high-magnification (as much as ×2000) examination of silicate minerals that have low levels of luminescence over large specimen areas (Laubach 1997; Milliken & Laubach 2000; Reed & Milliken 2003) (Figs 1–3). Sensitive photomultiplier-based CL systems, high magnification and stable SEM observing conditions provide clear resolution of both

zones within cements and cement-filled microfractures that cut grains and/or cement. This imaging permits construction of accurate microstructure maps that delineate grain and fracture boundaries and cement growth textures within fractures.

The detectors and processing used for these images record CL emissions in the range of ultraviolet through visible into near infrared and convert them to grey-scale intensity values. To allow textures not evident on panchromatic images to be revealed, colour images were acquired using filters, superposition of multiple images (Reed & Milliken 2003) and image manipulation using commercial image-processing filters. Most images were acquired using an Oxford Instruments MonoCL2 system attached to a Philips XL30 SEM operating at 15 kV.

Crack–seal texture and associated cement bridges

Crack–seal texture results from repeated small increments of extension across a discontinuity and cement precipitation in the resulting gaps (Hulin 1929; Ramsay 1980) (Figs 1 and 2). Grain and cement fragments, fluid inclusions and luminescence colour bands parallel to fracture walls define the texture. In some cases, inclusion trails form at boundaries of grain fragments and tend to parallel the wall-rock displacement direction. Sharp-sided boundaries between broken grains and cement record individual opening and sealing events.

In sandstone, quartz and other phases form isolated masses, or bridges, that span fractures and that range in size from individual crystals to areas of cement with dimensions of centimetres or more. Bridges can be categorized by the presence or

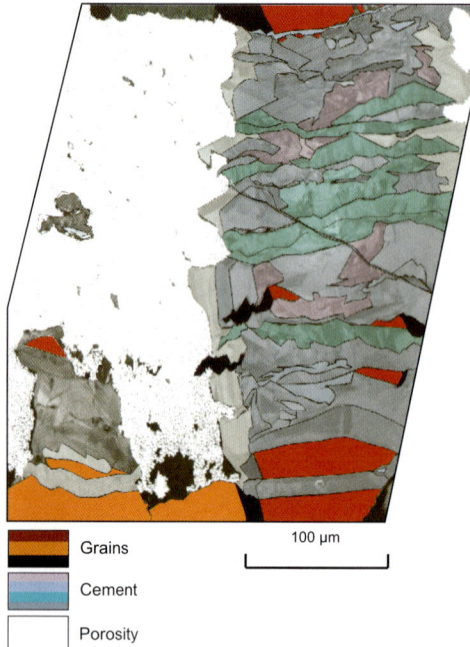

Grains

Cement

Porosity

100 µm

Fig. 3. Annotated scanned-CL image of bridge 7. Fracture trace is at a right angle to the long dimension of the bridge. Colour, texture, crystal outlines, and cross-cutting and overlapping relations separated cement zones and fractures. Figure 4 shows the fracture opening sequence.

absence of crack–seal texture. Bridge crystals lacking crack–seal textures either did not grow sufficiently to make a connection after fracture walls moved apart or formed in fractures that were quiescent (walls not moving). In some cases crystals merely grew across a fracture that subsequently did not widen (Fig. 1). Evidence of cement that precipitated in quiescent fractures is common. Such cements tend to overlap and locally surround bridges containing crack–seal structures (see fig. 1 in Laubach 1988). In sandstones the cements that form this latter type of bridge are frequently carbonate and sulphate minerals, not quartz.

The focus of this paper is on isolated bridges in sandstone composed of cements that precipitated while fractures were opening. These bridges are surrounded by fracture porosity or, later, post-kinematic cement. Bridges form where cement grows only locally across the fracture between widening increments. Repeated fracture creates gaps in the bridge that are subsequently filled with cement. Localization of new cement precipitation in these gaps results in much greater cement accumulation in the bridge compared with that of adjacent areas of the fracture wall. The resulting bridge structures can occupy a small footprint on the fracture wall, yet

cross from one fracture wall to the other with a nearly uniform dimension in the plane of the fracture. Because they occupy a small and discontinuous volume of the fracture, these bridges are not barriers to fluid flow.

Within bridges, quartz cement marking crack–seal texture comprises continuous-luminescence colour bands. In quartz, differences in CL intensity arise from slight variations in trace-element content or defect structure that characterize quartz of various origins (Sipple 1968). In addition to distinguishing grains entrained in crack–seal texture and cement (Fig. 2), contrasts in CL response can discriminate quartz cements deposited at different times in a rock's cementation history, helping to clarify structural events. For example, in Figure 1 early red-luminescent quartz coats grains and fills compaction fractures, whereas later blue-luminescent quartz fills tectonic microfractures and lines or, locally, bridges larger fractures.

Yet, crack–seal texture is frequently present only locally in fractures that contain bridges. A veneer of euhedrally terminated quartz overgrowths that lacks internal structure commonly lines fracture walls. The thickness of such veneers is typically a small fraction of crack–seal bridge thicknesses. Some euhedrally terminated crystals have crack–seal textures at their bases, succeeded by unfractured zoned crystals that developed after cement no longer bridged the opening fracture (see fig. 6 in Laubach 1997). As their width diminishes, many fractures show increased bridging, with partial–complete fill more prevalent near tips where apertures are smaller. Fractures that experienced episodic widening may also lack crack–seal texture if cement did not span the fracture.

Many bridges comprise quartz crystals elongated parallel to the *c*-crystallographic axis, the preferred rapid-growth direction for this mineral. The composition and crystallographic orientation of grains that make up fracture walls also influence cement patterns within fractures, locations where fracture porosity is retained and amount of fracture porosity. Residual fracture porosity tends to occur in association with rock fragments and feldspars that are not favourable substrates for quartz precipitation.

Core data show that bed-normal opening-mode fractures in sandstone have a wide range of sizes (Marrett *et al.* 1999). In terms of kinematic aperture, sizes range from microns to decimeters and more. Well-developed crack–seal texture is present in microfractures as narrow as 35 µm. Microfractures of this size and smaller tend to be sealed or to have only small, discontinuous areas of porosity. Fractures with kinematic apertures of millimetres commonly have crack–seal texture along at least part of their length. Larger fractures frequently have wide bridges with crack–seal texture and adjacent areas of porosity (Figs 1–4) or they may lack bridges entirely.

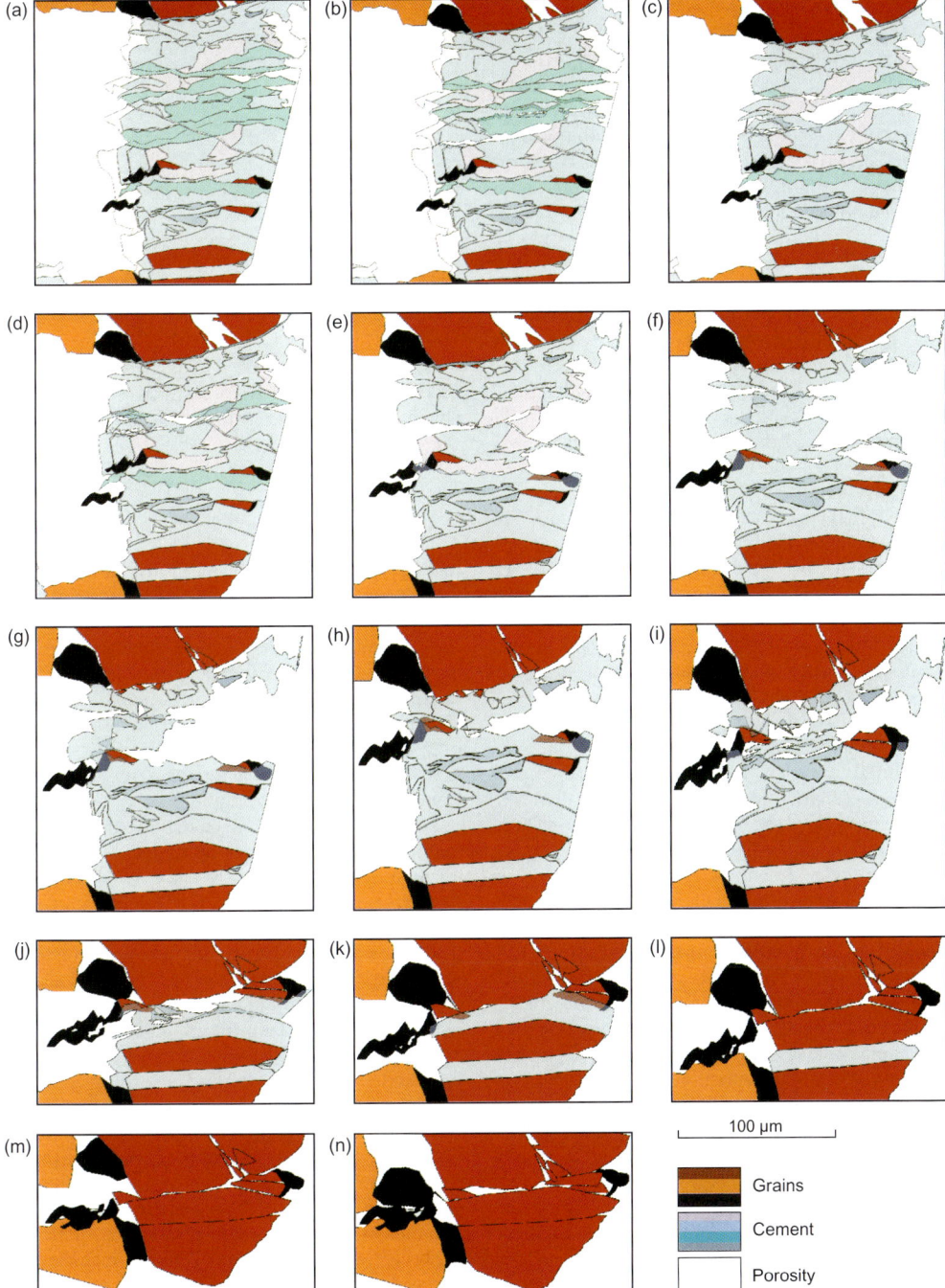

Fig. 4. Fracture opening restoration sequence (**a–n**) based on mapping cross-cutting cement and fractures on scanned-CL images, bridge 7, Upper Cretaceous Frontier Formation, Wind River Basin, Wyoming, depth 6242 m. Fracture trace is at a right angle to the long dimension of the bridge, which shows cement zones in (a) the present-day bridge; (n) is the prefracture, restored state. Animations of the opening sequence suggest that the fracture propagated from left to right in the reference frame of this diagram.

Bridge restoration

Of the 11 bridges examined in this study, 10 have relatively simple internal structures, exemplified by Figure 2. Visible within these bridges are wall-rock inclusions and uniform–textured luminescence bands. There are also areas, usually near or on the outer margins of bridges, with faceted crystals representing unimpeded crystal growth into pore space adjacent to the bridge. Luminescence bands may be continuous, with areas of euhedral crystal on bridge margins (upper part of bridge, Fig. 2a) marking crystal growth into open pore space from the edge of broken grains. Large, single crystal faces commonly mark outer edges of bridges. In some cases breakage of these crystal faces is evident in overlapping and cross-cutting zones of bands (Figs 2b and 3).

Within bridges, bands with uniform luminescence frequently cross-cut one another. The youngest bands may retain considerable fracture pore space lined with minute crystal faces (Fig. 2a, left side). In some cases cross-cutting relations among luminescence bands and crystal zoning along (or within) bridges give evidence of relative fracture timing (Fig. 2b, lower part of bridge). Textures in these bridges arise from repeated breakage and crystal growth on bridges with a simple initial structure, such as that shown in Figure 1. Even where bridges have broken in approximately the same location, cement accumulation leads to textural and luminescence differences that allow breaks to be recognized.

Mapping internal structures in bridges involves identifying and correlating fractures within bridges (cement-filled gaps) and determining their timing relative to other cement via cross-cutting and overlapping relations. Some patterns are readily identified because gaps are relatively isolated or they are bounded by prominent wall-rock inclusions (for example, much of Fig. 2b) or crystal faces. In areas with repeatedly broken cement and cross-cutting luminescence bands (Fig. 2a), high magnification and careful contact tracing are required for cross-cutting relations to be determined.

To identify fractures within bridges, we mapped grain and cement fragments derived from the wall rock and cement textures. We used the generally euhedral quartz cement on the bridge margin to find the latest fractures (for example, Fig. 2b). Consistent cement textures, abrupt luminescence band or crystal outline truncations, and euhedral crystal faces overlapped by later cement define cross-cutting relations. For example, in the bridge in Figure 3, areas marked in blue-green are interpreted to have formed during latest stages of bridge growth because bands with consistent texture and luminescence divide areas on the bridge margin marked by late euhedral overgrowths. However, the relative order of fracture and cements deposited during the

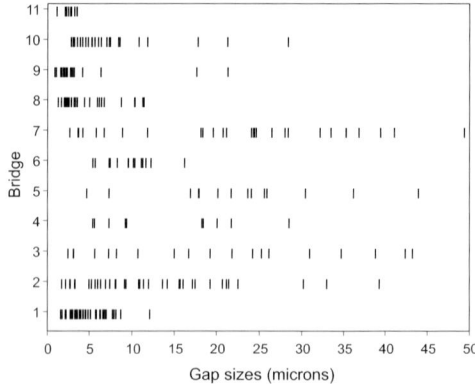

Fig. 5. Gap sizes v. bridge data set for gaps less than 50 μm. See Table 1 for sample depths. All filled gaps larger than 40 μm are from bridges 3 to 7, Frontier Formation at depths below 6242 m. Kinematic apertures are entire opening displacement spanned by the bridge.

middle stage of this bridge's growth (Fig. 4e–i) is ambiguous because repeated fracturing of cement created complex crossing relations. This ambiguity results in alternative permissible sequences of events during sequential restoration in this texturally complex cement bridge (Figs 3 and 4). However, fractures may not intersect, especially in wide bridges, leading to a lack of clear cross-cutting relations that may preclude unique identification of some relative fracture-timing relations.

As illustrated by bridge restoration (Fig. 4), gaps tend to localize in part of the bridge. However, the location of gaps does not necessarily conform to a progression from fracture wall to fracture interior or vice versa, and the locus of fracture may shift within the bridge. To trace a single rupture along a fracture would require the mapping of many bridges and the correlation of the fracture histories within each. Thus, gaps in adjacent bridges in the same fracture are not necessarily in the same position relative to the fracture wall (Fig. 5, bridges 4–6), or from one bridge to the next (Fig. 6).

Restorations show that euhedral crystal terminations form and are subsequently overgrown by other cement (and dissected by later fractures) as bridges develop (Fig. 4b, h). This pattern suggests that immediately after a fracture propagated, cement in the bridge was not in contact with the opposite fracture wall and thus could not have been supporting a normal stress. We found no textural evidence that cement precipitation and fracture opening were synchronous in the sense that solid cement was in continuous contact.

Restoration of bridges suggests that their breadth does not increase greatly through time, even though their cross-fracture dimension increases markedly,

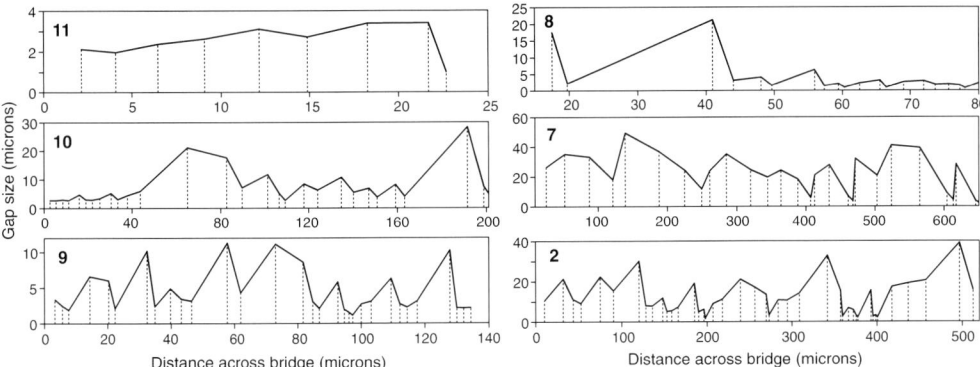

Fig. 6. Gap size v. distance across bridge. Dotted vertical lines mark the centre point of filled gap. The height of the dotted line is the width of the gap. Diagrams show spatial distribution of gap size; sequential restoration provides evidence of sequence of gap development (cement volume) that can be tied to cement history, fluid inclusion temperature and pressure data, and, thus, thermal and burial history and models of fracture growth (Lander *et al.*, 2002; Olson 2004). Numbers refer to bridges listed in Table 1. Note that scales differ for each profile.

in most cases matching the dimension of the overall fracture kinematic aperture. The main cause of increase in bridge volume is addition of cement on broken surfaces that are aligned normal to the opening direction and parallel to the main fracture trace. Accumulation of cement in the bridge is a by-product of repeated renewal of fracture surface area within bridges.

Competing fracture opening and cement-precipitation rates are responsible for patterns recorded by crack–seal textures (Fig. 6). There are several possible explanations for these contrasting opening patterns. Some may reflect the mechanics of fracture opening and the structural geology of specific fracture arrays. Others may merely represent complex breakage patterns in fractures with only small amounts of cement. For example, gap distributions may depend on bridge shape, crystal orientation, and bridge and substrate composition, as illustrated by three bridges from the same fracture that have different patterns of breakage (Fig. 5, bridges 4–6).

Bridge 7 records a history that is consistent with growth of a simple opening-mode fracture. Large initial opening increments record the propagation of the fracture tip (Figs 4 and 6). These are succeeded by smaller openings localized first near one wall of the fracture (upper wall in Fig. 4) and subsequently in the centre of the bridge. These smaller openings presumably track lengthening of a fracture with an opening displacement concentrated near the fracture tip. Some fractures show a similar progression (Fig. 6, bridge 8), but others have more complex patterns that might record interaction and linkage of fractures (Fig. 6, bridges 2 and 9). To test these concepts evidence is required at various positions along a fracture trace or from several fractures in an array.

Geomechanical modelling shows that fracture-opening histories can be complex even for simple fracture patterns (Olson 2004). The record from cement bridges deciphered using scanned-CL mapping provides the evidence for documenting and interpreting these opening histories.

Opening amounts

Isolated quartz bridges that contain crack–seal texture occur in opening-mode fractures that otherwise may be open and that lack appreciable cement. Unlike previous observations of extensively mineral filled veins in metamorphic rocks, these structures provide evidence of fracture-opening amounts from diagenetic environments that was previously lacking. Crack–seal patterns in bridges show that the fractures we studied have an involved opening history. Yet, there is little or no evidence of this history outside of the bridges, where a micron-scale veneer of faceted quartz crystals lines fracture walls.

Inspection identifies individual fractures in bridges (cement-filled gaps) marked by luminescence bands created by incremental fracture opening. We also used maps of fractures and cement zones within bridges to restore bridge development sequentially (Figs 2–4). Sequential restoration helps identify individual fractures (gaps) in areas of complex texture and also reveals the timing of gap development.

Gap sizes and locations were recorded along lines parallel to bridge axes (normal to fracture walls). Although this procedure does not accurately capture the cement volumes accumulated in each gap, it minimizes possible over or undercounting of the number and size of filled gaps in texturally complex areas with ambiguous cross-cutting relations and gives an accurate measure of individual opening steps.

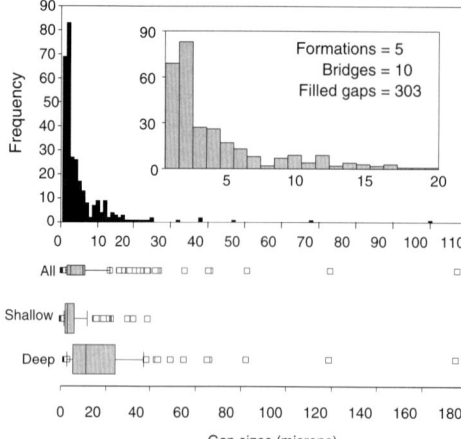

Fig. 7. Frequency distributions of gap sizes (filled opening increments) in bridges. Bridge 11, a microfracture, is included in box plots but not the histogram. The histogram shows the size distribution for gaps less than 120 μm; the inset histogram shows the distribution for gaps less than 20 μm. The first box plot shows the mean and range for the entire data set; second and third box plots show bridge data separated into relatively shallow (bridges 1, 2, 8, 9 and 11) and deep (bridges 3–7 and 10) samples.

Gaps in bridges created by fracture widening with contemporaneous cementation range from less than 1 μm to more than 1 mm, but nearly all are less than 20 μm and most are less than 5 μm (Fig. 7). The average gap size is about 10 μm, but gap sizes appear to have a log-normal size distribution (Fig. 7). The significance of this distribution is uncertain. It is probably truncated at both small and large sizes. At small sizes, small opening increments may not produce recognizable textural features with our current imaging methods. At large sizes, bridges record only opening amounts that could be spanned by subsequent cement growth rather than the entire range of permissible opening increments.

For all sample sets, the average gap size is less than the average thickness of quartz cement on monocrystalline quartz grains within adjacent host rocks. This difference is consistent with slow, continuous cement accumulation on fracture walls outside of bridges.

The largest minimum gap sizes are in the deepest samples and in the outcrop sample of Jurassic sandstone from NE Mexico, the burial history of which suggests likely fracturing at great depth (*c.* 5 km). The opening patterns in these fractures probably record diverse strain histories, possibly in the context of changing ambient conditions. Nevertheless, although the data set is small, overall average gap

sizes are somewhat larger for deeper samples (Fig. 7). The largest filled gaps are in the deepest samples, and the average gap size for deeper samples (18.9 μm) is also larger than for shallower samples (5.4 μm). These apparent differences between deep and shallow samples are consistent with a scenario in which cement grew faster relative to fracture opening in the deeper samples, allowing larger gaps to be spanned for a given opening history. Wider cemented gaps found in deeper samples could be a result of faster precipitation rates at these depths owing to higher temperatures.

Discussion and conclusions

Our observations are the first evidence of opening amounts from joint-like fractures formed in diagenetic environments (depths of 1–6 km, temperatures of >80–200 °C). Previous studies of arrest lines on fracture surfaces documented that such fractures experienced episodic growth. Our study shows that it is possible to recover fine detail of opening histories from fractures that retain little other evidence of their structural development. Fractures containing bridges may superficially even appear to lack mineralization.

Information on opening patterns for such fractures is useful for comparison with mechanical models of fracture pattern development (Olson 2004). Bridge structures also provide a link between the rock's mechanical and diagenetic histories. Cements are more readily associated with thermal and burial history than are fractures; crack–seal bridges may therefore provide a route to better dating of fracture formation and, thus, to improved models of fracture development. We found suggestive evidence that differences in the average size of fractures spanned by cement may provide an indication of the depth of fracture formation.

The patterns we observe in crack–seal texture and bridge patterns are consistent with a structural diagenetic model of bridge growth in which the chemical environment is dominated by host-rock composition, and cement accumulation on new fracture surface areas is temperature sensitive (Lander *et al.* 2002). In conjunction with diagenetic models that predict cement accumulation as a function of time and temperature (Lander & Walderhaug 1999), patterns of cement accumulation associated with fracture opening can potentially indicate rate of fracture opening as a function of time.

Partly supported by US Department of Energy Contract No. DE-FC26-00BC15308, by associates of the Fracture Research & Application Consortium: Saudi Aramco, ChevronTexaco, Devon Energy Corporation, Ecopetrol, Marathon Oil, PEMEX Exploración y Producción and IMP, Petroleos de Venezuela, Petrobras, Repsol-YPF, Shell

International Exploration & Production, Schlumberger, Tom Brown, Inc., Williams Exploration & Production, and Geocosm LLD. We are grateful to associates for samples used in this study, to J. Gale, R. Marrett and K. Milliken for discussion, and C. Onasch and T. Engelder for helpful review comments.

References

GALE, J. F. W., LAUBACH, S. E., MARRETT, R., OLSON, J. E., HOLDER, J. & REED, R. M. 2004. Predicting and characterizing fractures in dolostone reservoirs: using the link between diagenesis and fracturing. Geological Society, London, Special Publications, **235**, 117–192.

HULIN, C. D. 1929. Structural control of ore deposition. *Economic Geology*, **24**, 15–49.

LANDER, R. H. & WALDERHAUG, O. 1999. Predicting porosity through simulating sandstone compaction and quartz cementation. *AAPG Bulletin*, **83**, 433–449.

LANDER, R. H., GALE, J. F. W., LAUBACH, S. E. & BONNELL, L. M. 2002. Interaction between quartz cementation and fracturing in sandstone. *American Association of Petroleum Geologists Annual Convention Program*, **11**, A98–A99.

LAUBACH, S. E. 1988. Subsurface fractures and their relationship to stress history in East Texas Basin sandstone. *Tectonophysics*, **156**, 37–49.

LAUBACH, S. E. 1997. A method to detect fracture strike in sandstone. *AAPG Bulletin*, **81**, 604–623.

LAUBACH, S. E. 2003. Practical approaches to identifying sealed and open fractures. *AAPG Bulletin*, **87**, 561–579.

LAUBACH, S. E., REED, R. M., OLSON, J. E., LANDER, R. H. & BONNELL, L. M. 2004. Co-evolution of crack–seal texture and fracture porosity in sedimentary rocks: cathodoluminescence observations of regional fractures. *Journal of Structural Geology*, **26**, 967–982.

MARRETT, R., ORTEGA, O. O. & KELSEY, C. 1999. Power-law scaling for natural fractures in rock. *Geology*, **27**, 799–802.

MILLIKEN, K. L. & LAUBACH, S. E. 2000. Brittle deformation in sandstone diagenesis as revealed by scanned cathodoluminescence imaging with application to characterization of fractured reservoirs. *In*: *Cathodoluminescence in Geosciences*. Springer, New York, Chap. 9, 225–243.

OLSON, J. E. 2004. Predicting fracture swarms – the influence of subcritical crack growth and the crack-tip process zone on joint spacing in rock. *In*: COSGROVE, J.W. & ENGELDER, T. (eds) *The Initiation, Propagation, and Arrest of Joints and Other Fractures*. Geological Society, London, Special Publications, **231**, 73–87.

RAMSAY, J. G. 1980. The crack–seal mechanism of rock deformation. *Nature (London)*, **284**, 135–139.

REED, R. M., & MILLIKEN, K. L. 2003. How to overcome imaging problems associated with carbonate minerals on SEM-based cathodoluminescence systems. *Journal of Sedimentary Research*, **73**, 326–330.

SIPPLE, R. F. 1968. Sandstone petrology, evidence from luminescence petrography. *Journal of Sedimentary Petrology*, **38**, 530–554.

WALDERHAUG, O. 1996. Kinetic modeling of quartz cementation and porosity loss in deeply buried sandstone reservoirs. *AAPG Bulletin*, **80**, 731–745.

Growth of ductile opening-mode fractures in geomaterials

PETER EICHHUBL*

Rock Fracture Project, Department of Geological and Environmental Sciences, Stanford University, Stanford, CA 94305-2115, USA
*Present address: Texas A&M University – Corpus Christi, College of Science and Technology, 6300 Ocean Drive, Corpus Christi, TX 78412, USA

Abstract: Opening-mode fractures in clinker and opal-CT chert spheroids form by growth and coalescence of pores, and are associated with extensive textural and compositional changes in the host material. Extensive inelastic deformation outside the immediate vicinity of fracture tips characterizes these fracture processes as ductile. Fracture formation in clinker is concurrent with high-temperature combustion alteration of diatomaceous mudstone. Fracture formation in chert spheroids is associated with the opal-CT to quartz transition in the same host material during early marine diagenesis. In both cases, growth of elongate pores is attributed to the combined effects of diffusive-fracture growth and flow by solution–precipitation creep. Pore growth and coalescence occur preferentially ahead of fracture tips along two directions oblique to the mean macroscopic fracture direction. This growth process, referred to as side-lobe damage, is interpreted to reflect the shear-stress dependence of pore growth by solution–precipitation creep. The tendency for oblique fracture growth is suppressed by global stress and strain-boundary conditions forcing the fracture along a characteristic zig-zag propagation path that is macroscopically perpendicular to the loading direction. These examples of ductile fracture demonstrate that macroscopic fracture formation is not uniquely associated with damage processes by microfracture at low-temperature 'brittle' subsurface conditions. Instead, fracture is a deformation process that can occur due to various inelastic-deformation mechanisms under diverse crustal environments, which include high-temperature conditions.

The localization of deformation in the brittle upper crust commonly results in displacement discontinuities that are observed as faults and opening-mode fractures. Significant progress has been made in recent years in understanding the formation of single fractures and of fracture systems by applying concepts of linear elastic fracture mechanics (Ingraffea 1987; Pollard & Segall 1987; Pollard & Aydin 1988; Whittaker et al. 1992; Engelder et al. 1993). Linear elastic fracture mechanics approaches rock failure as the propagation of pre-existing flaws and the coalescence of multiple interacting fractures (Segall 1984; Du & Aydin 1991; Fleck 1991; Kemeny & Cook 1991; Olson 1993; Renshaw & Pollard 1994). The linear elastic fracture mechanics approach is based on the underlying assumption that inelastic deformation associated with the breakage of bonds by inter- and transgranular fracture is limited to a small zone ahead of the macroscopic fracture tip. For large classes of engineering materials such as ductile metals, polymers and ceramics at high homologous temperature, failure is associated with extensive inelastic deformation (Kinloch & Young 1983; Thomason 1990; Anderson 1995). Processes associated with creep failure of ceramics at high homologous temperatures and slow loading rates (Dalgleish et al. 1985; Wiederhorn & Fuller 1985; Robertson et al. 1991) appear particularly relevant to failure under geological conditions. Creep failure in ceramics results from the nucleation, growth and coalescence of pores by predominantly diffusive-deformation mechanisms (Wilkinson 1981; Hutchinson 1983; Evans & Blumenthal 1984; Wilkinson et al. 1991; Krause et al. 1999). Processes similar to those of creep failure in ceramics are potentially significant to failure of Earth materials in chemically reactive environments.

Two examples of opening-mode rock fracture under geological conditions are presented in this study where fracture formation is associated with pore growth and coalescence, as well as significant inelastic deformation well outside the immediate vicinity of fracture tips. Inelastic deformation is accompanied by a mineralogical and textural re-organization of the host material, which in both cases is diatomaceous mudstone. The conditions, however, under which fractures form in these two examples are entirely different. Fractures in the first example formed under high-temperature–low-pressure conditions during natural *in situ* combustion of hydrocarbons and the associated alteration of diatomaceous mudstone to clinker. The second example is observed in chert spheroids that are contained in diatomaceous mudstone where fractures are inferred to form during and subsequent to spheroid growth under early marine-diagenetic conditions. Emphasis is placed here on the fracture geometry and its implications on fracture-growth processes. Based on the extent of inelastic deformation associated with fracture in both cases, fractures will be designated as ductile. The term ductile is used here following Rutter (1986) to account for the distributed inelastic deformation associated with fracturing. The term ductile is not

From: COSGROVE, J. W. & ENGELDER, T. (eds) 2004. *The Initiation, Propagation, and Arrest of Joints and Other Fractures.* Geological Society, London, Special Publications, **231**, 11–24. 0305-8719/04/$15 © The Geological Society of London 2004.

intended to imply a specific stress–strain behaviour, which cannot be observed in outcrop, nor any specific deformation mechanism. This use differs from the common usage in engineering and experimental rock fracture mechanics where ductility refers to the ability of a material to withstand inelastic strain (Handin 1966; Stouffer & Dame 1996). In engineering fracture mechanics ductile fracture is usually equated with crystal–plastic deformation in metals (Wilsdorf 1983), whereas creep failure in ceramics is considered a brittle process (Sabljic & Wilkinson 1995). Based on inferred fracture-growth processes for clinker and chert spheroids, and on analogue processes in ceramics, I will develop criteria that delineate ductile rock fracture growth from brittle rock fracture.

Ductile fracture associated with combustion alteration of siliceous mudstone to clinker

Clinker occurs in Miocene diatomaceous mudstone at several locations in southern California resulting from natural *in situ* combustion of hydrocarbons (Bentor & Kastner 1976). At Orcutt oil field south of Santa Maria, three or four zones of clinker and associated alteration haloes are exposed in a quarry (Cisowski & Fuller 1987; Lore *et al.* 2002). Clinker forms subvertical tabular bodies 1–2 m-wide that are brecciated in their core. The continuously exposed 5–20 m-wide transition from unaltered mudstone to clinker provides insight into the mineralogical and textural changes associated with combustion alteration (Eichhubl & Aydin 2003). Unaltered diatomaceous mudstone is composed of biogenic opal-A and smectite, with minor opal-CT, kaolinite, illite, quartz and feldspar. With increasing alteration, tests of diatoms composed of opal-A are dissolved, and cristobalite, illite and haematite become the predominant mineral phases. Across a distinct boundary the mineralogical composition changes to tridymite, cordierite and calcic plagioclase resulting in the hardened or sintered, yet porous, appearance of clinker. Bedding, weakly expressed in unaltered mudstone by a preferred fissility, is obliterated in clinker. Based on the mineral associations and textural characteristics suggesting the presence of a partial melt phase, Eichhubl *et al.* (2001) inferred combustion temperatures of about 1100 °C. Compositional changes are associated with distinct changes in rock texture. Siliceous mudstone has approximately 60% of predominantly submicron-sized porosity. Clinker has 28% porosity, with individual pores predominantly >1 μm (Eichhubl *et al.* 2001).

Clinker is characterized by an orthogonal–polygonal system of up to 5 cm-long unfilled fractures with blunt fracture tips, large apertures and high intersection angles (Fig. 1a). In the centre of high-temperature alteration, these fractures are sufficiently interconnected to brecciate the formation. Under close macroscopic inspection and under the microscope, the formation of macroscopic fractures is observed to result from growth and coalescence of elongate pores (Fig. 1b, c). Pore growth from small spherical pores to larger elongate pores is inferred based on an increase in length to width ratio with an increase in pore long dimension (fig. 9 in Eichhubl & Aydin 2003). Pore coalescence resulted in remnant shapes of pores in macro- and microscopic fractures, and in remnants of failed ligaments between adjacent pores (Fig. 1d). Remnants of failed ligaments form blunt protrusions along irregular fracture walls (arrows in Fig. 1d). An increased abundance of elongate pores ahead of fracture tips indicates that fractures grew by preferred pore growth and coalescence ahead of fracture tips (Fig. 2). In addition to preferred growth and coalescence of pores at individual fracture tips, coalescence of circular pores is observed away from factures as well as along the sides of fractures where pore coalescence contributed to the widening of fractures.

Ductile fracture in opal-CT spheroids

Chert spheroids are concretion-like features of oblate ellipsoidal shape found in diatomaceous mudstone (Taliaferro 1934; Kolodny 1969; Brueckner *et al.* 1987; Behl 1992; Behl & Garrison 1994). Chert spheroids, collected at World Minerals Inc. (formerly Johns-Manville) Celite diatomite quarry in South Lompoc, California, reach diameters of up to 30 cm and are composed of opal-CT (Behl 1992). They are found in diatomaceous mudstone of the Miocene Monterey Formation that is preserved at this location in its least-altered opal-A stage of silica diagenesis. Unlike the chalk-like consistency of surrounding opal-A diatomite, chert spheroids are hard with a waxy luster on broken surfaces.

Chert spheroids are composed of concentric opal-CT shells that are separated by concentric spheroidal fractures (Fig. 3). The concentric shells are, on average, 5–10 mm thick (Fig. 4b). Spheroidal fractures have apertures of 1–5 mm, being wider along equatorial regions of chert spheroids, and narrower towards top and bottom regions where fractures become discontinuous. Bedding is preserved within the chert spheroids (Fig. 4a). Spheroidal fractures are parallel to bedding at the top and bottom of spheroids, and cut bedding at high angles in mid-sections of spheroids. Walls of spheroidal fractures are quite straight, and well defined in top and bottom sections of chert spheroids (Fig. 4e, f), but are irregular and amoeboid in mid-sections where they merge into a maze of irregular-shaped pores (Fig. 4c, d). Based on

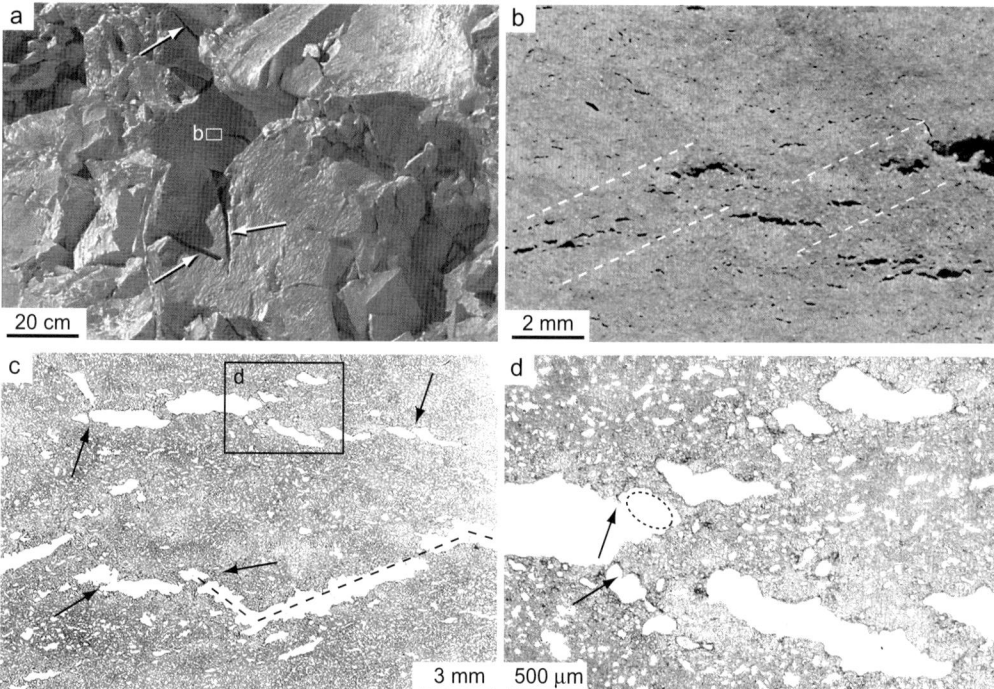

Fig. 1. Ductile opening-mode fracture in brecciated clinker that resulted from natural *in situ* combustion of hydrocarbons in diatomaceous mudstone of Upper Miocene Sisquoc Formation, Red Rock Canyon S Orcutt, California, USA. (**a**) Outcrop view of brecciated clinker. Arrows point at blunt-tipped opening-mode fractures. (**b**) Close-up view of the tip region of opening-mode fracture in (a), exposed on a natural fracture surface. Small voids and fractures are distributed throughout the sample but are concentrated at the tip of the larger fracture. Notice en échelon arrays (outlined by dashed lines) of smaller fractures ahead of the larger fracture tip. (**c**) Photomicrograph of (b). Fractures are characterized by zig-zag propagation paths (dashed lines) and blunt tips. Pores are distributed throughout the section but are more abundant and more elongate at the tip of larger pores and fractures. Arrows point at thinned or failed ligaments between pores. (**d**) Detail of (c). Higher abundance of elongate pores ahead of larger pores and fractures suggests that fractures form by growth and coalescence of pores. Ellipse outlines the remnant shape of a coalesced pore. Arrows point at a failed ligaments. Petrographic thin sections were prepared after vacuum epoxy impregnation to preserve details of pore structure.

image analysis of vertically oriented thin sections, space occupied by fractures and microscopic (>1 μm in diameter) pores amounts to 25% of the total image area along equatorial transects and to 18% along vertical transects of a particular spheroid. Pores and spheroidal fractures are cemented with banded chalcedonic and microflamboyant quartz. Larger fractures are only partially cemented by quartz that lines fracture walls and leaves remaining open fracture space. Bottom sections of fractures are filled with fragments and fine-grained opal-CT detritus that predates fracture-lining quartz cement. This detrital infill indicates that fracture-lining cement post-dates fracture opening.

The presence of quartz as a pore- and fracture-filling mineral phase in chert spheroids suggests that pore growth and fracture are associated with the transformation of metastable opal-CT to quartz. This transformation is part of the opal-A to opal-CT to quartz transformation sequence (Behl 1992) that is characteristic of biogenic silica undergoing burial diagenesis (Isaacs 1982; Williams & Crerar 1985). In detrital-rich sections of the Monterey Formation, the opal-CT to quartz reaction occurs at 65–80 °C, with higher transformation temperatures associated with a lower detrital content (Keller & Isaacs 1985). Behl & Garrison (1994) found that in detrital-poor diatomites containing chert spheroids, the opal-CT to quartz reaction occurred at temperatures as low as 36 °C and thus under early marine burial diagenetic conditions. Following Stein & Kirkpatrick (1976) and Williams & Crerar (1985), these silica transformation reactions involve dissolution of the metastable precursor phase (opal-A and opal-CT) and precipitation of the more stable phase (opal-CT and quartz, respectively) out of an aqueous solution.

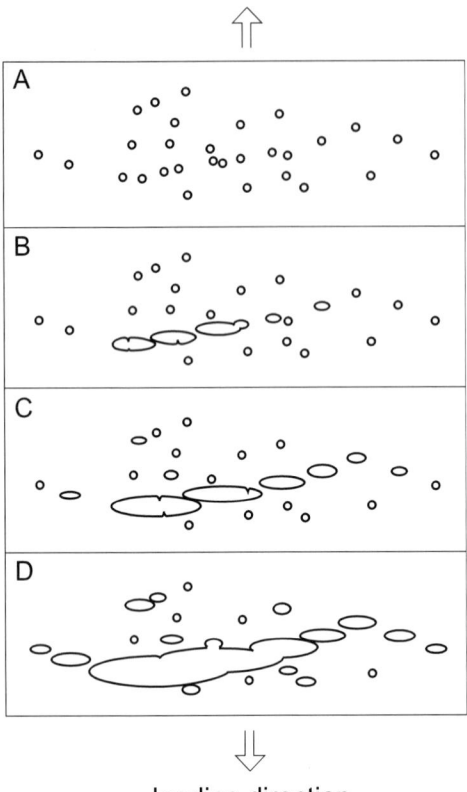

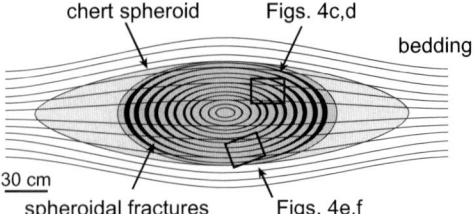

Fig. 3. Schematic cross-section across a chert spheroid (after Behl 1992). The location of detailed figures in Figure 4 and their relation to bedding are indicated.

loading direction

Fig. 2. Sequence of fracture growth by pore growth and coalescence. Fractures are inferred to grow with their long axis perpendicular to the maximum principal tensile or least principal effective compressive stress, σ_3. Pore shapes are depicted to reflect pores in clinker, but a similar growth process is inferred for fracture in chert spheroids.

Discussion

Controls on pore growth

Ductile fracture in clinker and chert spheroids involves two concurrent processes: (1) growth of pores; and (2) coalescence of pores along preferred directions. In clinker, these pores remained unfilled whereas pores in chert spheroids are filled with quartz. Based on a comparison of specific bulk volume of clinker and less altered mudstone, Eichhubl *et al.* (2001) inferred that the gain in micro- and macroscopic pore space associated with ductile fracture correlates with an equal loss in submicron porosity and the formation of denser mineral phases. This correlation suggests that larger pores and fractures grew at the expense of smaller ones, driven by

the difference in pore radius, and thus in chemical potential, between large and small pores. Large pores with higher radius of concave (negative) curvature, and thus higher chemical potential, are expected to be sites of preferred dissolution or melting relative to surfaces of small pores that are sites of preferred precipitation. This process is analogous, but opposite in sign, to grain coarsening by Ostwald ripening where surfaces of grains with larger radius of *convex (positive)* curvature are preferred sites of precipitation relative to the surfaces of smaller grains that undergo preferred dissolution (Lasaga 1998). Based on the inferred presence of a partial melt during fracture formation in clinker, Eichhubl & Aydin (2003) suggested that pore growth resulted from solution–precipitation and diffusion of ions in a partial melt phase.

A similar process of pore growth with associated matrix densification can be invoked for the formation of concentric fractures in chert spheroids. Following Behl (1992), submicron (<1 μm) porosity is reduced from 80% in opal-A diatomite to 15% in opal-CT chert. Assuming conservation of silica, this reduction in pore volume corresponds to a 30% reduction in radius of a spherical body. The predicted 30% reduction in radius exceeds the 25% fracture and microscopic (>1 μm) pore space measured in equatorial sections of chert spheroids. The volume of concentric fractures in chert spheroids can thus be attributed to the reduction in submicron porosity associated with the opal-A to opal-CT transformation. The remaining 5% porosity could be accounted for by vertical compaction of the spheroid. Flattening of the chert spheroids is apparent from their oblate shape and from the difference in measured porosity between equatorial and vertical transects. The close correspondence of porosity reduction associated with opal-CT chert formation with the amount of fracture and >1 μm pore space created in chert spheroids necessitates that the silica cement filling these fractures and pores was transported into the concretion. A similar conclusion was reached by Behl (1992), based on a comparison of bulk density measurements of opal-CT chert and

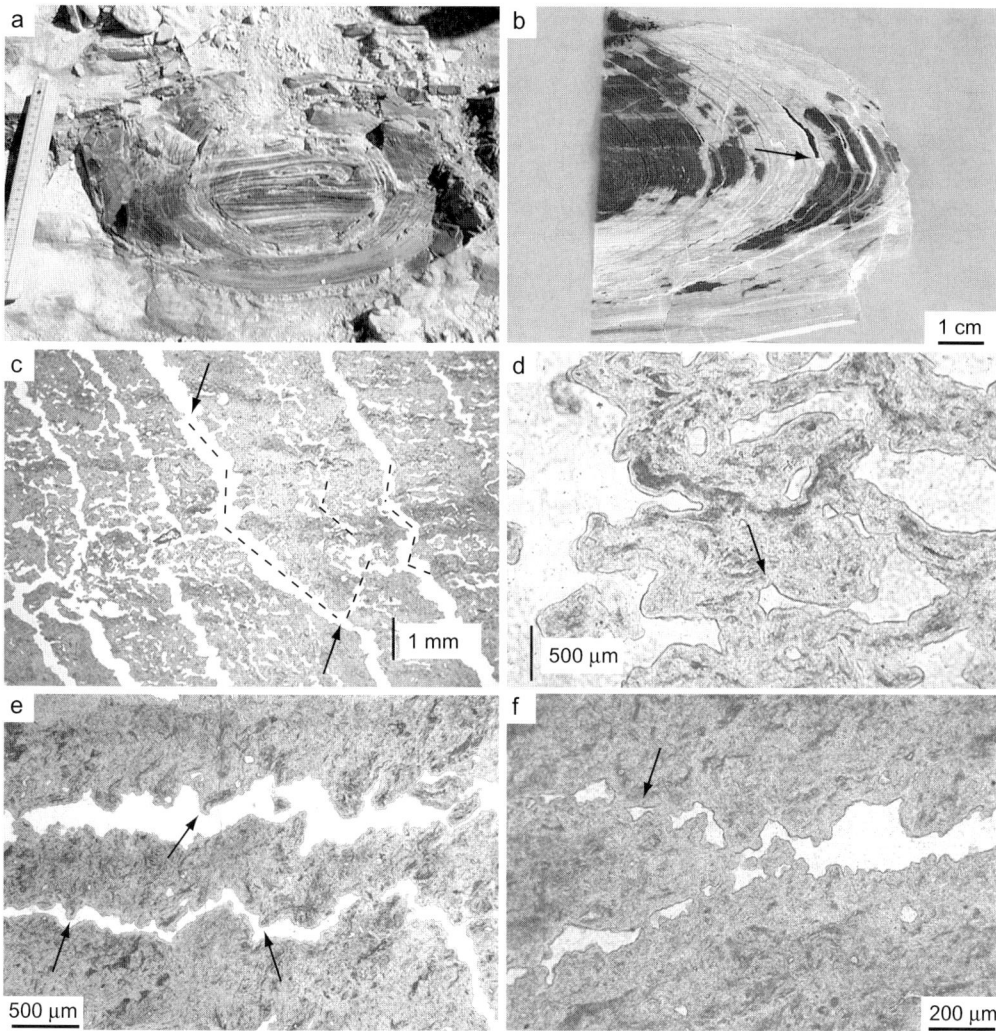

Fig. 4. Ductile fracture in chert spheroids, Miocene Monterey Formation, Celite (World Minerals Inc.) quarry South Lompoc, California, USA. (**a**) Outcrop view of a chert spheroid contained within opal-A diatomite. Bedding (with microfold in the core of the spheroid) is subhorizontal. (**b**) Quarter section of a chert spheroid. Concentric fractures are wider along equatorial sections and narrow towards the top and bottom of the spheroid. Arrow points at incomplete geopetal infill of spheroidal fracture. (**c**) Photomicrograph from the equatorial region of the spheroid in (b). Note irregular shape of spheroidal fractures and the irregular porous texture of opal-CT matrix between the fractures. Fractures and pores are filled with clear chalcedonic and microflamboyant quartz. Note the bifurcation of fractures (lower arrow) and zig-zag geometry (dashed lines) of coalescing pores. Pore coalescence resulted in remnants of broken bridges (e.g. at upper arrow). (**d**) Detail of irregular pore structure in the opal-CT chert matrix in (c). Pore space is filled with clear chalcedonic and microflamboyant quartz. Pores in opal-CT chert have a characteristic inward convex shape (arrow). (**e**) Photomicrograph from the basal section of the chert spheroid in (b). Fractures follow zig-zag propagation path. Remnants of failed ligaments between pores form blunt protrusions (arrows) along irregular fracture surfaces. (**f**) Voids aligned ahead of a crack tip in basal section of (b). Various stages of void coalescence suggest that fracture grew by void growth and coalescence. Arrow points at pore with notably inward convex shape.

opal-A diatomite. The diffusive or advective addition of silica in an aqueous solution is consistent with the findings of Stein & Kirkpatrick (1976) and Williams & Crerar (1985) that the opal-A to opal-CT transformation constitutes a solution–precipitation process and not a solid-state diffusion process.

Based on similar processes during reactive-liquid sintering of ceramics, Eichhubl *et al.* (2001) proposed that the elongation of larger pores in clinker and their coalescence along linear trends occurred in response to a tensile sintering stress. Sintering results from the thermodynamic tendency of porous or granular media to minimize the surface free energy of solid and liquid surfaces. Sintering thus leads to densification of a porous or granular medium and the tendency of the medium to contract (German 1996). If the sintered body is prevented from contracting freely, a tensile-sintering stress builds up. Under isotropic strain-boundary conditions in an isotropic material, sintering results in a reduction in mean total stress. In a general deviatoric stress state characteristic of subsurface conditions, macroscopic opening-mode fractures would then form perpendicular to the least compressive principal stress once the tensile-failure strength of the material is exceeded. In both clinker and chert spheroids, the constraint preventing free contraction is provided by the bonding of the sintered material with the adjacent unaltered diatomaceous mudstone. The roughly tabular shape of clinker bodies thus resulted in orthogonal–polygonal sets of fractures. Sintering within ellipsoidal chert spheroids resulted in a radial-symmetric sintering stress that is reflected in the concentric arrangement of spheroidal fractures.

The orientation of long axes of elongate pores in clinker parallel to the mean macroscopic fracture direction suggests that the initially spherical pores grew preferentially in the plane perpendicular to the least principal stress. This elongation direction is different from pore growth in ductile metals or polymers where plastic flow elongates initially circular pores in the direction of tensile loading (Rice & Tracey 1969; McMeeking & Hom 1990). Growth of elongate pores perpendicular to the loading direction is associated with creep fracture of ceramics where pore growth is dominated by diffusive-mass transfer (Wilkinson 1981; Hutchinson 1983). The elongation of pores perpendicular to the least principal stress is likely to result from enhanced rates of mass transfer and pore enlargement due to the concentration of tensile stress at the tips of elongate pores (Fig. 5). This positive feedback between tensile stress and mass transfer is characteristic of diffusive-fracture growth and of stress-corrosion fracture (Dutton 1974; Wiederhorn 1978; Atkinson & Meredith 1987). The tendency to form elongate pores due to the stress amplification at pore 'tips' is opposed by the tendency to minimize the surface free energy of pores by diffusive-mass transfer from pore surfaces of high curvature to areas of low curvature. This process tends to blunt the fracture tip and to form spherical pores by a process I refer to as diffusive-pore spheroidization (Fig. 5). The tendency of elongating pores perpendicular to the loading direction by diffusive-fracture growth is also opposed by flow that will tend to elongate pores parallel to the loading direction. Flow in clinker is likely to result from solution–precipitation creep and possible grain-boundary sliding associated with partial melting and the formation of high-temperature phases (Eichhubl & Aydin 2003). The shape of elongate pores thus reflects a balance between rates of pore elongation and fracture-tip sharpening as a result of stress-enhanced reactions, and of blunting due to flow and diffusive-pore spheroidization (Fig. 5). Without the last two effects, pores would evolve into sharp-tipped flaw-like structures rather than elongate voids with blunt tips.

Compared to clinker, pores in chert spheroids are less elongated and more equidimensional, as well as more irregular in shape. The more equidimensional shape of pores suggests that pore spheroidization was more effective compared to pore growth in clinker. Pore surfaces in chert spheroids are typically inward convex (arrows in Fig. 4d and f), contributing to the irregular pore shape. Within high-porosity equatorial regions of chert spheroids, the opal-CT–quartz texture resembles an emulsion (Adamson & Gast 1997), with opal-CT matrix effectively coagulating and forming islands within quartz 'pore' cement. This texture suggests that pore shape is strongly controlled by surface forces of opal-CT to quartz contacts. Despite these strong effects of surface forces on pore geometry, contributions of diffusive-fracture growth and of creep are still noticeable, however, resulting in the fracture-like coalescence geometry of pores and the zig-zag fracture geometry, as discussed in the following section.

In both examples of ductile fracture, the effect of bedding on the preferred orientation of pore and fracture growth is apparently weak. In clinker, this is attributed to the obliteration of any sedimentary preferred fabric during high-temperature mineral neoformation. Within chert spheroids, bedding is observed through variations in pigmentation of the silica phases (Fig. 4a). The concentric shape of fractures suggests, however, that fracture shape is dominantly controlled by the radial-symmetric stress field induced by constrained sintering with little regard to a bedding-parallel stress or strength anisotropy.

In contrast to creep fracture in alumina where pores nucleate by cavitation along crystal facets (Wilkinson 1981; Luecke & Wiederhorn 1999), pores in both clinker and chert are inferred to be inherited from the porous diatomaceous protolith. Pore nucleation is thus not considered here as a process separate from pore growth.

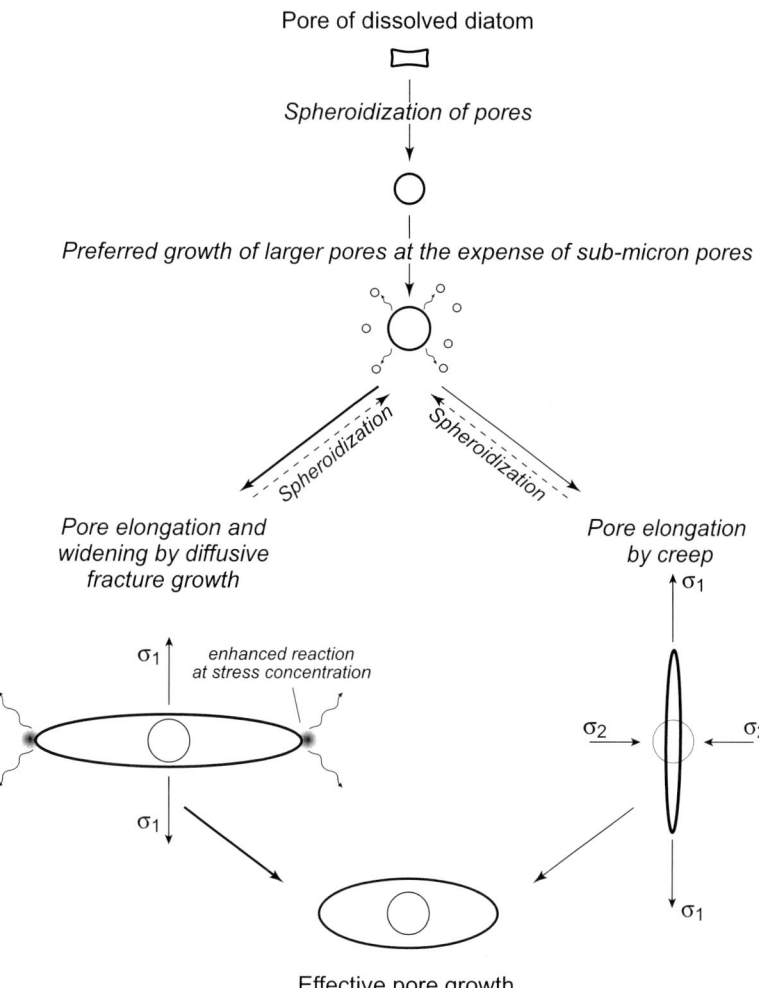

Pore of dissolved diatom

Spheroidization of pores

Preferred growth of larger pores at the expense of sub-micron pores

Spheroidization

Spheroidization

Pore elongation and widening by diffusive fracture growth

Pore elongation by creep

σ_1

σ_1

enhanced reaction at stress concentration

σ_2

σ_2

σ_1

σ_1

Effective pore growth

Fig. 5. Processes controlling pore growth. Mass transfer associated with sintering tends to minimize surface free energy leading to pore spheroidization. Diffusive-fracture growth results in pore elongation perpendicular to the maximum principal tensile stress due to enhanced reactions and mass transfer at tensile stress concentrations. Void growth due to solution–precipitation creep results in pore elongation parallel to the maximum principal tensile stress. The final pore shape is considered to be a result of all three processes.

Controls on pore coalescence

Continued deformation in clinker and chert spheroids results in pore coalescence (Fig. 2). In both materials, fracture growth by pore coalescence is inferred based on various stages of apparent pore coalescence preserved ahead of fracture tips. Pore coalescence in clinker takes place by the thinning of ligaments between pores and occasional ligament rupture, with remnant pore shapes clearly recognizable and ligaments forming blunt protrusions along the irregular fracture surfaces (Fig. 1d) (Eichhubl & Aydin 2003). A similar process is inferred for opal-

CT spheroids, although remnants of failed intra-pore ligaments (arrows in Fig. 4c) are more blunt. This suggests that diffusive-mass transfer in spheroids is more effective in rounding off failed ligaments, consistent with a stronger effect of surface forces on pore shape.

Pore growth and coalescence in clinker and chert spheroids occur preferentially obliquely ahead of fracture tips resulting in fracture growth along directions oriented 25–40° with respect to the mean macroscopic fracture direction (Figs 1d and 4f). Preferred void growth and coalescence along oblique directions, referred to as side-lobe damage

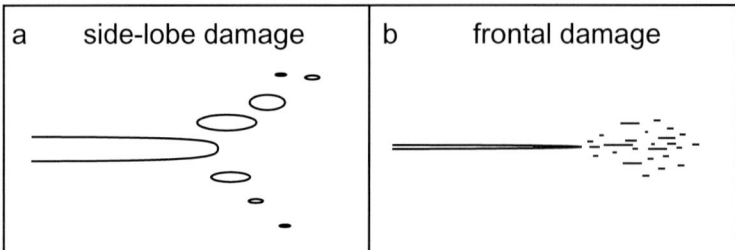

Fig. 6. Damage distribution at fracture tips, loaded in tension perpendicular to the fracture: (**a**) side-lobe damage (after Evans & Blumenthal 1984), shown here to reflect the pore geometry of clinker; (**b**) frontal damage.

(Fig. 6a) (Blumenthal & Evans 1984; Evans & Blumenthal 1984), is characteristic of ductile fracture in metals and of creep rupture in ceramics. Based on the spatial coincidence of side-lobe damage with the distribution of maximum shear stress ahead of fracture tips (Fig. 7a), Evans & Blumenthal (1984) explained side-lobe damage by the dependence of void growth on shear stress. Void growth in metals involves plastic flow by crystal–plastic deformation mechanisms which is usually considered to be dependent on the equivalent stress, $\sigma_e = \sqrt{\tfrac{1}{2}[(\sigma_1 - \sigma_2)^2 + (\sigma_1 - \sigma_3)^2 + (\sigma_2 - \sigma_3)^2]^{1/2}}$ (Hill 1950). For solution–precipitation creep, Paterson (1995) proposed a model that depends on the maximum shear stress, $\tau_{max} = (\sigma_1 - \sigma_3)/2$. Based on the inference that pore growth in clinker and chert spheroids resulted from solution–precipitation creep, contours of maximum shear stress are shown in Figure 7a for a penny-shaped crack loaded by a unit tensile stress acting perpendicular to the fracture. Contours in maximum shear stress are contrasted in Figure 7b with contours of normal stress, σ_{11}, acting perpendicular to the fracture. The normal stress perpendicular to the fracture would be expected to control the propagation of brittle opening-mode fractures consistent with linear elastic fracture mechanics (Anderson 1995). In Figure 7 it is noted that both τ_{max} and σ_{11} have two local maxima forming two lobes off the fracture axis, although the two maxima of the maximum shear stress τ_{max} are significantly more pronounced.

Preferred side-lobe damage ahead of fracture tips is consistent with the shear-stress dependence of void growth by solution–precipitation creep. In contrast, brittle opening-mode fracture by cleavage, trans- and intergranular fracture is typically associated with frontal damage (Fig. 6b), as observed by acoustic emissions during fracture testing (Labuz *et al.* 1987) and damage mapping (Hoagland *et al.* 1973; Nolen-Hoeksema & Gordon 1987). Because of the local minimum in maximum shear stress along the fracture axis ahead of the fracture tip,

frontal damage and coplanar fracture propagation are not expected for damage mechanisms that are controlled by the concentration of shear stress. Although flaw nucleation around voids (Sammis & Ashby 1986) or sliding cracks (Horii & Nemat-Nasser 1986) depend on shear stress for incipient flaw growth, the influence of shear stress on flaw growth decreases with increasing flaw length. Frontal damage localization characteristic of brittle opening-mode rock fracture is thus likely to be ultimately controlled by normal stress, which is consistent with experimental results (Schmidt 1980; Ouchterlony 1983). Because opening-mode fracture growth by stress corrosion and diffusion is, to a first approximation, controlled by normal stress (Dutton 1974; Wiederhorn 1978; Atkinson & Meredith 1987), side-lobe damage observed in clinker and opal-CT chert suggests that creep is involved in pore growth in both materials.

Side-lobe damage creates a propensity for ductile opening-mode fractures to grow along paths that are oblique to the loading direction. Crack growth by side-lobe damage is locally controlled by the stress distribution at the fracture tip and by dependence of the damage mechanism on the stress state. The macroscopic propagation direction, however, is controlled by the global stress and strain-rate fields (Anderson 1995). This is illustrated by the characteristic 'cup-and-cone' shape of failed cylindrical specimens of ductile metals loaded in uniaxial tension (Thomason 1990). In the centre portion of the specimen, where radial stresses perpendicular to the loading direction and therefore stress triaxiality are high, ductile fractures zig-zag around a mean direction perpendicular to the macroscopic loading direction producing a dimpled 'cup'. Towards the sides of the necking specimen, where radial stresses approach zero and stress triaxiality is low, the material fails in oblique opening-mode along planar zones that are oblique to the macroscopic loading direction resulting in the cone shape of the failed specimen. These deformation experiments thus suggest that zig-zag propagation

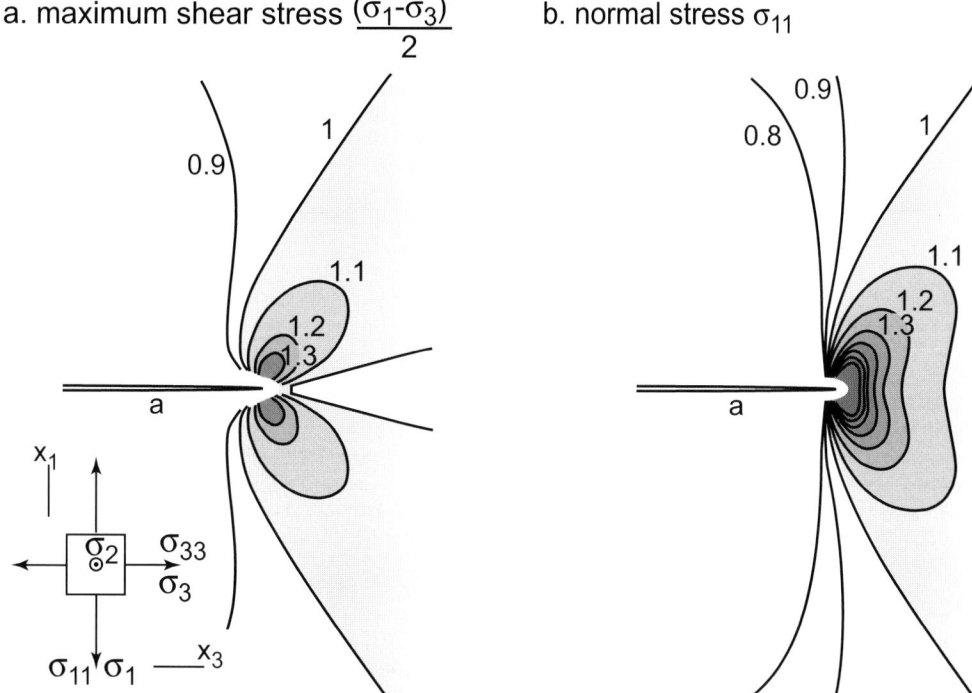

a. maximum shear stress $\frac{(\sigma_1 - \sigma_3)}{2}$

b. normal stress σ_{11}

Fig. 7. Contours of (**a**) normalized maximum shear stress and (**b**) of normalized normal stress acting on planes parallel to the fracture. Shear stress is concentrated in two distinct lobes that are oblique to the fracture tip and that are separated by a distinct local minimum along the fracture axis. Normal stress is characterized by a single concentration ahead of the fracture tip with two maxima at 60° with respect to the fracture axis. Contours obtained with the boundary-element code Poly3D (Thomas 1993) for a penny-shaped fracture of unit radius a, loaded by a unit tensile stress, σ_1, perpendicular to the fracture, an intermediate principal stress, $\sigma_2 = 0.5\,\sigma_1$, parallel to the fracture and perpendicular to the observation plane, and a least principal stress, $\sigma_3 = 0$. Remote loading directions and the designation of coordinate axes are shown in inset. Tension is considered positive, Poisson's ratio $= 0.1$

paths with short segments reflect high stress triaxiality and paths with long segments low stress triaxiality.

The zones of oblique opening-mode failure in uniaxial tension experiments of ductile metals, usually referred to as deformation bands, are zones of localized pore growth and coalescence (Thomason 1990) that can be considered analogous to oblique zones of pore growth and coalescence in ceramics (Dalgleish & Evans 1985). It is suggested here that the same analogy holds for pore growth along oblique zones in clinker and chert spheroids. Following Hill (1961, 1962) deformation bands can be considered as stationary velocity discontinuities in an initially homogenous plastic velocity field. Compatibility requirements of the velocity field limit the orientation of such stationary velocity discontinuities to planes along which components of the strain-rate tensor that are contained within the plane are zero and other strain-rate tensor components can be non-zero (Thomason 1990). For a Cartesian coordinate system chosen with the (x_2, x_3) plane parallel to the stationary velocity discontinuity (Fig. 8), zero strain-rate tensor components are $\dot{\varepsilon}_{22}$, $\dot{\varepsilon}_{33}$, and $\dot{\gamma}_{23}$. In strain-rate Mohr space, the angle Ψ between the deformation band and the direction of maximum principal strain rate, $\dot{\varepsilon}_1$, is thus given by the intercept of the $(\dot{\varepsilon}_1, \dot{\varepsilon}_3)$ strain-rate circle with the shear strain-rate axis $\frac{1}{2}$ (Fig. 8). For volume-conserving deformation (i.e. no volume dilation, $\Delta_v = 0$) such as dislocation slip, deformation bands are inclined at 45° to the direction of maximum principal strain rate, $\dot{\varepsilon}_1$, under plane-strain conditions ($\dot{\varepsilon}_1 = -\dot{\varepsilon}_3$, $\dot{\varepsilon}_2 = 0$) and at 54.7° for axisymmetric-strain conditions ($\dot{\varepsilon}_2 = \dot{\varepsilon}_3 = -\frac{1}{3}\dot{\varepsilon}_1$) Thomason 1990). With increasing volume dilation perpendicular to the band, the angle between deformation band and the direction of maximum principal strain rate increases (Rudnicki & Rice 1975), approaching 90° for uniaxial strain-rate conditions $\lim_{\dot{\varepsilon}_2,\dot{\varepsilon}_3,\to 0} (\dot{\varepsilon}_1 \gg \dot{\varepsilon}_2 = \varepsilon_3)$ (Fig. 8) (Dalgleish & Evans 1985). The angle of oblique fracture growth in clinker and chert spheroids of 25–40° relative to the mean macroscopic fracture

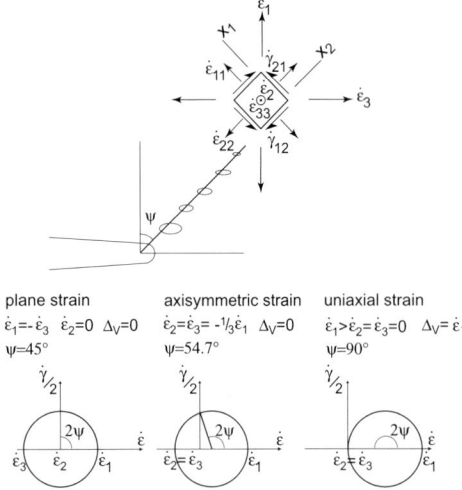

Fig. 8. Dependence of deformation-band orientation on strain-rate boundary conditions (after Thomason 1990). Deformation bands are predicted to follow directions along which $\dot{\varepsilon}_{22} = \dot{\varepsilon}_{33} = 0$. For isovolumetric deformation ($\Delta_V = 0$), deformation bands are predicted at angles of 45° for plane strain and at 54.7° for uniaxial strain with respect to the direction of maximum principal strain rate, $\dot{\varepsilon}_1$. With increasing positive dilation in the $\dot{\varepsilon}_1$ direction and $\dot{\varepsilon}_2 = \dot{\varepsilon}_3 \to 0$, the angle between deformation band and $\dot{\varepsilon}_1$ approaches 90°.

direction, corresponding to $\Psi = 50$–65°, is consistent with the observation of extensive pore growth and resulting dilation along zig-zag fracture paths.

The macroscopic direction of fracture arrays, controlled by global stress and strain-boundary conditions, is presumably not specific to either ductile or brittle–elastic fracture. It should be noted, however, that elastic interaction among fractures in en échelon arrays favours a consistent left- or right-stepping direction (Du & Aydin 1991) over alternating stepping that results in zig-zag arrays characteristic of ductile fracture.

Significance of ductile rock fracture

Growth and coalescence of elongate pores along preferred directions independent of a pre-existing textural anisotropy indicate that the directionality of these processes is controlled by stress- and strain-boundary conditions. This directionality of the growth process, in conjunction with the resulting macroscopic fracture-like discontinuity, characterizes these processes as a fracture phenomenon. Based on the observation of extensive pore growth and coalescence in the surrounding material outside the tip regions of fractures, these processes are

described as a ductile fracture process following the definition given earlier.

In both examples of ductile fracture presented in this study, inelastic deformation is inferred to result from solution–precipitation creep and mass transfer in the presence of a liquid phase. Ductile fracture by pore growth thus demonstrates that the formation of macroscopic fractures is not necessarily associated with brittle fracture on a grain scale. The geological occurrence of ductile fracture need not be limited to solution–precipitation creep, however. I speculate that, depending on subsurface conditions, material composition and texture, and deformation rate, ductile fracture may be associated with any inelastic-deformation mechanisms characteristic of Earth materials including diffusion and dislocation creep, and cataclastic or granular flow (Knipe 1989; Tullis 1990; Paterson 1995). In contrast, stress corrosion and diffusive fracture due to localized chemical reactions and diffusive-mass transport at the fracture tip (Atkinson & Meredith 1987) are considered here as processes of brittle subcritical fracture. Subcritical refers to fracture propagation at loading stresses below those required for failure by predominantly mechanical bond rupture (Atkinson & Meredith 1987), and is frequently designated 'slow' fracture as opposed to 'fast' – although not necessarily catastrophic – critical fracture in experimental fracture mechanics (e.g. Quinn 1990). The distinguishing parameter between ductile and brittle fracture is thus the extent of inelastic deformation and chemical mass transfer involved in fracture. Based on these criteria, Eichhubl & Aydin (2003) considered ductile fracture as an end member in a series spanning from brittle–elastic fracture and stress-corrosion fracture to large-aperture and fully ductile fracture.

Fracture-mechanisms maps can be constructed based on deformation-mechanism maps (Rutter 1976) and fracture-mechanism maps for engineering materials (Fields & Fuller 1981; Quinn 1990) that predict the occurrence of rock-fracture processes as a function of temperature and loading rate or stress for any given material composition (Atkinson & Meredith 1987). Figure 9 depicts a generalized map for ductile fracture and for subcritical and critical brittle fracture as a function of temperature and loading stress. The shape of the fracture-mechanism curves is based on the inferred temperature conditions for fracture formation in clinker and opal-CT chert, and on experimentally obtained curves for ceramics by Quinn (1990). Although speculative, the relative position of the fracture curve with respect to the triple junction among the fields represented by solution–precipitation creep diffusion creep and dislocation creep is based on the inference that the uniaxial tensile strength, σ_T, for critical brittle fracture at room temperature will usually exceed the stress magnitude of the triple junction

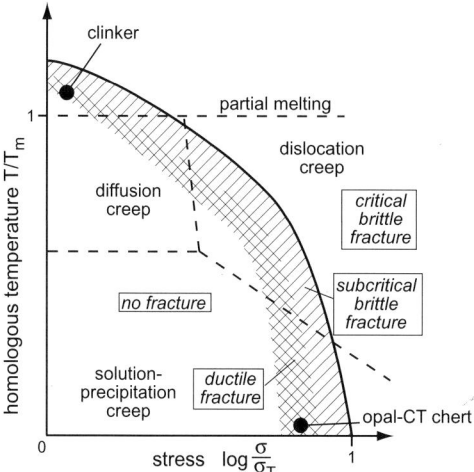

Fig. 9. Fracture-mechanism map in space of normalized stress σ/σ_T, where σ_T is the uniaxial tensile strength for critical brittle failure at room temperature, v. homologous temperature T/T_m, where T_m is the melting temperature. Fracture-mechanism fields are based on the inferred temperature conditions for ductile fracture in clinker and chert spheroids. The general shape of fracture curves follow Quinn (1990). The position of fracture curves relative to the deformation-mechanism fields is schematic only and will vary depending on material and pore fluid composition, and strain rate. Deformation-mechanism fields for fine-grained quartzite after Rutter (1976).

under subsurface conditions. Rutter (1976) estimates the stress of the deformation mechanism triple junction at about 1 MPa for fine-grained quartzite, whereas σ_T is in the range of 10–40 MPa for crystalline rocks (Jaeger & Cook 1979).

Figure 9 illustrates that fracture processes are not necessarily confined to low-temperature environments but are expected in high-temperature regimes provided that suitable loading conditions are met. Brittle fracture was observed in deformation experiments of partially melted granite by van der Molen & Paterson (1979), Dell'Angelo & Tullis (1988) and Rutter & Neumann (1995). Structures that share textural geometric properties of ductile fractures as described in this study have been produced recently in partial melting experiments by Bruhn *et al.* (2000) and Wark & Watson (2002). These experiments suggest that ductile fracture is potentially significant for melt migration, consistent with the observed geometry of melt-segregation veins (Clemens & Mawer 1992) and with theoretical models of melt migration in fractures (e.g. Sleep 1988). At a larger scale, ductile void growth was invoked by Regenauer-Lieb (1999) to explain the localization of intraplate volcanism.

Although inferred temperatures for ductile-fracture formation in clinker and chert spheroids

differ by >1000 °C, both examples formed at shallow burial depth and thus at low effective confining pressure. Under intermediate–deep subsurface conditions, formation of macroscopic ductile opening-mode fractures will require high pore-fluid pressure to obtain effective tensile-stress conditions. In addition, ductile fracture involving chemical-mass transfer such as solution–precipitation creep will require a protolith material that is sufficiently chemically reactive under given burial conditions and fluid environment. Chemically reactive conditions are most favourable in prograde diagenetic and metamorphic environments. Because reaction-controlled-deformation mechanisms are dependent on loading rate, ductile-fracture processes will be favoured by slow, interseismic rates of tectonic loading.

Conclusions

Fracture formation in clinker and opal-CT chert spheroids is characterized by pore growth and coalescence. Pore growth is accompanied by extensive textural and compositional reorganization of the host material. Pore growth is inferred to result from combined effects of diffusive-pore growth and solution–precipitation creep. Pore elongation in the macroscopic fracture direction results from diffusive fracture growth where mineral reactions and diffusive-mass transfer are enhanced by the tensile stress concentration within the tip regions of elongate pores and fractures. Flow by solution–precipitation creep, on the other hand, tends to elongate pores parallel to the direction of maximum tensile loading. Distortion of pore shape by these two processes is opposed by the thermodynamic tendency to minimize surface curvature thus tending towards spherical or isometric pore shapes. In clinker and, to a lesser extent, in chert spheroids diffusive-pore growth is the dominant mechanism resulting in pore elongation parallel to the mean macroscopic fracture direction. The shear-stress dependence of pore growth by creep is inferred to cause preferred pore growth within two lobes adjacent to fracture tips. Preferred side-lobe damage provides a propensity of the fracture to grow along two oblique directions with respect to the macroscopic fracture direction. The macroscopic fracture direction is controlled by global stress- and strain-boundary conditions, forcing the fracture to zig-zag along a mean direction perpendicular to the least compressive principal stress.

Ductile fracture growth in clinker and opal-CT chert demonstrates that macroscopic fracture formation is not necessarily associated with brittle fracture on a grain scale. Rather, fracture growth can involve a variety of deformation mechanisms depending on rock composition, temperature, stress and strain rate. Although both examples of ductile fracture formed at shallow depth and thus at low confining stress, temperature conditions and mineral reactions were different suggesting that ductile-fracture-growth processes can occur in chemically reactive environments and at slow loading rates under various burial conditions that include high-temperature environments.

Financial support for this study was provided by the industrial affiliates of the Stanford Rock Fracture Project. The author also received half-time support through grant DE-FG03-94ER14462 from the US Department of Energy, Office of Basic Energy Sciences, The Chemical Sciences, Geosciences, and Biosciences Division. This work benefited from discussions with A. Aydin, D. Pollard, R. McMeeking, N. Davatzes and R. Behl. A. Aydin, D. Pollard and GSL reviewers, J.-P. G. and S. Laubach, are thanked for detailed reviews.

References

ADAMSON, A. W. & GAST, A. P. 1997. *Physical Chemistry of Surfaces*, 6th edn. Wiley-Interscience, New York.

ANDERSON, T. L. 1995. *Fracture Mechanics. Fundamentals and Applications*, 2nd edn. CRC Press, Boca Raton, FL.

ATKINSON, B. K. & MEREDITH, P. G. 1987. The theory of subcritical crack growth with application to minerals and rocks. *In*: ATKINSON, B. K. (ed.) *Fracture Mechanics of Rock*. Academic Press, London, 111–166.

BEHL, R. J. 1992. *Chertification in the Monterey Formation of California and deep-sea sediments of the West Pacific*. PhD thesis, University of California, Santa Cruz.

BEHL, R. J. & GARRISON, R. E. 1994. The origin of chert in the Monterey Formation of California (USA). *In*: IIJIMA, A. (ed.) *Proceedings of the 29th International Geological Congress*, C. VSP, Utrecht, 101–132.

BENTOR, Y. K. & KASTNER, M. 1976. Combustion metamorphism in southern California. *Science*, **193**, 486–488.

BLUMENTHAL, W. & EVANS, A. G. 1984. High-temperature failure of polycrystalline alumina: II, creep crack growth and blunting. *Journal of the American Ceramic Society*, **67**, 751–759.

BRUECKNER, H. K., SNYDER, W. S. & BOUDREAU, M. 1987. Diagenetic controls on the structural evolution of siliceous sediments in the Golconda allochthon, Nevada, U.S.A. *Journal of Structural Geology*, **9**, 403–417.

BRUHN, D., GROEBNER, N. & KOHLSTEDT, D. L. 2000. An interconnected network of core-forming melts produced by shear deformation. *Nature*, **403**, 883–886.

CISOWSKI, S.M. & FULLER, M. 1987. The generation of magnetic anomalies by combustion metamorphism of sedimentary rock, and its significance to hydrocarbon exploration. *Geological Society of America Bulletin*, **99**, 21–29.

CLEMENS, J. D. & MAWER, C. K. 1992. Granitic magma transport by fracture propagation. *Tectonophysics*, **204**, 339–360.

DALGLEISH, B. J. & EVANS, A. G. 1985. Influence of shear bands on creep rupture in ceramics. *Journal of the American Ceramic Society*, **68**, 44–48.

DALGLEISH, B. J., SLAMOVICH, E. B. & EVANS, A. G. 1985. Duality in the creep rupture of a polycrystalline alumina. *Journal of the American Ceramic Society*, **68**, 575–581.

DELL'ANGELO, L. N. & TULLIS, J. 1988. Experimental deformation of partially melted granitic aggregates. *Journal of Metamorphic Geology*, **6**, 495–515.

DU, Y. & AYDIN, A. 1991. Interaction of multiple cracks and formation of echelon crack arrays. *International Journal for Numerical and Analytical Methods in Geomechanics*, **15**, 205–218.

DUTTON, R. 1974. The propagation of cracks by diffusion. *In*: BRADT, R. C., HASSELMAN, D. P. H. & LANGE, F. F. (eds) *Fracture Mechanics of Ceramics, Vol. 2: Microstructure, Materials, and Applications*. Plenum Press, New York, 647–657.

EICHHUBL, P., & AYDIN, A. 2003. Ductile opening-mode fracture by pore growth and coalescence during combustion alteration of siliceous mudstone. *Journal of Structural Geology*, **25**, 121–134.

EICHHUBL, P., AYDIN, A. & LORE, J. 2001. Opening-mode fracture in siliceous mudstone at high homologous temperature – effect of surface forces. *Geophysical Research Letters*, **28**, 1299–1302.

ENGELDER, T., FISCHER, M. P. & GROSS, M. R. 1993. *A Short Course Manual on Geological Aspects of Fracture Mechanics*. Geological Society of America, Boston, MA.

EVANS, A. G. & BLUMENTHAL, W. 1984. High temperature failure mechanisms in ceramic polycrystals. *In*: TRESSLER, R. E. & BRADT, R. C. (eds) *Deformation of Ceramic Materials II*. Plenum Press, New York, 487–505.

FIELDS, R. J. & FULLER, E. R. 1981. Crack growth mechanism maps. *In*: FRANCOIS, D. (ed.) *Advances in Fracture Research*. Pergamon Press, Oxford, 1313–1322.

FLECK, N. A. 1991. Brittle fracture due to an array of microcracks. *Proceedings of the Royal Society London*, **A432**, 55–76.

GERMAN, R.M. 1996. *Sintering Theory and Practice*. Wiley, New York.

HANDIN, J. 1966. Strength and ductility. *In*: CLARK, S. P. (ed.) *Handbook of Physical Constants*. Geological Society of America, Memoirs, **97**, 223–289.

HILL, R. 1950. *The Mathematical Theory of Plasticity*. Oxford University Press, Oxford.

HILL, R. 1961. Discontinuity relations in mechanics of solids. *In*: SNEDDON, I. N. & HILL, R. (eds) *Progress in Solid Mechanics, Volume II*. North-Holland, Amsterdam, 247–276.

HILL, R. 1962. Acceleration waves in solids. *Journal of the Mechanics and Physics of Solids*, **10**, 1–16.

HOAGLAND, R. G., HAHN, G. T. & ROSENFIELD, A. R. 1973. Influence of microstructure on fracture propagation in rock. *Rock Mechanics*, **5**, 77–106.

HORII, H. & NEMAT-NASSER, S. 1986. Brittle failure in compression: splitting, faulting and brittle–ductile transition. *Philosophical Transactions of the Royal Society London*, **A 319**, 337–374.

HUTCHINSON, J. W. 1983. Constitutive behavior and crack tip fields for materials undergoing creep-constrained grain boundary cavitation. *Acta metallurgica*, **31**, 1079–1088.

INGRAFFEA, A. R. 1987. Theory of crack initiation and propagation in rock. *In*: ATKINSON, B. K. (ed.) *Fracture Mechanics of Rock*. Academic Press, London, 71–110.

ISAACS, C. M. 1982. Influence of rock composition on kinetics of silica phase changes in the Monterey Formation, Santa Barbara area, California. *Geology*, **10**, 304–308.

JAEGER, J. C. & COOK, N. G. W. 1979. *Fundamentals of Rock Mechanics*, 3rd edn. Chapman & Hall, London.

KELLER, M. A. & ISAACS, C. M. 1985. An evaluation of temperature scales for silica diagenesis in diatomaceous sequences including a new approach based on the Miocene Monterey Formation, California. *Geo-Marine Letters*, **5**, 31–35.

KEMENY, J. M. & COOK, N. G. W. 1991. Micromechanics of deformation in rocks. *In*: SHAH, S. P. (ed.) *Toughening Mechanisms in Quasi-brittle Materials*. Kluwer, Amsterdam, 155–188.

KINLOCH, A. J. & YOUNG, R. J. 1983. *Fracture Behavior of Polymers*. Applied Science Publishers, London.

KNIPE, R. J. 1989. Deformation mechanisms – recognition from natural tectonites. *Journal of Structural Geology*, **11**, 127–146.

KOLODNY, Y. 1969. Petrology of siliceous rocks of the Mishash Formation (Negev, Israel). *Journal of Sedimentary Petrology*, **39**, 166–175.

KRAUSE, R. F., JR, LUECKE, W. E., FRENCH, J. D., HOCKEY, B. J. & WIEDERHORN, S. M. 1999. Tensile creep and rupture of silicon nitride. *Journal of the American Ceramic Society* **82**, 1233–1241.

LABUZ, J. F., SHAH, S. P. & DOWDING, C. H. 1987. The fracture process zone in granite: evidence and effect. *International Journal of Rock Mechanics and Mining Sciences & Geomechanics Abstracts*, **24**, 235–246.

LASAGA, A. 1998. *Kinetic Theory in the Earth Sciences*. Princeton University Press, Princeton, NJ.

LORE, J., EICHHUBL, P. & AYDIN, A. 2002. Alteration and fracturing of siliceous mudstone during in situ combustion, Orcutt field, California. *Journal of Petroleum Science and Engineering*, **36**, 169–182.

LUECKE, W. E. & WIEDERHORN, S. M. 1999. A new model for tensile creep of silicon nitride. *Journal of the American Ceramic Society*, **82**, 2769–2778.

MCMEEKING, R. M. & HOM, C. L. 1990. Finite element analysis of void growth in elastic–plastic materials. *International Journal of Fracture*, **42**, 1–19.

NOLEN-HOEKSEMA, R. C. & GORDON, R. B. 1987. Optical detection of crack patterns in the opening-mode fracture of marble. *International Journal of Rock Mechanics and Mining Sciences & Geomechanics Abstracts*, **24**, 135–144.

OLSON, J.E. 1993. Joint pattern development: effects of subcritical crack growth and mechanical crack interaction. *Journal of Geophysical Research*, **98**, 12 251–12 265.

OUCHTERLONY, F. 1983. Fracture toughness testing of rock. *In*: ROSSMANITH, H. P. (ed.) *Rock Fracture Mechanics*. Springer, Wien, 69–150.

PATERSON, M. S. 1995. A theory for granular flow accommodated by material transfer via an intergranular fluid. *Tectonophysics*, **245**, 135–151.

POLLARD, D. D. & AYDIN, A. 1988. Progress in understanding jointing over the past century. *Geological Society of America Bulletin*, **100**, 1181–1204.

POLLARD, D. D. & SEGALL, P. 1987. Theoretical displacements and stresses near fractures in rock: with application to faults, joints, veins, dikes, and solution surfaces. *In*: ATKINSON, B. K. (ed.) *Fracture Mechanics of Rock*. Academic Press, London, 277–349.

QUINN, G. D. 1990. Fracture mechanism maps for advanced structural ceramics – Part 1: Methodology and hot-pressed silicon nitride results. *Journal of Materials Science*, **25**, 4361–4376.

REGENAUER-LIEB, K. 1999. Dilatant plasticity applied to Alpine collision: ductile void growth in the intraplate area beneath the Eifel volcanic field. *Geodynamics*, **27**, 1–21.

RENSHAW, C. E. & POLLARD, D. D. 1994. Numerical simulation of fracture set formation: A fracture mechanics model consistent with experimental observations. *Journal of Geophysical Research*, **99**, 9359–9372.

RICE, J. R. & TRACEY, D. M. 1969. On the ductile enlargement of voids in triaxial stress fields. *Journal of the Mechanics and Physics of Solids*, **17**, 201–217.

ROBERTSON, A. G., WILKINSON, D. S. & CÁCERES, C. H. 1991. Creep and creep fracture in hot-pressed alumina. *Journal of the American Ceramic Society*, **74**, 915–921.

RUDNICKI, J. W. & RICE, J. R. 1975. Conditions for the localization of deformation in pressure-sensitive dilatant materials. *Journal of the Mechanics and Physics of Solids*, **23**, 371–394.

RUTTER, E. H. 1976. The kinetics of rock deformation by pressure solution. *Philosophical Transactions of the Royal Society London*, **A283**, 203–219.

RUTTER, E. H. 1986. On the nomenclature of mode of failure transitions in rocks. *Tectonophysics*, **122**, 381–387.

RUTTER, E. H. & NEUMANN, D. H. K. 1995. Experimental deformation of partially molten Westerly granite under fluid-absent conditions, with implications for the extraction of granitic magma. *Journal of Geophysical Research*, **100**, 15 697–15 715.

SABLJIC, D. B. & WILKINSON, D. S. 1995. Influence of a damage zone on high temperature crack growth in brittle materials. *Acta metallurgica et materialia*, **43**, 3937–3945.

SAMMIS, C. G. & ASHBY, M. F. 1986. The failure of brittle porous solids under compressive stress states. *Acta metallurgica*, **34**, 511–526.

SCHMIDT, R. A. 1980. A microcrack model and its significance to hydraulic fracturing and fracture toughness testing. *In*: *Proceedings of the 21st Symposium on Rock Mechanics*. University of Missouri, Rolla, MO, 581–590.

SEGALL, P. 1984. Formation and growth of extensional fracture sets. *Geological Society of America Bulletin*, **95**, 454–462.

SLEEP, N. H. 1988. Tapping of melt by veins and dikes. *Journal of Geophysical Research*, **93**, 10 255–10 272.

STEIN, C. L. & KIRKPATRICK, R. J. 1976. Experimental porcelanite recrystallization kinetics: a nucleation and growth model. *Journal of Sedimentary Petrology*, **46**, 430–435.

STOUFFER, D. C. & DAME, L. T. 1996. *Inelastic Deforma-*

tion of Metals. Models, Mechanical Properties, and Metallurgy. Wiley, New York.

TALIAFERRO, N. L. 1934. Contraction phenomena in cherts. *Geological Society of America Bulletin*, **45**, 189–232.

THOMAS, A. L. 1993. *Poly3D: A three-dimensional, polygonal element, displacement discontinuity boundary element computer program with applications to fractures, faults, and cavities in the Earth's crust.* MS thesis, Stanford University, Stanford.

THOMASON, P. F. 1990. *Ductile Fracture of Metals.* Pergamon Press, Oxford.

TULLIS, J. 1990. Experimental studies of deformation mechanisms and microstructures in quartzo-feldspathic rocks. *In*: BARBER, D. J. & MEREDITH, P. G. (eds) *Deformation Processes in Minerals, Ceramics, and Rocks.* Unwin Hyman, London, 190–227.

VAN DER MOLEN, I. & PATERSON, M. S. 1979. Experimental deformation of partially-melted granite. *Contributions to Mineralogy and Petrology*, **70**, 299–318.

WARK, D. A. & WATSON, E. B. 2002. Grain-scale channelization of pores due to gradients in temperature or composition of intergranular fluid or melt. *Journal of Geophysical Research*, **107**, ECV5-1–ECV5-15.

WHITTAKER, B. N., SINGH, R. N. & SUN, G. 1992. *Rock Fracture Mechanics – Principles, Design and Applications.* Elsevier, Amsterdam.

WIEDERHORN, S. M. 1978. Mechanisms of subcritical crack growth in glass. *In*: BRADT, R. C., HASSELMAN, D. P. H. & LANGE, F. F. (eds) *Fracture Mechanics of Ceramics, Vol. 4: Crack Growth and Microstructure.* Plenum Press, New York, 549–580.

WIEDERHORN, S. M. & FULLER, E. R. 1985. Structural reliability of ceramic materials. *Materials Science and Engineering*, **71**, 169–186.

WILKINSON, D. S. 1981. A model for creep cracking by diffusion-controlled void growth. *Materials Science and Engineering*, **49**, 31–39.

WILKINSON, D. S., CÁCERES, C. H. & ROBERTSON, A. G. 1991. Damage and fracture mechanisms during high-temperature creep in hot-pressed alumina. *Journal of the American Ceramic Society*, **74**, 922–933.

WILLIAMS, L. A. & CRERAR, D. A. 1985. Silica diagenesis, II. General mechanisms. *Journal of Sedimentary Petrology*, **55**, 312–321.

WILLIAMS, M. L. 1957. On the stress distribution at the base of a stationary crack. *Transactions of the American Society of Mechanical Engineers Series E – Journal of Applied Mechanics*, **24**, 109–114.

WILSDORF, H. G. F. 1983. The ductile fracture of metals: a microstructural viewpoint. *Materials Science and Engineering*, **59**, 1–39.

Age and depth evidence for pre-exhumation joints in granite plutons: fracturing during the early cooling stage of felsic rock

P. BANKWITZ[1], E. BANKWITZ[1], R. THOMAS[2], K. WEMMER[3] & H. KÄMPF[2]

[1] *Gutenbergstraße 60, 14467 Potsdam, Germany, (e-mail: epbank@web.de)*
[2] *GeoForschungsZentrum Potsdam, Telegrafenberg, 14473 Potsdam, Germany*
[3] *Göttinger Zentrum Geowissenschaften, Universität, Goldschmidtstraße 1–3, 37677 Göttingen, Germany*

Abstract: This chapter documents the fracture process associated with the early cooling stage of felsic magma. Characteristics of pre-exhumation joints include their spatial distribution in granite bodies, their fracture surface morphology, and geological and petrological evidence for the depth of fracture initiation. These characteristics allow inferences about the depth and the time of joint origin in the South Bohemian Pluton. The intrusion levels of currently exposed granites of the pluton were 7.4 km in the northern part and 14.3 km in the southern part.

Within the northern Mrákotín Granite (Boršov) early NNE joints propagated while the granite was at a temperature near the solidus, and, in part, magma was still being injected, post-dated by thin granite dykes along NNE joints. Evidence for the pre-exhumation initiation of these joints comes from the geochronological dating of these late-granite dykes (1–2 cm thick) at 324.9 Ma in age, which were creating their own rupture in the rock. The timing of the pluton emplacement at 330–324 Ma and the cooling ages of 328–320 Ma have been given by previous studies. From fluid inclusions within the late-granite dykes that occupy joint surfaces, the trapping depth of the analysed inclusions was calculated to be 7.4 km. Near the solidus H_2O separates during the crystallization of anhydrous phases. The associated increasing H_2O pressure can initiate the first cracks and can generate a small portion of new granitic melt, which forces the cyclic fracture propagation together with mobile, low-viscosity 'residual melt' input into the fracture.

The determination of the intrusion level and time at which the dykes began cooling provide evidence for the joint initiation at a depth of 7.4 km, which was connected with the level and process of final emplacement and early cooling of the Mrákotín Granite long before the main exhumation. At the earliest, the erosion of the upper rock pile, 7.4 km in thickness, started significantly after generation of the early joint sets. The NNE-trending joints are persistent in orientation throughout the South Bohemian Pluton, but the joint-surface morphology varies in all subplutons and occupies all sections of the stress intensity v. crack-propagation velocity curve (Wiederhorn–Bahat curve).

Granite bodies can fracture in various characteristic ways relative to the density of fractures, the number of joint sets and their distribution, and the type of fractographic features on joint surfaces. Some granite plutons are dominated by only one orthogonal subvertical joint system (e.g. in the Hercynian South Bohemian Pluton, Czech Republic; A in Fig. 1A), whereas other plutons contain two or more subvertical joint systems of different generations (e.g. the Cadomian Lausitz Pluton, Germany; D in Fig. 1).

The density of fractures in the Earth's crust decreases with depth (e.g. Price 1966; Ramsay & Huber 1987). This observation does not exclude fracture initiation within plutons at depths below 7 km. We consider two problems: (1) age and depth for propagation of pre-exhumation joints; and (2) the fracture mechanism during the early cooling stage in granite plutons.

(1) Evidence for propagation of NNE joints at the depth of granite emplacement in the South

Bohemian Pluton (SBP) will be discussed. Jointing just after emplacement, during early stages of cooling, may be associated with a temperature above the solidus of granite. Late-granite dykes, creating natural hydraulic subvertical joints, have well-developed fractographic features and date the timing of these NNE-trending joints (Table 1: Boršov, Mrákotín Granite, SBP).

For this purpose the ages of the granite dykes that created early joints were determined by K/Ar dating on micas. These ages (Table 2) are roughly the same as the timing of the SBP emplacement (zircon, monazite and Rb–Sr dating, Table 4). In addition, the depths of emplacement of several plutons (SBP, Czech Republic; Erzgebirge granite plutons, Germany) were calculated by fluid and melt inclusion analysis (Table 3). Depth and time are of importance for the interpretation of the fractographic features with regard to the conditions under which the joints originated.

From: COSGROVE, J. W. & ENGELDER, T. (eds) 2004. *The Initiation, Propagation, and Arrest of Joints and Other Fractures.* Geological Society, London, Special Publications, **231**, 25–47. 0305-8719/04/$15 © The Geological Society of London 2004.

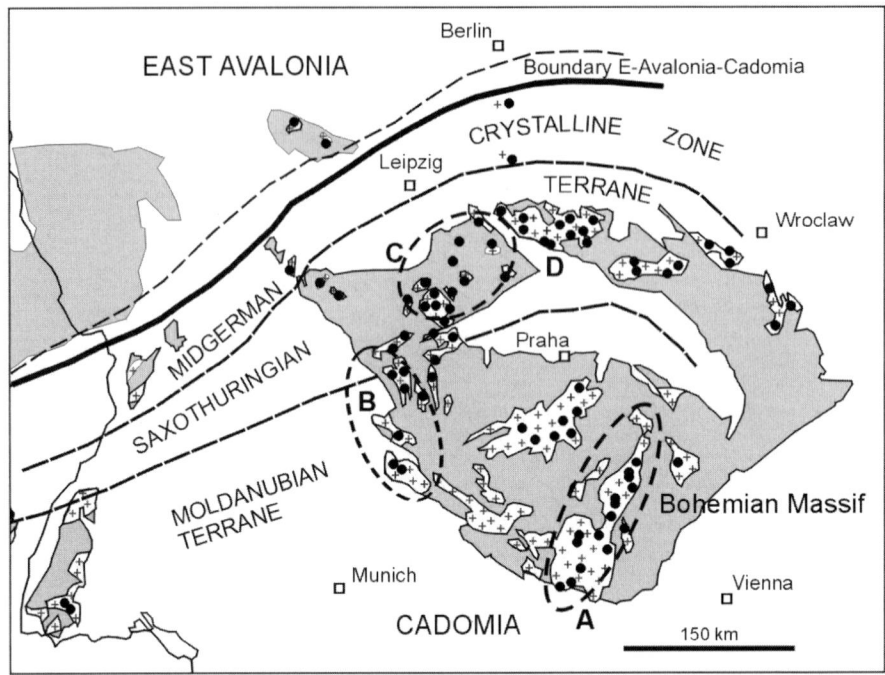

Fig. 1. Late to post-Variscan granites of the Bohemian Massif: A, South Bohemian Pluton (SBP); B, Oberpfalz plutons (surroundings of the KTB); C, Erzgebirge plutons; D, Lusatian Granite Massif: Cadomian–early Cambrian in age. Dots are the localities of sites of investigation.

Table 1. *Succession of early main joints in granite types of the SBP*

Quarry	Early joints First sets (mean values)	Second sets (mean values)	Later joints Third sets
Mrákotin Graníte (SBP-N)			
Boršov	subhorizontal NNE: 025°/090° + ESE: 110° to 120°/090° fringe cracks: 060°/090°	NNE + ESE 042°/040°	ENE, *c.* 120°/090°; subhorizontal;
Mrákotín	subhorizontal to W-dip NNE: 025°/085° + ESE: 120°/090°		
Řásná	subhorizontal to W-dip NNE: 010°/090° + ENE: 080°/090°	E–W	N–S
Mislotice	NNE	NNW to N–S	
Sumrakov	NNE: 020°/090° + ESE: *c.* 115°/090		
Olši	NNE		
Weinsberg Granite (SBP-S)			
Friepeß	NNE: 025°/085° fringe cracks: 060°/090°	ESE: *c.* 120°/090°	
Hartelberg	ENE: 080°/080° ESE: 120°/085° NNE: 018°/090° + ESE: 115°/090 fringe cracks: 060°/090°		
Hartberg	NNE: 010°/090° + ESE: 105°/090		
Windhaag	NNE: 030°/088° + ESE: 100°/085		

Table 2. *Age determinations of late-granite dykes along NNE-trending joints of the Mrákotín Granite (Boršov, Czech Republic) and of the Weinsberg Granite itself (Friepeß, Austria)*

Ar isotopic abundance (%)	Spike isotopic composition (%)	Decay constants (1/a)	Potassium	Standard temperature pressure (STP)	Molar volume (ml)	Atomic weight (g mol^{-1})
^{40}Ar: 99.6000	^{40}Ar: 0.0099980	lambda e: 5.810 × 10^{-11}	^{40}K: 0.011670%	0 °C; 760 mmHg	22413.8	total Ar: 39.9477
^{38}Ar: 0.0630	^{38}Ar: 99.9890000	lambda β: 4.962 × 10^{-10}	K$_2$O/K: 0.8302	Normal atmosphere (DIN 1343)		^{40}Ar: 39.9624
^{36}Ar: 0.3370	^{36}Ar: 0.0009998	lambda L$_{tot}$: 5.543 × 10^{-10}		273.15 K; 1013.25 mbar		total K: 39.1027

Sample	Spike (no)	K$_2$O (wt%)	^{40}Ar* (nl g^{-1})STP	^{40}Ar* (%)	Age (Ma)	2σ error (Ma)	2σ error (%)
Muscovite: Boršov (CZ) FK 1–2, 5	2767	10.37	119.0700	97.96	**324.9**	6.7	2.1
Muscovite: Friepeß (A) FK 12	2810	10.54	121.0500	97.83	**318.9**	6.7	2.1

Laboratory: Centre of Geoscience, Göttingen, Germany.

Table 3. *Trapping data of fluid inclusions in quartz of a late-granite dyke, intersecting the Mrákotín Granite (FK-1), and in quartz of the Weinsberg Granite (FK-11). Both granites are granite types of the SBP*

Sample	Salinity (NaCl equiv.%)	Homogenization temperature (°C)	n	Trapping temperature (°C)	Trapping pressure (kbar)	Depth (km)
FK-1	5.8 ± 0.6	375 ± 6	17	620	1.89	7.4
FK-11	13.8 ± 0.2	332 ± 10	20	662	3.65	14.3

n, number of determinations.

(2) The fracture process itself, the source and the driving stresses of early joints, together with their typical fractographic features, will be discussed. We focused our investigations on the SBP in the eastern part of the Bohemian Massif where emplacement was >7 km (A in Fig. 1A). In addition, we consider data from plutons in the western part of the massif (Fig. 1B, the surroundings of the KTB – which stands for Kontinentales Tiefbohrprogramm der Bundesrepublik Deutschland, or German Continental Deep Drilling Programme– Continental Deep Drilling, Bavaria; Fichtelgebirge and Oberpfalz plutons), which correspond remarkably with those from the SBP (A in Fig. 1). At least, the results were compared with joints of the Erzgebirge granites (C in Fig. 1) which intruded high up into subvolcanic crustal levels. We attempt to relate particular fractographic surface features of joints in the SBP with joints in other granite plutons of different emplacement depth, looking for features

unique to early jointing in deeply emplaced plutons.

Methods

Geological investigations

This study is based on field work carried out by the authors concentrating in the area of two plutons (northern Mrákotín Granite, Czech Republic, and Weinsberg Granite, Austria; Fig. 2). We studied the spatial organization of granite joints together with their fractographic surface pattern and evaluated the timing, in an attempt to recognize features unique to the deep fracture process. For this purpose the earliest joint sets were identified that could be initiated at the depth of granite emplacement. In each case we determined the succession of the joint sets (Table 1), using their fractographic features. Finally, we focused our study on the NNE-trending joints of the SBP.

Table 4. *Intrusion and cooling age data from the granite types of the SBP and, in addition, age data of (A) early granite joints with dykes (Boršov) and (B) the Weinsberg Granite itself (localities in Fig. 2)*

Type of granite	Intrusion/cooling ages	Method	Authors
Eisgarn Complex (2-mica granites)			
	330±6.5–320 Ma	Rb–Sr	Scharbert (1998)
① Čiměř Granite (N) and	**327±4 Ma**	U–Pb monazite	Friedl *et al.* (1996)
① **Mrákotín granite**	**326±1.5 Ma**	Rb–Sr	Frank (1994)
① Eisgarn Granite s.s.	**324±8 Ma**		Gerdes *et al.* (1998)
	① Cooling 328–325 to 320 Ma decreasing from north to south	^{40}Ar–^{39}Ar muscovite	Scharbert (1998)
• (A) JOINTS WITH DYKE	**• Late-granite dykes created joints:** **>324.9±6.7 Ma**	**K–Ar muscovite**	**Wemmer, this study**
① Čiměř Granite (S)	Cooling *c.* 312 Ma		Scharbert (1998)
Late suites: granite porphyries		Rb–Sr	Scharbert (1998)
– coarse grained	323±16 Ma	Rb–Sr	Scharbert (1998)
– fine grained	317±6 Ma	Rb–Sr	Scharbert (1998)
– Josephsthal type	318±3 Ma		
② Homolka Granite, etc.:			Breiter & Scharbert (1998)
– in the northern body	320 Ma		Breiter & Scharbert (1998)
– in the southern body	*c.* 314 Ma		
③ Granite bodies in the axial part of the SBP: Zvůle, Melechov and Čěřínek stock	(Fig. 2: deep-seated, ring-shaped, coarse-grained, gravity minima)		
Weinsberg Complex	328± 6Ma 327±5 & 323±4 Ma, 318 Ma	Rb–Sr whole rock Zircon and U–Pb monazite	Finger & von Quadt (1992) Friedl *et al.* (1996)
Weinsberg Granite I and & Weinsberg Granite II	Cooling 327-325–320 Ma, decreasing from north to south at the border	Muscovite	Scharbert (1998)
• (B) GRANITE Weinsberg I (from a NNE-joint)	**• Cooling muscovites** **318.9±6.7 Ma**	**K-Ar muscovite**	**Wemmer, this study**
	Cooling 313–308 Ma, in the central parts	Muscovite	Scharbert (1998)
Small stock granites	Cooling 316 and 320 Ma	Ar–Ar	Scharbert (1998)
Nebelstein greisen	311.6±1.4 Ma 313.3±1.6–315.8±1.6 Ma 312.2±1.5 Ma	Rb–Sr Ar–Ar muscovite Ar–Ar muscovite Ar–Ar biotite	Scharbert (1998)
Mauthausen group			
Freistadt granodiorite	303±2 Ma	Monazite U–Pb and	Finger *et al.* (1996);
Pfahl granitoids	305±10 Ma	microprobe technique	Friedl *et al.* (1992);
Mauthausen granite			Gerdes *et al.* (1998)

Bold type indicates our new data (A and B, this study), and the studied granite.

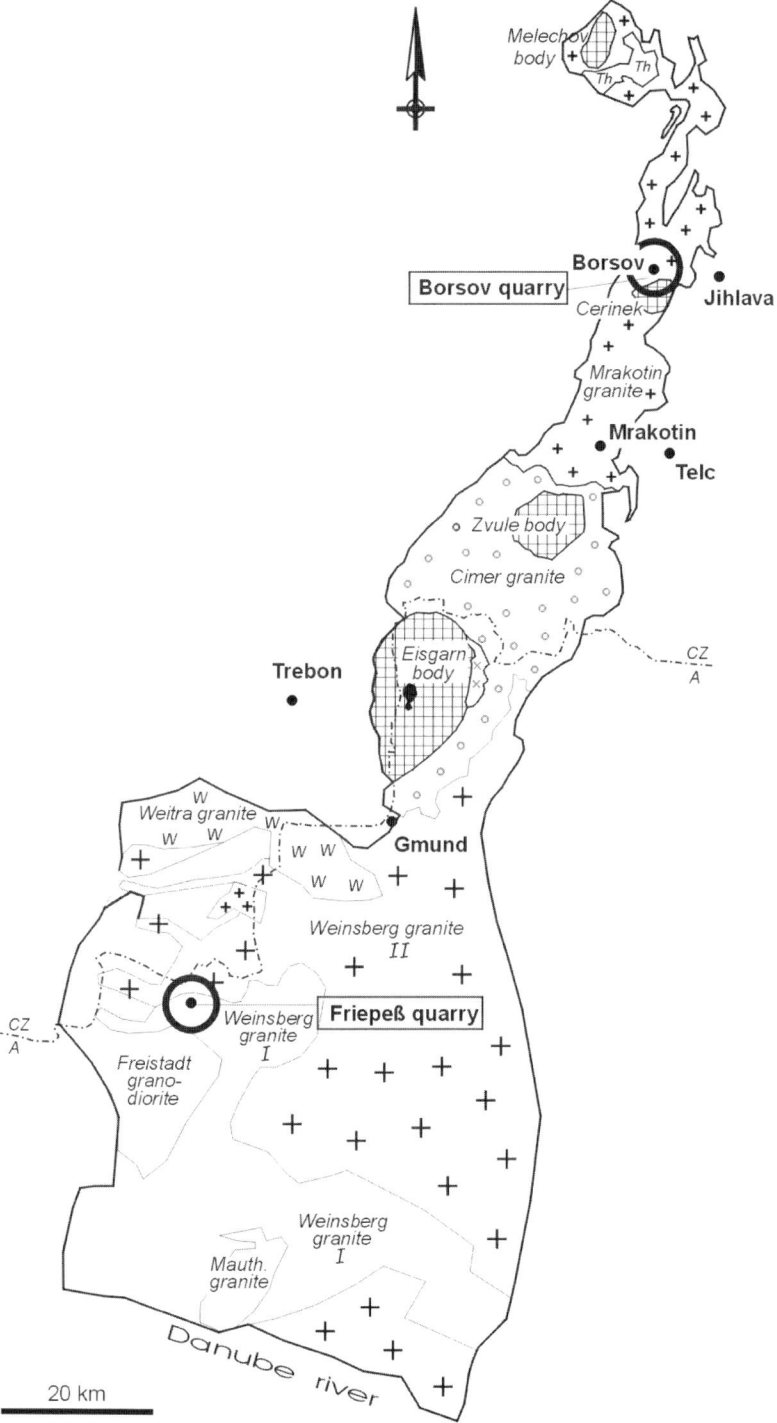

Fig. 2. Simplified geological map of the SBP, which is exposed throughout the SE Czech Republic and on to the Danube river in Austria (modified after Geological Survey of Czechoslovakia 1968; Gerdes *et al.* 1998; Breiter 2001). The SBP consists of a number of intrusions similar in age (Table 4). Sample locations for age determination of early joints are encircled.

Main steps of the study

- Each suitable joint was documented: strike and dip of the parent joint planes, undulations, fringe cracks, their twist spacing and step height (Engelder *et al.* 1993), twist and tilt angle, and other fractographic features, e.g. axis of plumes and main direction of propagation.

- In addition, all available geological, petrological, chronological (succession of joints) and geophysical data (regarding the thickness of granite bodies and their roots) were collected, in order to enhance our interpretation of joint propagation.

- Finally, we selected two quarries for comparison, one from the northern Mrákotín Granite (Boršov) and one from the southern Weinsberg Granite (Friepeß). The estimate of age and depth at these two locations within the SBP (Fig. 2) involved:
  - documentation of: (1) NNE-trending fractures related to the granite dykes (late melts) within the Mrákotín Granite; and (2) of early joint surfaces within the Weinsberg Granite, coated by muscovite, chlorite and biotite indicating a cooling temperature of *c.* 300–400 °C at the time of precipitation from the hydrothermal fluid;
  - age dating of the late-granite dykes (sample A in Table 4) and related joints to date indirectly the joint formation (K. Wemmer). The late-granite melt formed very thin granite dykes along some joints, probably 1–2 cm thick, the dyke thickness at the former opposite wall of the joint is unknown. In addition, the age of the Weinsberg granite (sample B in Table 4) was dated;
  - calculation of the level of granite emplacement in the southern part of the SBP and of the depth of late-granite dykes in the northern part (R. Thomas), based on pressure–temperature (*P*–*T*) conditions obtained from fluid inclusion studies (see the subsection on 'Thermometric investigation of granite samples'). The former depth of the recently exhumed granites indicates the maximum depth at which the earliest joints could have been formed.

Analytical procedure for K/Ar dating of granitic rocks

For dating of the late-granite dykes in the Mrákotín Granite and the Weinsberg Granite, the argon isotopic composition was measured in a Pyrex glass extraction and purification line coupled to a VG 1200 C noble gas mass spectrometer operating in static mode. The amount of radiogenic ^{40}Ar was determined by the isotope dilution method using a

highly enriched ^{38}Ar spike from Schumacher (Schumacher 1975). The spike is calibrated against the biotite standard HD-B1 (Fuhrmann et al. 1987). The age calculations are based on the constants recommended by the IUGS quoted in Steiger & Jäger (1977).

Potassium was determined in duplicate by flame photometry using an Eppendorf Elex 63/61. The samples were dissolved in a mixture of HF and HNO_3 according to the technique of Heinrichs & Herrmann (1990). CsCl and LiCl were added as an ionization buffer and internal standard, respectively.

The analytical error for the K/Ar age calculations is given on a 95% confidence level (2σ). Details of argon and potassium analyses for the laboratory in the University of Göttingen are given in Wemmer (1991).

Mica preparation. Purified micas were ground carefully in pure alcohol to remove altered rims that might have suffered a loss of Ar or K. In this way, only the fresh cores of the muscovites were analysed. The excellent quality of the separates from the South Bohemian Pluton is confirmed by the K_2O content of more than 10%.

Thermometric investigation of granite samples

For the discussion concerning at what depths early fractures were formed, first we had to determine the range of depths at which joints form. The subvertical joints are cross-cut by subhorizontal sheet fractures, which formed later at shallower depth. Sheet fractures are known to occur to depths of several tens to hundreds of metres. The depth of jointing must have been between the depth of granite emplacement, as the maximum, and below 1 km, as the minimum. Therefore, the *P*–*T* conditions and the depths of intrusion were calculated using thermometric investigations.

We have studied fluid inclusions ($n = 37$) in quartz of samples from: (1) the late-granite dyke intersecting the Mrákotín Granite in the northern part of the South Bohemian Pluton (SBP-N; Boršov quarry; Fig. 2); and (2) the Weinsberg Granite in the southern part (SBP-S; Friepeß quarry). The fluid inclusions are small, with a mean diameter of about 5 μm. They are randomly distributed and are not connected to trails or healed cracks, thus indicating the cooling process of the granite melt itself. Texturally, they represent the earliest fluids trapped as inclusions, probably formed during the α–β inversion of the quartz (α, high-temperature quartz; β, low-temperature quartz).

For the determination of phase transitions (melting temperature of ice, homogenization temperature) a calibrated LINKAM THMS 600 heating and freezing stage, together with a TMS92 temperature programmer and the LNP2 cooling system, was used. The salinity of the fluid inclusions (as NaCl-equivalent concentration) was calculated from the ice-melting temperature using the system H_2O-NaCl.

The isochors of the fluid inclusions were calculated with the model of Zhang & Frantz (1987) using the salinity and the liquid–vapour homogenization temperature of the fluid inclusions. Under the assumption that the fluid inclusions were formed at the α–β inversion of the quartz, we obtain as a first approximation the trapping conditions from the point of intersection of the isochors with the P–T relation of the inversion temperature (Yoder 1950). The following data were obtained: salinity, homogenization temperature and trapping temperature and trapping pressure. From the trapping data the minimum intrusion depths (depth of the sampled granite) can be estimated using a lithostatic model with a rock density of 2600 kg m^{-3}.

The results can be compared with a number of thermometric data (n = 1386) from the melt inclusions in several Erzgebirge granites (eastern Germany; Thomas & Klemm 1997). These data define the solidus temperature and the pressure at the solidus, from which the depth during the emplacement of the plutons can be determined.

Case study: South Bohemian Pluton

Geological setting

The South Bohemian Pluton (SBP, Fig. 2), a large late-orogenic batholith within the Variscan orogenic belt (c. 6000 km^2), is not a continuous granitoid mass, but a composite and multiple intrusion with isolated subtypes (Čméř, Mrákotín, Eisgarn, Weinsberg, Freistadt, Mauthausen types, etc.; Breiter & Koller 1999). The main stage of the pluton history started according to Breiter (2001) with the intrusion of a high-K peraluminous Th- and Zr-rich melt, which crystallized as a medium–coarse-grained porphyritic Čméř-type granite in the south, and fine–medium-grained non-porphyritic Mrákotín-type granite in the north of the Czech part of the SBP. Products of this intrusive phase built the largest part of the SBP. Further fractionation of the Čméř-type granite melt produced the Eisgarn-type granite *sensu stricto* (Table 4) in the central part of the pluton between the towns of Nová Bystřice and Gmünd, and the Weinsberg Granite between Gmünd and the Danube river (Austria). The SBP intruded Moldanubian high-grade metamorphic rocks, mainly cordierite–biotite paragneisses.

The SBP extends over a distance of 150 km

between Havlíčkův Brod and the Danube river, and further south is recognized from drillings below the Molasse Zone of the Alps (Scharbert 1998). The root zone of the SBP occurs in its southern part between the Danube and the Pfahl faults. According to Büttner (1997) the granitoids were emplaced at mid–upper-crustal levels into hot country rocks (c. 15–18 to 7–10 km, c. 450–650 °C) shortly after the thermal peak of regional metamorphism. P–T paths in the country rocks suggest that granite formation, low-P–high-T metamorphism and extensional thinning was preceded by a phase of intense crustal thickening, which occurred during late Palaeozoic continent–continent collision (Matte *et al.* 1990; Gerdes 1997; Gerdes *et al.* 1998).

Advantage of the study area

As a case study SBP was selected for two purposes: (1) to demonstrate particular fractographic features in granites (penny-shaped, circular or elliptical joints sharply bordered by rib marks as separation lines against the fringes) that are rare in other plutons; and (2) to discuss joint formation at deep crustal levels in the early cooling history of the pluton. For several reasons the SBP was favoured to discuss these problems: the geochemical and mineralogical composition of the subplutons, their succession, and age of emplacement and cooling are well known. An attempt to recognize the pluton structure was made by Gnojek & Přichystal (1997), giving the first comparison of the Czech and Austrian sides of the pluton based on geophysical, mainly gamma-ray, measurements. Breiter & Scharbert (1998) compiled all existing gamma-spectrometric data, and accessible structural and chemical data, defining the area of extent of the individual granite intrusions. Petrological and geochemical features of main granite intrusions were described in detail by Breiter & Koller (1999), and the geochronology by Scharbert (1998).

The petrographically homogeneous and fine–to medium-grained Mrákotín Granite was most suitable for this study. Conspicuous, frequent penny-shaped NNE-trending joints were concentrated in the Boršov quarry (Kavex Enterprise; depth of mining 40 m). Further advantages of the SBP are:

- fracture surface features are well developed and preserved. Most of the granites experienced little or no later deformation (Breiter & Sokol 1997). The early fractures maintained the original shape during exhumation;
- they often indicate an unrestricted propagation of the fracture. They can bear all fractographic features known from any geological rock at one individual fracture;

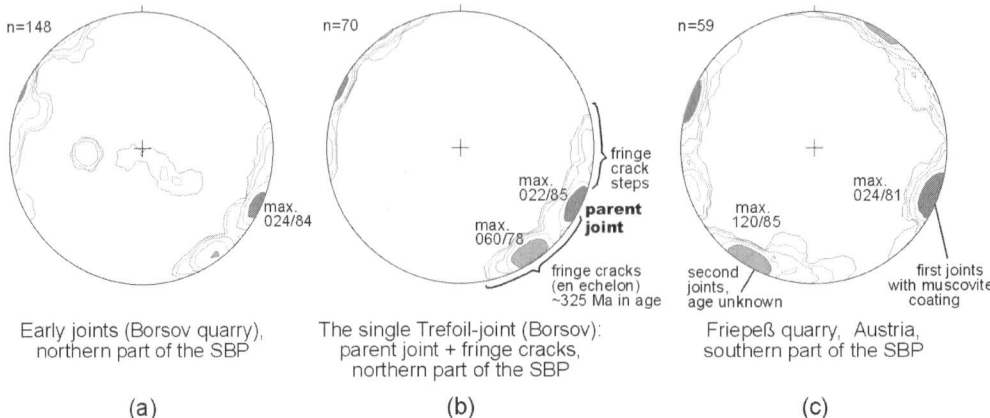

Early joints (Borsov quarry),
northern part of the SBP

(a)

The single Trefoil-joint (Borsov):
parent joint + fringe cracks,
northern part of the SBP

(b)

Friepeß quarry, Austria,
southern part of the SBP

(c)

Fig. 3. Main joints in granites of the SBP. From the northern part of the SBP: (**a**) early joints in the Boršov quarry (Mrákotín Granite); (**b**) measurements of the surface of one single joint consisting of a parent joint, several weak undulations and fringe cracks (en echelon), as well as their steps beween them (Trefoil joint, Boršov quarry, Fig. 4f); (**c**) southern part of the SBP (Friepeß quarry, main joints; Fig. 5).

- The early generated NNE–SSW-trending joint sets of various subplutons of similar age (Table 4) are persistent in their orientation over the whole length of the SBP (extension of 150 km);
- these joints have most probably been associated with the cooling history of the granites.

In the two quarries with sample locations, K/Ar ages of the muscovites in the samples corroborate the early jointing both of the muscovites within the late-granite dykes at the Boršov quarry (Mrákotín Granite) and the muscovites precipitated from hydrothermal fluids on the NNE-trending joints at the Friepeß quarry (Weinsberg Granite).

Determination of joint successions (SBP)

The joints of the main granite types within the SBP (Fig. 2) are present in the same orientation and succession in many quarries (Fig. 3). The relative timing of the joint set sequence was determined by means of fractographic surface features, butting and cross-cutting relationships. Fringe-crack propagation at the end of fractures stop at earlier existing joint planes or are modified along intersection lines with older fractures. Older and younger joints can be easily distinguished. The time interval between fracture events was not evaluated.

An early fracture system, which consists of ESE- and NNE-trending subvertical joint sets, is very common in the SBP. A few early subhorizontal joints (several decametres long) were exposed at the bottom of the quarries. These older horizontal joints differ from subhorizontal sheet fractures that formed very late by unloading at the erosion level, corre-

sponding with the surface morphology. Sheet fractures cross-cut older vertical joints. The first three fracture sets (subhorizontal, subvertical ESE and NNE), which dominate by their size and persistence of orientation, developed more or less synchronously with changing succession from one location to the next. The old subhorizontal set and the ESE set have only rarely modified or stopped the propagation of the predominant NNE joints that are most frequent, trending parallel to the central axis of the pluton.

The two key localities for our study of the SBP joints (Fig. 2) are the Boršov quarry within the northern Mrákotín Granite (Czech Republic) and the Friepeß quarry within the southern Weinsberg Granite (Austria). The NNE joints (Fig. 3a–c) correlate across the region, but it is only at the Boršov quarry that many of them display a penny-shaped parent joint (Fig. 4b–e). In any case, the NNE joints were the prominent fractures. They are the only fracture surfaces related to thin late-granite dykes (Boršov quarry) or coated by dense phyllosilicate cover between the plumose ridges (Friepeß quarry). The crossing joints (*c*. 120°/90°) are mostly younger than the main *c*. 025°/90° fractures and, essentially, they are without phyllosilicate coating or dykes. The only difference between the joint succession in the northern Mrákotín Granite (Fig. 3a, b) and the southern Weinsberg Granite (Fig. 3c) is the timing of the cross-cutting joints: in the southern part of the SBP they are clearly younger than the NNE joints, whereas in the northern part single wide-spaced cross joints were formed interacting with the formation of the NNE-strike joints. However, the main bulk of cross-joints are also younger in the northern granite, but they are less frequent than in the south-

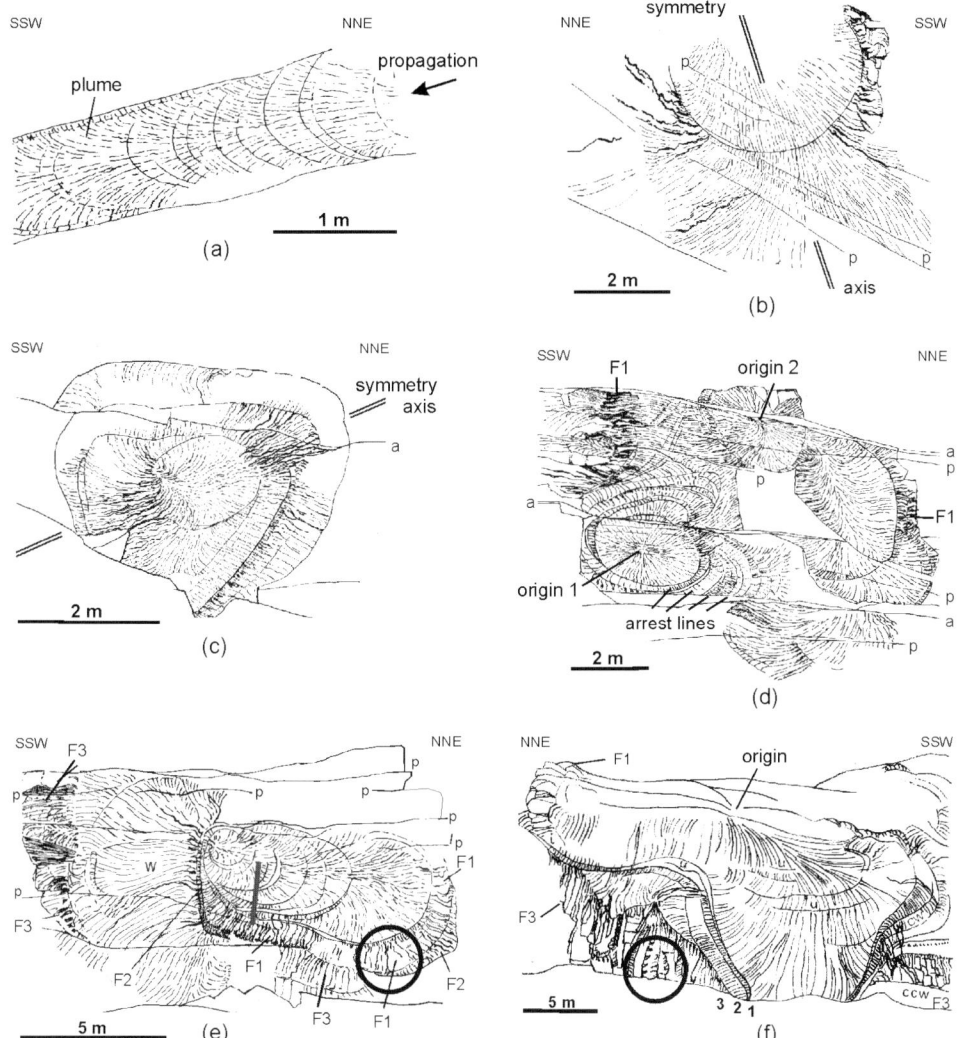

Fig. 4. Boršov quarry (SBP), granite joints with penny-shaped fractures (size between 2 and 30 m) and undulations (rib marks). (**a**)–(**c**) Undulations covered with a plume. (**d**) Two circular fractures that interacted during further propagation and formed a joined plume and fringe on the left side. (**e**) The initiation started in the central circular fracture, then the joint propagated laterally only to the right. After reaching fringe 2 (F2), the joint propagates only towards the left (overall length 15 m). (**f**) Large 'amoeboid' central parent joint with tongue-like propagation in each direction, the upper part of the joint is erased. (e) and (f), outer fringes are sample locations for age dating and depth determination (circles).

ern quarries. This means that we have to be aware that joints which developed at different times can be associated with the same azimuth.

Another point of interest is the relation between the azimuths of main fracture planes and of the en echelon fringe cracks. One example is given in Figure 3b, demonstrating one single joint (Fig. 4f; length >30 m, present-day height >20 m) together with fringe cracks (length >4 m) and associated steps. The two maxima do not represent two individ-

ual, independently formed sets of joints, as they were produced during one fracture process. Evidence for this comes from the gradual en echelon fringe cracks that start at the fringe root zone with a very low twist angle, which increases up to 38° after a short distance of *c.* 30–50 cm. The maximum 022°/85° contains measurements from the large parent joint and the maximum 060°/78° only uses data from the outer part of the en echelon fringe cracks.

(a)

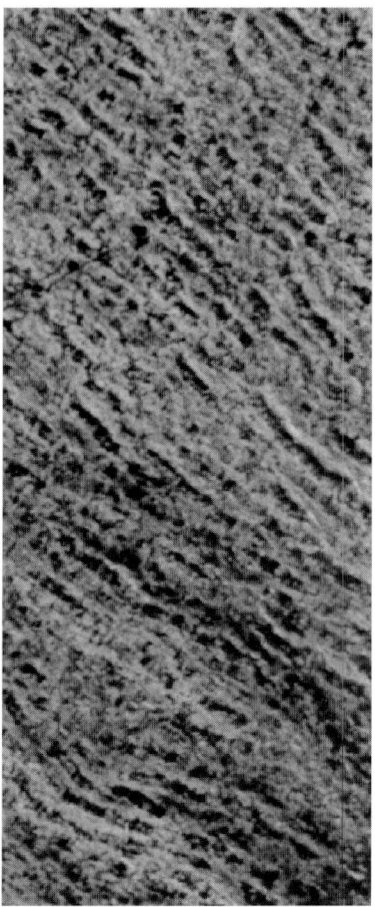

(b)

Fig. 5. SBP, Friepeß quarry, Austria. (**a**) NNE-trending joint with a large plume structure. On the left edge of the photograph is part of a smooth ESE joint surface without a thick plume structure and without coating. (**b**) Detail from the left photograph of a plume structure.

Fractography of joints in the South Bohemian Pluton

The surfaces of the investigated joints in the SBP are covered with well-developed surface morphology (Bahat *et al.* 2003). In general, these fractographic features are the same in all plutons as they are known to occur in all brittle materials, but vary in the fineness and sharpness of their surface morphology from one individual subpluton to another. In addition, the characteristic surfaces of the NNE-trending joints (Figs 4 and 5) can also vary within one of the granite types of the SBP. But the joint orientation of the NNE set is remarkably persistent through the whole pluton (Fig. 6). One speciality are the quasi-circular parent joints with sharp boundaries (which look like 'suns') that occur only at the Boršov quarry and which differ significantly from the joint-surface markers in other plutons (e.g. B–D, Fig. 1), except for a few joints in the Central Bohemian Pluton and in the Bohus Granite (Sweden).

Although dependent on the coarseness of the granitic rock and the quasi-homogeneity and quasi-isotropy of many granite types, granite joints can grow to an enormous size (>30–100 m). Often, the shape of a single granite joint covers the entire height and length of a wall within quarries. There, complete joint surfaces can display fractographic features that represent each stage of joint propagation (Bankwitz & Bankwitz 2000).

Northern SBP (example: Boršov quarry)

Joint surfaces exhibit evidence of origin, propagation and termination. The walls of the quarry consist of NNE-trending joints, and the most conspicuous fractographic features are complete quasi-circular or elliptical parent joints (with sun-like appearance; several metres in size) with undulations, which are known to occur in sedimentary rocks. Dominant surface features are plumes (representing the propa-

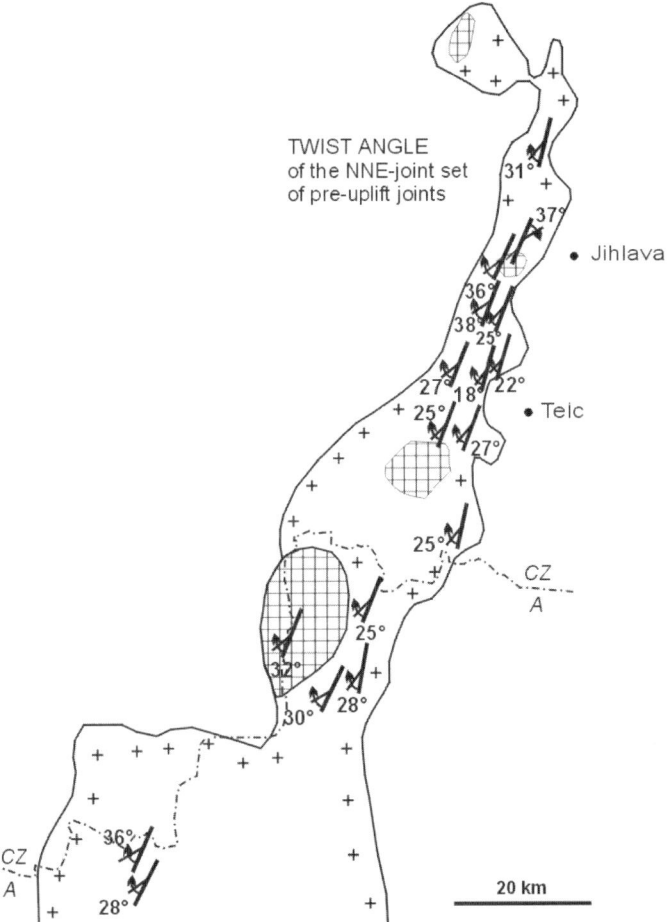

Fig. 6. South Bohemian Pluton (as in Fig. 2) and the trend of the early initiated NNE joints in different granite types. Thick lines illustrate the mean values of the dominant NNE set in quarries with suitable fractures for this study. Numbers give the mean rotation angle of the fringe cracks. The sense of rotation is always the same for this joint set.

gation of the crack front) that cover the penny-shaped fracture and, without interruption, the undulations (Fig. 4). Generally, the final fringe is tilted at a low angle and composed of twisted en echelon segments.

At the Boršov quarry the three initial fracture sets are widely spaced (mostly *c.* 2–5 m, but with some exceptions). The NNE-trending set is more densely developed, in part in bundles of joints that occur widely spaced (2–10 m) within the quarry. Many of the vertical joints propagated from penny-shaped parent joints, which ranged in size from 2 m to 30 m. The more or less concentric surface undulations (rib marks) indicate a cyclic propagation of the fracture front suggesting natural hydraulic joints. Several circular joints form composite fractures at the quarry face, which assumes that some joints have a faster

propagation front. Sometimes, a single propagation front is misaligned relative to the neighbouring one, resulting in them forming composite fractures (e.g. Fig. 4d). Large sections of the quarry faces were opened along such composite joint sets (e.g. Mrákotín; Bahat *et al.* 2003).

Southern SBP (example: Friepeß quarry)

Here two orthogonal joint sets of different age are predominant: NNE- and ESE-trending main joints occur regularly, spaced approximately 1–2 m apart. A difference between the first formed NNE joints and the younger 120° joints is striking (Fig. 5): only the NNE-trending joints demonstrate large plumes with a deep relief at the surface, which occurs as rough ridges and depressions. The surfaces are coated by a dense muscovite–chlorite–biotite cover that was generated in the depressions between the plume ridges (Fig. 5b). The phyllosilicates are the cause of a dark colour of the NNE joints in contrast to the bright surfaces of the ESE set. The ESE set surfaces are relatively smooth, or occur with only rare and smoothly developed wide-spaced single plume step traces (Fig. 5a, left boundary). Ruf *et al.* (1998) observed the same difference in joint-surface morphology between orthogonal sets. At the Friepeß quarry, the central part (0.5–1 m) of the non-penny-shaped ESE joint surfaces comes close to being a mirror surface. In a recent paper Savalli & Engelder (2004) interpret this difference as a manifestation of crack-propagation velocity, with the smoother surface representing Region I subcritical crack propagation where the ESE joint cuts smoothly through the rock and was not affected by local inhomogeneities. Compared to this, the earlier formed NNE joint set, which predates the ESE set, was formed more quickly and did not develop a mirror surface.

The muscovite- and chlorite-enriched coating is seen to cover only the cm-wide depressions between the higher ridges of the large plume, and not the whole joint surface, therefore a 'light–dark' striped pattern occurs on the joint. This phenomenon has also been recognized in other plutons of central Europe.

It should be noted that the preferred orientation of the dominant early 025° joint set was remarkably persistent within the entire SBP (Fig. 6). The NNE trend, and particularly the same sense of fringe-crack rotation throughout the SBP, indicates the rotation of a regional remote stress (e.g. Younes & Engelder 1999), or the increasing influence of the superimposed regional remote stress on the local stress within the crystallizing plutonic body during fracture propagation. However, if the regional stress field then rotated, this would be evidence of subcritical crack propagation during the rotation of the stress field, and would indicate very slow crack growth.

Because the pluton consists of several subplutons, the consistent azimuth of the earliest subvertical joints was not expected. The alignment of these NNE joints over large regions is indicative of a regional stress field at the time of fracturing. Such 'granite rift' is reported to be common due to thermoelastic loading after intrusion of these granites (Engelder & Fischer 1996). The same orientation in several granite types of similar age (Table 4) points to a longer-lasting constant remote stress field that controlled the propagation of the early joints.

The conformable fringe rotation (Fig. 6) cannot immediately be considered as evidence for a second stage of joint propagation after rotation of the stress field as many of the available joints show a gradual propagation from the parent joint into the root zone of the fringe. Only after a rotation of up to 10 cm did a larger twist angle develop, but without a visible break of the propagation.

Age dating of the NNE–SSW joint set (SBP)

Geologically, the timing of joint initiation is difficult to determine; generally, dating is relative, and is based on butting and cross-cutting relationships. In the case of the main NNE joint set in Boršov, the thin late-granite dykes along the joints can be used for K–Ar age determination of muscovites. This research was carried out at the University of Göttingen by K. Wemmer.

The dated samples were taken at two localities within the SBP (Fig. 2): (1) in the northern part (Boršov quarry, Czech Republic) from the surfaces of two large NNE-trending joints: the outer fringes (Fig. 4e, f) of the 'trefoil joint' and the 'bilateral asymmetric joint' (Bahat *et al.* 2003); and (2) in the southern part of the SBP (Friepeß quarry, north of Freistadt, Austria) from the normal granite of a NNE-trending joint.

The muscovites within the late-granite dykes that created the two studied joints of the Boršov quarry are 324.9 ± 6.7 Ma in age, confining the time of joint formation. The muscovites of the Weinsberg Granite in the Friepeß quarry have an age of 318.9 ± 6.7 Ma (Table 4). Essentially, the new data suggest that the fracturing was roughly synchronous with the emplacement and cooling of the Hercynian SBP at 330–318 Ma (Friedl *et al.* 1996, Scharbert 1998).

For the northern part of the SBP, because the intruding dyke was at a higher temperature ($>600\,^{\circ}$C) than that of the dated muscovites (350–400 °C), the muscovite cooling age of 324.9 ± 6.7 Ma should be considered the minimum age for the initiation of the NNE joints. In the southern part of the pluton the muscovite cooling age of the Weinsberg Granite is given as 318.9 ± 6.7 Ma. Fracturing was possible under the conditions at that time, and joints must have existed before the phyllosilicate coating of the surfaces plumes was formed. Probably, these data point to the timing of the hydrothermal fluid crystallization, which is restricted to the NNE joint set. Within the Friepeß quarry, ESE joints formed later and post-date any deposition of the hydrothermal activity. The precipitation along NNE joints may correspond temporally with the Greisen Formation and other additional late intrusions of the Weinsberg Complex (Table 4) in southern parts of the SBP.

Calculation of the intrusion depths (SE, north and west Bohemian Massif)

Hercynian SBP (330–318 Ma), SE part of the Bohemian Massif (BM)

Samples with the analysed fluid inclusions (Mrákotín Granite, Boršov quarry; A in Fig. 1 and 2) were taken from the same location as the samples for age dating (Fig. 4e, f). Fluid inclusions of the age-dated late-granite dyke itself were investigated regarding their depth of generation. The samples for age dating and determination of the intrusion depth of the Weinsberg Granite (Friepeß quarry) also came from the same location.

From the *P–T* conditions of the trapping data of fluid inclusions (Table 3), the minimum intrusion depths can be estimated using a lithostatic model with a rock density of 2600 kg m^{-3} (last column of Table 3). In the Mrákotín Granite the temperature of homogenization in the fluid phase within 17 fluid inclusions amounts to $T = 620\,°C$. That is the trapping temperature of the inclusions under pressure conditions of $P = 1.89$ kbar. Under the assumption of a lithostatic pressure at this crustal level, the depth at the time of inclusion formation within the samples amounts to 7.4 km. For the inclusions in the Weinsberg Granite, the level of generation was much deeper at 14.3 km with associated higher temperature and a trapping pressure of 3.65 kbar. In addition, we consider results from other plutons of the Bohemian Massif for comparison of the cooling and exhumation history of the eastern part of the SBP with those of the north and west part of the BM.

Hercynian Erzgebirge plutons (325–318 Ma; Förster et al. 1999), northern part of the BM

A different microthermometric method using a heating–quenching technique was applied to analyses of 1386 melt inclusions in quartz, apatite and feldspar of several plutons (e.g. Eibenstock, Kirchberg, Bergen and Ehrenfriedersdorf granites) of the Erzgebirge anticlinal zone, which is part of the Variscan Orogen in Germany (C in Fig. 1). According to the deep seismic reflection profile (Dekorp Research Group: Behr *et al.* 1994), most of the samples belong to a shallow granite 'sheet' within the upper crust without deep roots. The largest pluton is the Eibenstock Granite, which has an exceptional extension from the surface down to 10 km. Smaller bodies intruded like diapirs into the level above the granite plate. Some bodies of the shallow-intruded Erzgebirge Granites are geochemically highly evolved and enriched in ore minerals; in part, they are of subvolcanic nature. The estimated trapping pressures of melt inclusions in different plutons range between 1.1 ± 0.1 and 1.0 ± 0.1 kbar, and correspond to intrusion levels of about 3 km associated with trapping temperatures of 562 ± 2– $577 \pm 7\,°C$. Geological evidence for shallow intrusion levels comes from vitreous glass inclusions that occur together with normal crystallized melt inclusions in granitic quartz of several Erzgebirge plutons. Glass inclusions demonstrate rapid cooling or quenching that is not compatible with a deep intrusion level (Thomas 1989; Thomas & Klemm 1997).

Therefore, the first fractures could have been initiated for granite cooling below 560 °C at a depth of about 3 km, where the investigated inclusions of the Erzgebirge samples were trapped.

Hercynian Fichtelgebirge (325, 305 and 290 Ma) and Oberpfalz plutons (320–290 Ma), western border of the BM (Siebel 1998)

Melt inclusions from several Hercynian plutons of the KTB surroundings were also analysed (B in Fig. 1), but the calculated intrusion depths of 3–6 km disagree with the petrological evidence for a deeper intrusion of 9–12 km related to mineral equilibria (e.g. Falkenberg, Flossenbürg, Leuchtenberg and Steinwald granites; Zulauf *et al.* 1993). This assumption corresponds with results from structural analysis, and with calculations of denudation and degradation of the crust that indicate >12 km uplift and erosion in the south and SW part of the Bohemian Massif (Suk 1994). Nevertheless, from melt inclusion data, only a depth of formation of about 3 km could be obtained.

Discussion of timing and depth of joint generation

The new data give evidence for the granite depth–joint time relation within the northern and southern part of the SBP (Boršov and Friepeß quarries). The depth of 7.4 km is the crustal level of intrusion (Boršov quarry). Subsequently, more than 7 km were eroded or eliminated by extension of the crust: 7.4 km was neither the bottom nor the top of the vertical extent of the Mrákotín Granite. We suspect that the location of the sample belongs to the upper part of the granite, which formerly was 7.4 km more deeply seated.

In the southern part of the SBP (Friepeß quarry) the depth of the sampled granite is 14.3 km. The same depth range was estimated by Büttner (1997) and corresponds with results from Suk (1994) given in Figure 7. Despite of the removal of the *c.* 14 km-thick overburden, geophysical data suggest that the

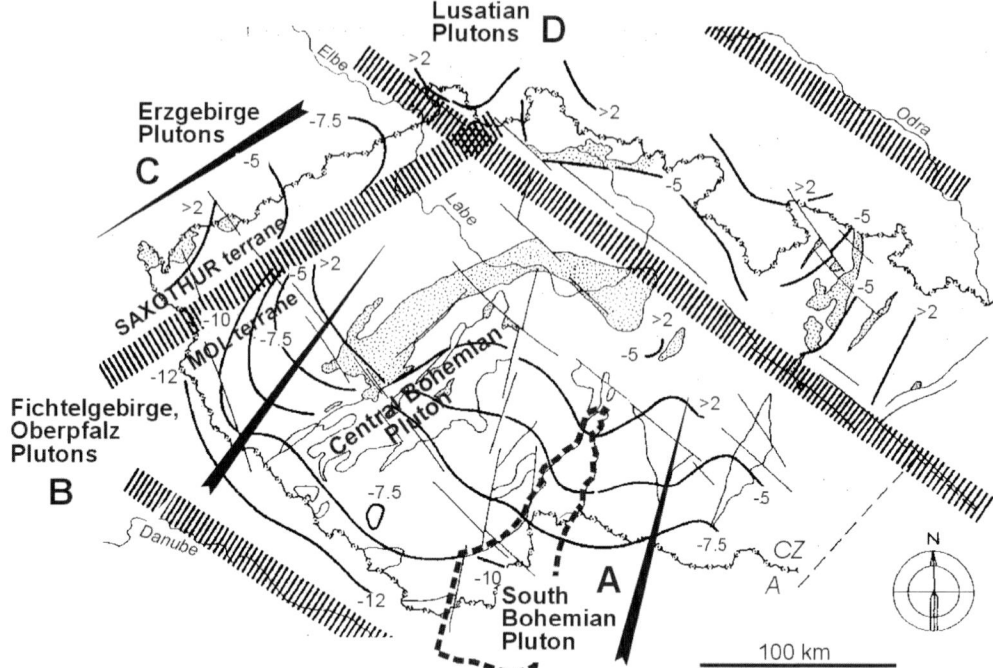

Fig. 7. Isohypses (height contours, in km) of the denudation level of the crystalline units (in km) at the present erosion level in the Czech part of the Bohemian Massif (modified according to Suk 1994). Significant exhumation occurred along the southern boundary. SAXOTHUR, Saxothuringian; MOL, Moldanubian.

present lower boundary of the Weinsberg Granite is at a depth of *c.* 12 km. Subsequent smaller intrusions into the SBP, such as the Eisgarn Granite and the Zvůle, Čeřínek and Melechov granite types (Table 4; hatched area in Fig. 2), also show a present-day rooting of between 12 and 15 km, which is indicated by circular negative anomalies of the gravity field associated with the axial trend of the SBP. The distribution of these coarse-grained granites coincides with the gravity minima (Meurers 1992; Breiter & Scharbert 1998). Deep roots of the pluton in these stock-shaped structures can help to explain these anomalies. The present depth of *c.* 15 km suggests that the lower boundary of these granite types was primarily at about 30 km within a much thicker crust than occurs now.

Correlation of joint initiation with pluton history

The existing intrusion and cooling ages of the main subplutons within the SBP are shown in Table 4. The intrusion of the Eisgarn complex, including the Mrákotín Granite, was determined to have occurred at 330–324 Ma. Mrákotín Granite is the local name of the Čiměř Granite in the more northern intrusion

area, and further fractionation of the Čiměř melt produced the Eisgarn Granite *sensu stricto* in the centre of the SBP (Breiter 2001). The U–Pb monazite age of 327 ± 4 Ma (Friedl *et al.* 1996) could be the most accurate age date. Ar/Ar muscovite cooling ages of this first granite intrusion phase of the SBP range between 328–325 and 320 Ma, decreasing from north to south (Scharbert 1998).

Joint samples, SBP-N. The determination of muscovites within the late-granite dykes along the Boršov joints (Mrákotín Granite) 324.9 Ma in age (K–Ar) corresponds with the muscovite cooling range in the granite body itself (Table 4) (Scharbert 1998), taking into account the uncertainties of the Ar–Ar age determination. The age of the granite dykes agrees remarkably well and suggests the sequence: intrusion of the Mrákotín Granite at 327 ± 4 Ma; cooling at 327–325 Ma; fracturing before or at 324.9 Ma by the late-granite dyke intrusion with muscovite cooling at 324.9 Ma.

After the initiation of fractures the cooling of the SBP continued up to 320 Ma, and further up to 312 Ma in the southern part of the Čiměř Granite. At the same time the younger Homolka Granite intruded (320–314 Ma) and, only after that event, the deep-rooting granite bodies in the axial part of the SBP intruded (e.g. Zvůle Granite; Breiter 2001). These sequences of intrusions give evidence for only weak exhumation during the cooling stage between 327 and 325 Ma, and supports the joint initiation at c. 7 km depth. A recognizable exhumation has only occurred since 310 Ma, as is known from the Bohemian Massif further to the NW. Evidence for these data comes from apatite-fission tracks (measured to the west of the SBP where analogous intrusions exist) that only started at 310–280 Ma due to the late-Variscan exhumation (Wagner & Van den Haute 1992; Menzel & Schröder 1994). The exhumation was not a continuous process: during the first period about 3.3 km of exhumation has been estimated to have occurred (Coyle & Wagner 1995); the second period of exhumation occurred in the Lower Triassic; and the third period in Upper Cretaceous–Tertiary times.

The relationship between depth, temperature and timing of the magmatic processes and the early fracture processes is informative. The temperature of the intruding late-granite dyke at a depth of 7.4 km was determined to be 620 °C at 324.9 Ma. The cooling temperature of the muscovites within the host granite of the dyke was about 350–400 °C at 325 Ma (K–Ar age; Table 4) (Scharbert 1998). The host granite had a lower temperature than the intruding dyke, which means that the granite was emplaced at c. 800 °C at a deeper level and >2 Ma earlier at 327–328 Ma, and was not exhumed to 7.4 km until 325 Ma. The first fractures (e.g. Fig. 4e, f) occurred about 324.9 Ma when the late-granite dykes created these joints by natural hydraulic fracturing of the Mrákotín granite body at a depth of 7.4 km. The fluid inclusions of the dykes were trapped at a temperature of 620 °C. It should be noted that the K–Ar age determination of the late-granite dykes is also connected with the cooling temperature of 350–400 °C of the muscovites within the dyke. The age of 324.9 Ma indicates the time when the late-granite dyke was cooled down from 620 to 350 °C (dot in Table 4: sample A). This fact gives evidence for an injection of the late-granite melt earlier than 325 Ma. The late-granite dyke intrusion, which created the joints, occurred some time before 325 Ma, because time is required to lower the temperature within the dyke from 620 to 350 °C (i.e. 270 °C). These data concerning the temperature and timing of the individual processes confirm the sequence of events: (1) cooling of the Mrákotín granite melt down to the solidus; and (2) injection of the rest melt into the solid phase of the Mrákotín Granite, thus forming late-granite dykes that create NNE-trending hydraulic fractures. These events occurred more or less simultaneously.

Joint samples, SBP-S. In the southern part of the SBP only the age of the Weinsberg Granite itself at the Friepeß quarry, sampled from one NNE joint, could be determined. The new K–Ar cooling age obtained from muscovites of the granite is 318.9 ± 6.7 Ma (dot in Table 4: sample B) and fits into the existing age data within Table 4. The listed zircon and U–Pb monazite intrusion ages (Friedl et al. 1996) and the cooling ages are equal to those of the Mrákotín Granite. The cooling ages range between 327 and 320 Ma at the border and along large shear zones, and 313–308 Ma in the central parts of the pluton, indicating a slow cooling (Scharbert 1998). The locality of the quarry belongs neither to the central part nor to the border area or to a shear zone. Therefore, this muscovite age is placed between the two end members of the cooling range. Hydrothermal fluids, probably associated with the magmatic process itself, are the source of the dense joint coating on the NNE set (Fig. 5). Tuttle & Bowen (1958) stated that in natural systems the escaping water near the solidus conditions, by crystallization of anhydrous phases, is introduced into the surrounding rocks where it may be the agent responsible for recrystallization, granitization and the hydrothermal ore deposits. This process and the cooling of the muscovites as the crystallizing phases are simultaneous in time. Therefore, we suspect roughly the same age of about 318 Ma for the precipitation of the muscovites on the NNE joint surfaces. The younger orthogonal ESE joint surfaces are without phyllosilicate coating.

The level of fluid inclusion formation in the Weinsberg Granite (Friepeß quarry) was determined by inclusion analysis at 14.3 km. The intrusion depth was estimated by different authors to be between about 12 and 15 km (e.g. Büttner 1997). These values correspond with the calculated trapping level of the fluid inclusions and exceed the estimated exhumation of this region (Fig. 7) (Suk 1994).

Comparison with fractures in plutons of the western BM

The NNE joint set of the Boršov quarry, in part with granite dykes or quartz–feldspar infilling, is not the only one Variscan fracture type. Other granite dyke rocks (Table 4: Late suites; granite porphyries) are products of the Čímeř–Eisgarn melt evolution. They intruded at 323 ± 16 Ma (to 318 ± 3 Ma), between the phase of the Eisgarn Granite emplacement and the phase of the intrusion of small bodies at the rim of the Eisgarn Granite (Fig. 2). The fractures of the granite porphyry dykes are similar in age to the early NNE joint set of 324.9 ± 6.7 Ma. Generally, such N–S-trending granite dykes form a 30-km long zone and indicate the beginning of the intense late-Variscan extension, which enabled the penetration of small portions of the residual melt into the uppermost parts of the SBP and into its gneiss mantle (Breiter & Scharbert 1998; Breiter 2001).

In the west part of the Bohemian Massif (Fichtelgebirge and Oberpfalz, B in Fig. 1) the older group of granites intruded between 325 and 310 Ma (Carl & Wendt 1993; Stettner 1994). The earliest late-Variscan fracture tectonics in this part of the BM was recognized by Zulauf (1993): subvertical hydrothermal dykes (quartz-feldspar or quartz-epidote) indicate the formation of early fractures and the beginning of crustal extension during the period 310–300 Ma. The fracture system occurs also within the granite plutons. Lamprophyric dykes 305–295 in age (Kreuzer *et al.* 1993) discordantly intersect the older dykes, again indicating the late-Variscan crustal E–W extension at about 310 Ma. These dyke events give evidence for early formed deep-seated fractures during the early cooling stage of the granites.

Fractures that were comparable in time and the associated dykes were recognized in the western (Oberpfalz, KTB surroundings) and the eastern (South Bohemian Pluton) parts of the Bohemian Massif, starting a little earlier in the eastern part. In both areas the first fractures were driven by pegmatoid (quartz-feldspar) or granitoid melt, and provide evidence of fracturing at deep levels and a very short time after the granite flood.

Discussion of periods and amounts of granite exhumation

Significant exhumation did not immediately follow the first granite-type emplacement (Mrákotín Granite) and ongoing crystallization. Repeated intrusions of more or less younger granite types into the somewhat older first subplutons provide evidence for further deep-crustal levels. Arguments for slow exhumation come from the slow cooling (Table 4) due to the granite intrusion into hot metamorphosed rocks (Gerdes *et al.* 1998), and from sedimentary geology.

Special investigations of the thickness and facies of younger sedimentary deposits, their detritus and source, and the study of structure and mineral parageneses, as well as fission track analyses, in granites provide the reconstruction of the cooling and exhumation periods (Peterek *et al.* 1994; Suk 1994; Coyle & Wagner 1995). The knowledge of this history supports the understanding of depth and time of the granite joint initiation within the SBP.

During late- and post-Variscan times the crustal exhumation of the southern half of the Bohemian Massif (BM) was similar in the western and SE part (A and B in Fig. 1). The contour lines (isohypses) of the basement denudation level (at the present erosion level; Suk 1994) trend parallel to the border of the massif: more than about −12 km near the

southern border. The exhumation decreases towards the NE below −5 km. Peterek *et al.* (1994) and Coyle & Wagner (1995) carried out a more precise reconstruction for the western part of the BM with three main stages of exhumation: 3.3 km during Lower Permian times; >3 km in the early Triassic (245–230 Ma); and after a long, nearly non-tectonic period, >4 km starting from the beginning of Upper Cretaceous (*c.* 100–60 Ma) up to Tertiary times; at 70–60 Ma regional exhumation occurred. The granites discussed here were not yet exposed during Upper Cretaceous time, granite pebbles in such sediments are rare and do not belong to the present-day exposed granites (Peterek *et al.* 1994).

Initiation process of deep-seated early joints within solidifying felsic magma

The question arises of how the late granitic melt injected into the quasi-circular joints. Backed up by fluid and melt inclusion analysis we are able to discuss the mobilization of small portions of melt at local randomly distributed locations within the cooling pluton. The cooling of the granite melt led to the separation of water from the anhydrous phase near the solidus and to an increase of pressure (Fig. 8) and so the formation of associated small cracks, and then, finally, to a secondary melt generation that took place within the initial small cracks, which then drove further propagation.

We also consider the shape and fractographic surface pattern of the joints, and the depth and time of their initiation, together with the processes within the cooling magma of the SBP (Mrákotín Granite and its late-granite dykes, and Weinsberg Granite; Fig. 8a) and various Erzgebirge granites (Fig. 8b). The NNE joint set in Boršov demonstrates characteristic surface structures, indicating a quasi-circular fracture propagation with many ribs (Fig. 4). This points to a cyclic growth of the central part of the fractures that show a clear boundary to the outer fringes at their edges. Evidence for natural hydraulically formed fractures comes from the associated dykes (Secor 1965; Secor & Pollard 1975; Engelder & Lacazette 1990). The joint fringes are often built up by very regularly distributed and straight en echelon cracks with large overlaps. For this outer part of the joints we have to take into consideration additional driving force effects (Engelder & Fischer 1996). The wavy central parts of these joints look like patches, driven from the centre by fluids to the boundary against their outer fringe, in spite of their size (a few metres to decametres). Within other plutons, such as the Rönne Granite on Bornholm Island, where numerous pegmatite dykes occur, fluid-driven joints with dykes had been initiated somewhere below and then propagated radially

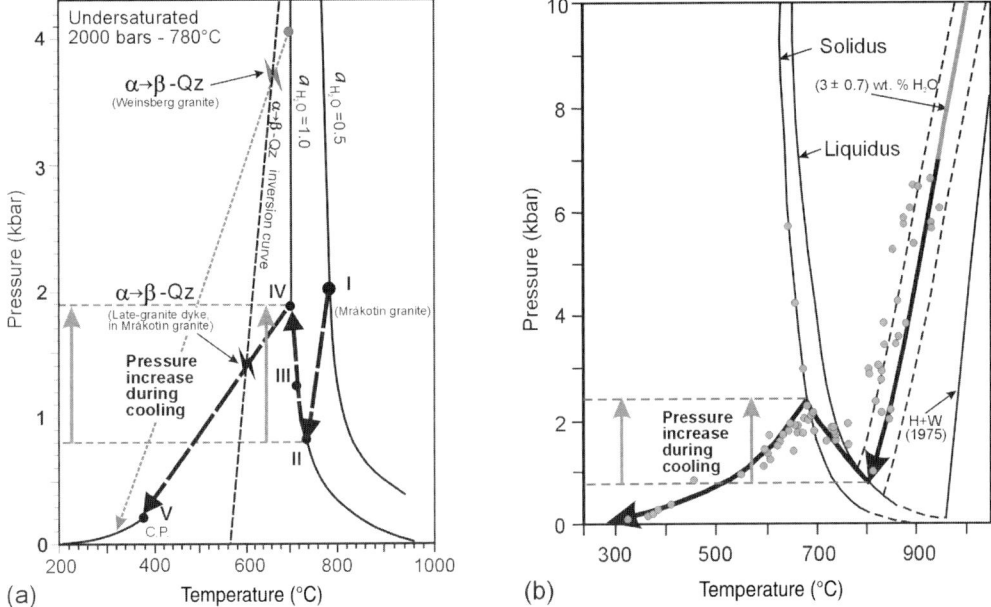

Fig. 8. *P–T* paths of the cooling granites during crystallization derived from fluid and melt inclusion studies (in part according to Thomas 1994; Student & Bodnar 1996). (**a**) Cooling in two subplutons of the SBP; lightning-symbols indicate the formation and trapping of fluid inclusions in quartz at the α–β transformation, which is connected with a volume change of about 1%. The trapped fluid inclusions represent minimum *P–T* conditions, because of the time difference between solidus crystallization and the α–β transformation of quartz. a_{H_2O}, water activity. The bold lines marked by $a_{H_2O} = 1$ and $a_{H_2O} = 0.5$ represent the water-saturated and water-undersaturated solidus curves, respectively, according to Johannes & Holtz (1996). Dots: I, melt inclusions trapped at water-undersaturated conditions, points II–IV represent the *P–T* conditions of the melt inclusions in water-saturated melt near the solidus. C.P., critical point of the salt–H_2O system. The line between IV and V represents an isochore of the fluid, trapped in quartz as a fluid inclusions or as fluid subsystem in melt inclusions (Thomas 1994). Near point V the isochore intersects the liquid–vapour curve. At the intersection point a vapour bubble forms in the originally homogeneous liquid at cooling. In the reverse direction of the process the bubble disappears on heating. (**b**) Data from various Erzgebirge plutons that show a pressure increase in the cooling range between liquidus and solidus (according to Thomas 1994). H + W, the line represents the dry solidus according to Huang & Wyllie (1975). The line marked by 3 ± 0.7 wt% corresponds to the liquidus curve of the system quartz–albite–orthoclase–H_2O with a water content of 3 wt% (according to Johannes & Holtz 1996). The grey dots are data points obtained from melt and fluid inclusions.

upwards and, in part, subhorizontally. There, the fluids or rest melts came from deeper parts of the pluton with continuing higher temperatures, migrating upwards. This is not the case, however, for the investigated NNE joints in Boršov. Feeding fissures outside of the natural termination of the joints were not recognized. So, from where did the small melt portion come?

In the case of the NNE-trending joints in Boršov, there had been some sort of anatectic melting near the initiation point of these joints, from where the samples were taken (Fig. 4e, f). The late-granite melt entered the point-like porosity of the microcrack at the initiation stage of the fracture, an eutectic melt is generated first and, from these 'pocket' of eutectic melt, a joint is driven under at very low effective stress. These are natural hydraulic fractures (Savalli & Engelder 2004). The late-granite melt of the dykes

has a spot-like appearance within the Mrákotín Granite and propagated radially, analogous to spot-like circular parent joints, which have a final size of a few metres at the quarry face.

The diagrams in Figure 8 demonstrate the development of *P–T* in a primarily water-undersaturated granitic melt under cooling conditions at the site of the granite emplacement. The *P–T* path is not simply an exhumation path, it is a cooling path. The cooling could not be caused by exhumation, as the heat was given to the country rock. The crystal growth of the melt (path to the solidus: point I to point II) indicates cooling but not necessarily exhumation, as the melt would crystallize even if under a constant rock pile. These diagrams were derived from the study of natural fluid and melt inclusions in quartz of different granites (Thomas 1994; Student & Bodnar 1996; Student 2002).

A melt inclusion forms when a small globule of magma is trapped in a growing crystal, such as quartz, and if this crystal as a container is able to preserve the original trapping conditions. If so, then accurate information can be obtained about ancient magmatic systems from the study of such melt inclusions (Roedder 1984). Thermometric measurements of melt inclusions are based on the generally recognized condition that the medium contained in an inclusion was an homogeneous phase (melt) at the time of trapping and heterogenized only during subsequent cooling without any gain or loss of material. Usually, the homogenization can be experimentally reproduced by heating the inclusion-bearing sample on a microscope heating stage and by the observation of all phase changes in the inclusion (partial homogenization of the fluid phase, the beginning of melting) up to the total homogenization corresponding to the trapping conditions. Thus, the melt inclusions can be applied to natural samples to constrain trapping pressures and temperatures. The different melt and fluid inclusions were trapped at different time and under different P–T conditions of the cooling melt, thus indicating varying moments of the cooling history and thereby demonstrating the path.

The evolution of granite magma during ascent, emplacement and cooling can be modelled by the behaviour of melt inclusions during trapping and by following cooling in a very simplified way. The generation of fluid inclusions then reflect the early post-magmatic history, probably formed during α–β inversion of the quartz (Yoder 1950), which is connected with a density change and the formation of microcracks. The point of α–β inversion of the quartz is marked by lightning-symbols in Figure 8a. The black lines denoted by $a_{H_2O} = 1$ and $a_{H_2O} = 0.5$ represent the water-saturated and water-undersaturated solidus curves, respectively, according to Johannes & Holtz (1996).

The analysed fluid inclusions in quartz (Table 3) were formed and trapped at the lightning-symbols along the P–T path in Figure 8a. The path (bold line) between point I and point IV indicates the Mrákotín granite melt evolution. For the intersecting late-granite dyke from Boršov (SBP-N) the coordinates of the fluid inclusions are $T = 607\,°C$ and $P \approx$ 1.4 kbar, located on the P–T path between points IV and V, corresponding to a depth of about 7.4 km, and the timing of this was at 324.9 Ma. This dating (minimum age) was obtained from the cooling age of the muscovites within the late-granite dyke associated with temperatures of about 350–400 °C (near point V in Fig. 8a), some time after the trapping of the inclusions at 620 °C. Within the Weinsberg Granite (Friepeß quarry; SBP-S) the fluid inclusions indicate trapping conditions of $T = 662 \pm 8\,°C$ and P $= 3.65 \pm 0.33$ kbar, located on the grey fine-hatched line in Figure 8a, and a level of inclusion trapping at 14.3 $\pm$ 1.3 km, taking into consideration lithostatic conditions.

The P–T path of the fluid inclusions (Fig. 8a) represents the cooling and crystallization of the granite melt (bold line) from the starting point, point I (representing complete homogenization of inclusions at the water-undersaturated solidus curve), down to point II. During this period water separates by crystal growth within the residual melt. Near the water-saturated solidus ($a_{H_2O} = 1$) this process produces a strong increase in H_2O pressure ('overpressure': points II–IV) of about 1.2 kbar, which can initiate fractures. Evidence for such overpressure comes from numerous measurements of the increasing trapping pressure of inclusions between points II and IV, as is demonstrated by data dots in the range between 680 and 800 °C in Figure 8b.

The increase of pressure from point II to point IV is a physical consequence of the crystallization process: the reaction, OH-bearing melt $\rightarrow$ crystals + H_2O vapour, takes place with an increase in volume and therefore with an increase of pressure. The rise in pressure between points II and IV was also experimentally and theoretically shown by Student & Bodnar (1996) and Student (2002) for the case of primary water-undersaturated granitic melts. Starting from point I, during cooling the inclusion crystallizes and the internal pressure inside the inclusion decreases until the remaining melt reaches water saturation (point II). Continued cooling leads to an increase in the internal pressure in the inclusion to the solidus conditions at point IV. The line between points IV and V represents an isochore of the fluid, trapped in quartz as fluid inclusions or as a fluid subsystem in melt inclusions (Thomas 1994). Near point V the isochore intersects the liquid–vapour curve. At the intersection point a vapour bubble forms in the originally homogeneous liquid on cooling; in the reverse direction of the process the bubble disappears on heating.

Burnham (1979) demonstrated that for constant volume conditions during crystallization of a felsic melt enormous pressures can be generated, theoretically up to several 10 kbars. In addition, evidence for the pressure increase during cooling and water separation comes from numerous measurements along this path by Thomas (1994), as is demonstrated in Figure 8b for the Erzgebirge Pluton, and by Student & Bodnar (1996).

The amount of the overpressure depends on the mechanical behaviour of the melt system. The latest melt phase can be liquid and mobile at very low temperatures (c. 500 °C, 1 kbar) if the volatile (H_2O, B, F) and semi-volatile (P_2O_5, rare alkalies) content is high enough. In the case of a quasi-closed system, fissures occur as the result of the changing density of the melt during crystallization. Such fissures will be filled with fluid or with the rest of the intergranular melt, because the opening space of the crack is able to connect with the entire granite body. The process is similar to the well-known fissure generation in the roof of a granite body and the development of 'Stockscheider' pegmatites. Experiments show that these 'residual melts' are characterized by a high water content, low density, low viscosity (similar to water) and a high mobility (Thomas & Webster 2000; Thomas *et al.* 2000, 2003). Possibly, the residual melt represents a supercritical fluid that became separated from the crystallizing granite.

This separated supercritical fluid, together with the pressure increase near the solidus, can lead to a reactivation and repeated melting that a new granitic or pegmatitic melt generates. We conclude that a causative sequence of local processes within the granite during cooling suggests the source and conditions of early joint initiation in the way that Figure 9 demonstrates.

This simplified succession of local processes involves several conditions under which fracturing can occur. In addition, the density and volume change during crystallization (thermoelastic contraction) can lead to further propagation of cracks. This is common in cooling igneous rocks. We propose a new concept of a 'magma self-made' *in situ* joint initiation caused by overpressure and residual melt or new melt generation within the closed system of the cooling granite, evolving rather like spots within the solidifying magma. Hence, the associated resulting fractures are spot-like distributed penny-shaped features with concentric undulations (Fig. 4). The regularly orientated propagation of these spot-like centres in a NNE–SSW direction requires additional forces.

Evidence from fluid and melt inclusion measurements for evolving overpressure near the solidus conditions provides arguments against the concerns that fracturing cannot occur in deeply buried rocks. Engelder & Lacazette (1990), Engelder & Fischer (1996) and Savalli & Engelder (2004) have treated several aspects of the role of fluids for the fracture process. We have considered here the additional aspect of magma-involved overpressure and associated local events.

Driving forces and stress conditions (SBP)

In suitable cases the type of joint driving forces (e.g. pluton-related local stress or regional stress related to tectonic events) the orientation of the main stress axis and, if existing, the anisotropy of stresses can be reflected by the joint-surface morphology and the spatial organization of joints.

The fractographic pattern of the discussed joints, which were generated at 7 km-deep crustal levels, does not simply reflect typical features of thermo-elastic contraction of the hot rocks, driving the fracture process as it is known from mafic volcanic rocks. Nor do the early granite fractures at Boršov represent random orientation or distribution comparable with those in mafics. Rather, the joint pattern demonstrates the imprint of additional joint-driving forces.

Engelder *et al.* (1993) distinguished four driving forces: (1) joint-normal stretching; (2) thermoelastic contraction; (3) fluid-driven fracturing; and (4) axial splitting. The fluid-driven joints are called natural hydraulic fractures. Evidence for an internal fluid drive is the incremental propagation of joints (Lacazette & Engelder 1992), common in layers of various sediments (Bahat 1991; Engelder & Fischer 1996) and is seen exposed with finest features in the Boršov quarry (Fig. 4). However, the fracturing process at Boršov is more specific.

Fluids are common in magmatic rocks, for example after water separation from the crystallizing melt near the solidus (Fig. 8), nonetheless, fluid-driven fracturing may operate together with other driving forces. Indicators for fluid-driven joints are the quasi-circular 'sun'-like penny-shaped joints associated with wave-like undulations, as can be seen in the Boršov quarry (Bankwitz *et al.* 2001; Bahat *et al.* 2003). The frequency of their occurrence is peculiar, and seldom developed in granites of central Europe. Only a few joints of the same type were recognized in the Central Bohemian Granite (340 Ma), the Bohus Granite (950 Ma) and the Sunne Granite (1650 Ma) in Sweden.

Because of the coeval processes of fracture and melt influx (as late melt injections producing the joint) it is to be supposed that the jointing started early after reaching the solidus of the melt and after the development of an overpressure (Fig. 8). The related new generation of a small amount of a local melt at many places within the granite, or small portions of residual melt within the solidifying granite entering at the initiation point of a crack, was the driving force of the first propagation stage. Many joints finished after several more or less concentric undulations (Fig. 4a–d), a few of them mismatches together propagating in each direction.

This stage of early fracturing demonstrates slow subcritical fracture propagation. Savalli & Engelder (2004) propose that joints driven by fluid pressure are stable and thus propagate in region III of the subcritical crack propagation regime, as documented by Atkinson & Meredith (1987). Stable propagation takes place because the driving fluid is slowly introduced to the joint surface. For the first fracturing stage this seems to be the case with the Boršov joints. Savalli & Engelder (2004) studied mechanisms controlling the shape of slow propagating joints in sediments and assumed that the stress intensity (K_1) related to the joint formation dictates the rupture

---

SUCCESSION OF EVENTS WITHIN SOLIDIFYING AND FRACTURING FELSIC MAGMAS

**Cooling of the magma to the solidus P-T conditions
at the emplacement level**
(evidence: melt & fluid inclusions)

**Increase of the H_2O pressure near the solidus
by crystallization of anhydrous phases**
(evidence: P-T conditions of melt & fluid inclusions)

Fracture initiation *(age & depth evidence)*
▼

**Cyclic propagation by input of a mobile low-viscosity "residual melt"
into the natural hydraulic formed fracture
slow subcritical crack growth**
(evidence: spot-like penny-shaped joints & late-granite dyke)
▼▼

Associated generation of small portions of new granitic melt
(evidence: overpressure near the solidus)

Continuing slow subcritical crack growth
(evidence: fractographic features & late-granite dyke)
▼

**Reaching a critical point of fracture propagation,
further autonomous crack propagation occur**
(fractographic evidence: fringe with straight en echelon cracks)

**Crystallization of the new melt or residual melt as late-granite dykes
along the fracture**
(evidence: inclusions & age determination)

Fig. 9. Model of early deap-seated fracturing in solidifying granite magma summarized from fractographic investigations, and from fluid and melt inclusion studies.

---

front velocity during joint growth. A K_1 dependent velocity, v, was found during slow, long-term loading in the presence of chemically reactive pore fluids (Wiederhorn 1972). The penny-shaped joints of Boršov are located in region III of the stress intensity (K_1) v. crack-propagation velocity curve.

But one peculiarity of the NNE joint set of the Boršov quarry is the variety of joint shapes and surface morphology (Bahat *et al.* 2003), thus indicating variations in crack growth. All parent joints developed first of all in region III, the subcritical crack-propagation regime. Some formed very smooth surfaces, which propagated according to Savalli & Engelder (2004) in region I of the subcritical regime. Some large individual joints generated wide fringes

with well-developed en echelon cracks, propagating dominantly vertically. Ahead at the parent joint boundaries the fracture process became autonomous in many cases and produced branching into several joint surfaces (Bahat *et al.* 2001, 2003) that indicates propagation in the critical regime. After the gradual propagation into the fringe, the rapid increase in velocity and intensity of fracture produced the segmentation into fringe cracks. Then again, the crack propagation slowed down and was driven, in some cases, by late-granite dykes. The vertical propagation of the en echelon fringe cracks indicates vertical compression due to the rock pile that was 7 km in thickness. The spatial organization of the en echelon fringe cracks demonstrates axial splitting (Engelder *et al.* 1993).

The orientation of these NNE-trending joints is regular and persistent throughout the SBP over a distance of more than 100 km. The propagation of the joint does not only depend on pluton-related stress, but, in addition, it depends on the regional remote stress. The straight orientation reflects an outer tectonic control. The rotation of the fringe cracks shows the same regularity regarding the degree and direction of the rotation through the entire pluton (see Fig. 6) similar to the joints from the Appalachian Plateau, as reported by Younes & Engelder (1999).

From fractographic features and melt-conditions research we expect different forces to be operating together during the formation of the NNE-trending joints at Boršov:

(1) spot-like crack initiation by overpressure near the solidus;
(2) cyclic propagation under fluid load creating natural hydraulic fractures, slow subcritical crack growth;
(3) imprint of regional remote stress establishing the consistent joint orientation;
(4) further slow propagation under fluid load by residual melt or small portions of a newly generated melt due to overpressure;
(5) rotation of the regional stress field;
(6) joint propagation into the transition zone and to the fringe;
(7) increase in propagation velocity, autonomous fracturing behind the parent joint boundary, critical crack growth vertically and segmentation of the fringe, imprint of the axial load;
(8) slowing down of the propagation of the fringe cracks; subcritical crack growth under continuing fluid load in the case of the late-granite dykes;
(9) development of en echelon fringe cracks under axial load + regional remote stress, vertical propagation of the fringe cracks;
(10) axial splitting + mixed-mode (I, II, and III) propagation.

Engelder & Fischer (1996) also consider a combination of joint-loading configurations as the most likely forces during the propagation of a natural joint and, in addition, axial splitting may operate with elastic contraction. This is in a good agreement with our interpretations of the pre-exhumation joints found in the SBP.

Conclusions

From geological observations, age dating, and fluid and melt inclusion studies it was possible to conclude that large joints of the SBP were formed during a late stage of solidification of the magma. The final crys-tallization was associated with a pressure increase and with infiltration of fluids and mobile, low-viscosity residual melts into the solid rock creating natural hydraulic fractures. This contributed to the origin of the dominant NNE joints of the pluton.

T. Engelder is thanked for his time in reviewing an earlier version of this chapter and for the new ideas that he proposed.

References

ATKINSON, B. K. & MEREDITH, P. G. 1987. The theory of subcritical growth with applications to minerals and rocks. *In*: ATKINSON, B. K. (ed.) *Fracture Mechanics of Rocks*. Academic Press, Orlando, FL, 111–166.

BAHAT, D. 1991. *Tectonofractography*, Springer, Berlin.

BAHAT, D., BANKWITZ, P. & BANKWITZ, E. 2001. Changes of crack velocities at the transition from the parent joint through the en echelon fringe to a secondary mirror plane. *Journal of Structural Geology*, **23**, 1215–1221.

BAHAT, D., BANKWITZ, P. & BANKWITZ, E. 2003. Pre-uplift joints in granite: evidence for sub critical and post critical fracture growth. *Geological Society of America Bulletin*, **115**, 148–165.

BANKWITZ, P. & BANKWITZ, E. 2000. Granitklüftung – Kenntnisstand 80 Jahre nach Hans Cloos. *Zeitschrift für geologische Wissenschaften*, **28**, 87–110.

BANKWITZ, E., BANKWITZ, D. BAHAT, K., BREITER, K & KÄMPF, H. 2001. Excursion B, South Bohemian Pluton. *Exkursionsführer und Veröffentlichungen der GGW*, Berlin, **212**, 129–156.

BEHR, H.-J., DÜRBAUM, H.-J. & BANKWITZ, P. 1994. Crustal structure of the Saxothuringian Zone: Results of the deep seismic profile MVE-90 (East). *Zeitschrift für geologische Wissenschaften*, **22**, 647–769.

BREITER, K. 2001. South Bohemian Pluton. Overview. *Exkursionsführer und Veröffentlichungen der GGW*, Berlin, **212**, 130–141.

BREITER, K. & KOLLER, F. 1999. Two-mica granites in the central part of the South Bohemian Pluton. *Abhandlungen der Geologischen Bundesanstalt*, Wien, **56**, 201–212.

BREITER, K. & SCHARBERT, S. 1998. Latest intrusions of the Eisgarn pluton (South Bohemia, Northern Austria). *Jahrbuch der Geologischen Bundesanstalt, Wien*, **141**, 25–37.

BREITER, K. & SOKOL, A. 1997. Chemistry of the Bohemian granitoids: Geotectonic and metallogenetic implications. *Sbornik geologickych ved*, **31**, 75–96.

BÜTTNER, S. 1997. Die spätvariszische Krustenentwicklung in der südlichen Böhmischen Masse: Metamorphose, Krustenkinematik und Plutonismus. *Frankfurter Geowissenschaftliche Arbeiten*, **A16**.

BURNHAM, C. W. 1979. The importance of volatile constituents. *In*: YODER, H. S. (ed.) *The Evolution of the Igneous Rocks: Fiftieth Anniversary Perspective*. Princeton University Press, Princeton, NJ, 439–482.

CARL, C. & WENDT, I. 1993. Radiometrische Datierung der Fichtelgebirgsgranite. *Zeitschrift für geologische Wissenschaften*, **21**, 49–72.

COYLE, D. A. & WAGNER, G. A. 1995. Spaltspuren – ein Beitrag zur postvariszischen Tektonik und Abtragung des KTB-Umfelds. *Geowissenschaften*, **13**, 142–146.

ENGELDER, T. & FISCHER, M. P. 1996. Loading configurations and driving mechanisms for joints based on the Griffith energy-balance concept. *Tectonophysics*, **256**, 253–277.

ENGELDER, T. & LACAZETTE, A. 1990. Natural hydraulic fracturing. *In*: BARTON, N. & STEPHANSSON, O. (eds) *Rock Joints*. A. A. Balkema, Rotterdam, 35–44.

ENGELDER, T., FISCHER, M. P. & GROSS, M. R. 1993. *Geological Aspects of Fracture Mechanics*. Geological Society of America, Boston, MA.

FINGER, F. & VON QUADT, A. 1992. Wie alt ist der Weinsberger Granit? U/Pb versus Rb/Sr Geochronologie. *Mitteilungen der Österreichischen Mineralogischen Gesellschaft*, **137**, 83–86.

FINGER, F., BENISEK, A., BROSKA, I., *et al.* 1996. Altersdaten von Monaziten mit der Elektronenmikrosonde – Eine wichtige neue Methode in den Geowissenschaften. In: *Symposium Tektonik–Strukturgeologie–Kristallingeologie*, Geological Institute, University of Salzburg, 100–102.

FÖRSTER, H.-J., TISCHENDORF, G., TRUMBULL, R. B. & GOTTESMANN, B. 1999. Late collisional granite in the Variscan Erzgebirge, Germany. *Journal of Petrology*, **40**, 1613–1645.

FRANK, W. 1994. Geochronology and evolution of the South Bohemian Massif. A review. *Mitteilungen der Österreichischen Mineralogischen Gesellschaft*, **139**, 41–43.

FRIEDL, G., FRASL, G., VON QUADT, A. & FINGER, F. 1992. Neue U/Pb Altersdaten aus der südlichen Böhmischen Masse. *Frankfurter Geowissenschaftliche Arbeiten*, **A11**, 215.

FRIEDL, G., VON QUADT, A. & FINGER, F. 1996. Timing der Intrusionstätigkeit im Südböhmischen Batholith. In: *Symposium Tektonik-Strukturgeologie-Kristallingeologie*, Geological Institute, University of Salzburg, 127–130.

FUHRMANN, U., LIPPOLT, H. J. & HESS, J. C. 1987. Examination of some proposed K–Ar standards: $^{40}Ar/^{39}Ar$ analyses and conventional K–Ar data. *Chemical Geology (Isotope Geoscience Section)*, **66**, 41–51.

GEOLOGICAL SURVEY OF CZECHOSLOVAKIA 1968. *Geological Map of Czechoslovakia*, 1:500000. Geological Survey of Czechoslovakia, Praha.

GERDES, A. 1997. *Geochemische und thermische Modelle zur Frage der spätorogenen Granitgesteine am Beispiel des Südböhmischen Batholiths: Basaltische Underplating oder Krustenstapelung*. PhD thesis, Göttingen.

GERDES, A., WÖRNER, G. & FINGER, F. 1998. Late-orogenic magmatism in the southern Bohemian Massif – geochemical and isotopic constraints on possible sources and magma evolution. *Acta Universitatis Carolinae, Geologica*, **42**, 41–45.

GNOJEK, I. & PŘICHYSTAL, A. 1997. Ground geophysical and geological mapping in the central part of the Moldanubian pluton. *Jahrbuch der Geologischen Bundesanstalt, Wien*, **140**, 193–250.

HEINRICHS, H. & HERRMANN, A. G. 1990. *Praktikum der Analytischen Geochemie*. Springer, Berlin.

HUANG, W. L. & WYLLIE, P. J. 1975. Melting reactions in the system $NaAlSi_3O_8$–$KAlSi_3O_8$–SiO_2 to 35 kilobars, dry and with excess water. *Journal of Geology*, **83**, 737–748.

JOHANNES, W. & HOLTZ, F. 1996. *Petrogenesis and Experimental Petrology of Granite Rocks*. Springer, Berlin.

KREUZER, H., HENJES-KUNST, F., SEIDEL, E., SCHÜßLER, U. & BÜHN, B. 1993. Ar–Ar spectra on minerals from KTB and related medium-pressure units. *KTB Report*, **93–2**, 133–136.

LACAZETTE, A. & ENGELDER, T. 1992. Fluid-driven propagation of a joint in the Ithaca siltstone, Appalachian basin, New York. *In*: EVANS, B. & WONG, T.-F. (eds): *Fault Mechanics and Transport Properties of Rocks*. Academic Press, London, 297–370.

MENZEL, D. & SCHRÖDER, B. 1994. Geologische Kriterien zur Unterbau-Exhumierung im Naab-Gebirge. *KTB Report*, **94–3**, 179–184.

MATTE, PH., MALUSKI, H., RAJLICH, P. & FRANKE, W. 1990. Terrane boundaries in the Bohemian Massif: Result of large-scale Variscan shearing. *Tectonophysics*, **177**, 151–170.

MEURERS, B. 1992. *Korrigierte Bougueranomalie der Südlichen Böhmischen Masse*. Schwerpunktprogramm S47GEO. Präalpidische Kruste in Österreich, Salzburg.

PETEREK, A., HIRSCHMANN, G. *et al.* 1994. Spät- und postvariszische tektonische Entwicklung im Umfeld der Kontinentalen Tiefbohrung Oberpfalz (KTB). *KTB Report*, **94–3**, 123–148.

PRICE, N. J. 1966. *Fault and Joint Development*. Pergamon Press, Oxford.

RAMSAY, J. G. & HUBER, M. I. 1987. *The Techniques of Modern Structural Geology*, Volume 2. Academic Press, New York, 309–700.

ROEDDER, E. 1984. *Fluid inclusions*. Review in Mineralogy, **12**.

RUF, J. C., RUST, K. A. & ENGELDER, T. 1998. Investigating the effect of mechanical discontinuities on joint spacing. *Tectonophysics*, **295**, 245–257.

SAVALLI, L. & ENGELDER, T. 2004. Mechanisms controlling the shape of the rupture front as joints cut through layered clastic sediments. *Geological Society of America Bulletin*, submitted.

SCHARBERT, S. 1998. Some geochronological data from the South Bohemian Pluton in Austria: a critical review. *Acta Universitatis Carolinae, Geologica, Praha*, **42**, 114–118.

SCHUMACHER, E. 1975. Herstellung von 99,9997% ^{38}Ar für die $^{40}K/^{40}Ar$ geochronologie. *Geochronologia Chimia*, **24**, 441–442.

SECOR, D. T. 1965. Role of fluid pressure in jointing. *American Journal of Science*, **263**, 633–646.

SECOR, D. T. & POLLARD, D. D. 1975. On the stability of open hydraulic fractures in the Earth's crust. *Geophysical Research Letters*, **2**, 510–513.

SIEBEL, W. 1993. Geochronology of the Leuchtenberg granite and associated redwitzites. KTB Report, **93–2**, 411–416.

SIEBEL, W. 1998. Variszischer spät- bis postkollisionaler Plutonismus in Deutschland: Regionale Verbreitung, Stoffbestand und Altersstellung. *Zeitschrift für geologische Wissenschaften*, **26**, 329–358.

STEIGER, R. H. & JÄGER, E. 1977. Subcommission on Geochronology: Convention on the Use of Decay Constants in Geo- and Cosmochronology. *Earth and Planetary Science Letters*, **36**, 359–362.

STETTNER, G. 1994. Spätkaledonische Tektonik im Zusammenschluß des mittleren Paläo-europas. *Zentralblatt Geologie und Paläontologie*, **1**, 763–772.

STUDENT, J. J. 2002. *Silicate melt inclusions in igneous petrogenesis*. PhD dissertation, Virginia Tech, Blacksburg, VA.

STUDENT, J. J. & BODNAR, R. J. 1996. Melt inclusion microthermometry: Petrologic constraints from the H₂O-saturated haplogranite system. *Petrology*, **4**, 291–306

SUK, M. 1994. The role of the present erosion level in the interpretation of the Bohemian Massif. *KTB Report*, **94-3**, 191–199.

THOMAS, R. 1989. *Investigations of melt inclusions and their application to the solution of various problems of deposit geology and petrology*. Dissertation B, Bergakademie Freiberg, Germany.

THOMAS, R. 1994. Fluid evolution in relation to the emplacement of the Variscan granites in the Erzgebirge region: A review of the melt and fluid inclusion evidence. *In*: SELTMANN, R., KÄMPF, H. & MÖLLER, P. (eds) *Metallogeny of Collisional Orogens*. Czech Geological Survey, Prague, 70–81.

THOMAS, R. & KLEMM, W. 1997. Microthermometric study of silicate melt inclusions in Variscan granites from SE Germany. *Journal of Petrology*, **38**, 1753–1765.

THOMAS, R. & WEBSTER, J. D. 2000. Strong tin enrichment in a pegmatite-forming melt. *Mineralia Deposita*, **35**, 570–582.

THOMAS, R., FÖRSTER, H.-J. & HEINRICH, W. 2003. The behaviour of boron in a peraluminous granite–pegmatite system and associated hydrothermal solutions: a melt and fluid-inclusion study. *Contributions to Mineralogy and Petrology*, **144**, 457–472.

THOMAS, R., WEBSTER, J. D. & HEINRICH, W. 2000. Melt inclusions in pegmatite quartz: complete miscibility between silicate melts and hydrous fluids at low pressure. *Contributions to Mineralogy and Petrology*, **139**, 394–401.

TUTTLE, O. F. & BOWEN, N. L. 1958. *Origin of Granite in the Light of Experimental Studies in the System Na AlSi₃O₈–KAlSi₃O₈–SiO₂–H₂O*. Geological Society of America, Memoirs **74**.

WAGNER, G. A. & VAN DEN HAUTE, P. 1992. Fission track dating. Enke-Verlag, Stuttgart.

WIEDERHORN, S. M. 1972. Subcritical crack growth in ceramics. *In*: BRADT, R. C., EVANS, A. G., HASSELMAN, D. P. H. & LANGE, F. F. (eds) *Fracture Mechanics of Ceramics, Volume 2*. Plenum, New York, 613–646.

WEMMER, K. 1991. K/Ar-Altersdatierungsmöglichkeiten für retrograde Deformationsprozesse im spröden und duktilen Bereich – Beispiele aus der KTB-Vorbohrung (Oberpfalz) und dem Bereich der Insubrischen Linie (N-Italien). *Göttinger Arbeiten Geologie und Paläontologie*, **51**, 1–61.

YODER, H. S. 1950. High–low quartz inversion up to 10.000 bars. *Transactions of the American Geophysical Union*, **31**, 827–835.

YOUNES, A. I. & ENGELDER, T. 1999. Fringe cracks: key structures for the interpretation of the progressive Alleghanian deformation of the Appalachian plateau. *Geological Society of America Bulletin*, **111**, 219–239.

ZHANG, Y. & FRANTZ, J. D. 1987. Determination of the homogenization temperatures and densities of supercritical fluids in the system NaCl–KCl–CaCl₂–H₂O using synthetic fluid inclusions. *Chemical Geology*, **64**, 335–350.

ZULAUF, G. 1993. Brittle deformation events at the western border of the Bohemian Massif. *Geologische Rundschau*, **82**, 489–504.

The feedback between joint-zone development and downward erosion of regularly spaced canyons in the Navajo Sandstone, Zion National Park, Utah

CHRISTIE M. ROGERS* & TERRY ENGELDER

Department of Geosciences, The Pennsylvania State University, University Park, PA 16801, USA (e-mail: engelder@geosc.psu.edu)
*Present address: ExxonMobil, 233 Benmar, Houston, TX 77060, USA (e-mail: christie.m.rogers@exxonmobil.com)

Abstract – Large NNW-trending slot canyons cut into, but generally not entirely through, the approximately 600 m-thick Jurassic Navajo Sandstone at Zion National Park (ZNP). These canyons sit immediately above, and parallel to, joint zones and exhibit a regular spacing (*c*. 450 m). The joint zones, in particular, consist of vertical and steeply dipping joints that tend to dip towards the axis of the canyon. These regularly spaced, joint-localized canyons are confined to the Navajo, suggesting a stress-shadow origin for their configuration; however, this explanation does not predict closely spaced joints in joint zones at each canyon. To explain the development of the joint zones, we treat the canyons themselves as cracks. Early, widely-spaced, NNW-trending joints propagated into the top of the Navajo, and later preferential erosion along these joints initiated the pattern of canyons with a cross-sectional profile consistent with blunt edge cracks spaced at about 450 m. Analogous to edge cracks, the canyons subsequently concentrated tensile stress at their tips while subjected to regional extension. Concentration of canyon-tip tensile stress was sufficient to drive steeply dipping secondary joints, reflecting principal stress rotation in a process zone ahead of the canyon tip. Joint density in each joint zone increases as a consequence of a gravity-induced shear traction that drives vertical wing cracks from the tips of steeply dipping secondary joints. Exfoliation jointing along canyon walls also contributes to the widening of canyons. The preferential erosion of slot canyons follows the joint zones, and thus, a feedback loop is set up between the growth of secondary jointing in the canyon-tip stress concentration and the downward erosion of the canyon.

NNW-tending joint zones have preferentially eroded into regularly spaced slot-shaped canyons yielding Zion National Park's (ZNP) dramatic landscape (Gregory 1950; Eardley 1965). From the air, these linear, regularly spaced slot canyons take on the appearance of a joint set where canyon spacing is approximately equal to the *c*. 600 m-thickness of the flat-lying Navajo Sandstone (Fig. 1). This canyon fabric is analogous to bedding-confined joint sets possessing a spacing proportional to bedding thickness (Price 1966; Gross *et al.*, 1995). The analogy is appropriate because the slot canyons follow joint zones that cut through the Navajo Sandstone, a stratigraphic unit that appears to have acted as an extraordinarily thick mechanical unit (Gross 1995). A stress-reduction-shadow theory is often proposed as the mechanism explaining this one-to-one relationship between joint spacing and bed thickness (Pollard & Segall 1987; Narr & Suppe 1991). The question is how did canyon-related joint zones in the Navajo come to posses this same spacing–thickness relationship, particularly when the close spacing of joints in each zone is not consistent with a stress-shadow mechanism?

Groupings of closely spaced joints are known as *joint swarms* (e.g. Laubach *et al.* 1995; Hennings & Olson 1997), *joint clusters* (e.g. Olson 1993; Cooke *et al.*, 2000) and *joint zones* (Engelder 1987; Dyer 1988). Dyer (1988) describes joint zones at Arches National Park as 'individual, subparallel en echelon joints confined to a narrow zone, separated from adjacent zones by a characteristic distance . . . confined to a single lithologic interval' and 'generally perpendicular to bedding'. Although the NNW-trending joint zones at Zion are also narrow, defined by a characteristic distance and confined to the thick Navajo Sandstone, they are not predominantly composed of en echelon joints. In addition, there is a considerable population of steeply dipping joints included in the canyon-tip joint zones that are not perpendicular to bedding. The explanation for the joint-zone development at ZNP must take these characteristics into consideration.

Mechanisms proposed for closely spaced joint propagation include the propagation of en echelon arrays prior to shearing/faulting followed by oblique secondary fractures (Myers & Aydin 1998), fold-hinge joint localization (Fischer & Jackson 1999), formation of fold-limb splay cracks due to bedding-plane slip (Cooke *et al.* 2000), subcritical crack growth (Olson 1993), joint propagation in response to elevated stress in crack-tip process zones (Dyer 1988) and process-zone development associated with

From: COSGROVE, J. W. & ENGELDER, T. (eds) 2004. *The Initiation, Propagation, and Arrest of Joints and Other Fractures.* Geological Society, London, Special Publications, **231**, 49–71. 0305-8719/04/$15 © The Geological Society of London 2004.

Fig. 1. Satellite image taken over Zion National Park. Slot canyons, which have eroded form NNW-trending joint zones, are readily apparent (note the white coloration) (modified from Davis 1999; source Chevron).

dyke emplacement (e.g. Delaney *et al.* 1986). Of the mechanisms described above, joint zones at ZNP are best explained in terms of a process zone. The process zone consists of additional cracks, which develop within a zone of elevated tensile stress, concentrated about the tip of the original crack. Above a threshold, tensile stress can drive new cracks independent of, but in close proximity to, the original crack.

The objectives of this chapter are to document joint-zone development within the Navajo Sandstone at ZNP and to examine the potential mechanisms by which such joint zones can form. Ultimately, our analysis leads to a detailed mechanical explanation for the growth of ZNP's regularly spaced slot canyons as a feedback between joint-zone growth and differential erosion following in the wake of the joint zones.

Geology of ZNP

ZNP is located in SW Utah at the western edge of the Colorado Plateau, adjacent to the central Basin and Range subprovince (Fig. 2). The Navajo Sandstone within ZNP exhibits prominent, regularly-spaced canyons associated with joint zones in a relatively undeformed block of the Colorado Plateau between the Basin and Range-style Hurricane and Sevier normal fault systems (Fig. 3). The ZNP canyon network is an attribute of the Jurassic Navajo Sandstone (Fig. 4). The Navajo is a cliff-forming, eolian sandstone ranging in thickness from 550 to 670 m at ZNP (Biek *et al.* 2000). The Mesozoic sediments underlying the Navajo Sandstone, including the Moenkopi, Chinle, Moenave and Kayenta formations, are predominantly clastic continental deposits, varying in composition between mudstone and fine-grained sandstone (Fig. 5).

Colorado Plateau uplift and extensional unroofing (i.e. Biek *et al.* 2000) accounts for the overburden removal and subsequent headward erosion that uncovered the jointed Navajo and initiated the downward erosion of slot canyons along pre-existing NNW-trending joints. Based on the stratigraphic section of the Zion region (Hintze 1988), the average overburden above the Navajo is estimated at approximately 1400 m at the time of initiation of Miocene WSW regional extension in the neighbouring central

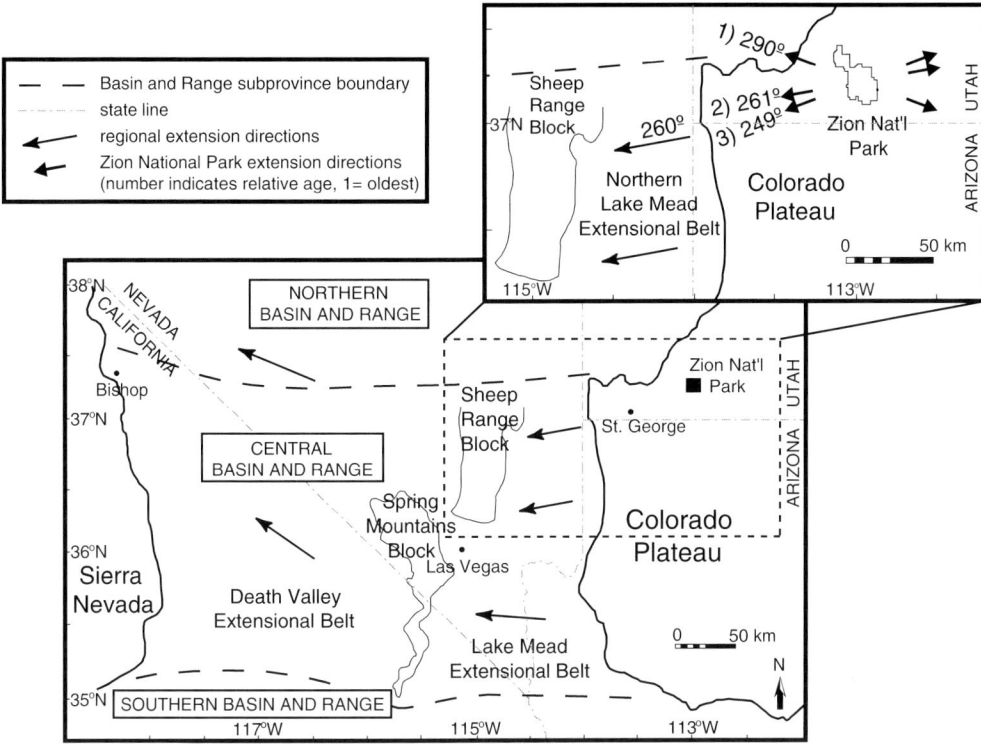

Fig. 2. The central Basin and Range subprovince with the relatively undeformed Colorado Plateau and Sierra Nevada to the east and west, respectively. Within the central Basin and Range, two highly extended regions, the Lake Mead and Death Valley extensional belts, differ in extension direction and are separated by the Sheep Range and Spring Mountains Blocks. Zion National Park is located at the western margin of the Colorado Plateau just east of the northern Lake Mead extensional belt (modified from Snow & Wernicke 2000).

Basin and Range subprovince. Aerial photographs and field checks indicate that the NNW-trending slot canyons and associated joint zones are typical for the Navajo Sandstone but not for the underlying Kayenta or the overlying Temple Cap formations. The orientation of the linear slot canyons suggests that regional jointing was critical to their origin during WSW Basin and Range extension imparted on the Colorado Plateau (Rogers *et al.* 2004). A 261°-trending extension apparently produced bed-confined, regularly spaced joints, and subsequent differential erosion following these early joints produced the NNW-trending slot canyons. This trend closely parallels the average extension direction of 260° identified for the northern Lake Mead Belt of the central Basin and Range subprovince (Snow & Wernicke 2000).

Field observations

Field observations were made in the southern half of ZNP where the canyon network is best exposed due to the headward erosion of the north and east forks of

the Virgin River. Horizontal exposures of joint zones were observed in the eastern half of the Park along Highway 9 (Fig. 4). Thirty-eight joint-perpendicular scanlines were taken in the Navajo Sandstone across these joint zones. Photographs of joint dip distributions across the joint zones were evaluated for trends in joint dip direction relative to position at slot-canyon tips in order to compare field observations with finite-element models (see the Appendix for the methodology used to sort out the distribution of joint dips). In addition, orientations/geometries reflective of wing-crack growth and exfoliation fracturing were identified in an effort to gain a further understanding of the evolution of the joint zones.

Overview

Slot canyons have two trends at ZNP, and they correlate with the strikes of joints in joint zones cutting the Navajo Sandstone. The NNW, *c.* 350°, set of canyons, spaced at approximately 450 m, provides the dominant topographic fabric within ZNP (Fig. 4).

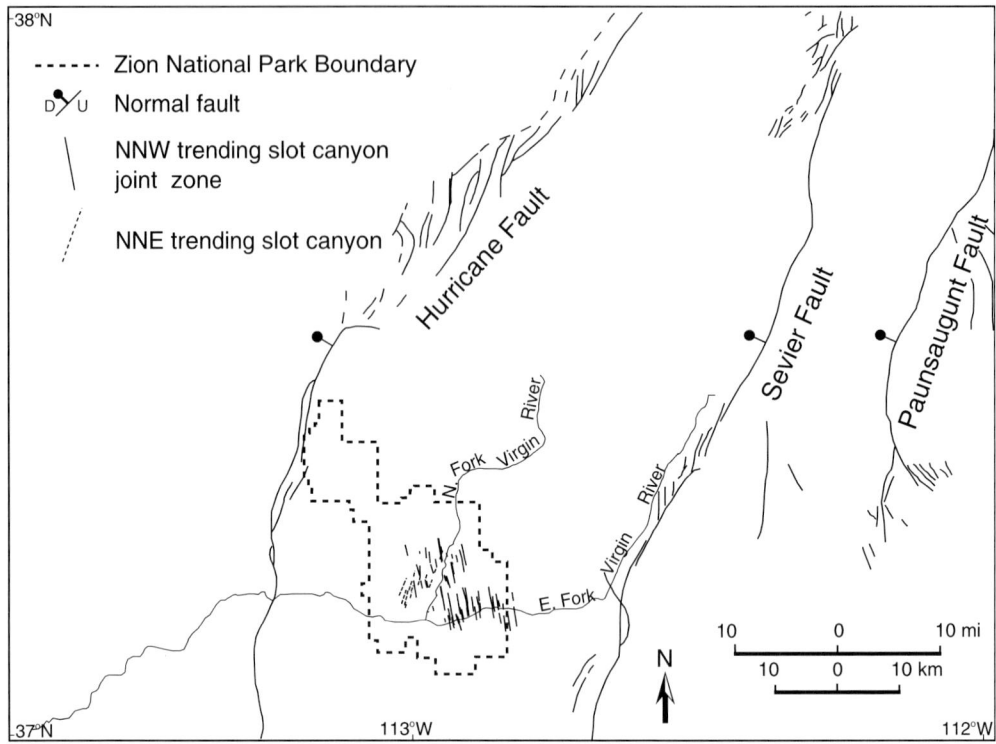

Fig. 3. Zion National Park is located in a relatively undeformed crustal block of the Colorado Plateau. This block is bounded by the Sevier and Hurricane high-angle, down to the west, Basin and Range style normal faults, to the east and west, respectively (modified from Davis 1999). Slot canyons located in the field area (Fig. 4) are represented.

NE-trending canyons located to the west of (and including) Zion Canyon, the widest and deepest (606 m) canyon in the field area, form a second set. In addition to the slot canyons, a third joint set, trending c. 340°, is found between slot canyons. These joints exhibit an average spacing of 22 m and do not preferentially erode into slot canyons. Joint-abutting relationships indicate that the 350°-trending canyons have eroded from joints that post-date those of the NE-trending canyons and predate those of the 340° orientation (Rogers *et al.* 2004). This sequence of jointing indicates that a counterclockwise rotation of the regional stress field has affected the Colorado Plateau in the vicinity of ZNP. This rotation may reflect gravity collapse of the Cordilleran thrust front that culminated in the WSW–ENE extension of the northern Lake Mead belt of the central Basin and Range about 10 Ma ago (Rogers *et al.* 2004).

The relationship between NNW-trending slot canyons and joint zones

The NNW-trending slot canyons are situated directly above one or more joint zones composed of closely spaced (sub)vertical joints. At many locations, the average joint orientation parallels the associated slot canyon. Cumulatively, the vector mean pole of data from all scanlines defines a joint trend of 171° (right-hand rule) (Fig. 6). In terms of distribution, these 351° (171°) joints are found exclusively in the vicinity of the slot canyons. Between slot canyons, the 351° joints are absent and the younger 320–340° joint set predominates.

Vertical outcrops reveal no stratigraphic offset on fractures within the zones. Although joint zones cut to the base of the Navajo Sandstone, individual joints do not. Local wing-crack growth associated with subvertical joints indicates a component of vertical propagation within joint zones. In addition, exfoliation joints occurring in canyon walls contribute to the jointing within the zones (Bahat *et al.* 1995) (Fig. 7).

Slot-canyon morphology is a function of the character of the associated joint zones. Single joint zones are associated with V-shaped, sharp-tipped, slot canyons (e.g. R slot, Fig. 7) while pairs of joint zones are associated with 'box' canyons (e.g. M slot, Fig. 8). These box canyons are V-shaped with a squared tip. Where the Navajo–Kayenta contact is

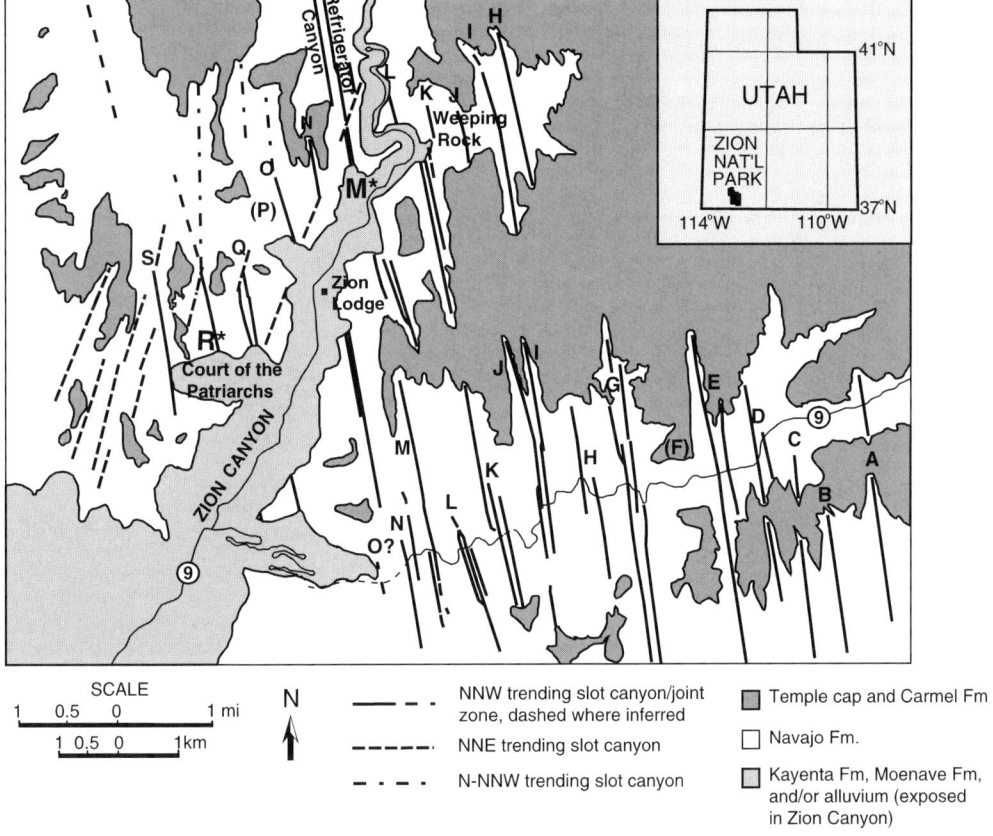

Fig. 4. Field area at Zion National Park. The NNW-trending slot canyons, confined to the Navajo Sandstone and labelled A–S, represent the locations of large-scale joint zone lineaments of particular interest to this investigation. The average orientation of joints that compose these joint zones indicates a 261° regional extension direction, an extension direction exhibited in the central Basin and Range subprovince immediately west of the Colorado Plateau at the latitude of ZNP (Fig. 2).

exposed the joint zones extend to, but not beyond, the base of the Navajo.

Joint zone dip distribution

Our hypothesis is that the erosional geometry of the slot canyons generates a local tensile stress concentration in the Navajo Sandstone below, but restricted to, the tip of the canyon. Furthermore, this stress concentration promotes joint propagation in the form of joint zones analogous to crack-tip process zones in the Navajo Sandstone of ZNP. As joints propagate normal to the local least principal stress, they serve as (palaeo)stress indicators. In the case of the slot canyons at ZNP, joint location and dip-angle distribution at slot-canyon tips should be indicative of the stress concentration that developed there, and by recognizing patterns, the

stress conditions that created these joints can be reconstructed.

Exfoliation jointing adds to the population of joints next to the walls of slot canyons (Fig. 7). In order to avoid the influence of exfoliation fractures in the joint dip distributions, the following observations concern jointing below the level of slot-canyon tips. The digitized joint population was split, east v. west, based on their respective position to the axis of the slot canyon. For our analysis of dip trends, K slot (Weeping Rock), M slot (Refrigerator Canyon) and R slot (Court of the Patriarchs) (Fig. 4) were chosen based on the quality of outcrop exposure below the level of their respective slot-canyon tips. The joint distributions were compared to a typical normal distribution for statistical skewness. A skewed distribution departs from a normal distribution in a particular direction, right or left, and in this case, east or west, indicating that one tail (side) of the distribution

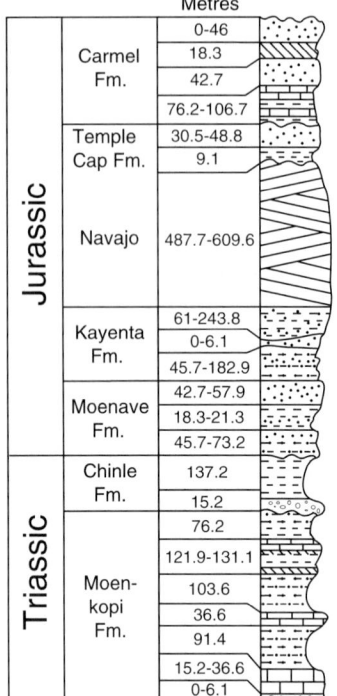

Fig. 5. Stratigraphic column of the Zion National Park area, Utah (Peterson & Pipiringos 1979; Marzolf 1983; Hamilton 1984; Hintze 1988).

exhibits a greater amount/range of data v. the other. This analysis assumes that a skewed distribution represents a measurable trend in dip direction for the population of individual joints within the zone. Each joint zone is briefly described below and followed by the statistical analysis that characterizes joint dip trends resulting from two sampling methods (described in the Appendix): (1) a collection of single joints measured tip to tip; and (2) a collection of 5 m-joint segments expected to account for dip deviations along a single joint due to a temporally changing stress field. This statistical information is summarized in Table 1.

Dip trends: K slot. A single joint zone characterizes K slot at the head of the ZNP's Weeping Rock trail (Figs 4 and 9). Jointing at K slot includes en echelon arrays expressed in V-shaped patterns, a characteristic that is less common at M or R slots. Approximately 120 m of Navajo Sandstone remains below the downward-eroding canyon, and, although the joint zone extends to the base of the Navajo, no *single* joint appears to cut through the remaining 120 m of sandstone.

Statistics derived from the single-joint sampling method suggests that the east half of the zone is skewed/biased toward W-dipping joints, and the

west half is skewed toward E-dipping joints. In comparison, the 5 m-segment sampling method suggests that both the east and west halves of the zone are skewed/biased toward W-dipping joints (Table 1).

Dip trends: M slot. Two joint zones define M slot, ZNP's Refrigerator Canyon, where it meets Zion Canyon (Figs 4 and 8). The associated slot canyon exhibits a 'box' morphology with approximately 120 m of Navajo Sandstone remaining below its base. Fractures within this pair of zones exist symmetrically below and define the box-canyon edges at the base of the canyon. Each zone extends to the base of the Navajo Sandstone. Although no *single* joint appears to cut the remaining Navajo below the zone to the west, a single joint does cut the Navajo below the zone to the east.

Both sampling methods reveal the same result. The eastern zone has a dip distribution biased toward W-dipping joints, and the western zone, has a dip distribution biased toward E-dipping joints.

Dip trends: R slot. A single joint zone controls the erosion of R slot in the Court of the Patriarchs (Figs 4 & 10). This slot canyon exhibits a tilted 'V' shape in cross-section and cuts 90% of the thickness of the

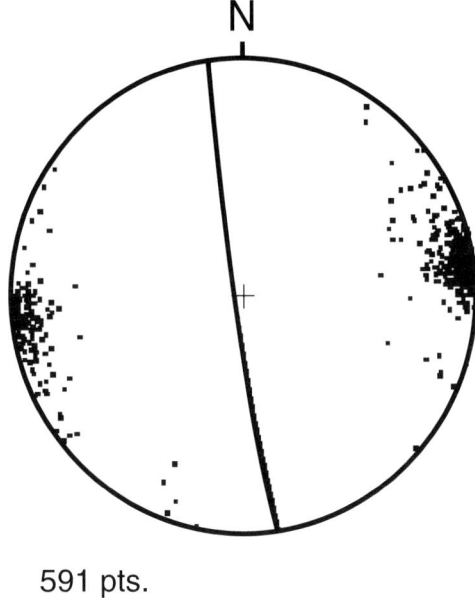

591 pts.

Joint Zones
171°/87°

Fig. 6. Lower-hemisphere stereonet plot displaying poles to joints measured in joint-perpendicular scan-lines across slot-canyon joint zones. The average strike and dip of the joints that compose the NNW-trending joint zones is displayed at 171°/87° (right-hand rule).

Fig. 7. View of the Navajo Sandstone at R slot looking north from the Court of the Patriarchs (Fig. 4). R slot is characterized by a single fracture zone and a 'V'-shaped slot-canyon morphology: examples of exfoliation jointing along canyon walls at R slot are highlighted. Exfoliation jointing acts to widen existing canyons (Bahat *et al.* 1995) and occurs in the final stage of joint-zone development at ZNP.

Fig. 8. Oblique view of the Navajo Sandstone at M slot, Refrigerator Canyon, looking north from Zion Canyon (Fig. 4). Two joint zones, evident from the intervening unfractured volume of rock, and a 'box canyon' ('V'-shaped canyon with squared tip) characterize M slot.

Navajo Sandstone where joints have been bypassed by the downward-eroding canyon. At the canyon tip a prominent joint cuts to the base of the remaining Navajo Sandstone.

Below the canyon tip, statistics derived from both sampling methods indicate that R slot joints are biased toward E-dipping joints on both sides of the canyon (Table 1). However, weaker confidence levels (<90%) for skewness are represented on the west side of the canyon by the single-joint sampling method and the east side of the canyon by the 5 m-segment sampling method.

Summary of dip-trend analysis. At each slot canyon, the following are true: (1) joints dip to the east and west regardless of which side of the canyon they are observed; (2) although these are skewed populations, the mode of each of these distributions is representative of vertical joints; and (3) positive kurtosis exhibited by each distribution indicates that the 'tails' of distribution cover a greater range of values

than that of a normal distribution. Although the mode of these distributions indicates an abundance of vertical joints, the skew and kurtosis values indicate that a considerable population of steeply dipping joints exists.

The whole-joint sampling method resulted in five out of six skewed distributions that show that joints below the level of slot-canyon tips, east v. west, have a tendency to dip towards each other/the axis of the

Table 1. *Summary of descriptive statistics for joint-dip distributions observed below canyon tips at K, M and R slots (Figs 9, 8 and 7, respectively). Joint populations were defined by position, east v. west, relative to the slot canyon. Two sampling methods were analysed for each population: (A) statistics for the distributions composed of whole joints; (B) statistics for the distributions composed of 5 m-joint segments. Normal distributions are described by skewness and kurtosis values of zero. Here, positive skewness denotes E-dipping measurements while negative values denote W-dipping measurements. In five or six cases of the whole-joint sampling method and four of six cases of the 5 m-segment sampling method, the skewness indicates that joints dip toward the slot-canyon axis. Positive kurtosis indicates that the 'tails' of the distribution cover a greater range of values v. a normal distribution and lends validity to the interpretation of a population of steeply dipping joints in addition to vertical joints.*

Joint zone	Number of observations	Mean	Standard deviation	Median	Skewness (Kurtosis)	Confidence level – skew
(A) Descriptive statistics – whole joint sampling						
K slot east	29	−88.7	9.64	88.5	−1.52 (2.65)	98% – West
west	101	−89.4	5.50	90.0	0.39 (1.80)	90% – East
M slot east	51	89.8	7.51	88.7	−0.59 (1.40)	90% – West
west	97	86.1	10.97	89.4	2.37 (5.97)	98% – East
R slot east	53	89.8	6.30	−89.6	1.19 (2.02)	98% – East
west	34	88.1	8.59	89.4	0.54 (0.04)	[<90% – East]
(B) Descriptive statistics – 5 m-joint segment sampling						
K slot east	64	90.0	8.25	89.4	−0.5 (0.79)	90% – West
west	188	−89.1	5.94	−89.9	−0.48 (1.68)	98% – West
M slot east	208	84.5	12.6	−88.2	1.74 (1.31)	98% – West
west	140	88.9	7.95	88.2	−0.59 (2.76)	98% – East
R slot east	167	89.5	9.05	90.0	0.24 (2.66)	[<90% – East]
west	283	−88.4	6.22	−87.9	0.49 (1.14)	98% – East

canyon. Furthermore, the comparisons of skewness and mean values complement each other. In the whole-joint sampling, not only does the skewness indicate a propensity for joints to dip toward the slot-canyon axis, but the values of the mean at each canyon also support this conclusion in a relative sense. The mean values for each analysed canyon reveal that the joints, east v. west of the canyon tip, have a tendency to dip towards each other. For example, at M slot, the east and west halves of the zone have means that are represented by E-dipping values (Table 1). However, the mean dip value of the west half of the zone dips less steeply (departs more from the vertical) than that of the east half of the zone, indicating the same *relative* dip relationship between the two halves of the canyon as expressed by the skewness. Both the mean dip and skewness

values indicate a relationship among joints below the canyon tip: joints from opposite sides of the canyon tend to dip toward each other. This relationship is true at K and R slots as well. We suspect that the sampling of vertical wing cracks has promoted this relative result and masked the absolute sense of dip relative to the canyon that is recognized by the skewness values.

In comparison, the 5 m-segment sampling method yields the same skewness result in four out of the six joint distributions, where only the result at K slot (west) differs between the two methods. This analysis, however, does not consistently reproduce the relative relationship between mean values observed in the whole-joint analysis. Understanding that the 5 m-segment data are flooded with vertical wing-crack measurements, we expect the dip trends in the distributions to weaken and so recognize that this method is more likely to measure the growth of wing cracks.

Models for slot-canyon development at ZNP

To explain the development of the joint zones, we presume that early, widely-spaced, NNW-trending joints propagated into the top of the Navajo and that later, preferential erosion downward along these joints initiated the pattern of canyons, which exhibits a cross-sectional profile consistent with blunt edge cracks spaced at about 450 m. Regarding the early NNW-trending joint set, the model considers two possibilities for the depth of penetration of individual joints into the Navajo: (1) the initial regional joints, produced by Miocene Basin and Range extension, cut the entire thickness of the Navajo Sandstone. Such through-cutting joints may be present at M and R slots. If true, the Navajo, as a mechanical layer, had no tensile strength during erosion of the slot canyons, and a gravity (body) load, alone, is the only possibility for a joint-driving stress ahead of (i.e. below) the slot-canyon tips where joint zones develop; (2) early regional joints only partially cut the Navajo, and the sandstone, as a mechanical layer, maintained tensile strength below the downward-eroding slot canyons. This appears to have been the case for K slot. Here, a slot-tip tension could have arisen from regional

Fig. 9. View of K slot joint zone in cross-section looking north from ZNP's Weeping Rock trail head (Fig. 4). A single joint zone characterizes K slot at this location. In addition, en echelon ('V'-shaped) joint patterns are evident here, yet not well developed at other slot canyons (e.g. M slot (Fig. 8) and R slot (Fig. 10)). No single joint is observed to cut the entire remaining section of Navajo Sandstone below the slot canyon.

Fig. 10. Detail of the joint zone at R slot looking north from the Court of the Patriarchs (Figs 4 and 7). (A) West half of R slot and (B) east half of R slot.

Table 2. *Range of material properties values assigned to the Navajo Sandstone during FRANC (finite-element) modelling. Values based on sandstone values reported by Birch (1966) and Fischer (1994)*

	Young's modulus (E)	Poisson ratio (v)	Density (ρ)	Loading
Model values	15–45 GPa	0.10–0.20	2.3 g cm^{-3}	Gravity body load and/or regional tension

extension and superimposed on the stress arising from gravity loading.

Finite-element models were constructed in order to provide insight into the development of the joint zones below the canyon tips at ZNP. Three slot-canyon geomorphologies were considered in the modelling process. Our initial working model incorporates a symmetric 'V'-shaped notch. Additional models include a symmetric 'V'-shaped notch with a squared tip (e.g. M slot; Fig. 8) and an asymmetric 'V'-shaped notch with a sharp tip (e.g. R slot; Fig. 7). These models were studied with and without the benefit of a tensile strength below the canyon tips.

The finite-element program FRANC (Wawrzynek & Ingraffea 1987) was used to generate a deformed mesh and calculate the resulting stresses at each

node. As the 350°-trending joint zones are exclusively linked to the slot canyons that have developed at Zion, each model was scaled to represent the configuration of a slot-canyon as observed in the field. Material properties, loading conditions and slot canyon geomorphologies were varied in order to understand how stress could be concentrated in the notch of the slot canyons.

Plane-strain simulations were constructed using a finite-element mesh scaled to represent the 600 m-thick Navajo Sandstone. The underlying Kayenta Formation is assumed to be in frictional contact with the Navajo and is included in the slot-canyon models. The loading conditions assigned to the models were designed to fit the tectonic environment that has affected the western edge of the Colorado Plateau in the vicinity of ZNP. Local tension and/or extensional displacements were set as boundary conditions in our models.

Material properties assigned to the Navajo Sandstone were derived from average sandstone properties (dry rock) reported by Birch (1966) and Fischer (1994) (Table 2). Variation in Poisson ratio (v) between 0.10 and 0.20 (Table 2) does little to change the stress during loading by extension, and, therefore, an arbitrary value of 0.15 was assigned to approximate the Navajo Sandstone. In contrast, variation in Young's modulus (E) from 15 to 45 GPa strongly influences the tensile stress at the canyon tip. A Young's modulus value of 15 GPa was found to offer the most realistic radius of stress in excess of the pre-

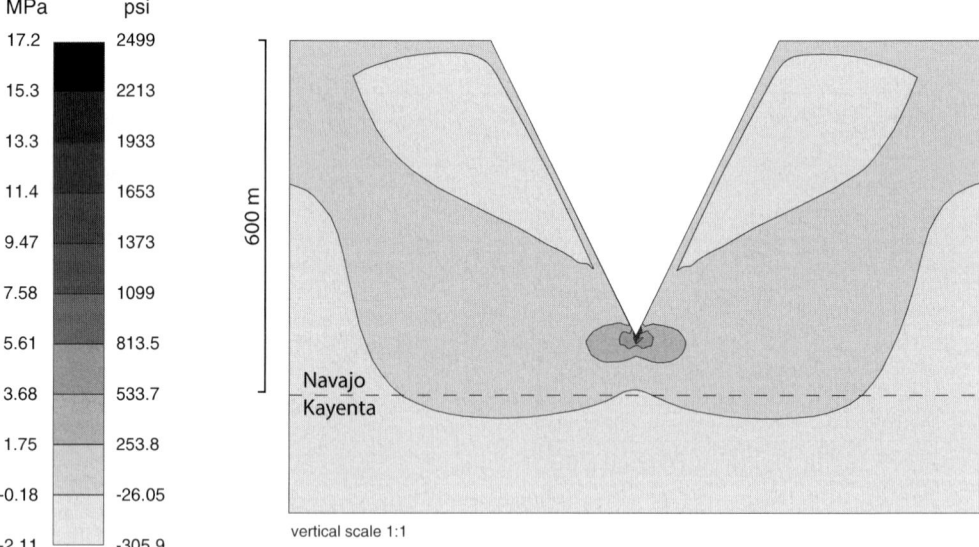

Fig. 11. FRANC (finite-element) model of an unconfined 500 m-deep Navajo Sandstone slot canyon subjected to a gravity body load (vertical scale = horizontal scale). Although highly restricted in lateral extent, the tensile stress concentrated at the very tip of the canyon exceeds the presumed tensile strength (5 MPa) of the Navajo. However, this tensile stress concentration does not account for the width of the observed joint-zone development at ZNP.

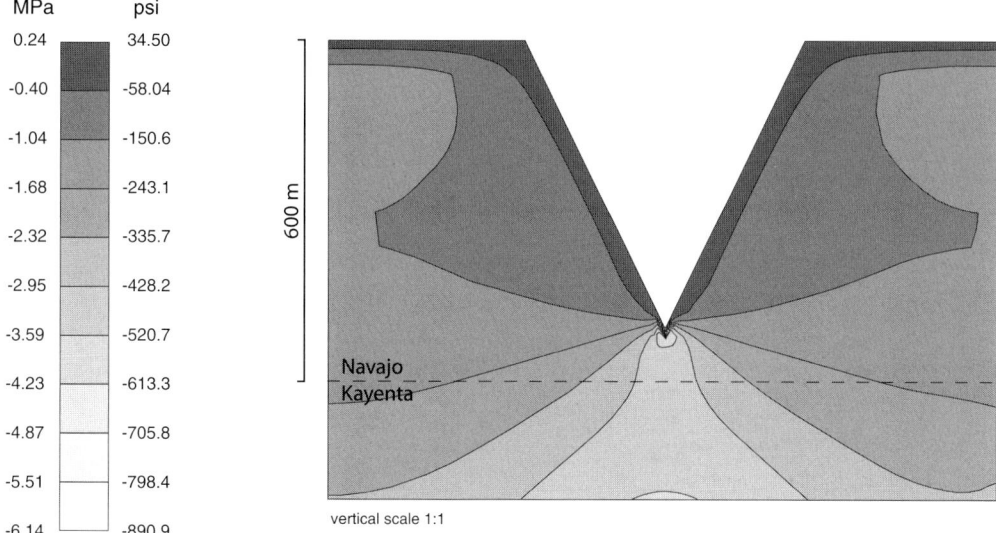

MPa	psi
0.24	34.50
-0.40	-58.04
-1.04	-150.6
-1.68	-243.1
-2.32	-335.7
-2.95	-428.2
-3.59	-520.7
-4.23	-613.3
-4.87	-705.8
-5.51	-798.4
-6.14	-890.9

600 m

Navajo
Kayenta

vertical scale 1:1

Fig. 12. FRANC (finite-element) model of confined 500 m-deep Navajo Sandstone slot canyon subjected to a gravity load (vertical scale = horizontal scale). The canyon tip is in a state of compression, while only minor tension develops just above the tip along the canyon walls. This tension is not predicted to exceed the tensile strength of the sandstone at any canyon depth in the Navajo.

sumed tensile strength of the rock, 5 MPa (tension is positive), available to joint-zone development. The Kayenta Formation was assigned constant values for thickness of 200 m (Hintze 1988), $E=12$ GPa, $v=0.10$ and, ρ, of 2.6 gcm^{-3}.

State of stress below slot-canyon notches

Our initial model represents a single slot canyon at ZNP with a simple symmetric V-shaped configuration (Fig. 11). Gravity body loads and horizontal tractions (i.e. regional extension) were simulated independently to understand the contribution of each to the state of stress at slot-canyon tips. In particular, the two loading conditions were simulated using combinations of four boundary conditions: an unconfined gravity load (no lateral boundaries) confined gravity load (zero displacement lateral boundaries), remote tension (displacement/ extension of lateral boundaries) and the mechanical interaction with the underlying Kayenta (welded bottom boundary). All have particular effects on the concentration of stress at the slot-canyon tip.

Each model was fixed at the mid-point of its base in the vertical direction. When a gravity body load was simulated independently, the base of the model was also fixed in the horizontal direction. In contrast, when remote tension was applied to a model, either alone or superimposed with gravity, the model was fixed on one side in the horizontal direction while the

rest of the model was allowed to slide horizontally (including the base) as the opposite side was 'pulled' to simulate the regional extension.

The gravity body load with no lateral constraints concentrates tensile stress at the canyon tip as long as the sandstone has a tensile strength; however, the width of the predicted zone of joint development at the canyon tip is too narrow to simulate the joint-zone widths observed at ZNP (Fig. 11). If, instead, a joint of zero strength below the slot tip cuts this model, then it does not support a tension under a gravity load. In these scenarios, joint zones at ZNP would be the result of exfoliation jointing at canyon walls, which does not account for the extent of (sub)vertical jointing in the zone *below* the downward-cutting canyon.

If a gravity body load is applied with zero-displacement lateral constraints, the canyon tip is in a state of compression even when the tip is cut by a vertical joint. Only after the simulated slot canyon has 'eroded' half way through the sandstone (approximately 300 m canyon depth) does minor tension appear in the vicinity of the canyon tip just *upward* along the canyon walls (Fig. 12). This is the only location of tension in the model, and although tensile stress at this location increases as canyon depth increases, it is not predicted to exceed the tensile strength of the sandstone. Again, joint-zone development would be a function of exfoliation jointing in the canyon walls, while no jointing is predicted below the canyon tip.

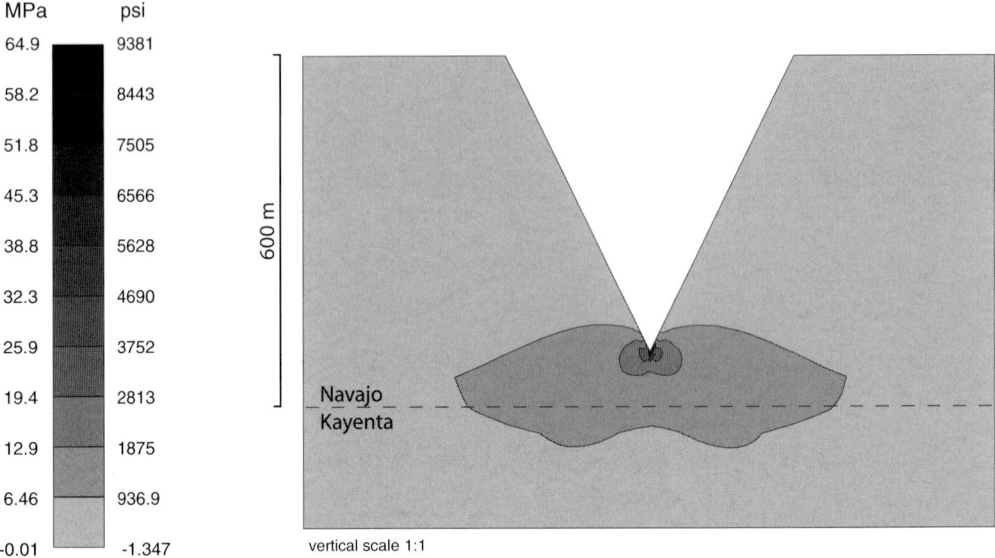

Fig. 13. FRANC (finite-element) model of a 500 m-deep Navajo Sandstone slot canyon subjected to tension equivalent to a 5×10^{-4} strain (vertical scale = horizontal scale). The tensile stress concentration at the canyon tip exceeds the presumed tensile strength (5 MPa) of the Navajo hundreds of metres from the tip of the slot canyon.

Remote tension results in a tensile stress concentration at the canyon tip as long as the Navajo Sandstone there is not cut by a joint and, thus, has a tensile strength. Tension is generated in the model by applying a displacement to one side of the model while fixing the other. Under strains of approximately 10^{-4}, the tensile stress concentration exceeds the tensile strength of the sandstone at a radius comparable to the widths of joint zones observed at ZNP (Fig. 13).

Finally, the underlying Kayenta Formation is assumed to be in frictional contact with the Navajo and, therefore, is included in these models. The influence of the Kayenta, as illustrated in the gravity, unconfined model, is such that an absence of Kayenta in the model results in a stress reduction within the Navajo at the canyon tip (Fig. 14A). Although a frictional contact with the Kayenta increases the tensile stress concentration at the canyon tip in the Navajo (Fig. 14B), slot canyons at ZNP do not penetrate into the Kayenta. This is observed in air photographs and in the field, and is also suggested by the gravity, unconfined model. As a Navajo slot canyon deepens, tensile stress concentrated at the tip is predicted to increase and then decrease in the final 100 m above the Kayenta contact (Fig. 14).

In each of the above models, where a net canyon-tip tensile stress concentration develops, stress trajectories indicate that joints on opposite sides of the canyon are predicted to dip toward each other and toward the axis of the slot canyon below the level of

the canyon tip (Fig. 15). Because this same trend in dip is observed in the joints below the tips of the slot canyons in the field at Zion, we are encouraged that our models do approximate the state of stress ahead of the downward-eroding slot canyons at ZNP. In this manner, the pattern of stress trajectories modelled at the tips of slot canyons is analogous to that of a crack-tip stress field (Lawn 1993), and the joint zone below the slot canyons is analogous to new cracks propagating in a crack-tip process zone.

Regional tension plus gravity on a 'V'-shaped slot canyon

Our finite-element models suggest that gravity, alone (either unconfined or confined scenarios), generates a canyon-tip tensile stress concentration that is too local to account for the width of the joint zones at ZNP. However, as canyon walls at ZNP maintain steep cliffs (hundreds of metres of Navajo Sandstone), an overburden load is an important component in each model. Consequently, we superimpose a regional extension in order to generate a local tensile stress concentration that reaches outward from the canyon tip at the scale of ZNP joint zones. This necessitates that tensile strength is maintained below slot canyons and that initial regional joints cut only part way through the Navajo. Therefore, any hypothesis for early through-going joints in the Navajo is rejected implying that the taller joints that appear below the R and M slot canyons are not a remnant of

A)

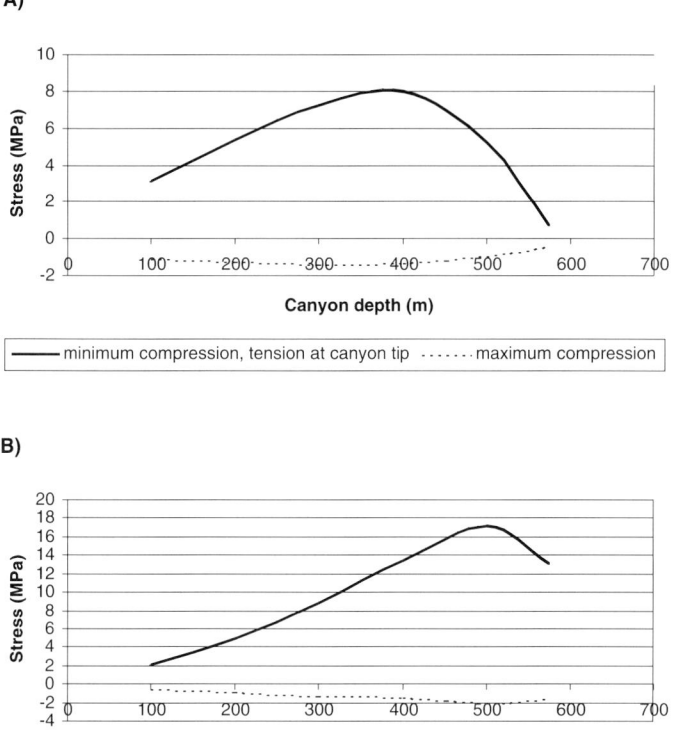

B)

Fig. 14. Graphs of FRANC (finite-element) model results from unconfined Navajo Sandstone slot canyon subjected to a gravity load: (**A**) the model without the underlying Kayenta Formation, i.e. no frictional contact; and (**B**) the model including a frictional contact with the Kayenta. Note that the tensile stress concentration at the canyon tip is predicted to increase when the Navajo is in frictional contact with the underlying Kayenta and, in both cases, peaks and then decreases as the Kayenta contact is approached by the deepening canyon.

a through-cutting joint dating from the Miocene time.

Simulating a regional extension at ZNP was accomplished by applying a uniform displacement to one side of the FRANC model. This displacement was increased until the 5 MPa (tension is positive) contour extended out a lateral distance from the slot-canyon tip that is comparable to the width of the joint zones at ZNP. With an extensional strain of 5×10^{-4}, our model (i.e. Fig. 16) simulates stress magnitudes and trajectories associated with a 575 m-deep canyon, approximating the depth of the slot canyons in the Court of the Patriarchs (Fig. 4). In this model everything under the depth of the slot-canyon tips is in tension, and an extensional strain of 5×10^{-4} generates a tensile stress in excess of the presumed tensile strength of the sandstone out to a radius of about 225 m (Fig. 17). Symmetric lobes of tension characterize the pattern of tension about the

symmetric V-shaped slot-canyon tip. In effect, the canyons are analogous to large-scale blunt edge cracks cutting downward into the Navajo Sandstone; therefore, the tensile stress pattern in our model is akin to a crack-tip stress field (Fig. 16).

Effect of canyon morphology

The same boundary conditions were applied to models for two additional slot-canyon geometries. The first geometry approximates the asymmetric V-shaped slot canyon at R slot (i.e. Figs 4 and 7) and the second geometry approximates the box canyon at M slot, the V-shaped canyon with squared base (i.e. Figs 4 and 8). Although a joint below the tip of R slot and the east zone of M slot appears to cut from the canyon tip to the base of the Navajo, there is no direct evidence that these joints once cut the entire Navajo

A)

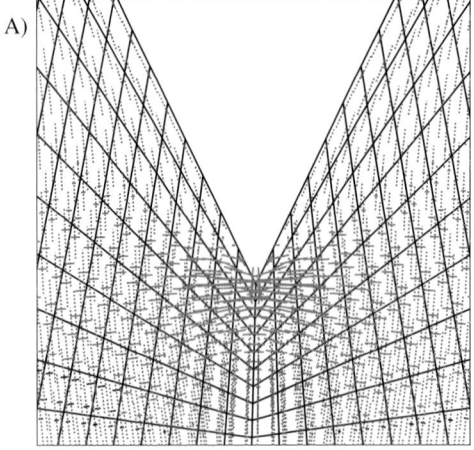

B)

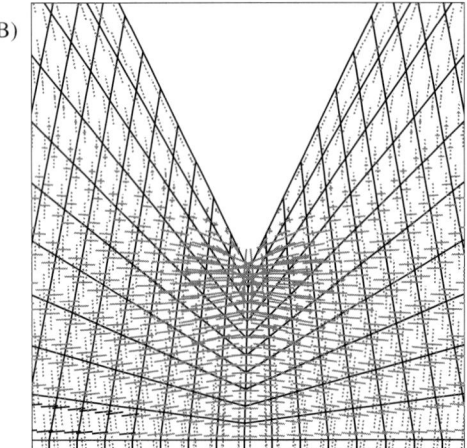

Fig. 15. Details of the FRANC finite-element mesh with stress trajectories below the level of the 500 m-deep slot canyon tip: (**A**) unconfined gravity load; and (**B**) combined gravity load plus regional extension. These images are focused at the canyon tip and distinguish individual stress trajectory bars: solid lines are tension trajectories and dotted lines are compression trajectories. Both model scenarios yield the same trend below the level of the slot canyon tip where joints are predicted to propagate perpendicular to the tension bars, dipping towards the canyon axis.

Sandstone prior to canyon development. Experiments with these two geometries included boundary conditions of gravity loading alone and regional extension superimposed on a gravity loading condition assuming a tensile strength below the slot canyons.

The asymmetric V-shaped canyon at R slot was simulated under the same boundary conditions as the symmetric V-shaped canyon (i.e. gravity loading plus regional extension of 5×10^{-4} strain). An asymmetric distribution of tension is predicted where the maximum predicted tension, located at the canyon

tip, is approximately 22 MPa (Fig. 18). The predicted zone of tensile failure (5 MPa contour) extends to a width of about 210 m underneath the more steeply dipping east wall. The 5 MPa contour under the less steeply dipping west canyon wall exhibits a radius of about 250 m. This result is consistent with the asymmetry in joint-zone width beneath the R slot canyon tip where the east half of the zone is about 250 m and the west zone is about 380 m in width (Fig. 7). The 5 MPa contour is predicted to extend to the base of the Navajo, as is observed in the field.

Although an asymmetric joint distribution is predicted when a superimposed gravity and regional extension are applied to the R slot model, the unconfined gravity model (i.e. no regional extension) better predicts the joint free 'shielded zone' observed just west of the R slot canyon tip (Figs 7 and 19). However, despite the better overall location of tensile stress concentration, a gravity body load, alone, does not result in a sufficient magnitude of tension to create the joint-zone width observed at R slot.

Under the same boundary conditions that predict a 225-m wide joint zone laterally away from the symmetric V-shaped canyon (i.e. gravity loading plus regional extension of 5×10^{-4} strain), the predicted zone of tensile failure (5 MPa contour) in the M slot model extends at least 380 m from the axis of the canyon (Fig. 20) with a maximum tension of approximately 32 MPa generated at the M slot canyon tip. This 5 MPa contour in the M slot model greatly exceeds the width, *c*. 75 m, of the joint zones observed in the field at M slot (Fig. 8), and the same contour appears to penetrate the Kayenta Formation, which does not contain the M slot joint zone in the field. In addition, the same M slot model does not predict the 'shielded' zone (i.e. no joint-zone development) under the centre of the canyon. The central shielded zone separates the two joint zones under the corners of the 'box' canyon at M slot. In an effort to match the 5 MPa contour with the field observations at M slot, we find that the unconfined gravity model offers a better solution for an M slot joint-zone radius, predicting joint development in the immediate vicinity of the canyon tip. However, the pattern of predicted tensile failure does not extend to the base of the Navajo nor indicate that a double joint zone at M slot would exist (Figs 8 and 21), indicating that additional complexities are inherent to the development of the double joint-zone lineaments at M slot.

Discussion

Closely spaced joint propagation is enigmatic in that it is not predicted by joint normal tractions; however,

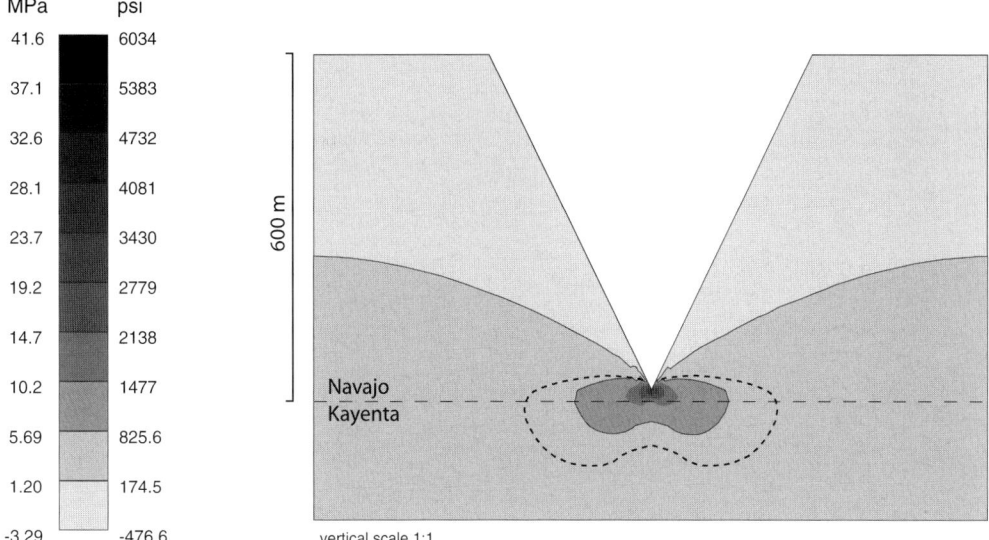

Fig. 16. Working slot-canyon FRANC model simulating the extent of joint-zone development associated with a canyon-tip tensile stress concentration at ZNP. Model parameters: 600 m of Navajo Sandstone (E=15 GPa, v=0.15, ρ=2.3 g cm^{-3}) in frictional contact with the underlying 200 m of Kayenta (E=12 GPa, v=0.10, ρ=2.6 g cm^{-3}) (vertical scale = horizontal scale) containing a symmetric, 'V'-shaped, sharp-tipped, 575 m-deep canyon with a superimposed gravity and regional extension (5×10^{-4} strain) loading condition. The dashed line represents the 5 MPa contour line, the extent to which joints are predicted to propagate in the Navajo due to the canyon-tip tensile stress concentration. This radius of joint propagation is comparable to the widths of joints zones observed at ZNP.

the dramatic NNW-trending, regularly spaced slot canyons of ZNP erode from distinct joint zones. The vertical and steeply dipping joints that compose the joint zones parallel the slot canyons and are spatially linked with them. To explain this observation we propose that there is a feedback between the downward erosion of the slot canyons at ZNP and the development of joint zones ahead of the downward-advancing tip of the canyons. This feedback has to generate joint zones below the canyon tip while leading to a regular spacing for the slot canyons that is roughly equivalent to the thickness of the Navajo.

Evolution of slot-canyon joint zones

We presume that the history of the NNW-trending slot canyons of ZNP goes back to the propagation of individual joints in intact Navajo Sandstone some time during the Miocene. At that time, the western edge of the Colorado Plateau stretched in response to extension in the northern Lake Mead Extensional Belt of the central Basin and Range subprovince. This extension was directed just south of west, an extension direction shared by the NNW-trending slot canyons of ZNP in the adjacent Colorado Plateau (Rogers *et al.* 2004) (Fig. 2). Later, erosion downward into these regional joints initiated ZNP's

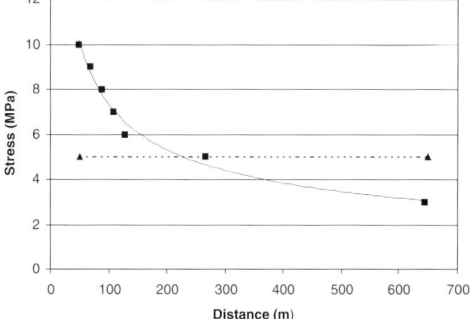

Fig. 17. Radius of tensile stress magnitudes extrapolated from σ_1 contours of the 575 m-deep canyon FRANC model (Fig. 16) subjected to a superimposed gravity load and remote tension (5×10^{-4} strain). The horizontal line represents the presumed tensile strength of the sandstone.

prominent NNW-trending slot-canyon network. Once initiated, these slot canyons acted as notches, concentrating tensile stress at their tips under a combination of gravity loading and minor regional extension.

This canyon-tip tensile stress concentration was responsible for closely spaced jointing in joint zones below canyon tips. The closely spaced jointing at

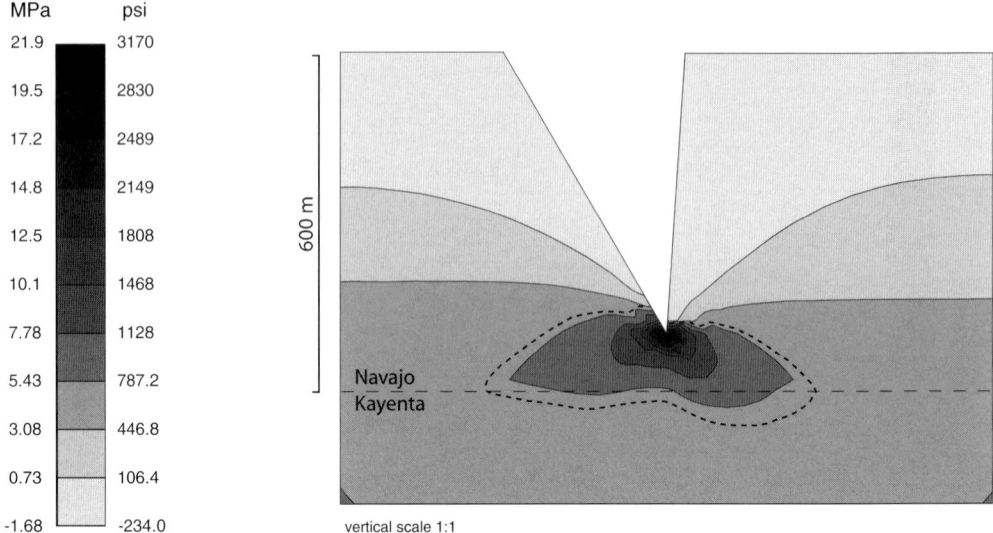

Fig. 18. Boundary conditions from the working model (Fig. 16) applied to an asymmetric 'V'-shaped, sharp-tipped canyon geomorphology simulating R slot at the Court of the Patriarchs (Fig. 7). The dashed line represents the 5 MPa contour line, the extent to which joints are predicted to propagate in the Navajo due to the canyon-tip tensile stress concentration.

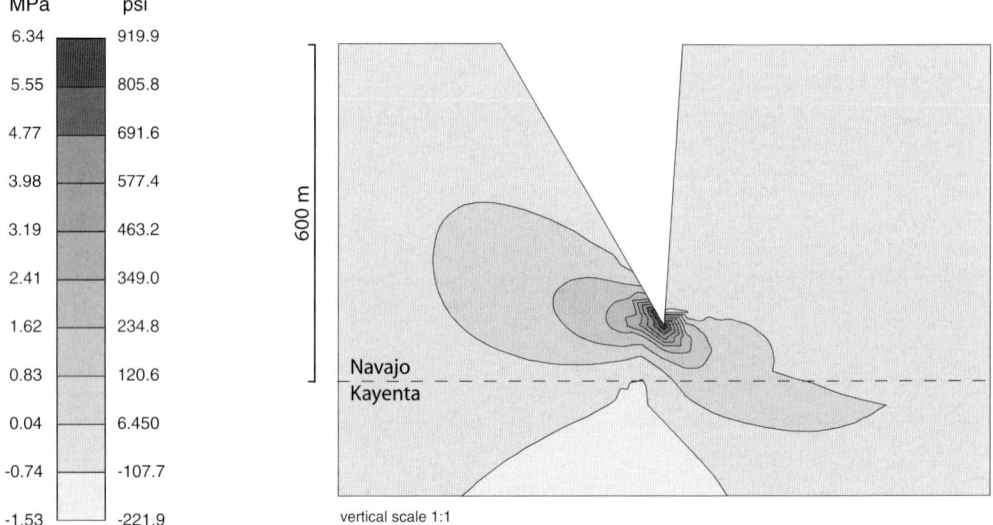

Fig. 19. Gravity body load applied to an unconfined FRANC model of R slot at the Court of the Patriarchs. Although, when compared to field observations, this model better predicts the pattern of tensile stress concentration below the canyon tip including the 'shielded' joint free zone offset toward the west, less steeply dipping cayon wall, it does not predict tensile failure at the observed joint zone width observed at R slot (Fig. 7).

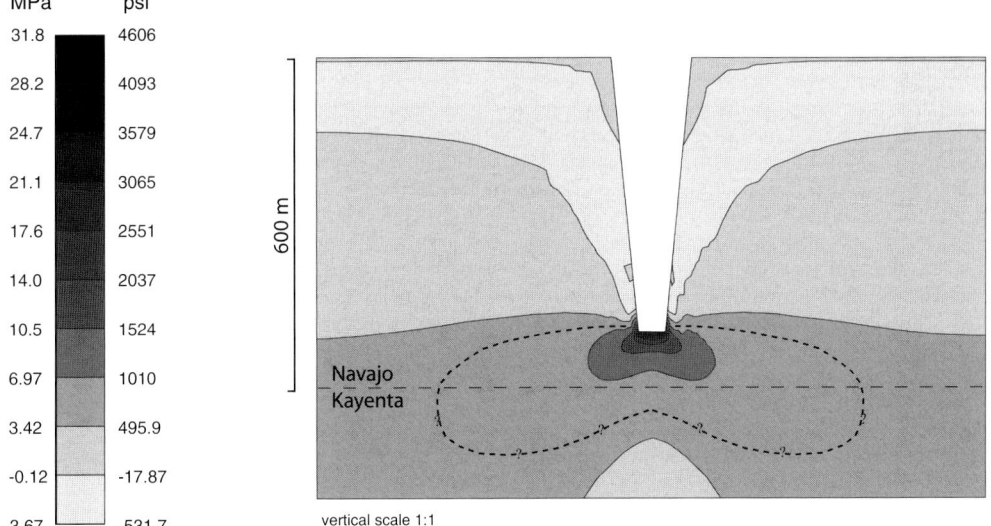

Fig. 20. Boundary conditions from the working model (Fig. 16) applied to a symmetric 'V'-shaped, squared-tip canyon geomorphology simulating M slot, Refrigerator Canyon. The dashed line represents the 5 MPa contour line, extrapolated into the Kayenta, indicating the extent to which joints are predicted to propagate in the Navajo and Kayenta due to the canyon-tip tensile stress concentration.

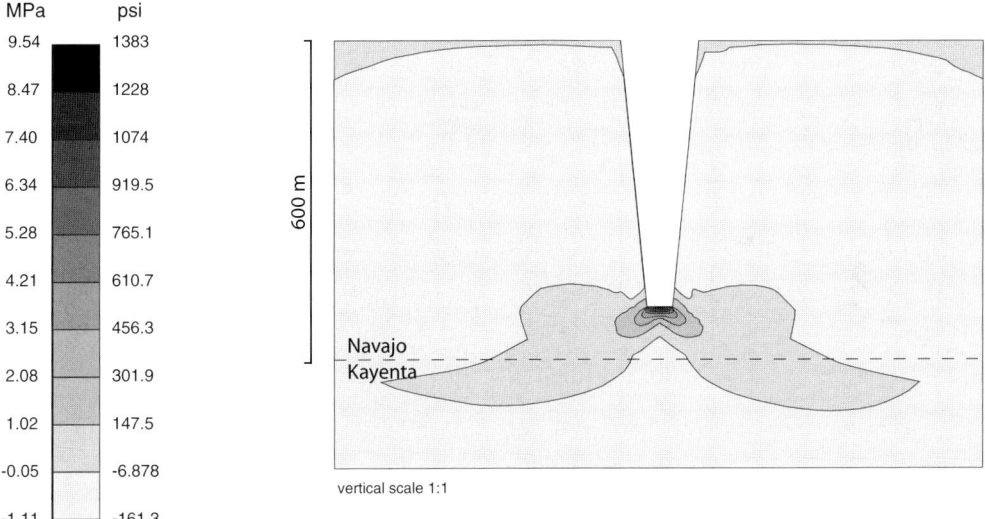

Fig. 21. Gravity body load applied to an unconfined FRANC model of M slot, Refrigerator Canyon. Although, when compared to field observations, this model better predicts a width of tensile failure in the Navajo comparable to the overall joint zone width observed at M slot (Fig. 8), it does not predict tensile failure to the base of the Navajo nor indicate that a double joint zone would exist.

slot-canyons tips is analogous to a crack-tip process zone where joints form in response to elevated tensile stress beyond the crack tip (Delaney *et al.* 1986). The slot-canyon 'notch' concentrates stress in a similar pattern to that generated by an edge crack in an elastic medium (Lawn 1993). A regional extension was necessary to generate a tensile stress concentration of sufficient magnitude and lateral extent to produce the observed joint zones. As the Navajo is assumed to be a confined system, this remote tension compensates for compression induced by the gravity load (weight) of the overlying Navajo Sandstone.

Joint-dip analysis performed from field photographs indicates that joints below the canyon tip have a tendency to dip toward the slot-canyon axis (Table 1). However, joints dip both east and west regardless of their position relative to the associated slot canyon. The models presented here represent stresses arising in a continuous medium where all trajectories dip toward the slot canyon. A jointed medium would not be expected to produce such a simple result, as the local stress field would continue to become more complex with the propagation of additional joints.

Following the relief of tensile stress at the canyon tip, the gravity load (overburden) then drives vertical wing cracks from the tips of the steeply dipping joints, therefore augmenting the canyon-tip joint zone. Joint-dip analysis reveals that the mode of each population of joints observed at slot canyons is vertical, the orientation of wing cracks propagating in response to gravity. In addition, the 5m-joint sampling statistics, which are biased towards measuring wing cracks, are weaker in the identification of a preferential joint dip direction within a particular canyon-tip joint population.

These wing cracks can exhibit a very close spacing as a response to a vertical, crack-parallel gravity load acting on inclined parent joints. This crack-parallel loading is distinct from the crack-perpendicular loading ((sub)horizontal tension) of original parent cracks (secondary joints) that propagated in response to elevated tensile stress in the canyon-tip process zone. This early crack-normal tension is already relieved by the propagation of the parent joints where spacing is expected to be governed by stress shadows. Subsequent wing-crack growth is required to infill and achieve a close joint spacing.

Ultimately, exfoliation jointing at canyon walls completes the sequence of jointing in the zones at ZNP slot canyons (Bahat *et al.* 1995). However, the development of a process zone under the canyon tip is an independent source of jointing from the exfoliation mechanism. Therefore, closely spaced joints found below the canyon tip are to be distinguished from exfoliation fracturing that occurs at canyon walls in direct response to lateral unloading while an overburden stress is still present.

Slot-canyon spacing

Our modelling of the slot canyons at Zion suggests several things about both the width of the joint zones and the spacing of the canyons. First, we conclude that the width of each canyon-tip joint zone(s) can be reproduced only when the 600 m bed of Navajo Sandstone maintains its tensile strength below slot canyons. A gravity-driven stress concentration at the tip of the slot canyons produces a zone of tensile stress concentration that is too narrow, in a laterally unconfined model, and of insufficient magnitude to cause tensile failure in a laterally confined model, to reproduce the joint zones at ZNP. Therefore, in order to generate a sufficiently far-reaching tensile stress contour, a superimposed regional extension is a necessary boundary condition in our models. This means that we must reject the hypothesis that the even spacing of the slot canyons at Zion is a consequence of a direct relationship with the initial propagation of through-going joints confined to the Navajo Sandstone in a manner similar to joints cutting the full thickness of stiff beds contained within shale layers. We reject the through-cutting joint hypothesis, because through-cutting joints would cause a local loss of tensile strength in the Navajo at the location of each eroding canyon. Such a lack of tensile strength pre-empts the possibility that a sufficient tensile stress concentration can develop at the leading edge of a downward-eroding slot canyon under a gravity load alone. Consequently, we cannot appeal to the classic stress-shadow theory as discussed by Gross *et al.* (1995) and many other authors to explain the relatively even spacing of the slot canyons at ZNP.

The density of joint-zone lineaments increases from west to east across the field area, while the depth of erosion of the slot canyons decreases, west to east, over the same area. If we measure the density of slot canyons cutting to the base of the Navajo Sandstone along the west side of the valley of Zion Canyon, we see that seven slots cut to within 100 m of the basal contact between slots M and S (Fig. 4) where Q slot counts for two data, P is missing and M is counted once. These slots project onto a profile of 4 km to give a density of 1.75 km^{-1}. If the stress-shadow model is used to explain the spacing of slot canyons, we predict a slot-canyon density of one per 600 m (average 600 m Navajo thickness), or 1.66 km^{-1}.

If we measure the density of slot canyons cutting through Highway 9 from D slot to N slot, we find that 18 joint zones/slots touch the road in Figure 4 in a profile distance of 6.5 km. This gives us a density of 2.76 km^{-1} or almost double the density found near the base of the Navajo Sandstone. In this distance there are several slot canyons that are composed of double joint-zone lineaments (i.e. D, E, G, H, K, L, M and N). In this same profile I and J have

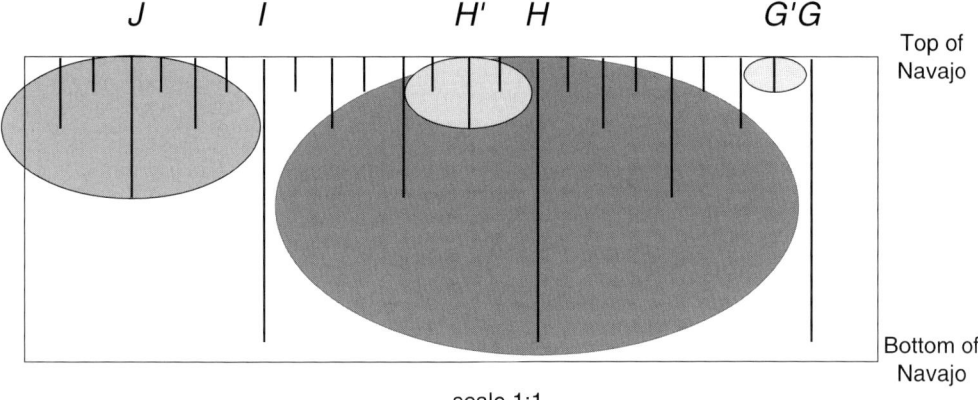

scale 1:1

Fig. 22. Hypothetical cross-section along Highway 9 from G slot to J slot (Fig. 4). Schematic of edge cracks (canyons) in the Navajo Sandstone analogous to Lachenbruch's thermally induced edge cracks in permafrost. The horizontal extent of stress shadows is comparable to edge-crack (canyon) depth and, as particular cracks propagate, they suppress the growth of neighbours due to their expanding stress shadow.

merged to form a canyon with a double lineament as well. Only F slot is missing. We believe that the presence of multiple canyons composed of double lineaments in the upper half of the Navajo Sandstone is a key pointing to the mechanism controlling the spacing of the slot canyons.

Lachenbruch (1961) presented an analysis of the development of thermally induced cooling cracks in permafrost. He pointed out that as the cracks grow into the permafrost, they may start as relatively close-spaced discontinuities. As they grow downward into the permafrost, their spacing increases. Lachenbruch (1961) believed that this growth pattern was a manifestation of the relief of tensile stress near the wall of cracks. The extent of the relief of tensile stress scales with the depth of the crack. This stress relief model of Lachenbruch (1961) is an early version of the stress-shadow model later quantified by Geyer & Nemat-Nassar (1982) in a series of experiments on the propagation of cracks from the edge of a glass plate as a consequence of a thermal shock. Pollard & Segall (1987) applied this model to joints that cut completely across a bed and Gross *et al.* (1995) provided greater detail for this model, which became known as the stress-shadow model. The distinction between the permafrost model and the later stress-shadow model is that the former model represents the growth of edge cracks from an unconfined boundary, whereas the latter involves the growth of internal flaws within a confined bed. In the edge-crack model (i.e. permafrost model of Lachenbruch), the faster growing cracks suppress the growth of adjacent cracks by the expansion of their stress shadow, a process referred to as crack-tip shielding (e.g. Olson 1993).

We can apply the edge-crack model to the devel-

opment of the slot canyons at ZNP. Although slot canyons are localized along pre-existing joints in the upper Navajo, they are themselves analogous to large cracks in the Navajo Sandstone and must be recognized as such in this explanation. Once slot-canyon 'cracks' begin to erode in the Navajo, they 'grow' downward due to a feedback loop between erosion and jointing arising from progression analogous to a process zone. The joint zones are more easily eroded, thus focusing erosion into a series of slot canyons. The slot-canyon 'cracks' concentrate a tensile stress downward from their tips, thus driving joint-zone development ahead of the downward-eroding slot canyons. Because the canyon walls become free surfaces, slot-canyon 'cracks' project stress shadows into the adjacent Navajo Sandstone at ZNP. This stress shadow will act to suppress the downward growth of closely spaced slot canyons, leaving only more widely spaced canyons to work further down into the Navajo.

Consider a hypothetical cross-section along Highway 9 from G slot to J slot (Fig. 22). In this profile we have demonstrated that the upper half of the Navajo Sandstone has many more joint zones than it should based on stress shadows developed about joint zones cutting the entire thickness of the Navajo. We denote G, H and I slots by joint zones that cut the entire thickness of the Navajo. In this hypothetical profile view we also add joint zones that are half the thickness of the Navajo, 25% of its thickness and 12.5% of its thickness. We also assign a spacing that is proportional to the depth of penetration for each size joint zone. Stress shadows are drawn about each size joint zone. This gives a profile for joint-zone penetration that is the same as that realized by Geyer & Nemat-Nassar (1982)

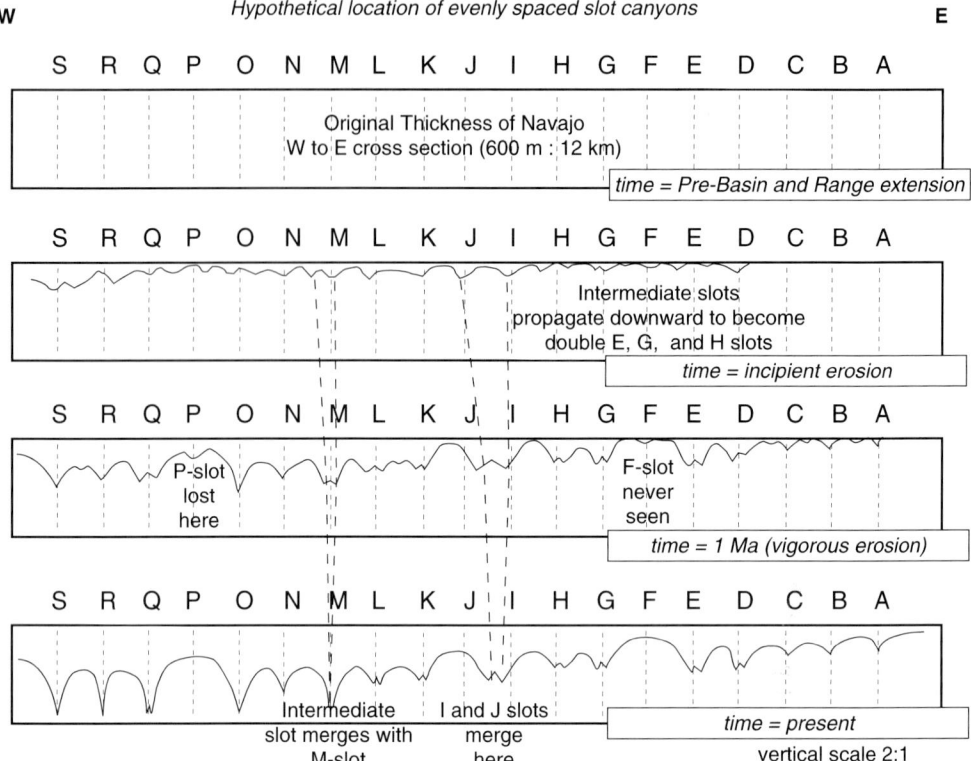

Fig. 23. Hypothetical set of cross-sections in a time sequence for the development of the topography at ZNP. We see the initial erosion gives a topography with many more slot canyons toward the top of the Navajo Sandstone. Canyons deepen due to the feedback loop between erosion and jointing in response to minor regional extension. These slot canyons either merge, form double joint zone lineaments or are suppressed in the present topography as stress shadows associated with canyon erosion develop.

experimentally and by Lachenbruch (1961) theoretically for the edge-crack model of stress-shadow development. On the Highway 9 profile we note that G slot has two joint-zone lineaments with G′ spaced about 72 m from G, and we note that H slot has a double lineament with H′ spaced about 140 m from H (Fig. 4). Along Highway 9, we note that and I and J approach each other. Finally, the 600 m spacing between slot canyons is denoted by an appropriately large stress shadow (Fig. 22). By this same model we predict that some double joint-zone lineaments (i.e. M slot) eventually merge to become one slot canyon.

Conclusions

Our thesis is that the eroding slot canyons within the Navajo Sandstone of ZNP are analogous to large, blunt-tipped 'edge cracks' and behave in the same fashion as cracks in permafrost (i.e. Lachenbruch 1961) and thermal-shock cracks (i.e.

Geyer & Nemat-Nassar 1982) where crack growth results in regular spacing that increases with depth due to stress-shadow shielding effects that suppress the growth of neighbouring cracks. This is seen in the pattern of canyons and joint-zone density in the Navajo Sandstone across ZNP. From this model, we can then construct a hypothetical set of cross sections in a time sequence for the development of the topography at ZNP. We see the initial erosion gives a topography with many more slot canyons toward the top of the Navajo Sandstone (Fig. 23). These slot canyons either merge, show double joint-zone lineaments or are suppressed in the present topography, depending on the depth of canyon erosion.

In summary, we find that stress-shadow theory does apply to Zion and that it is responsible for the even spacing of the slot canyons. However, our stress-shadow interpretation is based on the propagation of edge cracks (i.e. the eroding slot canyons) downward from the top of the Navajo Sandstone. We reject the hypothesis that early jointing extended the entire thickness of the Navajo Sandstone. Rejection of this hypothesis is consistent with the fact that we see little evidence for initial joints that propagated the full thickness of the Navajo Sandstone. In addition, rejection is also consistent with the observation that there are more double joint-zone lineaments within individual slot canyons near the top of the Navajo Sandstone than near the bottom. We see indications of slot-canyon development during a period of continued extension of the western edge of the Colorado Plateau. This extension was enough to place the tips of the slot canyons of ZNP into tension in a stress concentration zone, directly analogous to a crack-tip stress concentration, that may have extended more than 100 m into the adjacent rock ahead of the tip of the canyons.

This tensile stress concentration zone is responsible for initiating a 'process zone' of secondary jointing resulting in the joint zones observed at the base and ahead of each slot canyon. Following the slot-canyon-induced jointing, wing cracks develop from the tips of the secondary joints below the canyon under a gravity load, leaving exfoliation jointing to occur in a zone very near the canyon wall of the slot canyons. This exfoliation jointing is driven by a gravity load and is independent of the regional extension, which causes the joint zones ahead of the tip of the slot canyons. Cumulatively, these events result in the closely spaced jointing observed at slot canyons in ZNP.

This work was supported by Pennsylvania State University's Seal Evaluation Consortium (SEC), AAPG Grants-In-Aid, and a Krynine Grant from Pennsylvania State University. M. Nemcock and J. Cosgrove are thanked for reviewing an early version of this chapter.

Appendix: Photograph interpretation

Individual joints were digitized from field photographs of joint zones in cross-section, paying close attention to dip angle. Photographic interpretation was performed repeatedly. Multiple tracings (three–four) of the same photograph were compared and, in an effort to include only reproducible results, only those joints that appeared in each tracing were included in the final interpretation. Secondly, if a joint was recognized or suspected of trending in a direction other than that of the slot canyon, it was not included in the final tracing.

Scion Image computer software, developed at the Research Services Branch (RSB) of the National Institute of Mental Health (NIMH) (part of the National Institutes of Health (NIH)), was used to measure joint dip values directly from the cross-section photograph tracings. Scion Image is a public

Fig. A1. Joint sampling methods used for joint dip analysis by Scion Image software: (**A**) example of the whole joint (measured tip to tip) sampling; and (**B**) 5 m-joint segment sampling applied to M slot, Refrigerator Canyon.

domain image processing and analysis program used here to fit ellipses to line segments (joints). The program measures the angle of the long axis of the ellipse, which, in this case, is parallel to the joint (segment) length, thus providing the dip angle of digitized joints. Based on the average joint trend derived from scanline data (Fig. 6), all Scion Image data were corrected for the effects of apparent dip expressed in canyon walls prior to statistical analysis.

Two different joint sampling methods were analysed. Ellipses were first fit to the whole length of each digitized joint, measured from tip to tip (Fig. A1A). These data represent dip values of individual joints treating each as a planar feature without regard to fracture length. Secondly, all digitized joints were converted to 5 m-dashed line segments with minimal interdash intervals in order to approximate the full joint length (Fig. A1B). These data were expected to account for potential deviations in joint dip along a single joint, which may be attributed to a temporally varying stress field. In histograms these data were binned according to dip value, independent of joint identity and joint length.

Ultimately, the statistics and observations presented here are an accurate representation of trends in dip direction only, not the measured dip value. It was not possible to remove additional photographic distortions that affect these photographs (Wolf 1974) and, therefore, the accuracy of our measure of the absolute dip angles is unknown.

References

BAHAT, D., GROSSENBACHER, K. & KARASAKI, K. 1995. *Investigation of Exfoliation Joints in Navajo Sandstone at Zion National Park and in Granite at Yosemite National Park by Tectonofractographic Techniques.* Lawrence Berkeley Laboratory, **36971**.

BIEK, R. F., WILLIS, G. C., HYLLAND, M. D., & DOELLING, H. H. 2000. Geology of Zion National Park, Utah. *In*: SPRINKEL, D. A., CHIDSEY, T. C., JR. & ANDERSON, P. B. (eds) *Geology of Utah's Parks and Monuments.* Utah Geological Association Publications, **28**, 107–138.

BIRCH, F. 1966. Compressibility; elastic constants. *In*: Clark, S. P., JR. (ed.) *Handbook of Physical Constants.* Geological Society of America, Memoirs, **97**, 97–173.

COOKE, M. L., MOLLEMA, P. N., POLLARD, D. D. & AYDIN, A. 2000. Interlayer slip and joint localization in the East Kaibab Monocline. *In*: COSGROVE, J. W. & AMEEN, M. S. (eds) *Forced Folds and Fractures.* Geological Society, London, Special Publications, **169**, 23–49.

DAVIS, G. H. 1999. *Structural Geology of the Colorado Plateau Region of Southern Utah: With Special Emphasis on Deformation Bands.* Geological Society of America, Special Papers, **342**.

DELANEY, P. T., POLLARD, D. D., ZIONY, J. I. & McKEE, E. H. 1986. Field relations between dikes and joints: Emplacement processes and paleostress analysis. *Journal of Geophysical Research*, **91**, 4920–4938.

DYER, R. 1988. Using joint interactions to estimate paleostress ratios. *Journal of Structural Geology*, **10**, 685–699.

EARDLEY, A. J. 1965. Unpublished Open-file Reports. Park Library, Zion National Park Headquarters, Utah.

ENGELDER, T. 1987. Joints and shear fractures in rock. *In*: ATKINSON, B. (ed.) *Fracture Mechanics of Rock.* Academic Press, Orlando, FL, 27–69.

FISCHER, M. P. 1994. *Application of linear elastic fracture mechanics to solve problems of fracture and fault propagation.* PhD thesis, Pennsylvania State University, University Park, PA.

FISCHER, M. P. & JACKSON, P. B. 1999. Stratigraphic controls on deformation patterns in fault-related folds: a detachment fold example from the Sierra Madre Oriental, northeast Mexico. *Journal of Structural Geology*, **21**, 613–633.

GEYER, J. F. & NEMAT-NASSER, S. 1982. Experimental investigation of thermally induced interacting cracks in brittle solids. *International Journal of Solids and Structures*, **18**, 349–356.

GREGORY, H. E. 1950. *Geology and Geography of the Zion Park Region, Utah and Arizona.* United States Geological Survey, Professional Papers, **220**.

GROSS, M. R., FISCHER, M. P., ENGELDER, T. & GREENFIELD, R. J. 1995. Factors controlling joint spacing in interbedded sedimentary rocks: Integrating numerical models with field observations from the Monterey Formation, USA. *In*: AMEEN, M. S. (ed.) *Fractography: Fracture Topography as a Tool in Fracture Mechanics and Stress Analysis.* Geological Society, London, Special Publications, **92**, 215–233.

HAMILTON, W. L. 1984. *The Sculpting of Zion*, Zion National History Association.

HENNINGS, P. H. & OLSON, J. E. 1997. Relationship between bed curvature and fracture occurrence in a fault-propagation fold. *In*: *Annual Meeting Abstracts – American Association of Petroleum Geologists Annual Convention.* American Association of Petroleum Geologists and Society of Economic Paleontologists, Tulsa, OK, **6**, 49.

HINTZE, L. F. 1988. *Geologic History of Utah.* Brigham Young University Geology Studies, Special Publications, **7**.

LACHENBRUCH, A. H. 1961. The depth and spacing of tension cracks. *Journal of Geophysical Research*, **66**, 4273–4292.

LAUBACH, S. E., MACE, R.E. & NANCE, H. S. 1995. Fault and joint swarms in a normal fault zone. [Monograph.] *Proceedings of the Second International Conference on the Mechanics of Jointed and Faulted Rock*, **2**, 305–309.

LAWN, B. 1993. *Fracture of Brittle Solids*, 2nd edn. Cambridge University Press, Cambridge.

MARZOLF, J. E. 1983. Changing wind and hydrologic regimes during deposition of the Navajo and Aztec sandstones, Jurassic (?), Southwestern United States. *In*: BROOKFIELD, M. E. & AHLBRANDT, T. S. (eds) *Eolian sediments and processes, 11th International Association of Sedimentologists Congress, Developments in Sedimentology* [Collection Title], Elsevier, Netherlands, **38**, 635–660.

MYERS, R. D. & AYDIN, A. 1998. Fault damage distribution, evolution, and scaling in porous sandstones. *Geological Society of America, Abstracts with Programs*, **30**, 7, 63–64.

NARR, W. & SUPPE, J. 1991. Joint spacing in sedimentary rocks. *Journal of Structural Geology*, **13**, 1037–1048.

OLSON, J. E. 1993. Joint pattern development; effects of subcritical crack growth and mechanical crack interaction. *Journal of Geophysical Research, B, Solid Earth and Planets*, **98**, 12 251–12 265.

PETERSON, F. & PIPIRINGOS, G. N. 1979. Stratigraphic relations of the Navajo Sandstsone to middle Jurassic formations, southern Utah and northern Arizona. *US Geological Survey Professional Paper*, US Geological Survey, Reston, VA, B1-B43.

POLLARD, D. D. & SEGALL, P. 1987. Theoretical displacements and stresses near fractures in rock: with applications to faults, joints, veins, dikes, and solution surfaces. *In*: ATKINSON, B. (ed.) *Fracture Mechanics of Rock.* Academic Press, Orlando, FL, 277–350.

PRICE, N. 1966. *Fault and Joint Development in Brittle and Semi-brittle Rock.* Pergamon Press, Oxford.

ROGERS, C. M., MYERS, D. A., & ENGELDER, T. 2004.

Kinematic implications of joint zones and isolated joints in the Navajo Sandstone at Zion National Park, Utah: Evidence for Cordilleran relaxation. *Tectonics*, **23**, TC1007, 1–16.

SNOW, J. K., & WERNICKE, B. P. 2000. Cenozoic tectonism in the central Basin and Range: magnitude, rate and distribution of upper crustal strain. *American Journal of Science*, **300**, 659–719.

WAWRZYNEK, P. A. & INGRAFFEA, A. R. 1987. Interactive finite element analysis of fracture processes: An integrated approach. *Theoretical and Applied Fracture Mechanics*, **8**, 137–150.

WOLF, P. R. 1974. *Elements of Photogrammetry*. McGraw-Hill, New York.

Predicting fracture swarms – the influence of subcritical crack growth and the crack-tip process zone on joint spacing in rock

JON E. OLSON

Petroleum and Geosystems Engineering Department, University of Texas at Austin, 1 University Station C0300, Austin, TX 78746, USA (e-mail:jolson@mail.utexas.edu)

Abstract: Swarms or clusters represent an exception to the widely accepted idea that fracture spacing in sedimentary rock should be proportional to mechanical layer thickness. Experimental studies and static stress analysis do not provide adequate explanation for fracture swarm occurrence. The problem is re-examined numerically, accounting for the dynamics of pattern development for large populations of layer-confined fractures. Two crucial aspects of this model are: (1) the inclusion of three-dimensional effects in calculating mechanical interaction between simultaneously propagating fractures; and (2) the use of a subcritical crack-propagation rule, where propagation velocity during stable growth scales with the crack-tip stress intensity factor. Three regimes of fracture spacing are identified according to the magnitude of the subcritical index of the fracturing material. For low subcritical index material ($n = 5$) numerous fractures propagate simultaneously throughout a body resulting in irregular spacing that is, on average, much less than layer thickness. For intermediate subcritical index ($n = 20$) one fracture propagates at a time, fully developing its stress shadow and resulting in a pattern with regular spacing proportional to layer thickness. For high subcritical index cases ($n = 80$) fractures propagate in a fashion analogous to a process zone, leaving a fracture pattern consisting of widely spaced fracture clusters.

A common attribute of opening-mode fractures or joints (Pollard & Aydin 1988) in sedimentary rock is that observed fracture spacing is proportional to layer thickness (Ladeira & Price 1981; Narr & Suppe 1991; Gross *et al.* 1995; Wu & Pollard 1995; Bai & Pollard 2000*a*). Two-dimensional, plane-strain, static analysis demonstrates how the stress relief around a pre-existing joint can create a propagation 'exclusion' zone (Pollard & Segall 1987). The distance to which stress relief (or stress perturbation) extends from a joint can also be termed its 'mechanical interaction distance'. Any joints within another joint's area of perturbed stress will be mechanically influenced in some way, enhancing or hindering propagation as well as modifying the opening distribution (Pollard *et al.* 1982; Olson & Pollard 1989, 1991). Olson (1993) showed how this stress perturbation can develop in an areal sense as multiple joints grow in length and their stress shadows overlap, diminishing the stress available for additional parallel fractures to grow. Recent work (Bai & Pollard 2000a) has shown that in well-bonded, layered materials under crack-normal extensional loading (Fig. 1) the crack-normal stress between closely spaced, parallel joints (spacing less than or equal to layer thickness) actually becomes compressive. This surprising result dictates that increasing the remotely applied extensional strain will not promote the propagation of additional joints between the pre-existing ones, but will only cause the existing

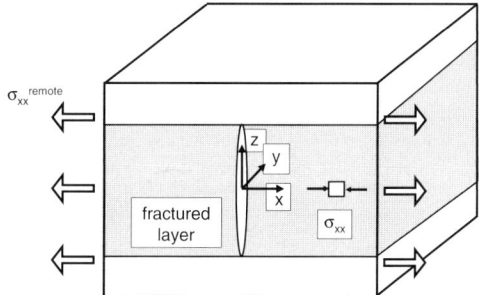

Fig. 1. A three-layer composite, where the middle layer is fractured. Typical plane-strain analysis for fracture spacing examines the variation of the crack-normal stress (σ_{xx}) with distance, x, from the fracture in the vertical x–z plane. The size of the stress relief or stress shadow scales with the height of the fracture, which is assumed to be equivalent to the mechanical layer thickness (Pollard & Segall 1987).

joints to open more to accommodate the added extension. Thus, a minimum spacing approximately equal to layer thickness is expected for parallel opening-mode fractures, and such a joint set is termed 'saturated', as there is no room for additional joints to grow (Rives *et al.* 1992; Wu & Pollard 1995).

The explanation that joint spacing scales with stress relief explains much of what is observed in

From: COSGROVE, J. W. & ENGELDER, T. (eds) 2004. *The Initiation, Propagation, and Arrest of Joints and Other Fractures*. Geological Society, London, Special Publications, **231**, 73–87. 0305-8719/04/$15 © The Geological Society of London 2004.

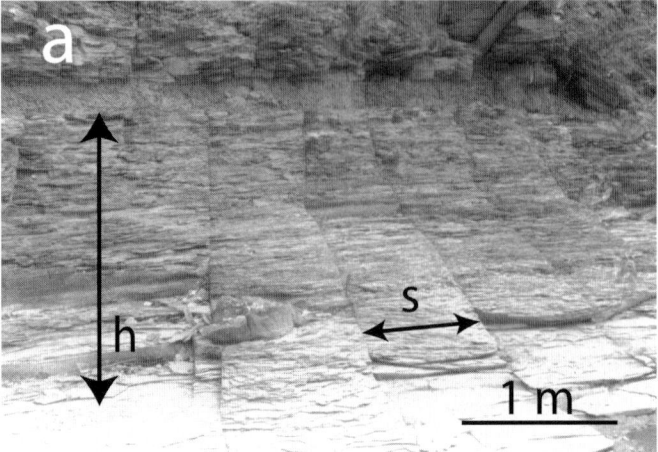

Fig. 2. Examples of fractures in rock with a spacing much less than layer thickness. (**a**) Regularly spaced joints in Devonian Huron Shale, NE Ohio, USA, where the average spacing, s, is less than one-third of the minimum fracture height, h. (**b**) A fracture cluster in the Triassic Wingate Sandstone on Comb Ridge, Utah, USA, where between three and five long fractures lie within a 20 cm-wide zone. The intercluster spacing is approximately 10 m, which is also the thickness of the fractured layer. (**c**) A large fracture swarm in the Cretaceous Frontier Sandstone at Oil Mountain, Wyoming, USA. These are cross-fold fractures with a spacing of less then 10% of layer thickness (Hennings *et al.* 2000).

fracture patterns in rock and other layered materials, but it cannot explain fracture swarms or other situations where local joint spacing is significantly less than layer thickness (Fig. 2). Bai & Pollard (2000b) proposed that joint spacing closer than bed thickness can be attributed to the vertical growth of flaws located near the intersection of fractures with layer boundaries. Their model predicted a minimum spacing to layer thickness ratio of approximately 0.3 based on the static analysis of a vertical cross-section. The description in this chapter is a slightly different approach analogous to the conceptualization of Wu & Pollard (1995), emphasizing the role of lateral fracture growth in the plane of bedding (propagation along the bed rather than vertically through it) as well as the time sequence of the growth process. The growth of all fractures from the flaw size to the macroscale is explicitly modelled using a

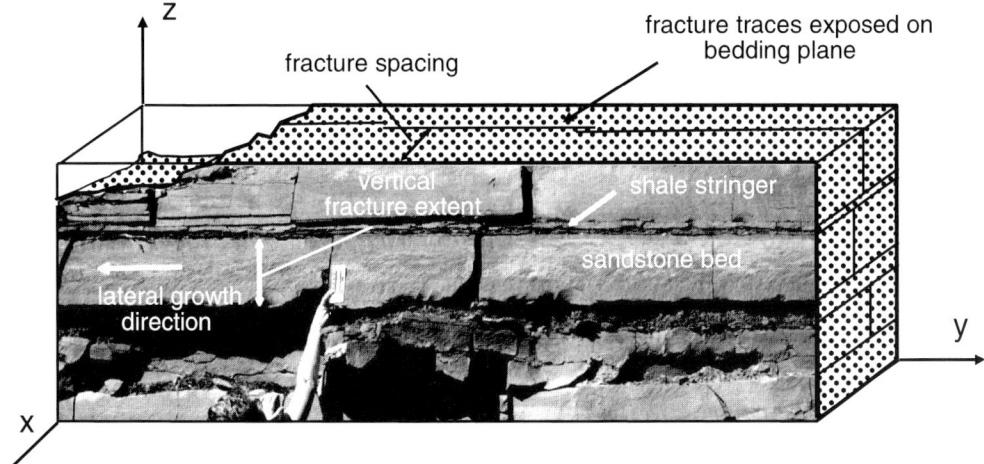

Fig. 3. Three-dimensional representation of a fractured roadcut exposure in the Triassic Chinle Formation at Hite Crossing, Utah, USA. The sandstone beds are separated by thin shale stringers that stopped the vertical growth of the fractures (see middle sandstone with a 10 cm scale). The plumose markings on the fracture surface for the middle bed indicate the fracture grew from right to left, reaching a length many times the height, indicating the dominance of lateral v. vertical fracture propagation.

subcritical crack-growth mechanism. The model addresses the mechanics of fracture swarms (and other closely spaced fractures) and shows how subcritical crack growth can be used to predict swarm occurrence. The modelling of time-dependent joint propagation and three-dimensional effects generates a wide range of spacing distributions, including spacing proportional to layer thickness, spacing much less than layer thickness, and the clustering of joints into swarms.

Fracture propagation in layered rocks

The importance of the lateral propagation of joints in layered rocks has been demonstrated by field observations (Pollard & Aydin 1988; Lacazette & Engelder 1992) and in the laboratory (Wu & Pollard 1995). Figure 3 represents a typical joint geometry from a roadcut example of interbedded sandstone and shale from the Triassic of SE Utah, USA. The photograph depicts the vertical outcrop face, which is a composite fracture surface made up of coplanar but non-continuous segments. The central brittle sandstone bed is bounded by thin, more ductile shale layers. The plumose structure on the joint in this layer can be used to determine that it propagated from right to left and was bounded in its propagation between the shale stringers indicated in the figure (Kulander & Dean 1995). The joint breached the entire thickness of the sandstone bed early in its growth history, and most of the subsequent propaga-

tion involved increasing the fracture length within the bed (lateral propagation), resulting in a fracture length that greatly exceeds the height.

A two-dimensional, cross-sectional analysis (that would be carried out in the x–z plane) would be inadequate for analysing the more dominant lateral fracture propagation along the y-direction. Only propagation in the vertical (z) direction can be adequately represented in such a cross-sectional geometry. However, most theoretical models in the literature that explain joint spacing are from the plane-strain perspective of a vertical, bedding-perpendicular cross-section, where length is assumed to be infinite, and the only variable fracture dimensions are height and opening (Hobbs 1967; Pollard & Segall 1987; Narr & Suppe 1991; Gross et al. 1995; Bai et al. 2000). The modelling approach described in this chapter, with the goal of understanding the variability of joint spacing and the occurrence of fracture swarms, presupposes that the fractures have already propagated vertically across a layer and focuses on the synchronous lateral propagation of numerous vertical fractures confined to a given bed, where bedding would be parallel to the x–y plane of Figure 3.

Subcritical crack growth

In order to analyse the simultaneous propagation of multiple opening-mode fractures, both a failure criterion and a propagation velocity model are required. Brittle fracture strength is influenced by environmental factors, such as relative humidity and chemical reactivity, that can weaken the bonds between material grains (Atkinson 1984; Swanson 1984). For instance, most rock and ceramic material exhibit maximum fracture resistance (termed critical fracture toughness or K_{Ic}) when tested in a vacuum, and that strength is significantly reduced in the presence of water or water vapour. Fracture propagation under critical conditions is unstable and occurs at velocities comparable to the elastic wave speed of the material (Lawn & Wilshaw 1975). Fracture propagation below the critical toughness, termed subcritical crack propagation, occurs at lower stress levels and much lower velocities. An important parameter from linear elastic fracture mechanics that quantifies the concentration of stress at the crack tip and the tendency for an opening-mode fracture to propagate is called the opening-mode stress intensity factor, K_I (Lawn & Wilshaw 1975). The K_I dependence of propagation velocity for subcritical growth is typically divided into three regions (Anderson & Grew 1977) (Fig. 4). In Region I, the log of propagation velocity varies linearly with the log of stress intensity factor, and this relationship is thought to be controlled by chemical reaction kinetics at the crack tip. In Region II, propagation velocity is roughly constant and is limited by the rate of delivery of the corrosive reactants to the crack tip. In Region III, propagation velocity accelerates to rupture speeds as the stress intensity factor approaches the fracture toughness of the material.

The importance of subcritical crack growth for geological situations is that it is a stable propagation mechanism (Segall 1984; Schultz 2000; Olson 2003) and, in Region I, the propagation velocity, v, is related to the opening-mode stress intensity factor at the crack tip, K_I, with an empirically quantifiable, power-law relationship (Atkinson 1984; Swanson 1984)

$$v = A \left(\frac{K_I}{K_{Ic}} \right)^n \qquad (1)$$

where K_{Ic} is the critical fracture toughness, n is the subcritical index and A is a proportionality constant. The power-law exponent, n, can vary widely depending on rock type and environmental conditions (such as dry v. wet). The higher the value of the subcritical index, the less important subcritical growth becomes, as very little propagation occurs before the fracture toughness is reached. For a given rock type, the subcritical index in Region I typically decreases with increasing water content in the environment (Atkinson 1984). Reported values vary from 20 or less for tests carried out on sandstone submerged in water (Atkinson 1984) to greater than 250 under dry conditions in carbonate (Olson et al. 2002). Work in clastic sedimentary rocks (Atkinson

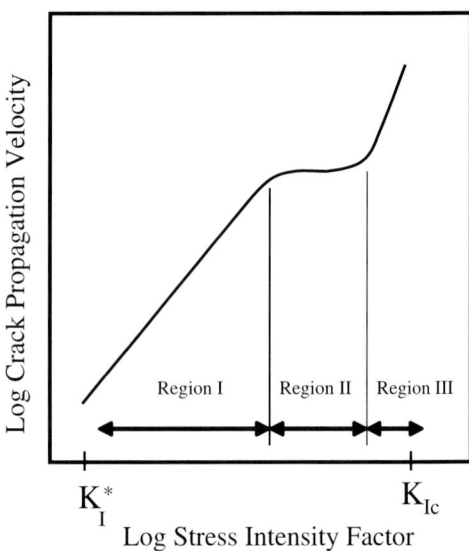

Fig. 4. A log–log plot of propagation velocity v. stress intensity factor (K_I) for subcritical crack growth. K_I^* is the minimum stress intensity factor below which there is no propagation. K_{Ic} is the fracture toughness of a material, at which propagation becomes critical. (See text for explanation of the three regions of behaviour.)

1987; Olson et al. 2002) suggests that grain size, grain mineralogy, cement type and porosity can influence the value of the subcritical index for a given environmental condition.

Two-dimensional, plane-strain modelling of the development of joint networks utilizing the mechanism of subcritical crack growth has shown that the value of the subcritical index, n, exerts a strong influence on the spatial arrangement and length distribution of fractures (Segall 1984; Olson 1993; Renshaw & Pollard 1994). Results in Olson (1993) demonstrated how the subcritical index controls the joint spacing to bed thickness ratio when modelling vertical propagation across a bed under plane-strain conditions. A very low subcritical index ($n = 1$) garnered a spacing to bed thickness ratio of 0.25, while a higher index ($n = 15$) resulted in a spacing to thickness ratio of 0.875. The Olson (1993) numerical results also demonstrated a mechanism for the growth of widely spaced clusters that had originally been postulated by Dyer (1983) for jointing in sandstone at Arches National Park in Utah, USA. The idea was that a propagating joint causes the stresses ahead of the tip to be more tensile, promoting the growth of nearby fractures in a manner similar to the process zone often observed around igneous dykes, where the intensity of dyke-parallel joints is found to be very high close to the dyke (Delaney et al. 1986; Pollard 1987).

Fracture propagation model

Fracture-pattern development is strongly influenced by the mechanical interaction between neighbouring fractures throughout the fracture growth history. This interaction is manifested by the opening or shearing of one fracture perturbing the stress field acting on other nearby fractures. Mathematically, the normal stress acting on an ith fracture element (σ_n^i) due to shearing and opening displacement discontinuities on the jth fracture element (D_s^j and D_n^j, respectively) can be represented by the equation (modified after Crouch 1976)

$$\sigma_n^i = \sum_{j=1}^{N} G^{ij} C_{ns}^{ij} D_s^j + \sum_{j=1}^{N} G^{ij} C_{nn}^{ij} D_n^j \qquad (2)$$

where C_{ns}^{ij} are the plane-strain, elastic influence coefficients giving the normal stress at element i due to a shear displacement discontinuity at element j, and C_{nn}^{ij} gives the normal stress at element j due to an opening displacement discontinuity at element j. An analogous equation can be written for shear stresses. The fundamental integral for determining the influence coefficients C is presented by Crouch (1976).

G^{ij} is a three-dimensional correction factor by which the plane-strain influence functions are multiplied, and it is given by

$$G^{ij} = 1 - \frac{d_{ij}^{\beta}}{[d_{ij}^2 + (h/\alpha)^2]^{\beta/2}} \qquad (3)$$

where d_{ij} is the distance between the centres of elements i and j, h is the fracture height (assumed equal to mechanical layer thickness), and β and α are empirically determined constants. The form of equation (3) was modelled after the analytical plane-strain equation for the normal stress, σ_{xx}, acting perpendicular to an uniformly loaded, isolated, vertical crack of finite height and infinite length (Pollard & Segall 1987). The solution for the fracture-induced σ_{xx} along the x-axis (at the mid-height point of the fracture where $z = 0$) can be written as

$$\sigma_{xx}(x) = \Delta\sigma_I \left[\left(\frac{|x|^3}{(x^2 + (h/2)^2)^{3/2}} - 1 \right) \right] \qquad (4)$$

where $\Delta\sigma_I$ is the mode I driving stress. The relationship between the correction factor in equation (3) and σ_{xx} is made evident by substituting G^{ij} into equation (4) with $\beta = 3$, $\alpha = 2$ and $d_{ij} = x$, resulting in

$$\sigma_{xx}(x) = \Delta\sigma_I(-G^{ij}). \qquad (5)$$

G^{ij} imparts a three-dimensional aspect to calculations in that it represents an influence of fracture height on fracture-induced stress, and the plane-strain influence functions, C, preserve the length dependency of the mechanics, resulting in an

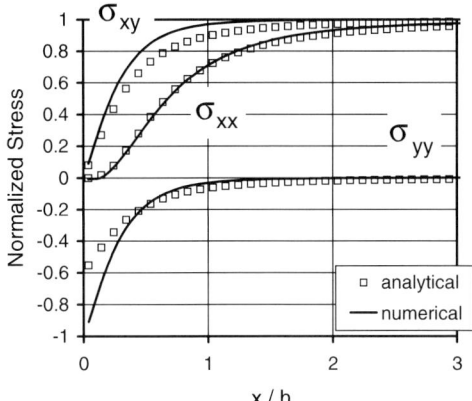

Fig. 5. Crack-induced stress components v. normalized distance, x/h, perpendicular to the fracture trend (see Fig. 1 for fracture orientation with respect to the coordinate system). Stress magnitudes are normalized by the magnitude of the mode I driving stress, $\Delta\sigma_I$, for σ_{xx} and σ_{yy}, and normalized by σ_{xy}^{remote} for σ_{xy}. The open squares represent the analytical solution for a plane-strain fracture with infinite length in the y-direction and finite height, h, in the z-direction. The solid curve is the approximate numerical solution for a three-dimensional crack that is very long in the y-direction and has a finite height, h.

approximate three-dimensional solution. The best values for β and α were empirically derived by comparing σ_{xx} from the plane-strain analytical solution to the boundary-element model result based on equation (3). Using $\beta = 2.3$ and $\alpha = 1$, the numerical approximation for the crack-normal stress component from a finite-height fracture is almost a perfect match to the analytical solution given in equation (4) (Fig. 5). The plane-strain solution for the crack-parallel stress component in the x–y plane, σ_{yy}, is a function of vertical stress and Poisson's ratio (Pollard & Segall 1987)

$$\sigma_{yy} = v(\sigma_{xx} + \sigma_{zz})$$

where

$$\sigma_{zz} = \Delta\sigma_I \left[\frac{|x|^3 + 2|x|(h/2)^2}{(x^2 + (h/2)^2)^{3/2}} - 1 \right].$$

The approximate numerical solution for σ_{yy} is less accurate than for σ_{xx}, although the general trend is reasonable. Finally, the shear stress in the x–y plane for the plane-strain configuration is given by (Pollard & Segall 1987)

$$\sigma_{xy} = \sigma_{xy}^{remote} \left[\frac{x}{(x^2 + (h/2)^2)^{1/2}} - 1 \right]$$

where σ_{xy}^{remote} is the shear stress acting on the fracture at a distance. The numerical approximation of the

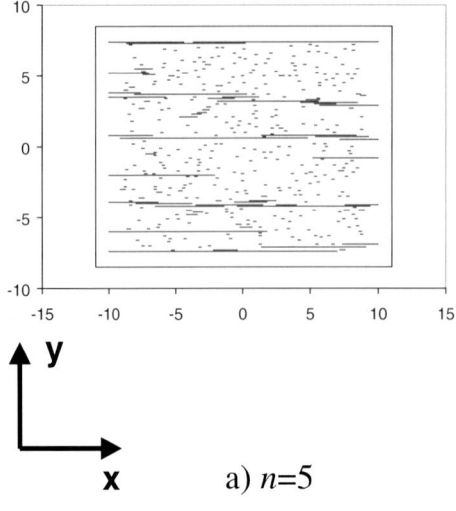

y

x

a) *n*=5

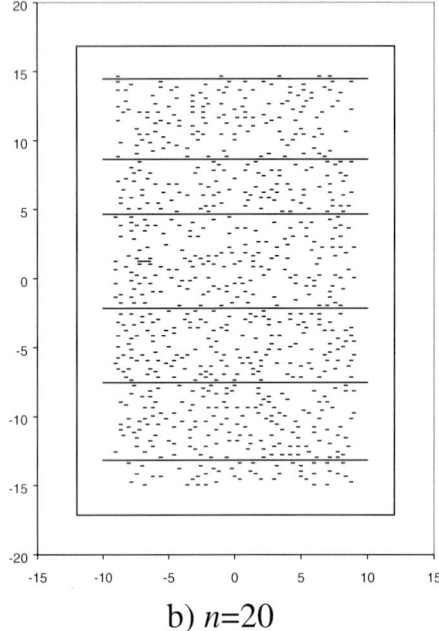

b) *n*=20

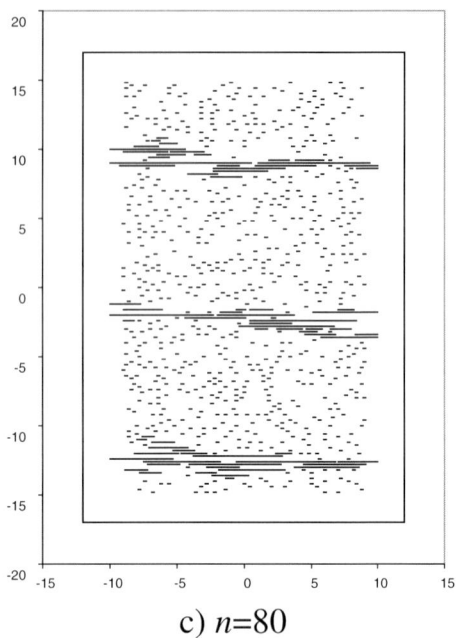

c) *n*=80

Fig. 6. Examples of subcritical fracture growth for subcritical indices of (**a**) $n=5$, (**b**) $n=20$ and (**c**) $n=80$. All simulations started with the same randomly located parallel flaws, a layer thickness of 8.0 m, a Young's modulus of 20 GPa and a Poisson's ratio of 0.25. Strain was imposed by normal displacement in the y-direction at a strain rate of 2.0×10^{-20} s^{-1} to an ultimate extension of 9×10^{-5}. The boundary conditions on the right and left boundaries of the body have zero normal displacement and all boundaries have zero shear stress. To reduce the computing burden for the $n=5$ case, a smaller representative body was simulated, changing all parameters appropriately to keep simulations equivalent.

shear stress shows some deviation from the analytical solution, but in general the accuracy is very good.

Variations in accuracy notwithstanding, however, the trends of all the stress components are correct and consistent with three-dimensional analysis. G^{ij} goes to zero as the ratio d_{ij}/h becomes large, such that fractures that are widely spaced relative to their height have no mechanical influence on one another. In this way, mechanical interaction distance scales with height for relatively long joints, consistent with the observation that joint spacing, as controlled by stress relief, scales with layer thickness for layer-spanning fractures. When the ratio d_{ij}/h is very small, G^{ij} approaches 1, and the solution reverts to the slot-like joint case, where length controls stress-shadow size and mechanical interaction distance, allowing the incorporation of lateral-propagation effects in the horizontal plane.

Although three-dimensional effects are taken into account for mechanical interaction of joints propagating in a given layer, height growth is not modelled. It is assumed that the initial starter flaws, no matter how short, extend across the full thickness of the layer. However, the mechanical influence of these microcracks is controlled by their shortest dimension (Olson 1993), which will be the length until considerable growth has occurred. As described in Olson (1993), individual fractures are represented by a series of equi-length boundary elements connected end to end. The boundary elements for these simulations were all 0.1 m long, and fracture propagation was based on the computed propagation velocity, as determined using equation (1). The time-step for each iteration was computed as the shortest amount of time required to add an entire element's worth of length to the fastest propagating crack. The growth increments of the more slowly propagating cracks in that same iteration typically represent only a fraction of the total element length. These partial growth increments are tracked and are treated as cumulative, so that after many iterations if a particular crack tip has accumulated enough growth to equal an entire boundary element, an element is added. No more than one element is added to any crack tip during any program iteration.

Numerical results

The simulation results demonstrate the role of subcritical index, layer thickness and initial flaw density on opening-mode fracture spacing and length. The initial set of simulations were run with 800 starter flaws, all with a length of 0.2 m and a height equal to the layer thickness. (There is no variation in fracture pattern from the top to the bottom of the layer, and fracture heights are constant and equal to layer thickness throughout the simulations.) The flaws were randomly located in a finite body with an x-dimension of 24 m, a y-dimension of 34 m and a layer thickness of 8 m. To prevent unwanted edge effects between the propagating fractures and the boundaries of the finite body, initial flaws were excluded from a 2 m-thick border around the body perimeter, and subsequent growth was excluded from a slightly thinner 1 m-thick border. Straight-crack propagation was imposed to simplify calculations, but such a geometry is reasonable if a strong horizontal stress anisotropy is assumed (Olson & Pollard 1989). Crack growth was induced by 10 equal increments of extension in the y-direction, spaced equally in time with a final strain magnitude of 9×10^{-5}. The strain was imposed at an average strain rate of 2×10^{-20} s^{-1}. (Because of the greater amount of fracture propagation for the $n = 5$ case, it was run for a body of half the area and half the flaws to reduce computing time and memory requirements. However, the strain history and initial areal flaw intensity were exactly the same.)

At the beginning of a simulation, each flaw has a slightly different value of K_I due to the random spatial locations of neighbouring flaws that mechanically interact with it. Side by side juxtapositions diminish K_I while en echelon and tip-to-tip arrangements enhance K_I (Olson & Pollard 1991). The flaw with the highest K_I propagates first, and the magnitude of the propagation velocity contrast between the higher and lower stress intensity cracks is defined by the power-law relationship of equation (1). Previous work has shown that for very low subcritical index values ($n < 10$) (Olson 1993; Holder et

al. 2001; Olson et al. 2001) many cracks propagate simultaneously and at roughly the same velocity. Even flaws that are initially close together relative to the layer thickness increase in length at a comparable rate, penetrating one another's propagation exclusion zones prior to the stress relief being fully developed.

Figure 6 shows the trace maps of the simulated fracture patterns and the boundary of the finite body on which the loading was imposed. The numerous shorter fracture segments in the plots represent starter flaws that did not grow. Fracture length and spacing can be inferred from these trace maps, but length can be misleading as modelled fractures sometimes meet tip to tip, thus merging into a single trace. In particular, the length distribution for the case of $n = 20$ (Fig. 7) appears inconsistent with the trace map of Figure 6b, as the trace map apparently shows only six different fractures, all with a length of about 20 m, while Figure 7 indicates a distribution of more than 30 fractures ranging in length from 0.3 to 12 m. In reality, each of the 20 m-long fractures in Figure 6b are made up of several shorter segments that meet tip to tip. The initial flaws were not placed entirely randomly, but their initial locations were rounded to the closest 0.1 m, forcing the starter crack centres to fall onto nodes of a 200×300 grid for the 20 m $\times$ 30 m fractured body. The reason for this constraint on starter flaw locations was to prevent numerical instabilities in the boundary element program arising from two crack elements coming too close to one another (Crouch & Starfield 1982). A more quantitative representation of the fracture pattern is found in Figures 7 and 8, where the cumulative frequency of length and perpendicular spacing are reported. These statistics are only compiled for fractures that grew – non-propagated starter flaws were removed from the analysed population. The length data include all fractures that propagated within the modelled area, while spacing is measured along a single scanline running perpendicular to the fracture direction in the middle of the fracture area.

Using $n = 5$, an irregularly spaced fracture pattern was generated with a large variety of fracture lengths and an average spacing much less than the bed thickness (Figs 6a, 7 and 8). Using a higher subcritical index of 20 significantly reduced the number of fractures that propagated and the amount of total fracture length created, but more longer fractures grew (Figs 6b and 7). The spacing became more regular and systematic, with an average spacing of almost 6 m – about 75% of the bed thickness (Fig. 8). This pattern development was characterized by fractures that grew one at a time as a consequence of a large contrast in relative velocity (as defined by equation 1) between fractures of even slightly different K_I values.

Finally, the subcritical index case of $n = 80$ gave some surprising results (Figs 6c, 7 and 8). Based on

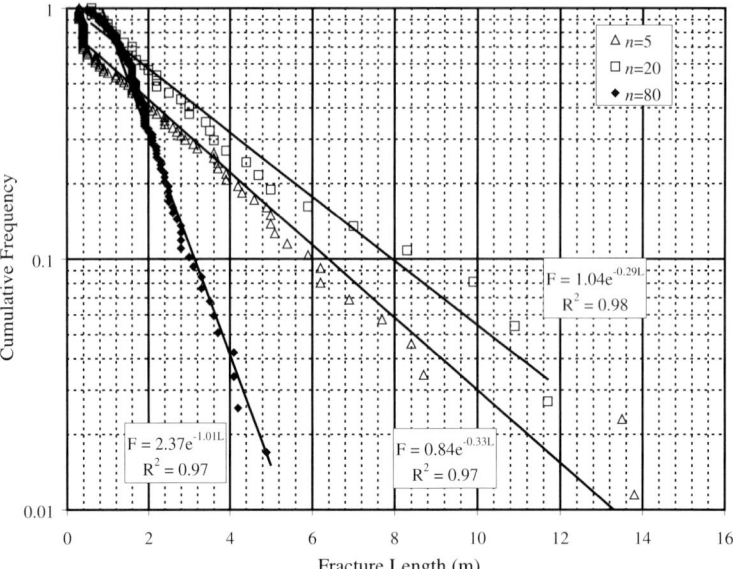

Fig. 7. Cumulative frequency of fracture length for the patterns from Figure 6. Each case can be well described with a negative exponential curve.

previous work (Olson 1993; Renshaw & Pollard 1994), it was expected that for very high values of n the propagation velocity contrast between neighbouring cracks would be even stronger than for $n = 20$, and the fracture pattern would be more sparse but still regularly spaced, developed by one fracture propagating at a time. Instead, fracture clusters formed that were made up of many closely spaced short fractures. The development of this pattern through time involved clusters of smaller fractures propagating simultaneously in the tip region of larger fractures, in a fashion similar to a process zone. These fracture clusters appear to reflect the same propagation mechanism as other geological examples of process-zone fracturing, such as m-scale joints in the process zone of km-scale dykes (Delaney et al. 1986; Pollard 1987) and mm-scale microcracks in the process zone of m-scale joints (Labuz et al. 1987; Nolen-Hoeksema & Gordon 1987). The main difference seems to be that in the joint/dyke and microcrack/joint cases, the process-zone fractures are orders of magnitude smaller than the main crack around which they form. In the simulations described in Figure 6 for $n = 80$, the 'process-zone' fractures and the 'main' crack are all about the same length.

A feature common to all of the length distributions from these simulations is that the cumulative frequency, F, is well described by a negative exponential function (Fig. 7). It has been proposed that a negative exponential distribution is theoretically required if fracture length is limited by crack–crack mechanical interaction (Olson et al. 2001), and these results lend support to that idea. The fracture-length distributions for different subcritical indices show that the $n = 5$ case is significantly different from the $n = 20$ case. They both have approximately the same negative slope on the semi-log plot (-0.33 for $n = 5$ and -0.29 for $n = 20$), with the main difference being that the $n = 20$ distribution is slightly shifted toward longer fractures. The clustered case of $n = 80$, however, shows a significant reduction in fracture lengths achieved, exemplified by the much steeper negative slope of -1.01. The conclusion is that clusters are made up of a large number of short fractures in en echelon arrangement spanning the width of the body.

The cumulative frequency distribution of spacing values (Fig. 8) shows significant and systematic differences between each of the subcritical index cases. Although the data are too sparse to make definitive statistical statements, there is a small degree of clustering (indicated by very small spacing values) for $n = 5$ that is not present in the data for $n = 20$. Also, the pattern for $n = 20$ has substantially higher fracture-spacing values on average than the $n = 5$ case, with median values of 6 and 2 m, respectively. The maximum spacing value for $n = 20$ is almost 7 m, close to the layer thickness, while the maximum spacing for $n = 5$ is only 3 m, less than half the layer thickness. For the $n = 80$ case, spacing is bimodal, with an average spacing of less than 0.5 m, dominated by intracluster spacings, and a maximum spacing of about 10 m representing the distance between clusters.

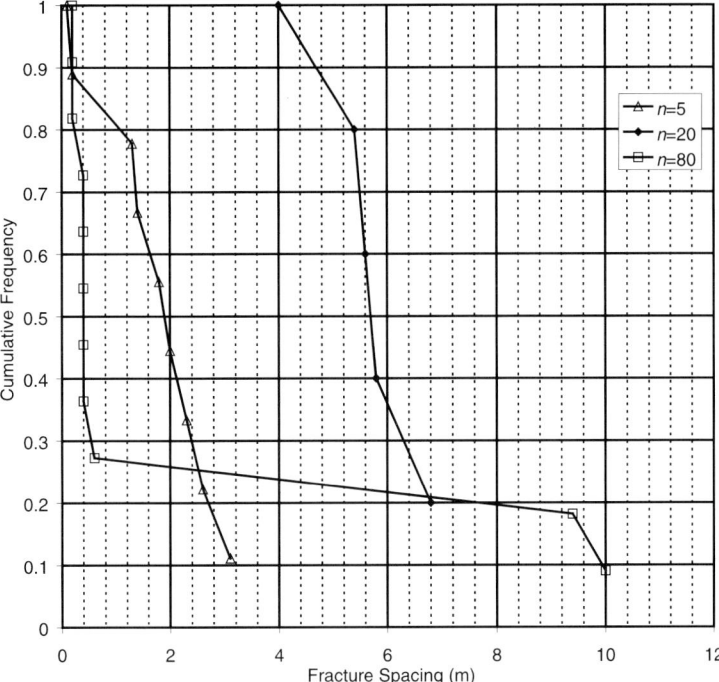

Fig. 8. Cumulative frequency of fracture spacing for the fracture patterns in Figure 6 based on a scanline at $x = 0.05$ m.

Based on these simulation results, three different types of fracturing were identified based on subcritical index value. (1) Materials with very low subcritical index can be expected to develop spacing that is substantially less than layer thickness. Because numerous fractures propagate simultaneously for low subcritical index, they can penetrate into one another's stress shadow before it becomes fully developed, resulting in fracture spacing that appears too close given final fracture height and length magnitudes. (2) Intermediate subcritical index values are most likely to result in fracture spacing roughly proportional to bed thickness. As one fracture grows at a time at this subcritical index, each fracture can fully develop its stress shadow before neighbouring fractures get a chance to compete for propagation energy, and the spacing scales with the size of the fully developed fracture's stress perturbation (Pollard & Segall 1987). (3) Finally, materials with very high subcritical index are likely to generate widely spaced clusters, and the spacing between clusters appears to scale with mechanical layer thickness.

Additional simulations were run for the case of $n = 80$ to demonstrate the effects of varying bed thickness and initial flaw density on fracture clustering (Figs 9–11). These simulations were run under the same strain boundary conditions used for Figure 6

but with only half the initial flaw density (400 instead of 800 flaws) and with bed thickness varying from 2 to 8 m. All of the trace maps show some degree of fracture clustering (Fig. 9), but clustering appears weakest for the thinnest bed ($h = 2$ m). However, even for the layer thickness of 8 m, the clustering is not as strong as the case of $n = 80$ using 800 initial flaws (Fig. 6c).

Even though the trace maps for the different bed thickness cases look substantially different, the fracture-length cumulative frequency data for all cases are very similar, again appearing to follow a negative exponential trend (Fig. 10). This suggests that subcritical index has a stronger effect on length distribution than does mechanical layer thickness. However, comparing the exponent for the negative exponential fit in Figure 11 (-0.3678) to the exponent for the $n = 80$ case in Figure 7 (-1.011) shows that the initial flaw density has a very strong influence on fracture-length distribution. The steeper slope for the higher fracture density case (Fig. 7) implies that more flaws cause additional hindrance to length growth, and the result is shorter fractures. The maximum fracture length for the higher flaw density case was 5 m, while for the lower density case it was almost 10 m. This strong dependence of length development on flaw density is similar to that found in Olson *et al.* (2001).

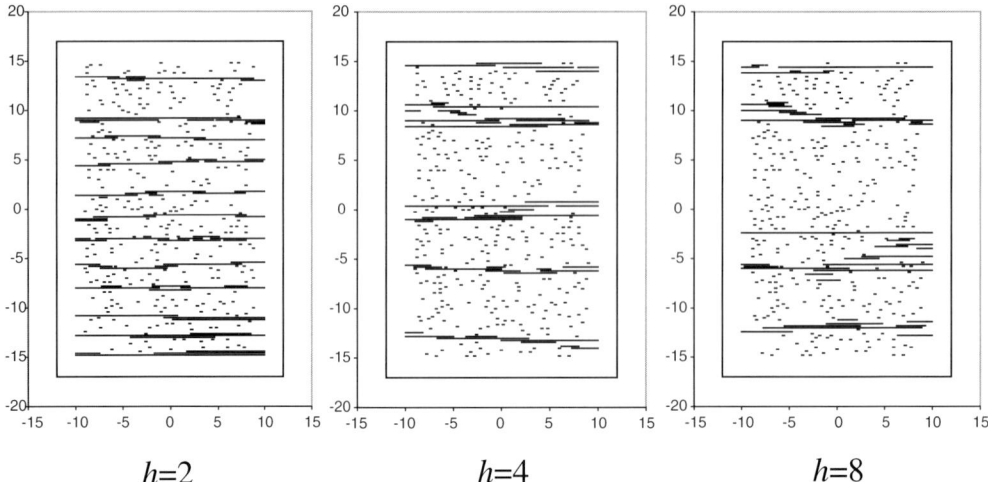

$h=2$ $h=4$ $h=8$

Fig. 9. Fracture trace maps for simulations run under similar conditions to those for Figure 6, except that all cases have a subcritical index of 80, there are 400 starter flaws instead of 800 and the bed thickness is varied from 2 to 4 to 8 m.

As expected, fracture-spacing distributions (Fig. 11) are strongly affected by bed thickness, even though the fracture-length distribution is relatively independent of it (Fig. 10). The maximum spacing increases with increasing layer thickness, but the median spacing has the reverse relationship. The low median value is interpreted to be an outgrowth of the effect of clustering on the spacing distribution. There is a strong change in slope of the cumulative frequency curves of fracture spacing for the cases of different bed thicknesses. From inspecting the trace maps in conjunction with the spacing data, it is evident that the larger spacings represent the distance between clusters (intercluster spacing) and the smaller values represent the distance between fractures within a cluster (intracluster spacing). For example, for the thin-bedded case ($h=2$), the data at a cumulative frequency of $F=0.78$ and above represent intracluster spacings of 0.2 m, while the data at cumulative frequency of $F=0.72$ and below, valued at 1.8 m or higher, are intercluster spacings (or in some instances the spacing between non-clustered fractures). Interpolating, the population break can be picked at a cumulative frequency of $F=0.75$. Thus, for the $h=2$ case, about 25% of the spacings represent measurements within a cluster. For the cases of $h=4$ and $h=8$, the break in the population can be picked at cumulative frequencies of $F=0.41$ and $F=0.45$, respectively. The thicker beds exhibit stronger clustering, with between 55% (for $h=4$) and 60% (for $h=8$) of the spacing measurements coming from within clusters.

Reiterating, the maximum spacing values in the distributions indicate the distance between clusters, and the dependence of intercluster spacing on mechanical layer thickness suggests it is related to the stress shadow around fractures. The spacing within clusters (intracluster) is most probably related to crack-tip-process zone and initial flaw spacing.

Process-zone mechanics and joint clusters

As noted in the discussion of the $n=80$ simulation, the mechanism of fracture clustering can be likened to a process zone propagating across the body. The reason for clustering to occur in the high subcritical index cases is related to the magnitude of the stress intensity factor when propagation occurs. As fracture-propagation velocities for material with very high n are initially very low due to the power-law nature of equation (1), propagation is delayed until more strain has accumulated. Consequently, when fracture growth finally occurs, it can be at stress intensity factor values that approach or exceed critical values. Because fracture-induced stress scales linearly with stress intensity factor in the near-tip region (Lawn & Wilshaw 1975), the tensile stress perturbation around the crack tip is increased by high K_I values and the propagation of flaws in the crack-tip region is enhanced for high n cases. The clusters or fracture swarms essentially record the movement of a process zone across the rock body and, because of the higher stress intensity factor values, the propagation mechanism is no longer described by subcritical region I, but by region II, III or critical propagation ($K_I \geq K_{Ic}$). However, even if fracture growth approaches critical conditions, high propagation velocities are not reached under typical

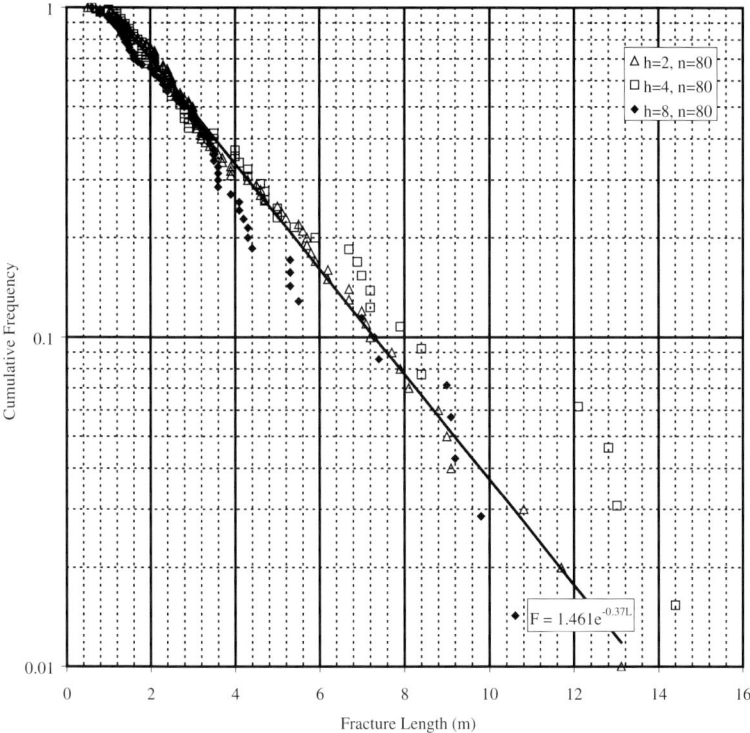

Fig. 10. Cumulative frequency of fracture length for the patterns of Figure 9. Note that all three patterns have markedly similar length distributions that follow a negative exponential shape.

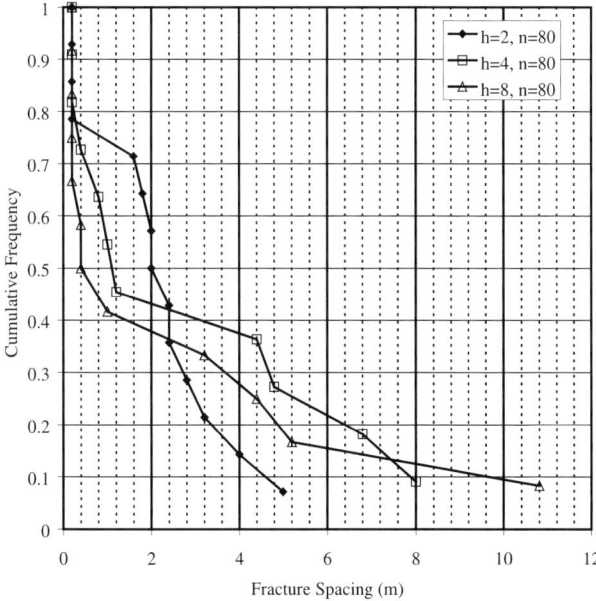

Fig. 11. Cumulative frequency of fracture spacing for the fracture patterns in Figure 9 based on a scanline at $x = 0.05$ m.

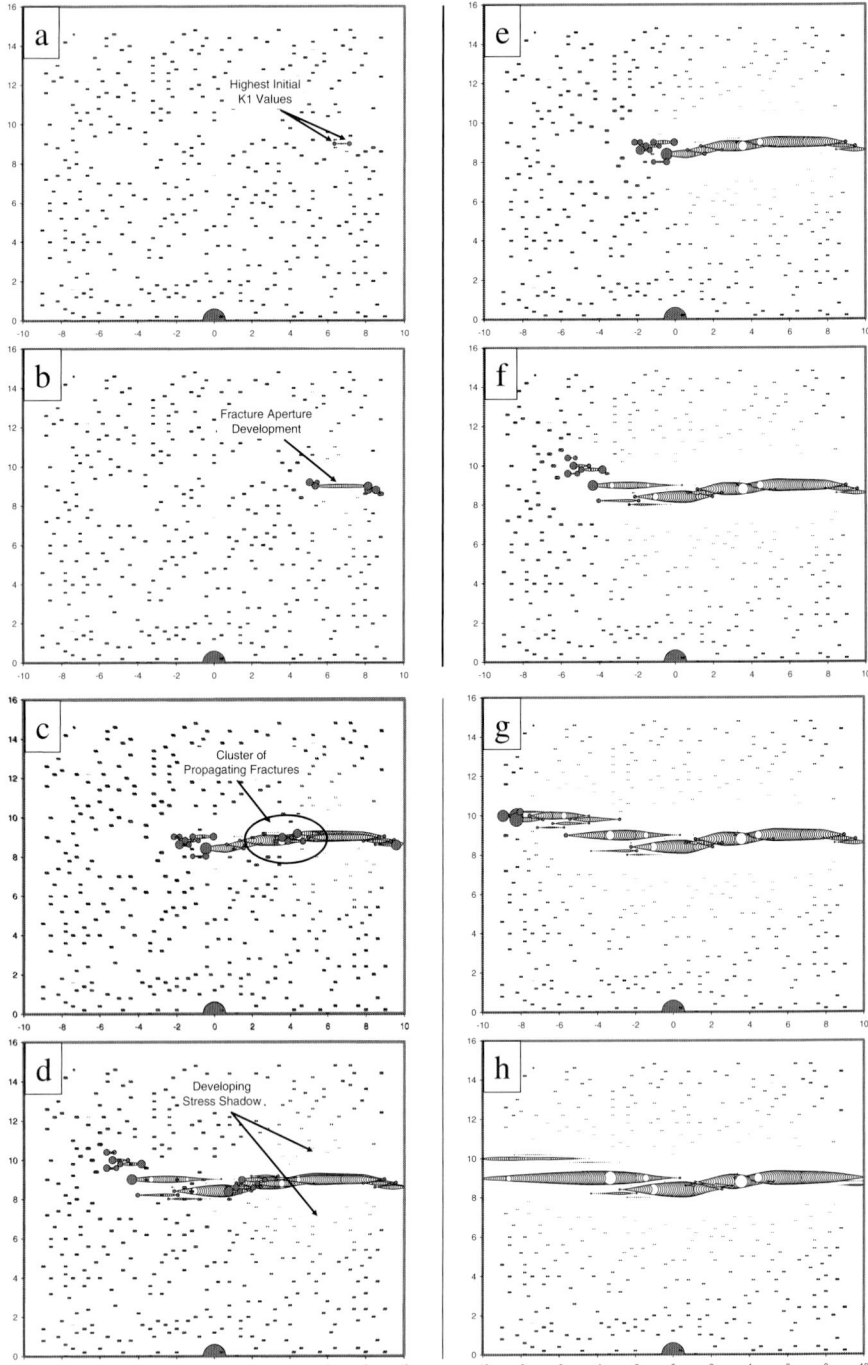

Fig. 12. Sequence of fracture aperture and stress intensity factor development for the cluster from Fig. 6 (case $n = 80$) in the vicinity of $y = 8$ m. The circle centred at $(0, 0)$ is the scale for K_I, showing the diameter appropriate for a value of 10 MPa-m$^{1/2}$. Each fracture segment modelled has an opening represented by a open circle with a proportionate diameter. The fracture-tip elements have a shaded circle representing the magnitude of K_I. Frames (a) – (h) show the propagation of a cluster across the body, the stress-shadow effects of growing fractures on other nearby fractures and the tapered fracture opening shape approaching the crack tips for interacting fractures.

geological conditions (Olson 2003) because strain-softening effects (Segall 1984) or fluid-flow restrictions (Engelder & Lacazette 1990; Lacazette & Engelder 1992; Renshaw & Harvey 1994) reduce the stress intensity factor with propagation, resulting in a quasi-stable process. The apparent increase in cluster size with increased mechanical layer thickness inferred from the spacing data of Figure 11 suggests that thicker beds also allow for stronger mechanical crack interaction and more stress elevation in the crack-tip region, which may cause the growth of more fractures over a wider area in the 'process zone'.

Looking at the sequence of development of fracture aperture and stress intensity factor in a growing cluster helps illustrate the cluster growth mechanism. Figure 12 shows the time sequence of fracture growth for the cluster located around $y = 8$ m from the $n = 80$ case in Figure 6, mapping fracture opening at each boundary element in the simulation and stress intensity factor at every crack tip. Fracture opening is represented by the diameter of the open circles located at each fracture patch centre. (Fracture aperture exaggeration is approximately $\times 450$ – the maximum aperture in the final plot of the sequence, Fig. 12h, is approximately 1.75 mm located at $x = 0$ m, $y = 8.4$ m.) The stress intensity factor is proportional to the diameter of the shaded circles that are located at the centre of every crack-tip element. The scale for stress intensity factor magnitude is the large, dark circle centred at (0, 0) on each plot, whose size represents a stress intensity factor value of 10 MPa-m$^{1/2}$. (There is no crack element at this location, it is merely a scale.)

In Figure 12a, only a small amount of fracture growth has occurred, starting at the cracks with the largest initial K_I values. None of the cracks in the simulation have any visible fracture aperture except for the cracks whose tips are indicated to have the highest K_I on the plot. After additional growth has occurred (Fig. 12b), the fracture aperture for the main crack is clearly visible, and the K_I values are rising for this growing crack as well as for some of its close neighbours. The fracture cluster growth (or the process zone) is well developed by Figure 12c, where several cracks are propagating ahead of the left tip of the main crack. It is interesting to note that the cracks in the cluster have elevated K_I values at both tips, suggesting that they are both propagating out ahead of the main crack (to the left) as well as back towards it (to the right). Eventually, the crack interaction of the overlapping en echelon crack tips hinders growth of the clustered fractures towards the right and the main crack towards the left, causing arrest at some tips and providing a limitation on length growth (Pollard *et al.* 1982; Olson & Pollard 1989).

Another aspect of crack interaction can be seen by comparing Figure 12a with Figure 12d. While there

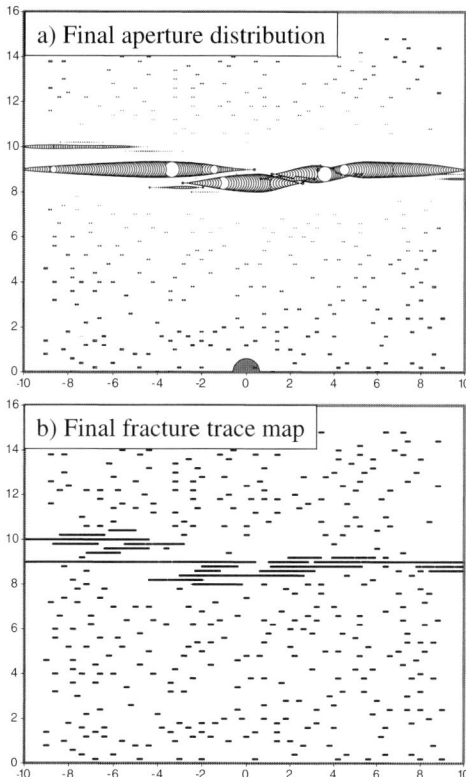

Fig. 13. Comparison of the final fracture aperture distribution and trace map of fractures that grew, showing that some fractures grow and then subsequently close or never reach an observable aperture.

is fracture-propagation enhancement ahead of the tip of the main propagating cracks, there is propagation hindrance to either side of the main body of the crack, as noted in Figure 12d. The propagation hindrance is exemplified by the diminished K_I values of the pre-existing flaws in that area compared to their initial state. Further propagation of the fracture cluster completely across the body is shown in Figure 12e–h. Further development of the stress-shadow or propagation-suppression zone is evident, as well as the tapered, non-elliptical displacement profiles at the tips for many of the overlapping, en echelon fractures.

Comparing the final fracture aperture map with a trace map (Fig. 13) shows that not all of the fractures that propagated have an appreciable aperture at the end of the deformation cycle. The figure clearly shows that, although there are many fractures that propagated in the cluster (Fig. 13b), there are a few dominant ones that have the largest aperture, and these are typically the longest fractures. Also, exaggerating the K_I scale for the initial and final

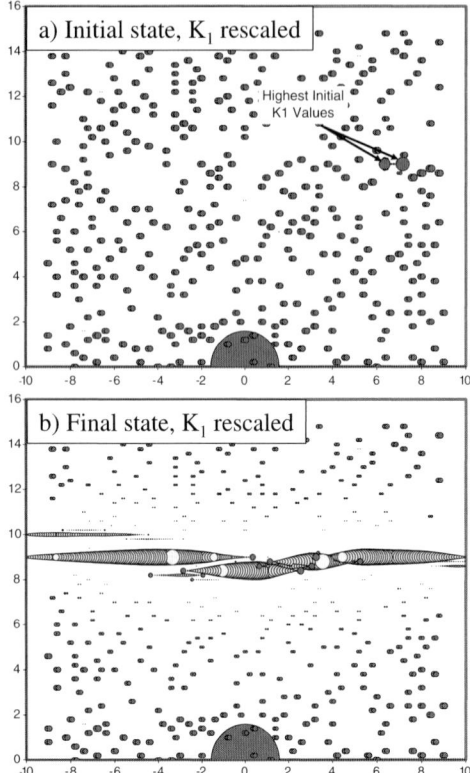

Fig. 14. Rescaled versions of Figure 12a and h, emphasizing the heterogeneity of K_I throughout the fractured body.

states emphasizes the stress-shadow or propagation-suppression effects around the propagated cluster (Fig. 14). This K_I suppression extends 4–6 m to either side of the open fracture zone ($h = 8$ m).

Conclusion

This work shows that the spacing of opening-mode fractures constrained to a single layer in a homogeneous elastic body depends not only on the static-stress distribution around the fractures but also on fracturing dynamics and the three-dimensional nature of lateral fracture propagation. These numerical fracture-propagation results are proposed to be analogous to the lateral propagation of joints confined by sedimentary layering. The plane strain, static analysis of stress relief or stress-shadow effects around a fracture imply a minimum fracture spacing approximately equal to layer thickness, but that static analysis represents only one of three newly identified fracture-propagation regimes delineated by variations in subcritical crack properties.

Regular fracture spacing approximately equal to mechanical layer thickness is attributed to subcritical crack propagation in rock with subcritical indices of intermediate magnitude ($n \cong 20$). Another regime of propagation behaviour exists for very low subcritical index materials ($n \cong 5$), where fractures tend to have spacing that is very irregular and much less than bed thickness. This spacing regime may be responsible for very large fracture swarms, such as pictured in Figure 2c. Finally, at very high subcritical index ($n \cong 80$), fracturing is also very clustered, but the clusters are widely spaced and the fracture growth probably occurs at critical stress intensity factor values. The fracture pattern for this high subcritical index regime resembles the fracture clustering in Figure 2b.

Although fracture height (as constrained by bed thickness) has an influence on fracture spacing, simulations show little impact of bed thickness on the cumulative frequency of fracture-length. The key parameters controlling fracture-length development appear to be subcritical index and flaw density in the body. Another unusual result related to bed-thickness effects was found in the high subcritical index cases ($n = 80$). Although increasing bed thickness increased the maximum observed spacing between fractures for a given pattern, which seems to be consistent with stress-shadow ideas, the average spacing of the distribution actually decreased as bed thickness was increased. This behaviour is attributed to the stronger mechanical-crack interaction for larger bed thickness cases, causing more intense fracturing in the near-tip region of a propagating crack. For a thin bed, the zone of elevated stress is less well developed around the propagating crack tip, and thus fracture clusters are less significant and median spacing is broader.

This material is based in part on work supported by the US Department of Energy under Award No. DE-FC26-00BC15308, the Texas Advanced Technology Program, and by the Fracture Research and Application Consortium (FRAC) of The University of Texas at Austin. The author thanks T. Engelder for helpful review comments.

References

ANDERSON, O. L. & GREW, P. C. 1977. Stress corrosion theory of crack propagation with applications to geophysics. *Reviews of Geophysics and Space Physics*, **15**, 77–104.

ATKINSON, B. K. 1984. Subcritical crack growth in geological materials. *Journal of Geophysical Research*, **89**, 4077–4114.

BAI, T. & POLLARD, D. D. 2000a. Fracture spacing in layered rocks: a new explanation based on the stress transition. *Journal of Structural Geology*, **22**, 43–57.

BAI, T. & POLLARD, D. D. 2000b. Closely spaced fracture in layered rocks: initiation mechanism and propagation kinematics. *Journal of Structural Geology*, **22**, 1409–1425.

BAI, T., POLLARD, D. D. & GAO, H. 2000. Explanation for fracture spacing in layered materials. *Nature*, **403**, 753–756.

CROUCH, S. L. 1976. Solution of plane elasticity problems by the displacement discontinuity method. *International Journal for Numerical Methods in Engineering*, **10**, 301–343.

CROUCH, S. & STARFIELD, A. M. 1982. *Boundary Element Methods in Solid Mechanics: With Applications in Rock Mechanics and Geological Engineering*. Allen and Unwin, London.

DELANEY, P. T., POLLARD, D. D., ZIONY, J. L. & MCKEE, E. H. 1986. Field relations between dikes and joints: emplacement processes and paleostress analysis. *Journal of Geophysical Research*, **91**, 4920–4938.

DYER, J. R. 1983. *Jointing in sandstones, Arches National Park, Utah*. PhD thesis, Stanford University.

ENGELDER, T. & LACAZETTE, A. 1990. Natural hydraulic fracturing. *In*: BARTON, N. & STEPHANSSON, O. (eds) *International Symposium on Rock Joints*. A. A. Balkema, Loen, 35–43.

GROSS, M. R., FISCHER, M. P., ENGELDER, T. & GREENFIELD, R. J. 1995. Factors controlling joint spacing in interbedded sedimentary rocks; integrating numerical models with field observations from the Monterey Formation, USA. *In*: AMEEN, M. S. (ed.) *Fractography: Fracture Topography as a Tool in Fracture Mechanics and Stress Analysis*. Geological Society, London, Special Publications, **92**, 215–233.

HENNINGS, P. H., OLSON, J. E. & THOMPSON, L. B. 2000. Combining outcrop and 3-d structural modeling to characterize fractured reservoirs: an example from Wyoming. *American Association of Petroleum Geologists Bulletin*, **84**, 830–849.

HOBBS, D. W. 1967. The formation of tension joints in sedimentary rocks: an explanation. *Geological Magazine*, **104**, 550–556.

HOLDER, J., OLSON, J. E. & PHILIP, Z. 2001. Experimental determination of subcritical crack growth parameters in sedimentary rock. *Geophysical Research Letters*, **28**, 599–602.

KULANDER, B. R. & DEAN, S. L. 1995. Observations on fractography with laboratory experiments for geologists. *In*: AMEEN, M. S. (ed.) *Fractography: Fracture Topography as a Tool in Fracture Mechanics and Stress Analysis*. Geological Society, London, Special Publications, **92**, 59–82.

LABUZ, J. F., SHAH, S. P. & DOWDING, C. H. 1987. The fracture process zone in granite; evidence and effect. *International Journal of Rock Mechanics and Mining Sciences & Geomechanics Abstracts*, **24**, 235–246.

LACAZETTE, A. & ENGELDER, T. 1992. Fluid-driven cyclic propagation of a joint in the Ithaca Siltstone, Appalachian Basin, New York. *In*: EVANS, B. & WONG, T.-F. (eds) *Fault Mechanics and Transport Properties of Rock*. Academic Press, London, 297–324.

LADEIRA, F. L. & PRICE, N. J. 1981. Relationship between fracture spacing and bed thickness. *Journal of Structural Geology*, **3**, 179–183.

LAWN, B. R. & WILSHAW, T. R. 1975. *Fracture of Brittle Solids*. Cambridge University Press, Cambridge.

NARR, W. & SUPPE, J. 1991. Joint spacing in sedimentary rocks. *Journal of Structural Geology*, **13**, 1037–1048.

NOLEN-HOEKSEMA, R. C. & GORDON, R. B. 1987. Optical detection of crack patterns in the opening-mode fracture of marble. *International Journal of Rock Mechanics and Mining Sciences & Geomechanics Abstracts*, **24**, 135–144.

OLSON, J. E. 1993. Joint pattern development: effects of subcritical crack-growth and mechanical crack interaction. *Journal of Geophysical Research*, **98**, 12 251–12 265.

OLSON, J. E. 2003. Sublinear scaling of fracture aperture versus length: an exception or the rule? *Journal of Geophysical Research*, **108**, 2413.

OLSON, J. E. & POLLARD, D. D. 1989. Inferring paleostresses from natural fracture patterns: A new method. *Geology*, **17**, 345–348.

OLSON, J. E. & POLLARD, D. D. 1991. The initiation and growth of en echelon veins. *Journal of Structural Geology*, **13**, 595–608.

OLSON, J. E., HOLDER, J. & RIJKEN, P. 2002. Quantifying the fracture mechanics properties of rock for fractured reservoir. *In*: *Proceedings of the SPE/ISRM Rock Mechanics in Petroleum Engineering Conference 2002*. Society of Petroleum Engineers, Richardson, TX, 421–432.

OLSON, J. E., QIU, Y., HOLDER, J. & RIJKEN, P. 2001. Constraining the spatial distribution of fracture networks in naturally fractured reservoirs using fracture mechanics and core measurements. *In*: *Proceedings – SPE Annual Technical Conference and Exhibition, 2001*. Society of Petroleum Engineers, Richardson, TX, 279–290.

POLLARD, D. D. 1976. On the form and stability of open hydraulic fractures in the Earth's crust. *Geophysical Research Letters*, **3**, 513–516.

POLLARD, D. D. 1987. Elementary fracture mechanics applied to the structural interpretation of dykes. *In*: HALLS, H. C. & FAHRIG, W. F. (eds) *Mafic Dyke Swarms; A Collection of Papers Based on the Proceedings of an International Conference, Mississauga, ON, Canada*, **34**. Geological Society of Canada, Toronto, ON, 5–24.

POLLARD, D. D. & AYDIN, A. 1988. Progress in understanding jointing over the past century. *Geological Society of America Bulletin*, **100**, 1181–1204.

POLLARD, D. D. & SEGALL, P. 1987. Theoretical displacements and stresses near fractures in rock: with applications to faults, joints, veins, dikes and solution surfaces. *In*: ATKINSON, B. K. (ed.) *Fracture Mechanics of Rock*. Academic Press, London, 277–350.

POLLARD, D. D., SEGALL, P. & DELANEY, P. T. 1982. Formation and interpretation of dilatant echelon cracks. *Geological Society of America Bulletin*, **93**, 1291–1303.

RENSHAW, C. E. & HARVEY, C. F. 1994. Propagation velocity of a natural hydraulic fracture in a poroelastic medium. *Journal of Geophysical Research*, **99**, 21659–21677.

RENSHAW, C. E. & POLLARD, D. D. 1994. Numerical simulation of fracture set formation: a fracture mechanics model consistent with experimental observations. *Journal of Geophysical Research*, **99**, 9359–9372.

RIVES, T., RAZACK, M., PETIT, J. P. & RAWNSLEY, K. D. 1992. Joint spacing: analogue and numerical simulations. *Journal of Structural Geology*, **14**, 925–937.

SCHULTZ, R. A. 2000. Growth of geologic fractures into large-strain populations: review of nomenclature, subcritical crack growth, and some implications for rock engineering. *International Journal of Rock Mechanics and Mining Science*, **37**, 403–411.

SEGALL, P. 1984. Formation and growth of extensional fracture sets. *Geological Society of America Bulletin*, **95**, 454–462.

SWANSON, P. L. 1984. Subcritical crack growth and other time- and environment-dependent behavior in crustal rocks. *Journal of Geophysical Research* **89**, 4137–4152.

WU, H. & POLLARD, D. D. 1995. An experimental study of the relationship between joint spacing and layer thickness. *Journal of Structural Geology*, **17**, 887–905.

Fracture-pattern variations around a major fold and their implications regarding fracture prediction using limited data: an example from the Bristol Channel Basin

MANDEFRO BELAYNEH & JOHN W. COSGROVE

Department of Earth Science and Engineering, Imperial College of Science Technology and Medicine, Royal School of Mines, Prince Consort Road, London SW7 2BP, UK (e-mail: m.belayneh@ic.ac.uk)

Abstract: This chapter focuses on the evolution of fractures during the inversion of the Bristol Channel Basin, and examines the lateral and vertical consistency of the resulting fracture network within the alternating Liassic limestones and shales. The study has two principal aims. The first is to determine the reliability of fracture systems deduced using more limited data from less well-exposed regions or unexposed regions sampled only by drilling, and the second is to assess whether the fractures are linked to a regional stress field or are the result of a local stress field controlled by the geometry and mechanism of formation of a fold.

The joint patterns were studied using a combination of scanline and window sampling, and the results indicate that there are considerable variations in the fracture systems between adjacent limestone beds and also lateral variation within the same bed. Although there is little doubt that the independent development of fracture patterns in adjacent limestone beds is facilitated by the intervening shale horizons, which allow them to become mechanically decoupled, the reasons for these variations are still unclear.

The aim of the present chapter is to examine the geometry and orientation of the fracture networks within the Liassic rocks of the south coast of the Bristol Channel Basin (BCB). The study area has undergone a relatively simple structural history involving basin opening that began in the Permian, inversion that occurred during the Cretaceous–Tertiary, and subsequent uplift and exhumation. The work focuses on the joint sets in rocks of the Lower Lias (and their related fracture networks) formed during inversion. Despite the simple structural history of the BCB, the fracture networks show remarkable variations from one area to another within an individual bed, and from one limestone bed to the next. These observations have profound implications regarding the determination of the geometry of fracture networks from limited data sets such as are obtained from boreholes. Joints by definition have little or no displacement across or along them (e.g. Price 1966; Blès & Feuga 1986) and this means that they are difficult to detect on a conventional seismic section. Consequently, the study of joint patterns in outcrops plays a key role in reservoir-development studies as it is the only method available for obtaining a realistic picture of fluid flow in the rock mass (i.e. reservoir). Whilst acknowledging that some joints form near the surface as a result of exhumation and the release of residual stresses, the authors argue that by studying natural analogues of fractured reservoirs exposed at the Earth's surface, such as the fractured and ampli-fied rollover fold that occurs in the Bristol Channel area, production planning and reservoir management can be achieved more efficiently and effectively in similar structural settings where there are proven, but unexposed, hydrocarbon provinces elsewhere in the world.

Joints in the BCB were formed at various stages in its evolution i.e. during burial and diagenesis associated with basin opening, during basin inversion, and as a result of uplift and exhumation. The fractures described in this chapter are exposed around two amplified rollover folds at Lilstock and Kilve (Figs 1 and 2). The timing and cause of these fractures is still a matter of dispute. For example, Rawnsley *et al.* (1998) argued that they were formed as a result of post-Alpine stress release. In contrast, Engelder & Peacock (2001), based on a study on the southern limb of the Lilstock buttress anticline, argued that the joints were related to folding during basin inversion. These ideas are discussed later in this chapter after the results of the study are presented.

The study area is situated on the southern margin of the BCB (Fig. 1); an area that has been the subject of many stratigraphic, palaeontological and structural investigations. The excellent cliff and foreshore exposures allow an almost complete three-dimensional (3D) characterization of the bedding and other structures in the Upper Permian–Triassic and Lower Lias succession. The literature on the area is substantial and only those works relevant to this study will be considered in this article. Kamerling (1979),

From: COSGROVE, J. W. & ENGELDER, T. (eds) 2004. *The Initiation, Propagation, and Arrest of Joints and Other Fractures.* Geological Society, London, Special Publications, **231**, 89–102. 0305-8719/04/$15 © The Geological Society of London 2004.

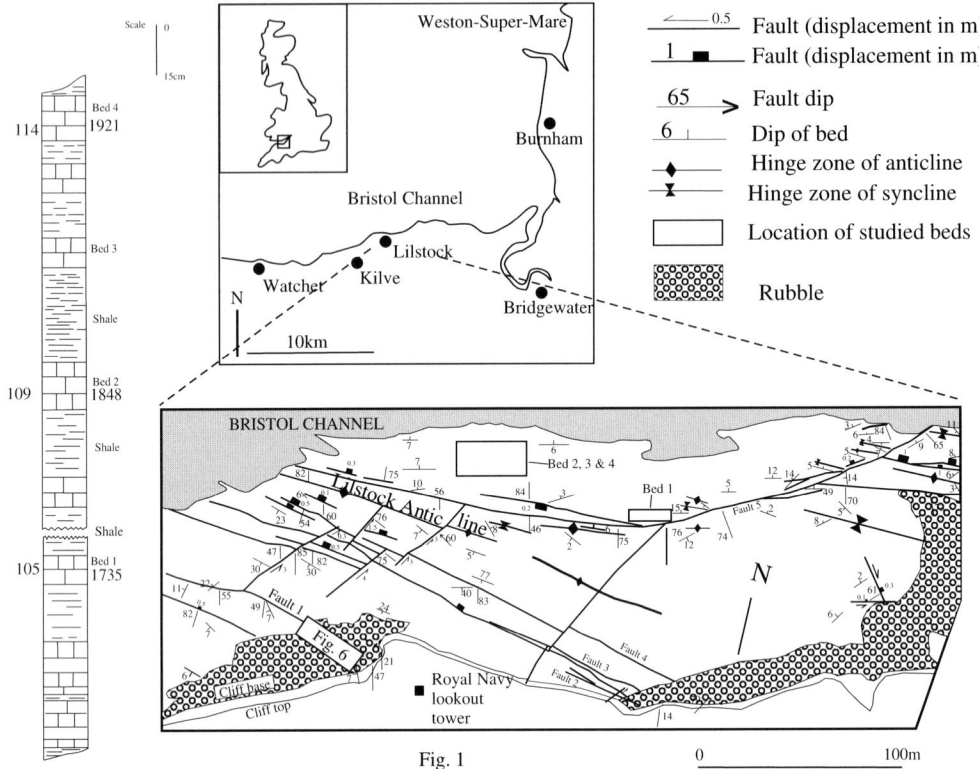

Fig. 1. Simplified structural map of Lilstock on the southern margin of the BCB showing the Lilstock anticline (from Rawnsley *et al.* 1998). The stratigraphic log shows the position of the studied limestone beds. Numbers to the right of the column are designation for the beds above an arbitrary datum, as given in Engelder & Peacock (2001) and those to the left relate to the BGS charting of these beds (Whittaker & Green 1983).

Whittaker & Green (1983) and Van Hoorn (1987) give a detailed description of the stratigraphy of the study area. The region has been used to study various structures including normal faults (Davison 1995; Nemčok & Gayer 1996), the development of relay ramps in association with normal faults (Peacock & Sanderson, 1991, 1994), inverted normal faults and thrust faults (Brooks *et al.* 1988; Chadwick 1993; Dart *et al.* 1995; Nemčok *et al.* 1995; Kelly *et al.* 1999) and strike-slip faults (Peacock & Sanderson, 1995; Willemse *et al.* 1997; Kelly *et al.* 1998). Veins from the area have been studied by Caputo & Hancock (1999) and Cosgrove (2001), and joints by Loosveld & Franssen (1992), Rawnsley *et al.* (1998), Al-Mahruqi (2001), Engelder & Peacock (2001) and Belayneh (2003, 2004). Peacock (2001) used some examples from the study area to illustrate the temporal relationship between joints and faults in the region, and Bourne & Willemse (2001) discuss fault-related opening-mode fractures at Nash Point on the northern margin of the BCB.

As a result of early lithification, carbonate rocks are particularly susceptible to brittle deformation (i.e. fracturing), which may take place at various times throughout their history. Secor (1965) pointed out that joints can form over a wide range of depths that extends from the surface to a depth of several kilometres, if ambient conditions promote tensile failure. An understanding of the impact of fractures on the properties of a rock mass (e.g. permeability, porosity and strength) is essential for workers in the oil, water, waste management, mining and rock engineering industries.

The efficient production of hydrocarbons from fractured reservoirs depends on a good understanding of the mechanisms controlling fracture development and that, where fractures are present, the permeability contrast between the fracture and the surrounding matrix can cause very heterogeneous fluid-flow patterns (Nelson 1985). Joints/fractures significantly affect the hydrological and mechanical properties of the rock mass, and the hydraulic properties of fractures may be altered by dissolution and/or mineralization – processes that may have either a positive or a negative effect on the permeability of a rock mass (Nelson 1985).

Fig. 2. Gently folded alternating limestone and shale beds, and early joints formed parallel to the fold axis around the hinge zone of Kilve anticline. The hammer at the right centre of the photograph is for scale.

Before discussing joint networks from the study area, it is important to define the terminology used in this article. A joint set is formed by a set of parallel joints. A joint network (or pattern) is formed by more than one joint set. For mutually orthogonal joints, the term 'cross-joint' (see Hodgson 1961; Gross 1993) has been proposed. It is not used in this work as it implies that the second joint set abuts the former joint set rather than cross-cutting it. Joint pattern is therefore preferred, and ranges from ladder, intermediate and grid patterns depending on the proportion of abutting to cross-cutting of later joints (see Rives *et al.* 1994).

In the following sections the geology and structural evolution of the area based on the study of the literature and the present authors' observations are outlined. Various fracture networks from around an amplified rollover anticline (the Lilstock anticline) are presented using 1D (scanline) and 2D (window sampling) data sets, and the problem of joint propagation across bedding interfaces is considered. Particular attention is paid to the effects of the interface properties and the properties of an incompetent bed separating two adjacent limestones on the propagation of joints between the two.

Geology of the study area

The Bristol Channel Basin is a rift basin initiated in the Permian–Early Triassic (Kamerling 1979; Van Hoorn 1987; Brooks *et al.* 1988; Nemčok *et al.* 1995; Rawnsley *et al.* 1998). Rifting continued throughout the Jurassic into the Early Cretaceous (Dart *et al.* 1995; Nemčok & Gayer 1996). The basin has a width of about 30 km, an E–W trend and stretches across southern England for 155 km. Extensive recent coastal erosion has exposed horizontal, gently, and up to 90°, dipping stratigraphic sequences on the foreshore. These excellent wave-cut platform exposures of Liassic carbonates provide up to several hundred square metres of expo-

sure of individual beds and are ideal for the study of fracture networks.

The Bristol Channel was formed in a Palaeozoic basement that consisted of Devonian sandstones and slates and Carboniferous limestones deformed during the Variscan orogeny as a result of the N–S closing of the Variscan Ocean (Dart *et al.* 1995). The oldest beds in the BCB are Permo-Triassic continental sediments. The major facies change that occurred at the end of the Triassic was the result of arid continental conditions giving way to marine conditions, as isolated continental rift basins became linked to each other and to the sea as Pangaea became progressively more fragmented. The first marine sediments deposited according to Dart *et al.* (1995) are the Westbury and Lilstock formations, which were formed in a shallow-marine lagoon or embayment. The Lilstock Formation is overlain by rocks of the Lower Lias consisting of rapidly alternating beds of grey mudstones (or grey/black bituminous shales) and light grey limestones of Lower Jurassic age (Hettangian and Sinemurian), locally known as the Blue Lias. It is the fracture patterns in these limestone beds that are the subject of this chapter.

Structures

In this section the structures observed along part of the southern shore of the BCB are described. The contact between the Triassic and Lower Lias horizons occurs along the coastline, approximately 5 km west of Lilstock (Fig. 1). Because of the differing rheological properties between the Triassic marl, which has only a poorly defined bedding fabric and is relatively homogeneous in composition, and the alternating micritic limestone beds and shales of the overlying Lower Lias, which have a marked mechanical anisotropy, the response to stress is distinctly different for these two rock types. The faults that occur in the marl occur as fault zones. In contrast, faults within the Liassic rocks are more localized and form discrete, sharp fault planes.

The earliest structure observed by the authors in the area is a conical-shaped feature approximately 5 m in diameter at its base, which occurs in the Blue Lias beds at GR ST155450. It is probably a mud volcano formed as a result of the local overpressuring of the sediments during the early stages of burial and diagenesis, and is made up of a mixture of brecciated limestone and shale.

The earliest tectonic structures observed within the micritic limestone beds are closely spaced, bed-normal (i.e. generally subvertical) microveins with a general WNW–ESE strike. They were formed as a result of regional extension, and are well developed in the area between Kilve and Lilstock. These high-density veins (Rawnsley *et al.* 1998) or microveins

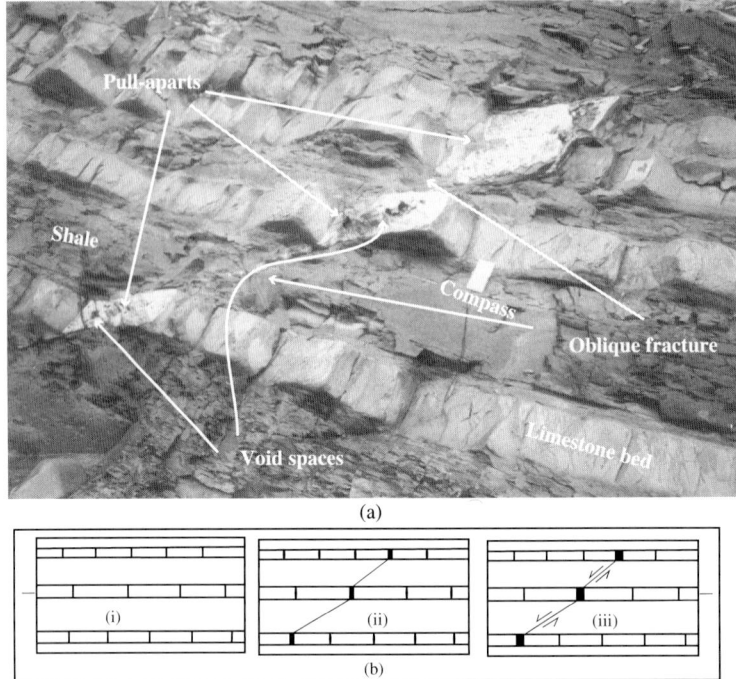

(a)

(b)

Fig. 3. (a) Pull-apart structures in southward-tilted alternating limestone and shale beds at Lilstock. Note the open cavities at the centre and lower left of the photograph with which calcite crystals have grown. These voids were formed at dilational jogs in the normal faults generated by the linking of approximately positioned originally vertical extensional fractures in the limestones, and probably indicate that the calcite grew in pre-existing open space created when the extension joints formed. The compass at the centre of the photograph is for scale. **(b)** Schematic diagram illustrating evolution of normal faults from: (i) randomly positioned joints in individual limestones; (ii) followed by linkage of individual joints by oblique fractures; and (iii) the development of normal faults and pull-apart structures.

(Caputo & Hancock 1999) form in zones 1–2 cm wide consisting of tens of microveins that taper as they approach the bedding-plane interface with the underlying and overlying shales. The height of the microveins is determined by the thickness of the limestone beds containing them. They are best exposed when erosion has exploited the contrast in properties between the microveins and the country rock. We propose that the microveins represent the earliest manifestation of a relatively diffuse extension, during which deformation took place throughout the beds before the onset of more focused displacement along discrete localized normal fault planes. The microveins are commonly oblique to, and may be cut by, veins that were formed during the major basin-opening event and also by later barren joints. Caputo & Hancock (1999) argue that they were formed by a 'crack-jump' mechanism, and inferred a cyclical build up and release of stress during their formation. When vein filling is weaker than the surrounding rock, then the next rupture occurs either within the vein or along the rock–vein interface and results in vein growth by the crack–

seal mechanism (Ramsay, 1980). Alternatively, if the vein material is stronger than the country rock, a new and independent fracture will be initiated within previously intact country rock. The term 'crack-jump' mechanism has been suggested by Caputo & Hancock (1999) for this latter process.

These layer-normal extensional veins nucleated in and were contained by individual limestone beds. Their present dip depends on the dip of the bedding, which on the moderately dipping southern limb of the Lilstock anticline is between 30° and 35° to the SSW (190°–195°, Fig. 3). As basin opening continued, further extension resulted in the oblique linkage of appropriately positioned extensional joints by oblique 'shear' fractures that developed in the shales. Progressive extension caused slip on these composite, stepped fractures that acted as normal faults resulting in the formation of the openings (pull-apart voids) in the limestone beds (Fig. 3a, b) (see Peacock & Sanderson 1992). The formation of a shear fracture by the linking of appropriately positioned extensional fractures has also been observed on a microscopic scale (see, e.g.,

Bieniawski 1967). When the bedding planes shown in Figure 3a are restored to the horizontal, the veins in the limestone beds become subvertical and the average dip of the composite fracture becomes approximately 60°, i.e. that of a classical normal fault.

As can be seen from Figure 1 the major structure in the area is the Lilstock anticline. It is situated on the southern margin of the BCB and the fold axis trends ESE–WNW, which coincides with the strike of N-dipping major basin-bounding faults. The fold has a long, gently dipping back limb (N-dipping) and a shorter moderately dipping southern limb. Rawnsley *et al.* (1998) argue that the fold relates to wall-rock deformation around an E–W striking normal fault (i.e. it is a rollover fold), formed during basin opening, that was tightened during the N–S Alpine compression. The present authors hold a similar view, arguing that if the Lilstock anticline was a simple rollover fold its northern limb would be subhorizontal. Its present gentle dip to the north is thought to represent the effect of fold amplification during basin inversion. The more steeply dipping southern limb dips towards the normal fault with which the fold is associated.

An identical fold occurs at Kilve, approximately 2.5 km west of Listock (Fig. 2). The Kilve anticline is one of the best-exposed folds in the area. The fold is gentle with a subhorizontal axis trending 100°, and limbs that dip to the NNE and SSW. It has a wavelength of approximately 50 m and a strike length of over 80 m. A set of joints run parallel to the fold axis and, as is discussed later, this is thought by the present authors to be related to the Alpine inversion of the basin. The fold has been displaced approximately 30 m by a 040°-striking sinistral, strike-slip fault.

A three-fold classification of the joints in the area has been proposed (Belayneh 2003). These are burial joints related to basin opening and subsidence, Alpine joints formed as a result of basin inversion during the Alpine orogeny and exhumation joints. The burial joints are mostly vein filled and the present authors have not made a systematic study of these structures. The post-Alpine joints were formed during uplift and erosion of the sediments in the basin and are short, linking longer joints formed during the Alpine inversion of the BCB and are relatively easy to identify (Belayneh 2003). The majority of the systematic joints in the study area are considered by the present authors to be Alpine in age and formed approximately parallel to the basin margin, i.e. in a general ESE–WNW direction. It is the temporal and spatial variation of these joint sets that form the main subject of this chapter. The authors have attempted to establish the chronology of these joint sets and to examine how the fracture networks that result from their superposition vary in

a bed at different positions around a fold and in adjacent beds at the same location. There is some controversy regarding the age of the various joint sets in the limestones of the study area. Rawnsley *et al.* (1998) have suggested that they are related to post-Alpine uplift and exhumation, whereas Engelder & Peacock (2001) provide evidence that at least one of the dominant joint sets (J_3 in block 4 of fig. 3 of Engelder & Peacock 2001) formed as a result of the amplification of the Lilstock anticline by flexural-flow folding during basin inversion. J_3 joints strike 285°–295°, i.e. subparallel to the Lilstock anticline, and abutting and cross-cutting relationships show that these joints post-date J_2 and J_1 (this work). The present authors agree with Engelder & Peacock (2001) that the systematic joints are linked to the Alpine inversion of the Bristol Channel, but, as is discussed later in this chapter, argue for a different mechanism of folding.

In addition to the reverse reactivation of the early normal faults during basin inversion related to the N–S Alpine compression of the BCB, NNW–SSE-striking dextral and NNE-SSW-striking sinistral strike-slip faults were also formed. They form a conjugate pair about an approximately N–S compression (Peacock & Sanderson 1995), and it has been suggested that these faults may be linked to reactivation during the Mesozoic and Tertiary of Variscan transfer faults with similar trend within the basement (Van Hoorn 1987; Brooks *et al.* 1988; Chadwick 1993).

Methods

Scanline and window sampling were carried out on the limestone beds in order to compare joint orientations and patterns at different locations around a fold and at the same location in adjacent limestone beds. Scanlines were used to examine the orientation of joints encountered along a particular direction, and these results were compared with similar studies at other locations around the folds and in the overlying and/or underlying limestone beds. The results were plotted either on stereographic projections or graphs. Window sampling involved careful mapping and photographing of areas of limestone bedding planes marked with a grid. A grid spacing of about 1 m was chosen. The photographs were then rectified using in-house software to enable the construction of undistorted 2D joint trace maps. As with the scanline studies, the window data were used to examine the variation of fracture patterns around the fold structure within a particular bed and to compare the joint patterns in adjacent limestone beds, and, thus, to determine the continuity of joints from one carbonate bed to the next through the intervening shale.

Results

The results of the field study on the Lilstock anti-cline have been subdivided into three groups relating to work carried out on the northern limb, the hinge area and the southern limb of the fold.

Joints on the gently dipping northern limb of the Lilstock anticline

Limestone bed 1. The sampled window on bed 1 (Figs 1 and 4a) is located in the vicinity of the E–W-and ENE–WSW-striking normal fault, shown as fault 5 in Figure 1. This bed may correlate with bed 105 (Whittaker & Green 1983) or 1735 in Engelder & Peacock (2001). The fault is synthetic to the main fault, against which the rollover fold formed and probably experienced reverse reactivation during the Alpine orogeny. The bed is 10 cm thick, situated on the gently dipping, northern limb of the Lilstock anticline and has a dip/dip direction of 05°/008°. The thicknesses of the overlying and underlying shales are 90 and 19 cm, respectively. There are two joint sets in this bed: an early set (J_1) striking approximately E–W and the second set (J_2) made up of N–S-striking short joints that abut against J_1 at approximately 90° to form ladder patterns. These 'short' joints are often referred to as cross-joints (e.g. Hodgson 1961; Gross 1993).

Limestone bed 2. The thickness of this limestone bed is 16 cm (Figs 1 and 4b). It has a dip/dip direction of 12°/004°, and the thicknesses of the overlying and underlying shale beds are 32 and 23 cm, respectively. This bed correlates with bed 109 of Whittaker & Green (1983) or 1848 of Engelder & Peacock (2001). Based on abutting relationships six joint sets have been identified. The earliest set (J_1) is made up of long straight joints striking between 125° and 130°. It correlates to J_1 in Figure 6a of Engelder & Peacock (2001). The second set (J_2) strikes between 110° and 115° (set 1 in Loosveld & Franssen (1992) or J_2 in Engelder & Peacock (2001). The curving of joints from this set as they approach the earlier set is commonly observed. The third joint set (J_3) strikes between 085° and 095°, and is approximately subpar-allel to the axis of the Lilstock anticline (Figs 1 and 4b). These joint sets (J_2 and J_3) abut against the earlier joint set to form oblique ladder geometries. Joint set J_3 correlates with set 2 in Loosveld & Franssen (1992) or J_4 in Figure 6a of Engelder & Peacock (2001). The fourth set (J_4) is poorly developed, and strikes between 065° and 070°. J_4 in this bed may correlate with J_6 in figure 6a of Engelder & Peacock (2001) or set 3–4 in Loosveld & Franssen (1992). The fifth set (J_5) strikes between 335° and 345°, and the sixth set (J_6) strikes approximately N–S (360° ± 10°).

Limestone bed 3. Bed 3 has a thickness of 10 cm and a dip/dip direction of 08°/360° (Figs 1 and 4c, d). The thicknesses of the overlying and underlying shale beds are 15 and 32cm, respectively. The under-lying carbonate bed is bed 1848 of Engelder & Peacock (2001). The joint networks shown (Fig. 4c, d) are from two localities on the same bed. Based on abutting relationships, four joint sets have been iden-tified. The first set (J_1) strikes NW–SE, the second set (J_2) strikes N–S, the third set (J_3) strikes NW–SE and the fourth set (J_4) is made up of short, N-S strik-ing joints. Non-systematic joints are also commonly observed in this bed.

Limestone bed 4. Bed 4 (Figs 1 and 4e, f) is the uppermost exposed limestone bed on the gently dipping northern limb of the Lilstock anticline and has a thickness of 15 cm. The underlying shale is 8 cm thick and the overlying shale is not exposed. This bed correlates with bed 114 of Whittaker & Green (1983) or 1921 in Engelder & Peacock (2001). Based on joint interactions, three joint sets have been identified. The strike of the first and second joint sets is 320° and 300°, respectively. The first set consists of long straight joints with wide apertures. The second set curves into parallelism with the first set as the joints approach the earlier joints (Fig. 4e). The last joints to form in this bed are short joints with variable orientations.

Measurements made along three scanlines with lengths of 92 m trending 100° on bed 2, 51.5 m trending 075° on bed 3 and 55.22 m trending 070° on bed 4 are summarized in Figure 5 and Table 1. As can be seen, the joint orientations in the four beds are different, indicating that the limestone beds were mechanically decoupled during joint formation.

Joints around the hinge zone

Alpine joints in the hinge region of the Lilstock fold (i.e. block 4 of fig. 3 & J_3 in fig. 7b of Engelder & Peacock 2001) are subparallel to the fold axis and the variation in joint pattern between adjacent limestone beds is less than on the limbs. The most common joint pattern is a ladder pattern (formed by an initial set of ENE–WSW-striking, long, subparallel joints and non-cross-cutting approximately N–S-striking joints of the second set). A general observation at Lilstock is that, although the thin limestone beds gen-erally show a smaller joint spacing than thicker beds, the variation is not systematic.

Another hinge zone that was studied crops out at Kilve, approximately 2.5 km west of Lilstock (ST 152448), where a gentle anticline in alternating lime-stone and shale beds is exposed. It has a horizontal fold axis trending 110° and a wavelength of over 50 m. The exposed strike length is more than 80 m. The

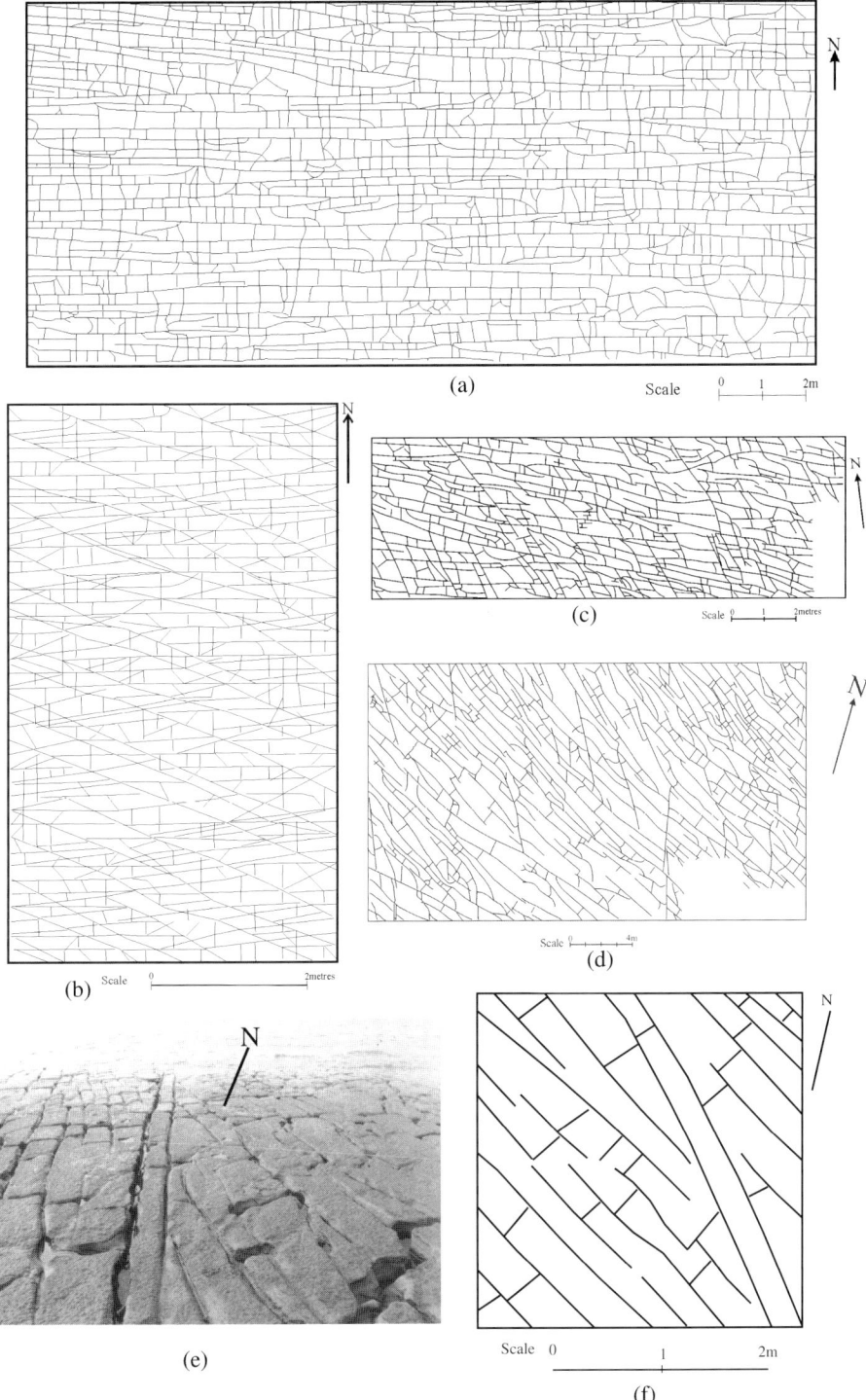

Fig. 4. Fracture patterns mapped on the gently dipping northern limb of the Lilstock anticline: (**a**) joint pattern in the lower bed (bed 1, Fig. 1); (**b**) joint pattern bed 2; (**c**) and (**d**) joint patterns in bed 3; (**e**) photograph showing joint pattern in bed 4; and (**f**) joint pattern mapped from the uppermost bed (bed 4 in Fig. 1).

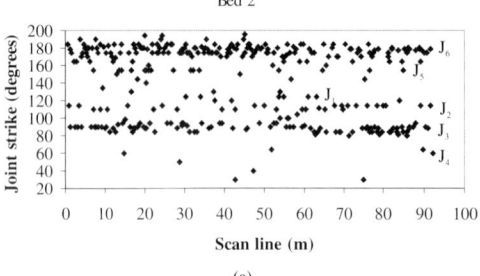

(a)

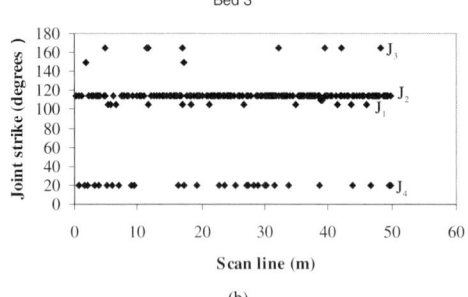

(b)

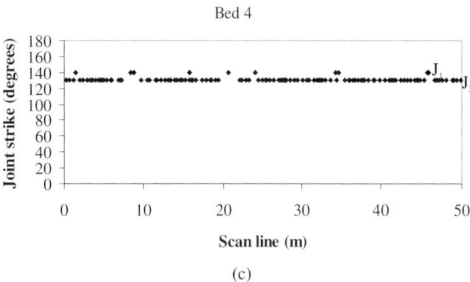

(c)

Fig. 5. Strike of joints along the scanline plotted against the three scanlines for beds 2, 3 and 4. The first subscript denotes that joint set number and the second represents bed number, respectively.

Table 1. *Summary of scanline and window sampling of joints on the northern limb of the Lilstock anticline*

Bed no.	Bed thickness (cm)	Joint set	Joint strike	Remark (age)
1	10	J_1	090°	Alpine
		J_2	N–S	Post-Alpine
2	16	J_1	125°–130°	Alpine
		J_2	110°–115°	Alpine
		J_3	085°–095°	Alpine
		J_4	065°–070°	Alpine
		J_5	N–S	Post-Alpine
		J_6	335°–345°	Post-Alpine
3	10	J_1	105°	Alpine
		J_2	165°	Alpine
		J_3	115°	Alpine
		J_4	020°	Post-Alpine
4	15	J_1	140°	Alpine
		J_2	130°	Alpine
		J_3	N–S	Post-Alpine

Joints on the moderately dipping southern limb of the Lilstock anticline

Because the exposures of bedding planes suitable for the construction of 2D joint trace maps on the southern limb of the anticline are limited, scanlines were used instead of window sampling. These beds are dipping to the south as shown in Figure 1. The scanlines were oriented approximately ESE–WNW (i.e. parallel to the strike of the beds). As was found on the northerly-dipping limb of this fold, differences in joint orientation and the geometry of the joint networks between adjacent limestone beds also occur on the short, southerly-dipping limb. The lower-hemisphere stereographic projections of systematic joint orientations from adjacent limestone beds are presented in Figure 6. As can be seen, the strike of joints in the adjacent (competent) limestone beds is highly variable.

joints in the hinge region strike approximately parallel to the fold axis and are thought to be fold related (i.e. Alpine in age). The second set is made up of short, non-cross-cutting fractures formed possibly during uplift. The fold has been displaced by a 040°-striking sinistral strike-slip fault with a horizontal displacement of about 30 m. Measurement on nine scanlines perpendicular to the fold axis were made on successive limestone beds in the hinge zone, the lowest 17 cm thick, the middle 20 cm thick and the upper 5 cm thick. The average joint spacing for the 17, 20 and 5 cm-thick limestone beds is 22, 25 and 20 cm, respectively, and show poor correlation between joint spacing and bed thicknesses.

Discussion

Some of the earliest formed fractures in the carbonate layers of the BCB are layer-normal microveins formed during basin opening when the beds were subjected to an approximately N–S layer-parallel extension. An abundance of fluid charged with $CaCO_3$ was drawn into these 'open' spaces, and the process of rupture and infilling by the crack-jump mechanism (Caputo & Hancock, 1999) produced a system of high-density veins throughout the carbonate beds. The microveins terminate at the limestone–shale contacts, which, as discussed briefly later, reflect the differences in mechanical properties of the two lithologies.

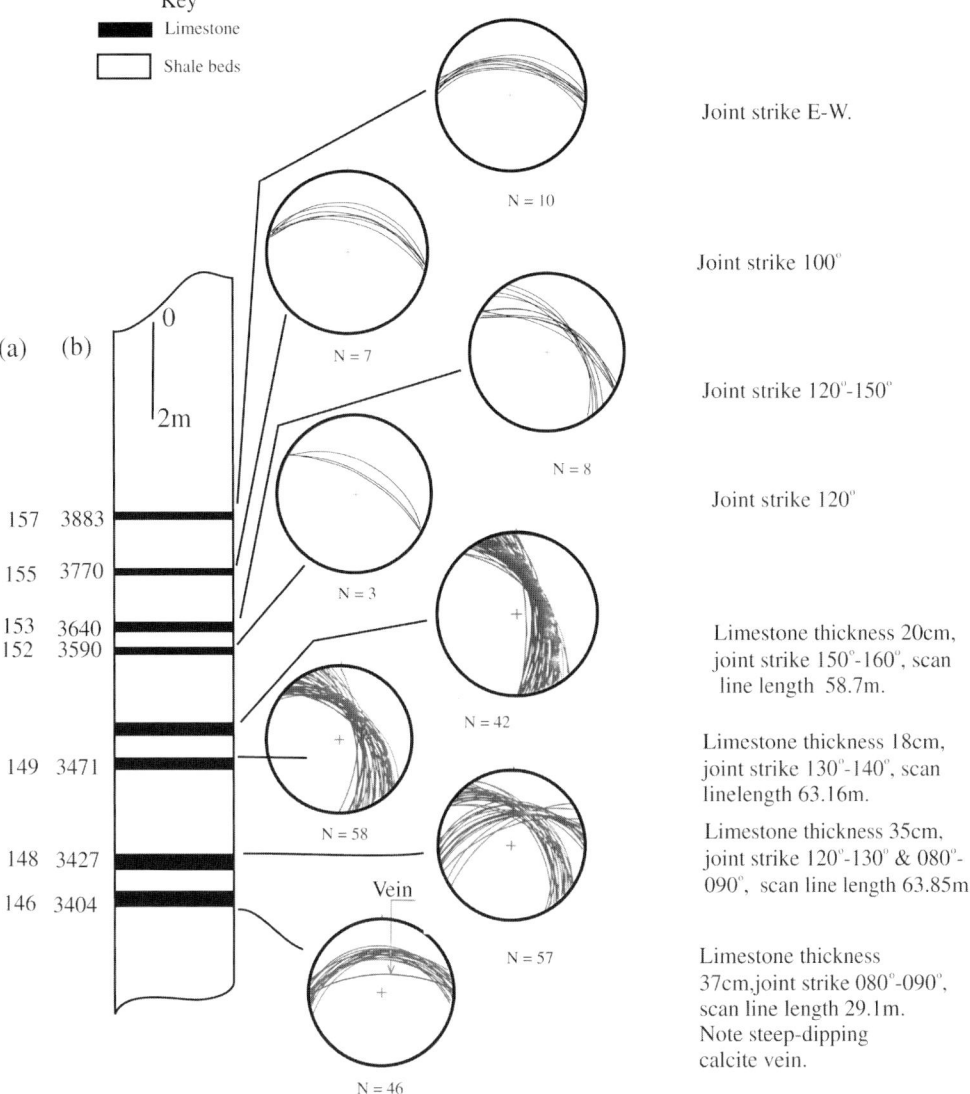

Fig. 6. Lower-hemisphere stereographic projection of the dominant joints on moderately dipping southern limb of the Lilstock anticline. It appears that each bed fractured independently of the overlying and underlying competent beds, and that there was therefore no mechanical communication between adjacent jointed beds. Numbers to the left of column in (**a**) relate to the BGS charting of these beds (Whittaker & Green 1983) and in (**b**) are designation for the beds above an arbitrary datum as given in Engelder & Peacock (2001).

As basin opening continued more vertical extension fractures formed in the carbonates some of which remained barren or only partially infilled with calcite. Some of these joints became linked to similar joints in the adjacent carbonates by the formation of oblique fractures in the intervening shales resulting in the formation of normal faults.

Another important fracture system formed during basin inversion that was initiated in the Late Cretaceous by the Alpine orogenic collision. This generated a new suite of structures that included the formation of several joint sets (the main focus of this chapter), the reverse reactivation of some of the normal faults and the amplification of the associated rollover folds, and the initiation of strike-slip faults.

Despite the relatively simple structural evolution of the region, namely basin opening, basin inversion and exhumation, detailed observations of the fracture networks on the limestone beds show clearly that fracture networks in adjacent carbonate horizons are often very different. This can be illustrated by comparing beds 1–4 in Figures 1 and 4a–f. In 'Limestone bed 1' the fracture network consists of an early joint set comprising long, E–W-striking joints and a second set that abuts against these at approximately 90° to form a ladder pattern (Fig. 4a). This joint pattern forms in a relatively unperturbed region of the fold and is similar to fractures generated experimentally during the cylindrical folding of a Perspex sheet covered with a thin brittle film (Rives & Petit, 1990). However, a progressive change in strike of the E–W-striking fractures is seen as they approach fault 5 (Fig. 1), reflecting the local stress perturbation it causes (see Rawnsley *et al.* 1992).

In contrast to bed 1, in 'Limestone bed 2' (Fig. 4b) the earliest joint set is made up of long joints striking between 125° and 130°. Most subsequent joint sets abut against this set; however, joints of the second set, J_2 (Fig. 5a), that strike between 110° and 115° are deflected into an orientation perpendicular to the early joints (J_1) as they approach them. The factors that determine whether a later joint is deflected into an orientation normal to, or parallel to, a pre-existing fracture as it approaches it have been discussed by Dyer (1988), who showed that the sense of rotation and change in the magnitude of the principal stresses near an existing fracture are functions of the orientation and the ratio of the magnitudes of the far-field principal stresses and the coefficient of friction across the joint. His work shows that younger joints curving into parallelism with older joints is indicative of a stress field in which the ratio of the maximum (σ_H^r) and the minimum (σ_h^r) principal horizontal far-field stresses lies in the range $-3 < \sigma_H^r/\sigma_h^r < -1/3$ (Fig. 4e, f). A large negative number means that σ_H^r is large and compressive, whereas σ_h^r is effectively tensile (Engelder 1993). For younger joints to curve into an orientation perpendicular to earlier joints as they approach them the ratio lies in the range $-1/3 < \sigma_H^r/\sigma_h^r < 1$. The joint pattern in this bed (Fig. 4b), which is characterized by 'oblique ladder patterns' and 'oblique ladders within ladders', is similar to that produced in the analogue experiments referred to above associated with non-cylindrical bending (Rives & Petit 1990).

Inspection of the fracture patterns in beds 1 and 2 (Fig. 4a, b) and in beds 3 and 4 (Fig. 4c, d and e, f, respectively) shows that each bed develops its own characteristic fracture network. Similarly, when the fracture patterns from different localities on the same bed are compared, e.g. Figure 4c and d, considerable differences often occur. The reasons for these differences are not clear.

Rawnsley *et al.* (1998) argued that the joints discussed above were formed as a result of post-Alpine stress release. However, based on a study on the southern limb of the Lilstock anticline, Engelder & Peacock (2001) argue that the joints were formed predominantly during the inversion of the basin and that they are linked to the amplification of the fold. They suggested that some of the structural complexity, expressed in the variation in fracture patterns, might be related to the mechanism of folding, which they considered to be by flexural-flow folding.

The strain distribution around conceptual folds, such as tangential longitudinal strain, flexural flow and flexural-slip folds, has been described by Ramsay (1967), Ramsay & Huber (1987) and Price & Cosgrove (1990). A consideration of these models indicates that flexural-flow folding would be expected to develop in a uniform, homogeneous anisotropic rock such as thick shale, and the state of strain in the bed would be homogeneous simple shear parallel to the bed boundary.

Because of the occurrence of alternating limestone beds and shales that characterize the rocks of the study area and the evidence provided by calcite slickensides, which occur at the junction of some of the limestone–shale contacts and which indicate that discrete slip has occurred along the bedding planes, Al-Mahruqi (2001) favoured the flexural-slip fold model. Here the shear is localized along discrete bedding planes and the regions between two adjacent slip planes are affected by a much lower bedding-parallel shear strain. The result is that the strain distribution within the folded sequence is that of heterogeneous simple shear. In both these models the maximum strains occur at the inflection points and increase with increase in limb dip. One would therefore expect to find the fractures concentrated on the limbs and therefore for there to be more fractures on the steeper dipping limb. However, the present authors note that in some beds (e.g. bed 2 shown in Fig. 1) it is the gently dipping limb that is the more fractured. We agree with previous workers that the majority of the systematic fractures were associated with basin inversion, but argue that the fractures are incompatible with the strain distributions described above. In contrast, in tangential longitudinal strain folding maximum strains develop at the hinge and zero strain occurs at the inflection point. The present authors argue that the fractures parallel to the fold axis that form in the hinge area of the fold are characteristic of tangential longitudinal strain folds, and that a composite fold model is indicated in which bedding-parallel slip occurs between individual layers during folding but where the individual layers fold by tangential longitudinal folding. In addition, experimental work using analogue materials and field observations of natural folds show that folds tend to have a periclinal geometry, i.e. they form elongate domes. The fracture patterns that might develop around these structures have been described by Stearns (1967) and Price & Cosgrove (1990). Although these models show a systematic variation in fracture orientation around the folds, it is still difficult to account for the fracture-pattern variation seen associated with the Lilstock anticline in terms of the variation in stress field around a pericline, regardless of the conceptual model of folding used. In addition it will be recalled that some of the subhorizontal to gently N-dipping beds (i.e. 'less strained' in this context) have more joint sets than their more steeply S-dipping counterpart, an observation incompatible with any of the fold models.

However, although the reasons for the differences in fracture patterns between adjacent layers are not clear, the mechanism by which they are permitted to develop is less problematic. In well-bedded successions, such as the alternating shales and limestone beds of the Liassic of the BCB, the weak interface between incompetent and competent beds and the drop in Young's modulus on passing from the limestone into the shale (see, e.g. Simonson *et al.* 1978) allows the adjacent limestone beds to become mechanically decoupled, and thus able to develop independent joint systems.

The problem of joint propagation across an interface in alternating competent and incompetent layers has been considered by a variety of workers including Helgeson & Aydin (1991) who worked on the turbidite sequences of the Devonian Genesee Group in the Catskill Delta Complex, Appalachians Plateau.

Their work, a combination of detailed field observation and finite-element analysis, indicates that if adjacent layers have similar mechanical properties and are not separated by a weaker layer (i.e. one with a much lower Young's modulus) joint segments confined by the layer interfaces are arranged 'in-plane' with each other, i.e. the two joint segments form part of the same plane and constitute a single composite joint. In contrast, they show that even a thin weak (inhibiting) layer between the two layers results in the formation of joints in the two layers in an 'out-of-plane' manner.

Later workers have confirmed their results. For example, Rijken & Cooke (2001), studying the Austin Chalk of Texas in an attempt to understand the propagation of fractures across an interface and through an intervening incompetent layer between two competent units, note that: (i) the majority of vertical fractures occur in chalk layers and abut against contacts with shale layers; and (ii) the amount of 'in-plane alignment' of fractures in adjacent chalk layers decreases with increase in thickness of the intervening shale layers. They concluded that both the resistance of the shale to fracturing and the thickness of shale layers act together to inhibit fracture propagation across the shale into the next chalk layer.

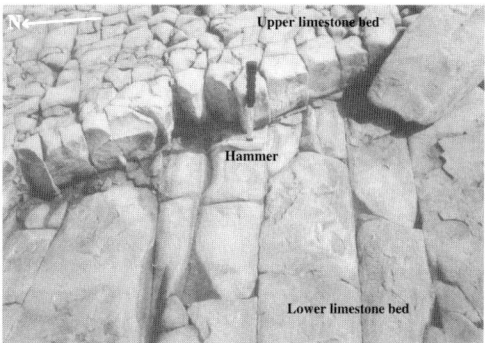

Fig. 7. A predominant Alpine joint set striking WNW–ESE, propagating either vertically up or down without significant offset, to form composite joints. The hammer at the middle of the photograph is approximately 45 cm for scale.

Al-Mahruqi (2001) investigated the propagation of fractures from a more competent to a less competent layer using finite-element analysis. He considered the effect of both welded and non-welded interfaces. When slip is allowed at the interface (non-welded contacts) the fracture is invariably arrested when it reaches the interface. In contrast, when the interface between the two layers is welded it was found that, once the fracture initiates within the competent unit, it is able to propagate across the interface into the shale unless the shale is extremely weak in which case the interface approximates mechanically to a free surface. Based on his study Al-Mahruqi (2001) concludes that the variation in joint patterns between adjacent competent beds in natural bilaminates made up of alternating competent and incompetent layers is probably related to the complex interplay between the remote stress field, local structures (folds, faults), material properties, joint interactions, bed thickness and basin history. Based on these results, it can be concluded that the lack of continuity of fractures across the limestone–shale interfaces in the study area indicates that: (i) the interface was not welded; and/or (ii) a strong contrast in material properties existed between the limestone beds and adjacent shales during the formation of the joints.

East of Lilstock, an upper 20 cm-thick limestone bed with a dip/dip direction of 07°/325° is separated from an underlying limestone by very thin shale (Fig. 7). Joints have propagated from one limestone bed to the other through the intervening shale without any significant offset. However, most of the limestone beds on the southern margin of the Bristol Channel coast are bounded top and bottom by ductile shale beds at least as thick as the limestone beds, and these inhibit the propagation of joints from one limestone to the next. Figure 8 is taken from

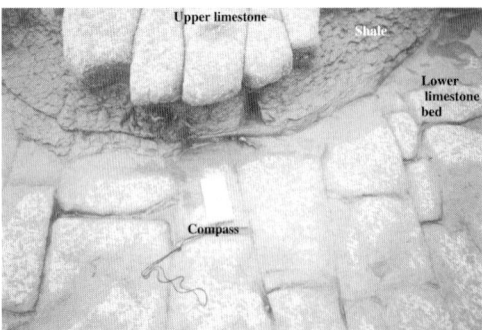

Fig. 8. Two limestone beds separated by a shale bed of approximately the same thickness. Note that some of the joints propagate from one competent bed to the next through the shale bed and others are arrested at the limestone–shale bedding interface.

Lilstock to show two limestone beds separated by shale of approximately the same thickness. It can be seen from the figure that some of the joints propagate from one limestone bed to the next through the shale without any offset and others are arrested at the limestone–shale contacts. Although a systematic variation between the thickness of the shale separating the adjacent competent limestone beds and the horizontal offset is very difficult to establish, the tendency for the horizontal offset of joints increases with increasing thickness of the intervening shale indicating the effectiveness of the shale beds in the decoupling of strain between adjacent limestones during jointing.

Conclusions

The aim of this chapter is to examine the geometry of the fracture networks developed in the Liassic carbonates of the BCB during its inversion and to quantify their variation using scanlines and window sampling. A specific structure was selected (the Lilstock anticline) and fracture networks were measured within the same bed at different positions around the fold and in different beds from the same locality. The results of the scanline measurements and window sampling show convincingly that major variations in joint patterns occur both laterally within the same bed at different positions around the structure and from bed to the next (e.g. beds 1–4 in Fig. 4, beds 2–4 in Figs 5 and 6).

The processes that allow different fracture systems to form in adjacent competent layers separated by weaker lithologies such as shale have been explored by previous workers and are reasonably well understood. As noted in the Discussion, this study and that of previous workers have shown that lack of cohesion along bedding planes and weak

interbeds both tend to inhibit fracture propagation from one competent bed to the next through the intervening incompetent layer and result in the joints that form being strata-bound. Thus, in a bilaminate of the type making up the Liassic rocks of the study area the shales act as horizons that permit the mechanical decoupling of the adjacent carbonates. However, although geologists have an understanding of the mechanical properties of a multilayer that allow decoupling of the carbonate beds and the formation of independent fracture networks in each, the reasons for these variations are less well understood.

Theoretically, if two adjacent limestone beds, separated by an intervening shale, experience the same stress history and occupy the same position on a major structure (e.g. the limb or hinge of a fold) they should develop the same fracture sets in the same order, and thus build up fracture networks with the same basic geometry and cross-cutting relationships. The individual fractures in the two beds may not align (i.e. may not be 'in plane') across the intervening shale but the orientation of the fracture sets and the geometry of the fracture network resulting from the superposition of the various fracture sets, should match. In some localities a correlation can be found, i.e. common joint sets can be recognized in adjacent limestone beds even though the fractures do not form 'in plane'. Engelder & Peacock (2001) argue for such a correlation on the steeply dipping, southern limb of the Lilstock anticline. However, it is clear from the study outlined in this chapter that the correlation is generally not apparent, as can be seen by examining the fracture patterns in Figure 4, which shows significant variations in four adjacent limestone beds (i.e. beds 1–4 in Fig. 1) on the gentle, N-dipping limb. Figure 6 shows fracture data from the steeply dipping southern limb. Although some correlation of the joint sets can be seen from bed to bed, the development of the different joint sets in each bed is significantly different.

The structural traps (faults, rollover folds) and lithologies (fractured carbonates) of the study area are small-scale examples of those typical of many fertile hydrocarbon provinces. It can be seen that even when the exposures are almost 100% and the geometry of the fracture network at a particular locality known, it is not generally possible to accurately predict the lateral variations of fractures in the same bed around a structure or the fracture patterns in adjacent competent beds. These observations have a significant impact on the problem of predicting fracture networks for regions where data are more limited either as a result of poor outcrop or because the region of interest is subsurface and the only data available are from boreholes. It is difficult to see how this difficulty of fracture network prediction can be overcome until the reasons for these large variations are more fully understood.

This work is part of PhD research of the first author at Imperial College and he would like to thank Mrs C. Thomas for financial support. This work has benefited from the discussions and field excursions during the Paul L. Hancock Memorial Meeting held at Weston-Super-Mare, UK, in August 2001. T. Engelder is thanked for a constructive review of the manuscript.

References

AL-MAHRUQI, S. A. S. 2001. *Fracture patterns and fracture propagation as a function of lithology.* PhD thesis, Imperial College, University of London.

BELAYNEH, M. 2003. *Analysis of natural fracture networks in massive and well-bedded carbonates and the impact of these networks on fluid flow in dual porosity modelling.* PhD thesis, Imperial College, University of London, UK.

BELAYNEH, M. 2004. Palaeostress orientation inferred from surface morphology of joints on the southern margin of the Bristol Channel Basin, UK. *In*: COSGROVE, J. W. & ENGELDER, T. (eds) *The Initiation, Propagation, and Arrest of Joints and Other Fractures.* Geological Society, London, Special Publications, **231**, 243–255.

BIENIAWSKI, Z. T. 1967. Mechanism of brittle failure of rock: Parts I & II. *International Journal Rock Mechanics, Mining Science and Geomechanics*, **4**, 395–423.

BLÈS, J. L. & FEUGA, B. 1986. *The Fracture of Rocks.* North Oxford Academic, Oxford.

BOURNE, S. J. & WILLEMSE, E. J. M. 2001. Elastic stress control on the pattern of tensile fracturing around a small fault network at Nash Point, UK. *Journal of Structural Geology*, **23**, 1753–1770.

BROOKS, M., TRAYNER, P. M. & TRIMBLE, T. J. 1988. Mesozoic reactivation of Variscan thrusting in the Bristol Channel area, UK. *Journal of the Geological Society, London*, **145**, 439–444.

CAPUTO, R. & HANCOCK, P. L. 1999. Crack-jump mechanism and its implications for stress cyclicity during extension fracturing. *Journal of Geodynamics*, **27**, 45–60.

CHADWICK, R. A. 1993. Aspects of basin inversion in southern Britain. *Journal of the Geological Society, London*, **150**, 311–322.

COSGROVE, J. W. 2001. Hydraulic fracturing during the formation and deformation of a basin: a factor in the dewatering of low-permeability sediments. *AAPG Bulletin*, **85**, 737–748.

DART, C. J., MCCLAY, K. & HOLLINGS, P. N. 1995. 3D analysis of inverted of extensional fault systems, southern margin of the Bristol Channel Basin, UK. *In*: BUCHANAN, J. G. & BUCHANAN, P. G. (eds), *Basin Inversion.* Geological Society, London, Special Publications, **88**, 393–413.

DAVISON, I. 1995. Fault slip evolution determined from crack–seal veins in pull-aparts and their implications for general slip models. *Journal of Structural Geology*, **17**, 1025–1034.

DYER, R. 1988. Using joint interactions to estimate palaeostress ratios. *Journal of Structural Geology*, **10**, 685–699.

ENGELDER, T. 1993. *Stress Regimes in the Lithosphere.* Princeton University Press, Princeton, NJ.

ENGELDER, T. & PEACOCK, D. C. P. 2001. Joint development normal to regional compression during flexural-flow folding: the Lilstock buttress anticline, Somerset, England. *Journal of Structural Geology*, **23**, 259–277.

GROSS, M. 1993. The origin and spacing of cross joints. *Journal of Structural Geology*, **15**, 737–751.

HELGESON, D. E. & AYDIN, A. 1991. Characteristics of joint propagation across layer interfaces in sedimentary rocks. *Journal of Structural Geology*, **13**, 897–911.

HODGSON, R. A. 1961. Classification of structures on joint surfaces. *American Journal of Science*, **259**, 493–502.

KAMERLING, P. 1979. The geology and hydrocarbon habitat of the Bristol Channel Basin. *Journal of Petroleum Geology*, **2**, 75–93.

KELLY, P. G., PEACOCK, D. C. P., SANDERSON, D. J. & MCGURK, A. C. 1999. Selective reverse-reactivation of normal faults, deformation around reverse-reactivated faults in the Mesozoic of the Somerset coast. *Journal of Structural Geology*, **21**, 493–509.

KELLY, P. G., SANDERSON, D. J. & PEACOCK, D. C. P. 1998. Linkage and evolution of conjugate strike-slip fault zones in limestones of Somerset and Northumbria. *Journal of Structural Geology*, **20**, 1447–1493.

LOOSVELD, R. J. H. & FRANSSEN, R. C. M. W. 1992. Extensional vs. shear fractures: implications for reservoir characterisation. *Society of Petroleum Engineers*, **SPE 25017**, 23–30.

NELSON, R. A. 1985. *Geologic Analysis of Naturally Fractured Reservoirs.* Gulf Publishing, Houston, TX.

NEMČOK, M. & GAYER, R. 1996. Modelling palaeostress magnitude and age in extensional basins: a case study from the Mesozoic Bristol Channel Basin, UK. *Journal of Structural Geology*, **18**, 1301–1314.

NEMČOK, M., GAYER, R. & MILIORIZOS, M. 1995. Structural analysis of the inverted Bristol Channel Basin: implications for the geometry and timing of fracture porosity. *In*: BUCHANAN, J. G. & BUCHANAN, P. G. (eds) *Basin Inversion.* Geological Society, London, Special Publications, **88**, 355–392.

PEACOCK, D. C. P. 2001. The temporal relationship between joints and faults. *Journal of Structural Geology*, **23**, 329–341.

PEACOCK, D. C. P. & SANDERSON, D. J. 1991. Displacement, segment linkage and relay ramps in normal fault zones. *Journal of Structural Geology*, **13**, 721–733.

PEACOCK, D. C. P. & SANDERSON, D. J. 1992. Effects of layering and anisotropy on fault geometry. *Journal of the Geological Society, London*, **149**, 793–802.

PEACOCK, D. C. P. & SANDERSON, D. J. 1994. Geometry and development of relay ramps in normal fault systems. *AAPG Bulletin*, **78**, 147–165.

PEACOCK, D. C. P. & SANDERSON, D. J. 1995. Strike-slip relay ramps. *Journal of Structural Geology*, **17**, 1351–1360.

PRICE, N. J. 1966. *Fault and Joint Development in Brittle and Semi-Brittle Rock.* Pergamon Press, Oxford.

PRICE, N. J. & COSGROVE, J. W. 1990. *Analysis of Geological Structures.* Cambridge University Press, Cambridge.

RAMSAY, J. G. 1967. *Folding and Fracturing in Rocks.* McGraw-Hill, New York.

RAMSAY, J. G. 1980. The crack–seal mechanism of rock deformation. *Nature*, **284**, 135–139.

RAMSAY, J. G. & HUBER, M. I. 1987. *The Techniques of Modern Structural Geology, 2: Folds and Fractures.* Academic Press, New York.

RAWNSLEY, K. D., PEACOCK, D. C. P., RIVES, T. & PETIT, J.-P. 1998. Joints in the Mesozoic sediments around the Bristol Channel Basin. *Journal of Structural Geology*, **20**, 1641–1661.

RAWNSLEY, K. D., RIVES, T., PETIT, J.-P., HENCHER, S. R. & LUMSDEN, A. C. 1992. Joint development in perturbed stress fields near faults. *Journal of Structural Geology*, **14**, 939–951.

RIJKEN, P. & COOKE, M. L. 2001. Role of shale thickness on vertical connectivity of fractures: application of crack-bridging theory to the Austin Chalk, Texas. *Tectonophysics*, **337**, 117–133.

RIVES, T. & PETIT, J.-P. 1990. Experimental study of jointing during cylindrical and non-cylindrical folding. *In*: ROSSMANITH, H. P. (ed.) *Mechanics of Jointed and Faulted Rock*. Balkema, Rotterdam, 205–211.

RIVES, T., RAWNSLEY, K. D. & PETIT, J.-P. 1994. Analogue simulation of natural orthogonal joint set formation in brittle varnish. *Journal of Structural Geology*, **16**, 419–429.

SECOR, D. T. 1965. Role of fluid pressure in jointing. *American Journal of Science*, **263**, 633–646.

SIMONSON, E. R., ABOU-SAYED, A. S. & CLIFTON, R. J. 1978. Containment of massive hydraulic fractures. *Society of Petroleum Engineers Journal*, SPE **18**, 27–32.

STEARNS, D. W. 1967. Certain aspects of fracture in naturally deformed rocks. *In*: RIECKER, R. E. (ed.) *Rock Mechanics Seminar: US Air Force Cambridge Research Laboratories, Contribution AD669375*, 97–118.

VAN HOORN, B. 1987. The south Celtic Sea/Bristol Channel Basin: origin, deformation and inversion history. *In*: ZEIGLER, P. A. (ed.) *Compressional intra-plate Deformations in the Alpine Foreland. Tectonophysics*, **137**, 309–334.

WHITTAKER, A. & GREEN, G. W. 1983. *Geology of the Country Around Weston-Super-Mare*. Memoirs of the Geological Survey of Great Britain Sheet 279, and parts of sheets 263 and 295.

WILLEMSE, E. J. M., PEACOCK, D. C. P. & AYDIN, A. 1997. Nucleation and growth of strike-slip faults in limestones from Somerset. *Journal of Structural Geology*, **19**, 1461–1477.

The index of hackle raggedness on joint fringes

DOV BAHAT[1], PETER BANKWITZ[2] & ELFRIEDE BANKWITZ[2]

[1]Department of Geological and Environmental Sciences, Ben Gurion University of the Negev,
Beer Sheva, Israel
[2]Gutenbergstraße 60, 14467 Potsdam, Germany

Abstract: In a previous laborious study we showed that considerably more new area is formed in hackle sections than in en echelon sections per given nominal area of fringe. In the present study we resort to an alternative quantitative method that would enable a relatively rapid correlation between the fracture mode of secondary fractures on joint fringes and their change in fracture morphology. We introduce a set of seven criteria that enable a geologist to place every fringe on the 'index of hackle raggedness' (IHR) along a scale between 1 and 10. This method is applied to the 'Mrákotin joint set' from a granite quarry in the Bohemian Massif of the Czech Republic. The results show an IHR range between 1.4 ± 0.7 and 6.8 ± 0.7. The gap between the top value of the scale, 10, and the value 6.8 probably suggests that the joints in this study did not form under the most dynamic conditions. Experimental results carried out on polymers show that dynamic fracture is better correlated with the dynamic stress intensity and additional fracture mechanics parameters than with fracture velocity. The demonstration that en echelon segmentation may be obtained at maximum fracture velocity, if mechanical constraints are imposed on the system by crystal anisotropy, possibly suggests that similar results may be obtained under other constraints that need to be elucidated.

Following the fractographic terminology by Bahat (1991, pp. 118–119) structures including mist, hackles, en echelon segmentation and branching occur beyond the mirror plane, on the joint fringe. These features are to be distinguished from striae (plumes) that occur within the mirror plane on the parent joint (Fig. 1a–d). The structures that form on joint fringes are summarized and compared in Table 1. Cuspate hackles often occur as substitutes for hackles on fringes of fracture surfaces that form by explosion (Bahat 1991, p. 180). This is a good criterion for the distinction between natural and man-made fractures. This terminology differs from the one used by other authors in that the term hackle is used in various applications, including 'twist hackle' that may occur on both the mirror plane and the fringe, and 'inclusion kackle', which is shown on the mirror plane (e.g. Kulander & Dean 1985, figs 2 and 7, respectively).

Recent studies have shown that some joint fringes consist exclusively of hackles (termed here 'hackle fringes'), while other fringes consist exclusively of en echelon cracks (termed here 'echelon fringes') (e.g. Bahat et al. 2002). These two fringe types constitute end-member fringes (Table 1). The general term 'fringe cracks' may represent both end members. Occasionally, end members may mix, to form a 'spectrum' on a given fringe.

The present study concerns a display of a 'spectrum' of different crack styles on fringes of adjacent joints that form a single set. This display is represented by fractographies of idealized end members of hackle cracks and en echelon cracks (Fig. 1a–d).

The explanation of the behaviour of the 'spectrum' involves four aspects: (1) the definitions of dynamic fracture v. quasi-static fracture; (2) the morphological characterization (including differences in crack areas) of the various crack styles; (3) reasons such as changes in the material behaviour or in the fracture mode (I v. III) for changes in crack area of the various crack styles; and (4) causes for changes in the fracture mode. These four aspects are interconnected. While it is quite challenging to elucidate their interrelationships, they are partly addressed in this study.

Whereas hackles form at high velocities (e.g. Kerkhof 1975) and are the 'dynamic end member' (at a velocity of $\geq 10^{-1}$ m s^{-1} in glass: Wiederhorn et al. 1974) (Fig. 1a), en echelon segmentation generally occurs at reduced velocities (e.g. Kulander et al. 1979; Müller & Dahm 2000) (at velocity of $\leq 10^{-1}$ m s^{-1} Wiederhorn et al. 1974), representing the 'quasi-static end member' in a fringe 'spectrum'. These definitions are based on crack velocities. At the end of this chapter we re-examine this issue.

The investigated fractures that we term the 'Mrákotin joint set' exhibit representatives from the two end members. Hackle fringes often show non-uniform widths compared to uniform widths exhibited by en echelon fringes (Fig. 1c) (Bahat 1998). Dynamic fracture in glass may result in repeated cycles of four zones. Initially, the primary mirror, mist and hackle, and fracture branching that initiates at the outer boundary of the hackle zone constitute the first cycle (beginning at M1 in Fig. 1d). These zones may then appear in the same sequence as

From: COSGROVE, J. W. & ENGELDER, T. (eds) 2004. *The Initiation, Propagation, and Arrest of Joints and Other Fractures.* Geological Society, London, Special Publications, **231**, 103–116. 0305-8719/04/$15 © The Geological Society of London 2004.

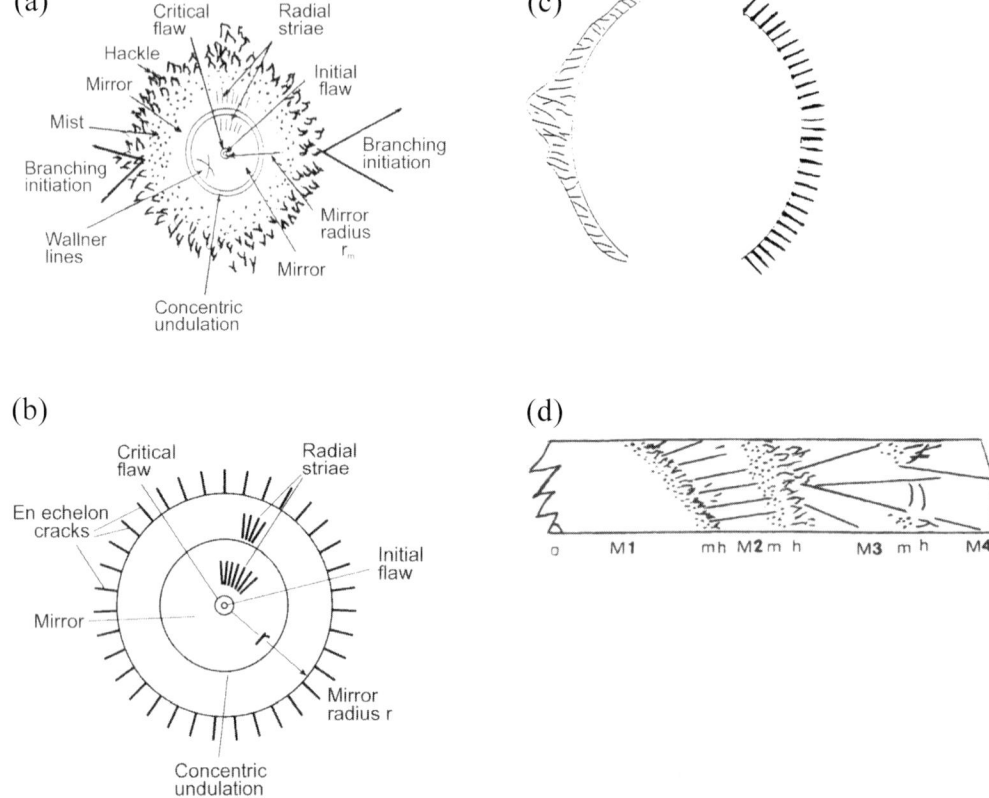

Fig. 1. (**a**) Schematic representation of a fracture surface: initial flaw (fracture origin), critical flaw, mirror plane, radial stiae, concentric undulations (ripple marks), Wallner lines, mist, hackle and the mirror radius, r_m. (**b**) Schematic representation of a fracture surface showing various fractographic elements. The mirror radius, r (arrow), is measured from the critical flaw to the inner boundary of the rim of the en echelon cracks (not scaled). (**c**) Diagram of mirrors with a hackle fringe of non-uniform width (left) and an en echelon fringe of uniform width (right). (**d**) Drawing of a piece of a fractured glass from a wall of a fractured bottle (3 mm in width), showing on the 'vertical split' and related features (Bahat *et al.* 1982) the origin, o, part of the primary mirror, M1, a thin rim of mist, m, at the mirror boundary, and a rim of hackle, h. Additional sequences occur three times, beginning with M2, M3 and M4, respectively (the fourth cycle is broken), Fracture branching initiates at the outer boundary of each mist or hackle rim, before the various M signs (further details have been given by Bahat *et al.* 2004).

secondary and tertiary cycles (beginning at M2 and M3). These cycles are caused by systematic fluctuations in crack velocities and local stress intensities, such that individual branches often start as mirrors (Bahat *et al.* 2004).

Following a geological section, the present chapter concentrates on the morphological characterization of the various crack styles on joint fringes. This is carried out by the introduction of a new classification method termed the 'index of hackle raggedness', whereby various fracture styles from the above spectrum may be indexed. In addition, several previous publications on fracture physics are cited in the Discussion, focusing on properties of hackles that were experimentally

obtained in different materials, so that the role of various fracture mechanic parameters and mechanical constraints in influencing the morphology of fringes can be shown.

Geological setting

The joint set crops out at the Mrákotin granite quarry (Bankwitz & Bankwitz 1984) in the Czech Republic part of the Bohemian Massif (Fig. 2a, b) that relates to the internal zone of the Variscan belt of Europe. The Bohemian Massif consists in its SE part predominantly of kyanite–sillimanite-bearing gneisses and schists of Late Proterozoic–Early Palaeozoic age

Table 1. *Main distinguishing features of en echelon cracks, hackle cracks and cuspate hackle cracks along joint fringes in rocks**

Feature no.	Feature	En echelon fringe cracks	Hackle fringe cracks	Cuspate hackle fringe cracks
1	General shape	Distinct planes overlapping like shingles, forming from a common centre	Cracks fan out from a common centre and combine flat planes subparallel to the parent fracture with striae and disoriented surfaces	Ridges with sharp summits fan out from a common centre
2	Crack orientation with respect to parent fracture	Uniform	Non-uniform	Approximately uniform
3	Crack orientation in a given outcrop	Uniform	Non-uniform	Non-uniform
4	Crack propagation	Subparallel to each other	Radial	Radial
5	Surface texture	Smooth	Rough	Rough
6	Neighbour steps or striae	Alternating with echelons present	Not well defined	Not developed
7	Plume markings on crack surface	Common	Occasionally	No
8	Length/width ratios	Approximately systematic	Non-systematic	Non-systematic
9	Fringe orientation with respect to parent fracture	Approximately coplanar	Inclined	Approximately coplanar

* Modified from Bahat (1991, table 3.3).

(Suk 1984; Matte *et al.* 1990). The metamorphism is of about 370–350 Ma. The Central Moldanubian Granite Pluton has post-tectonically intruded the core of an antiform composed of gneisses and schists (Benes 1971). Its exposed crest runs NNE–SSW through the Czech part of the Bohemian Massif to northern Austria down to the Danube river, over a distance of more than 170 km. The granite massif consists dominantly of two phases: the coarse-grained Eisgarn Granite and the fine-grained Mauthausen Granite. The studied Mrákotin Granite quarry (Fig. 2b) is part of the Eisgarn type. The intrusion age of the Mrákotin granite is 330 Ma, according to Matte *et al.* (1990). The Mrákotin Granite is dominated by two sets of joints: (1) cross-cutting joints, more or less E–W running; and (2) NNE–SSW-trending joints parallel to the structural crest, both of them formed before uplift processes. Our interest is focused on the NNE joints, which follow the long extension of the Central Moldanubian Pluton (Bankwitz *et al.* 2000; Bahat *et al.* 2001).

Joint exposure in the Mrákotin granite quarry

Our study relates to a joint set that is exposed in a zone of intense fracture about 30 m long and about 18 m high. The Mrákotin joint set consists of nine neighbour-distinct joints, all striking about 025° (Bahat *et al.* 2001). Joints A and B occur on one plane, joints C–H occur as a 'composite fracture' on another plane, which is further inside the rock body, and joint I forms a third plane still deeper in the rock. Spacing between the three planes varies from 5 to 20 cm, such that joint fringes partly touch neighbouring planes. Joints A–H, with the exception of C, contain mirrors that have one or two hackle fringes above and below them. Joint C, on the other hand, exposes only a fringe at its lower part that consists of en echelon segments (Fig. 3a, b).

Two large hackle fringes occur above and below the mirror of joint A (Fig. 3c). The length of the lower fringe is 9 m. The profile of joint A reveals an angular relation between the primary mirror and the outside surface of the fringe, producing the angle φ of 7° ± 3° (Fig. 3d). The upper part of the parent joint C is partially covered by joint B, but the lower boundary of the mirror of joint C is exposed (Fig. 3e). The visible part of joint C gives evidence for the propagation of only one fringe, below the mirror. This fringe consists of en echelon segments at an angular relation relative to the mirror. The upper fringe of joint F has a 'quasi-triangular' shape (Fig. 3f).

(a)

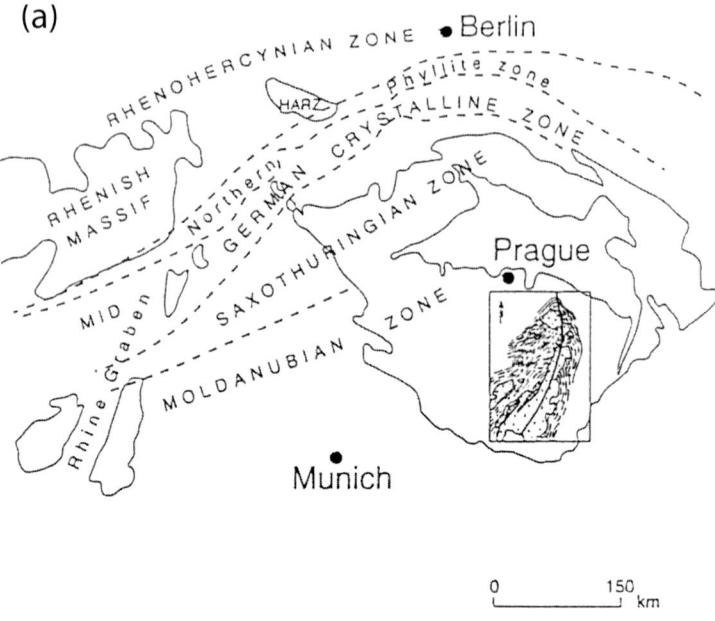

(b)

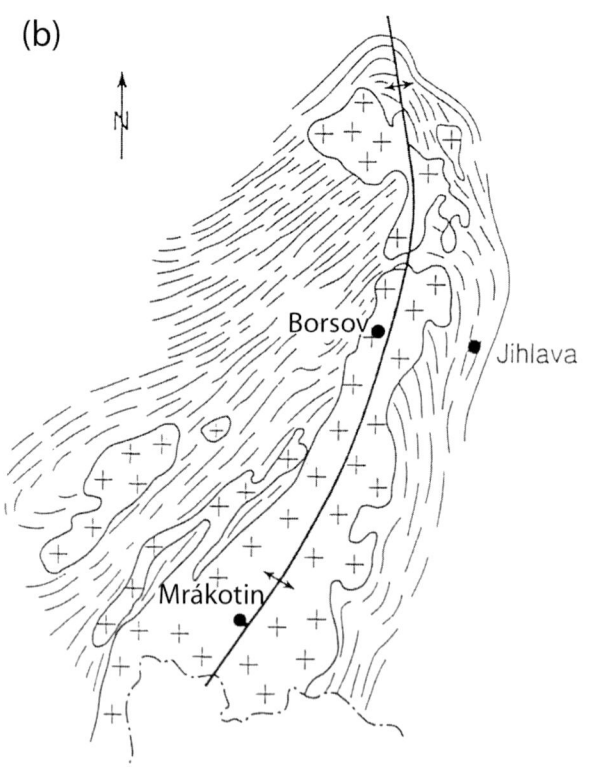

Fig. 2. (**a**) Simplified map of the central European Variscides (modified after Dallmeyer *et al.* 1995); (**b**) is framed. (**b**) A simplified geological map of the Moldanubian Pluton (marked+), the axis of the major fold and the orientation of foliation of gneiss country rock on various sides of the pluton (after Benes 1971).

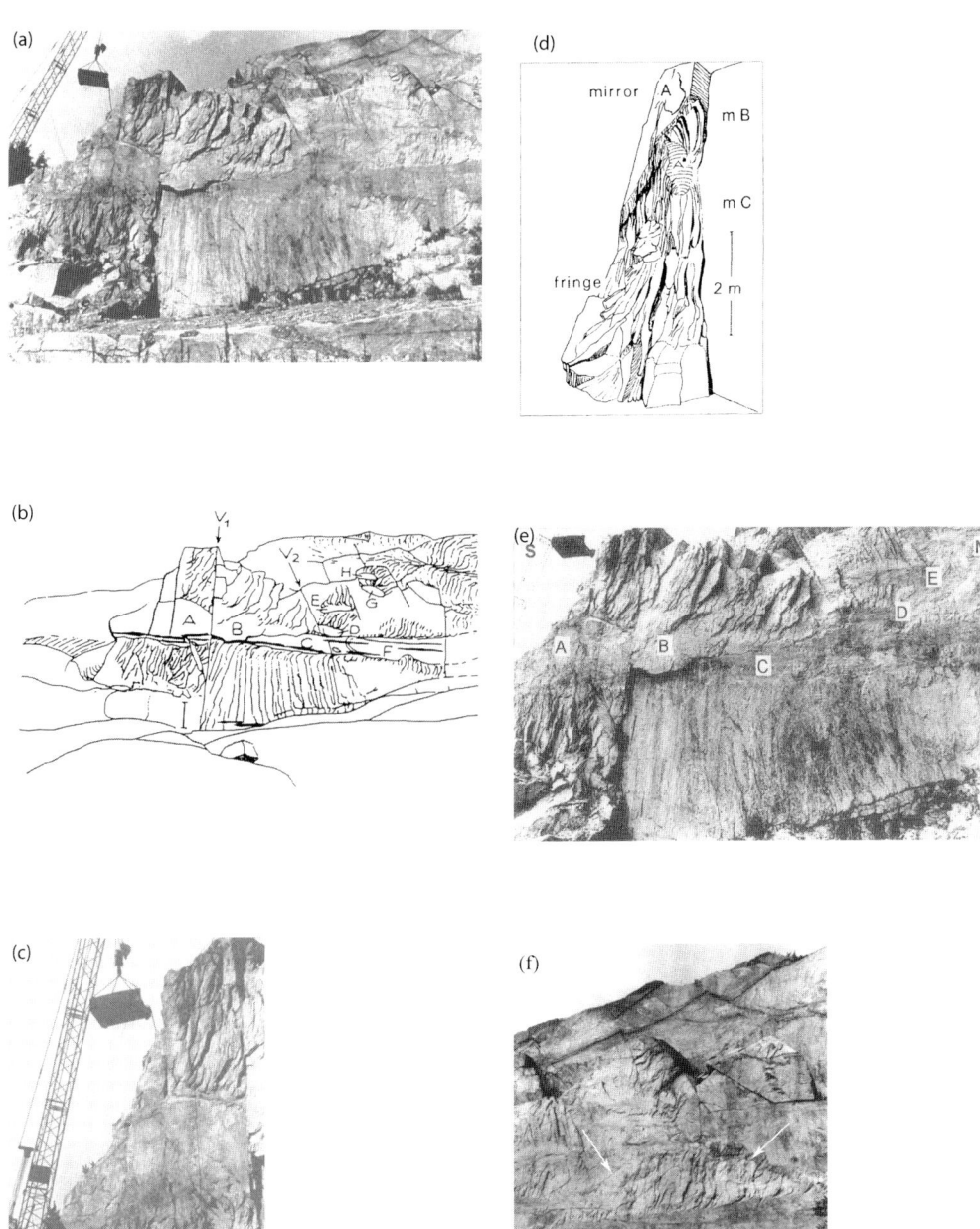

Fig. 3. (**a**) and (**b**) Photograph and diagram of the Mrákotin set of nine joints. The subvertical fracture V1 separates joints A and B, and V2 cuts joint D. The vertical scale bar is 2 m (a–f are after Bahat *et al.* 2001). (**c**) Photograph of joint A. Note an inclined secondary mirror above the 2 m scale. (**d**) Profile drawing of joint A. The vertical ruler is 2 m. (**e**) Photograph of joint C (the same scale as in c). (**f**) Joint F with its upper 'quasi-triangular' fringe at the centre of the photograph (between two legs shown by white arrows and the mirror boundary; see Fig. 5, criterion 3).

Introduction of the 'index of hackle raggedness'

We now briefly state our motivation for formulating the index of hackle raggedness (IHR). The IHR enlargement that is involved in the transition from en echelon fringe to hackle fringe generally correlates with the increase in fracture area (Bahat *et al.* 2002). Why is it important? A basic physical tendency of a cracked body is to minimize its stored 'strain energy'; that energy reflects the sum of the chemical bonds and energetic states in the material body. But the creation of a new crack area requires an investment in a different kind of energy, the 'surface energy', which relates to bond opening in the new area. Accordingly, the crack grows and its area increases if the release of strain energy equals, or is greater than, the consumed surface energy (Griffith 1920). It can be shown that often the crack growth results in an excess of energy, which is associated with an increase in the rate (per unit crack distance not per unit time) of energy release. This excess energy is expected to have various manifestations, including kinetic energy, which relates both to crack dimension and to crack velocity. However, experimental investigations show that crack velocities do not reach their ultimate theoretical (Rayleigh wave) speed. It is now thought that the reason for the limits in crack velocity relates to the development of secondary cracks that branch out from the primary crack as it propagates. Apparently, the energy that dissipates through the extra new area provided by this secondary cracking prevents the rise of crack velocity beyond certain limits (Sharon & Fineberg 1999). Fractographically, the secondary cracks appear in various shapes from hackle cracks to en echelon segments on fringes that surround mirrors of joint surfaces (Fig. 1a, b).

As the crack area is a fundamental parameter in fracture physics we seek to devise a method of correlating changes in crack area by tectonofractographic methods. The measurement and calculation of the 'new' crack area is quite cumbersome (Bahat *et al.* 2002) and therefore not always possible to accomplish. Here, we resort to an alternative quantitative method that would enable a relatively rapid correlation between the fracture mode of secondary fracture on joint fringes and the change in their fracture geometry. We use a scale of index of hackle raggedness in which index values increase along a gradual transition from en echelon fringe to hackle fringe.

Criteria of IHR and application to the Mrákotin joint set

We showed in a previous study that considerably more new area is formed in hackle sections than in en echelon sections per given nominal area of fringe

Table 2. *Summary of IHR indices for joints A, C and F*

Criterion no.	Criterion	A	C	F
1	Length variability	6	1	3
2	Rectangularity	6.5	2.5	3.5
3	Fringe width	n.m.	1.7	3.4
4	Fringe thickness	6.5	0.5	n.m.
5	Fringe angularity	5	0.5	n.m.
6	Crack dipping	10	0.5	2.5
7	Secondary mirror	n.m.	n.m.	n.m.
Sum		34	6.7	12.4
IHR (mean)		6.8	1.1	3.1

n.m., not measured.

(Bahat *et al.* 2002). Possibly, the transformation of some K_I to K_{III} (the tensile and shear stress intensities, respectively) in the transition from hackle to en echelon fringe involves an increase in heat loss, which is the portion of energy not used in the creation of new fracture surfaces in the latter fringe. Also, shear stresses promote the enlargement of the plastic zone in the material (Hahn & Rosenfield 1965; Tschegg 1983) and this enlargement is also associated with the increase in heat loss. Hence, the differences in fracture mode and in the material behaviour result in the creation of distinct crack morphologies in the two different fringes (Fig. 3c, e). Accordingly, the following IHR criteria are based on geometric distinction of crack features between the hackle and en echelon end members in the fringe.

We summarize in Table 2 the IHR properties of three joints, studying the lower fringes of joints A and C and the upper fringe of joint F (Fig. 3a–f). The key criteria in formulating the IHR are defined below and are applied to these three joints. However, before getting into the actual definition and procedure, we have to consider three limitations that qualify our rules. First, there is a need to establish which of the key seven criteria presented below is applicable to a given outcrop. Nature is quite selective in revealing good fracture markings and very often only small parts of the ideal fractography (Fig. 1a, b) are exposed. Occasionally, only one fringe is well displayed (the upper or lower). Therefore, in many outcrops only a certain number of these criteria would be useful, according to the quality of the crack exposure. Second, it is a requirement to apply independent estimations of the IHR index, ranging from 1 to 10, to each of the above criteria. Third, as the IHR index is new (even to the engineering literature), a lack of previous experience justifies us at this stage to assume an equal weight to the various criteria. Therefore, the IHR index for a given joint will be the mean of IHR values from all measurable parameters (criterion 3 is an exception, where sin α is converted to IHR index by multiplication by 10 (see below)). The total IHR error per joint, consisting mainly of measurement errors, is thought to be ±0.7.

Criterion 1. Length variability

In many fringes en echelon segments maintain uniform lengths if microcracks in root zones (Bankwitz 1966) and elsewhere on the segments are excluded. Hackles, on the other hand, show great length variabilities even among neighbouring cracks. Hence, increasing deviation of length ratios (of maximum to minimum length in a given fringe) from length uniformity (Fig. 1c) produces enlargement of IHR. The length ratios of fringe cracks for joints A, C and F are 6, 1 and 3, corresponding to the respective IHR indices.

Criterion 2. Rectangularity

En echelon segments approximate rectangular shapes whose aspect ratios can be readily measured. Hackles, on the other hand, deviate considerably from such shapes. Measurement of the deviations from straight lines (DSL) along the rectangle length and width can provide information regarding the departure of the fringe crack from its rectangular shape. We use a set of profiles (that somewhat resemble the scale of fracture-surface roughness constructed by Barton & Choubey 1977) that show a gradual DSL ranging from 0 to 10 (Fig. 4). The DSL value is the mean of combined deviations along the length and width of three fringe cracks, assessed manually. Thus, the DSL values of A, C and F are 6.5, 2.5 and 3.5, respectively.

Criterion 3. Fringe width

The tendency of en echelon fringes (to be distinguished from individual segments) to maintain uniform widths, compared to non-uniform widths of hackle fringes (Fig. 1c), has been repeatedly observed (Bahat 1998). This is also demonstrated by the 'quasi-triangle' shapes of the two fringes of joint F (Fig. 3f) compared to the approximate parallelism of the mirror and fringe boundaries at the lower side of joint C (Fig. 3e; part of the lower right side of the fringe has been removed by erosion). A measure of deviation from a uniform fringe width is given by the angle α (formed by the mirror boundary and a leg of the triangle) that increases with the 'triangularity' of the fringe. Unfortunately, this parameter cannot be measured with high confidence for joint A because only certain parts of the fringe lengths of joint A are revealed (sufficient for the study of criteria 1–5 but not for this param-

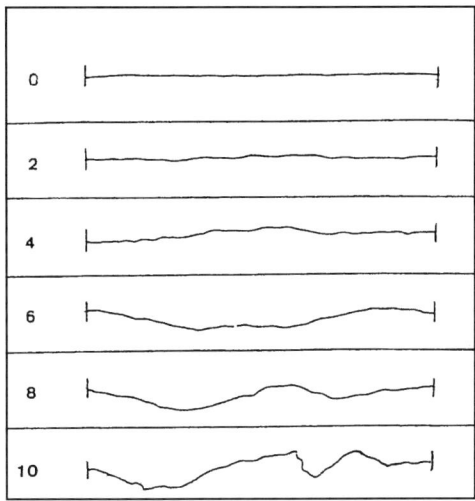

Fig. 4. A set of profiles that shows gradual increases of deviation from a straight line (DSL), ranging from 0 to 10.

eter) so that an appropriate index cannot be established for the latter joint. Accordingly, the sin α values for joints C and F were estimated to be 0.17 and 0.34 (α = 10° ± 2° and 20° ± 2°), respectively, and the corresponding IHR indices for these joints are 1.7 and 3.4.

Criterion 4. Fringe thickness

En echelon segments have approximately (quasi-two-dimensional) planar surfaces and display a uniform thickness along their lengths. However, hackles consist of multifracture surfaces superposing each other non-uniformly along the crack lengths, often leading to an increase in thickness of individual cracks and in the fringe thickness (FT) with distance from the mirror–fringe boundary. Therefore, FT is normally at a maximum at the far end of the fringe and at a minimum close to the mirror boundary. FT is abruptly reduced when fracture branching occurs, but there is no clear evidence that such a process actually took place in this outcrop. The change in FT often renders the hackle fringe to look quasi-three-dimensional (Fig. 3d), such that IHR increases with the ratio of maximum/minimum FT. Changes in thickness were measured on an enlarged profile of a photograph of joint A. These ratios for A and C were found to be 6.5 and 1 ± 1, respectively (no measurements for joint F are available).

Criterion 5. Fringe angularity

There are indications that the angle φ that the fringe (as a set of cracks, rather than an individual one) forms with the mirror increases with the 'fracture dynamics' of the process, i.e. roughly corresponding to the increase in fracture velocity. Bahat & Rabinovitch (2000) found that in the same rock (chalk) the ϕ varies from $0°$ to $30° \pm 5°$ on a natural joint, and from $30° \pm 8°$ to $50° \pm 8°$ on an artificial fracture along a roadcut (that was formed by explosion). This relationship may, however, change significantly from rock to rock (further details have been given by Bankwitz & Bankwitz 2004). Even so, we assume that such a relationship does occur in granites (maintaining similar petrographic properties). It was found that $\phi = 5° \pm 2°$ for joint A, while $\phi = 0° - 1°$ for joint C (it could not be measured for joint F); we therefore assign IHR values of 5 and 1 ± 1 to joints A and C, respectively.

Criterion 6. Crack dipping

Deviation of individual fringe cracks from a uniform dip also increases the IHR. Individual en echelon segments in a fringe commonly dip in the same direction as the mirror, i.e. the 'normal direction', forming zero to small $+\phi$ angles. However, occasionally, different hackle flakes may also dip into the mirror plane, i.e. in the 'opposite direction', forming $-\phi$. For determining the IHR value we combine the maxima of $+\phi$ with those of $-\phi$ for different cracks on the same fringe to give a single angle ϕ_f representing a fringe. Our results show that the ϕ_f values for joints A, C and F are $10° \pm 5°$ ($+\phi = 7° \pm 3°$ plus $-\phi = 3° \pm 2°$), *c.* $0.5°$ and $2°-3°$, respectively. Accordingly, the corresponding IHR indices are 10, 1 and 2.5.

Criterion 7. Secondary mirror

Gross deviations from 'normal fringe morphologies' require special attention. They are very rare on en echelon fringes and rare on hackle fringes. Quite intriguing is the the rarity of 'secondary mirrors' in natural exposures (Fig. 3c, d) compared to their frequent appearance in dynamically fractured glass (Fig. 1d). It is presently impossible to construct a scale for secondary mirrors. Quantification of this

parameter remains a challenge for future study. Qualitatively, however, this distinction supports the relatively high IHR value assigned to joint A that exhibits a secondary mirror.

Discussion

The seven criteria

Criteria 1–6 are summarized diagramatically in Figure 5. Note that criteria 1–3 relate to changes observed parallel to the mirror, while criteria 4–6 concern parameters perpendicular to the mirror. Criterion 7 requires measurements both parallel and perpendicular to the mirror. We did not use deviations of fringe-cracks from a uniform strike as an IHR criterion, because we unexpectedly found it not to be diagnostic. The gap between the top value of the scale, 10, and the value of 6.8 found for joint A probably suggests that joint A was not formed by the most dynamic conditions, and that the 10 IHR value would be obtained under more extreme dynamic conditions, such as intense fracture of glass in the laboratory (e.g. Bahat *et al.* 1982, fig. 2).

Hackle formation in glass

Previous views about the mist and hackle zones maintain that they are 'identical in appearance but different in scale' (e.g. Mecholsky 1991). More recently, however, Rabinovitch *et al.* (2000a) suggested a fundamental difference between these zones, as explained below. The secondary cracks (SC) start to grow from an 'existing flaw' (Lawn 1993, p. 13) when a critical stress intensity is attained in front of the tip of the primary crack (PC) cutting the glass (Fig. 6). The SC move under a changing stress field caused by the PC, whose distance from the SC is continuously changing. Hence, the SC start from zero velocity and, asymptotically, reach v_t (the terminal, or maximum, crack velocity). In the mist zone the PC overtakes the SC almost instantly after

Fig. 5. Diagrammatic illustrations for the criteria of index of hackle raggedness, where L and W are length and width of the fringe crack, and α and ϕ are the angles of fringe triangularity and fringe angularity, respectively (the latter can also be used to calculate the crack dipping). Criterion 1. Length variability. On mirror A hackles vary considerably from short, $L=1$, to long, $L=3$, lengths, while on mirror C all lengths of en echelon segments are about the same. Criterion 2. Rectangularity. On mirror A hackles vary considerably in shape, whereas on mirror C en echelon segments are approximately rectangular. Criterion 3. Fringe width. Deviation of the fringe of joint F from width uniformity is given by sin α. Criterion 4. Fringe thickness. An approximately uniform thickness of the fringe of joint C (looking along the length of the mirror and inspecting the profile of the fringe) and an increasing thickness of the fringe of joint A with distance from the mirror boundary (see Fig. 3d). Criterion 5. Fringe angularity. The angle ϕ that forms between the mirror and fringe is small on joint C and large on joint A. Criterion 6. Crack dipping. The angle ϕ forms by combining the dips of hackle flakes in the 'normal' and 'opposite' directions with respect to the orientation of the mirror, forming the angle $+\phi$ $-\phi$, respectively. This angle is small on joint C and large on joint A.

Criterion 1

Mirror A

Mirror C

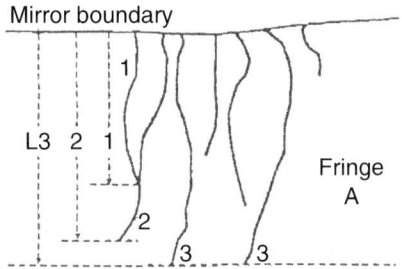

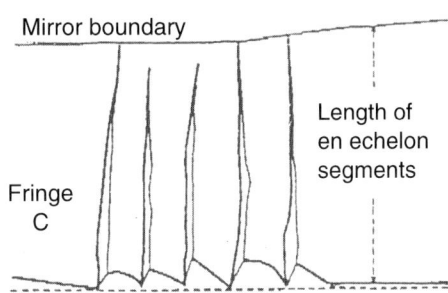

Criterion 2

Mirror A

Mirror C

Criterion 3

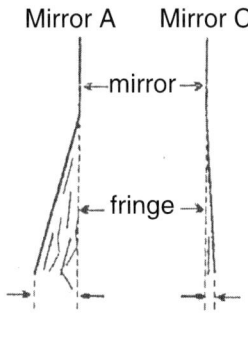

Hackle-fringe

En echelon-fringe

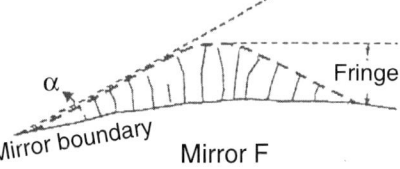

Criterion 4

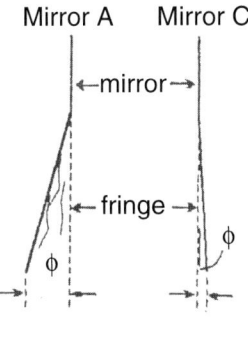

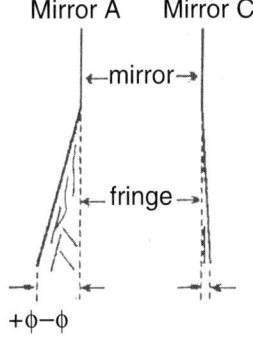

(a)

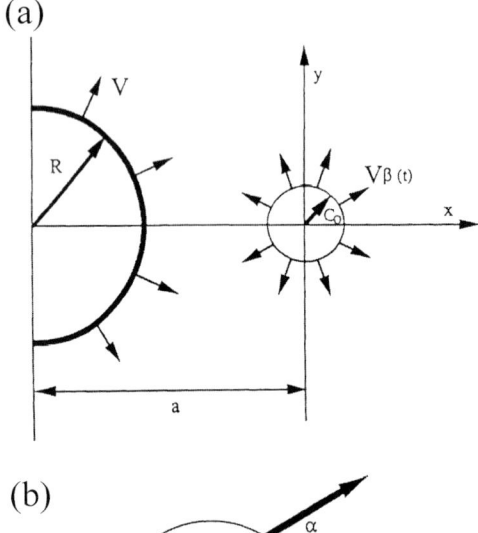

(b)

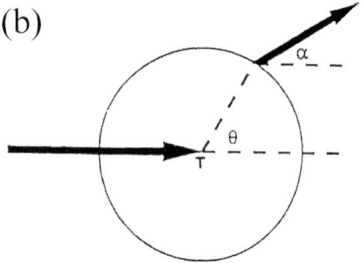

Fig. 6. (a) A schematic diagram of the geometrical set-up of the primary and secondary cracks. A secondary crack of length $2c_0$ at a distance a from the origin of the primary crack starts to grow at time $t=0$ (above). **(b)** Side view of the primary crack, T is the tip (below) (after Rabinovitch *et al.* 2000*a*).

they begin to grow (Fig. 6). Therefore, the SC in this zone are short (for details see Rabinovitch *et al.* 2000*b*). However, towards the end of the mist zone the growing SC become longer before the PC overtakes them, and they manage to attain larger sizes and rougher morphologies. Apparently, the transition to the hackle zone is related to the inability of the PC to catch up with the SC, which then grow separately, compared to those SC that are, to a large extent, 'swallowed up' by the PC and form the mist (Rabinovitch *et al.* 2000*a*). Previous authors made the (erroneous) assumption that the SC move, from their very incipience, with the same v_t of the PC, resulting in parabolic or hyperbolic shapes all through. Such shapes actually appear in the hackle zone but not before. Hence, hackle formation is associated with the distinct attainments of v_t by both the primary and the secondary cracks. This *association*, however, does not imply a *dependence* of hackle formation on v_t. It has been suggested that hackle formation is, instead, stress intensity dependent (see an extended discussion on this topic in Bahat *et al.*

1982). This dependence is, perhaps, even more intricate than the latter, as shown below.

Hackle formation in polymers

Arakawa & Takahashi (1991) compared fractographic parameters to several mechanical ones in polymers. Fractography was measured by two methods, pit density and surface roughness λ. They scanned the fracture surfaces by a scanning instrument at an interval of 0.5 mm with a needle (10 μm in tip radius) in the direction perpendicular to the direction of crack propagation, and defined surface roughness as:

root mean square roughness (RMS):
$$[1/L \int_0^L |f(x)|^2 dx]^{1/2} \qquad (1)$$

where L is the scanned length and $f(x)$ is the roughness height at the point x. A good correlation was found between changes in pit density and surface roughness. Arakawa & Takahashi (1991) employed the shadow optical method of caustics to evaluate K_d and K_c, the dynamic tensile stress intensity factor during crack propagation and the tensile stress intensity factor for arresting crack, respectively, where a Cranz–Shardin-type high-speed camera was used. K_d was evaluated by:

$$K_d = (2\sqrt{2\Pi}/3Z_0 dc \; \eta^{3/2}) \, (\phi/3.17)^{5/2} \qquad (2)$$

where ϕ is the caustic diameter, Z_0 is a distance between the specimen and the image plane, d is the specimen thickness, c is stress optical constant for dynamic conditions and η is a convergency factor for incident light rays.

Arakawa & Takahashi (1991) showed that the curve linearly correlating crack velocity, v with K_d may be divided into three regions of different slopes in Hommalite-100 (Fig. 7a). There is a gradual increase in K_d with v in the low-velocity region A, where λ remained relatively smooth. In region B there is a rapid increase in K_d along very short intervals of v and a corresponding increase in λ. The curve slope in region C exhibits an additional rapid increase in K_d and an extremely rough surface that reached its peak prior to branching. The latter slope clearly shows that branching did not occur at the peak crack velocity but at the maximum value of K_d, i.e. branching is K_d-dependent. The dotted curve connects the peak velocity points, separating the accelerating and decelerating areas in the diagram (see below).

Arakawa & Takahashi (1991) show that the maximum λ is best correlated with Gv. The physical meaning of Gv is energy per unit crack width per unit time, and it is considered to relate to the flow rate of energy into the crack tip region. A very good fit among the maxima of the curves of λ, G and Gv occurs for epoxy (araldite D), while it is shown that

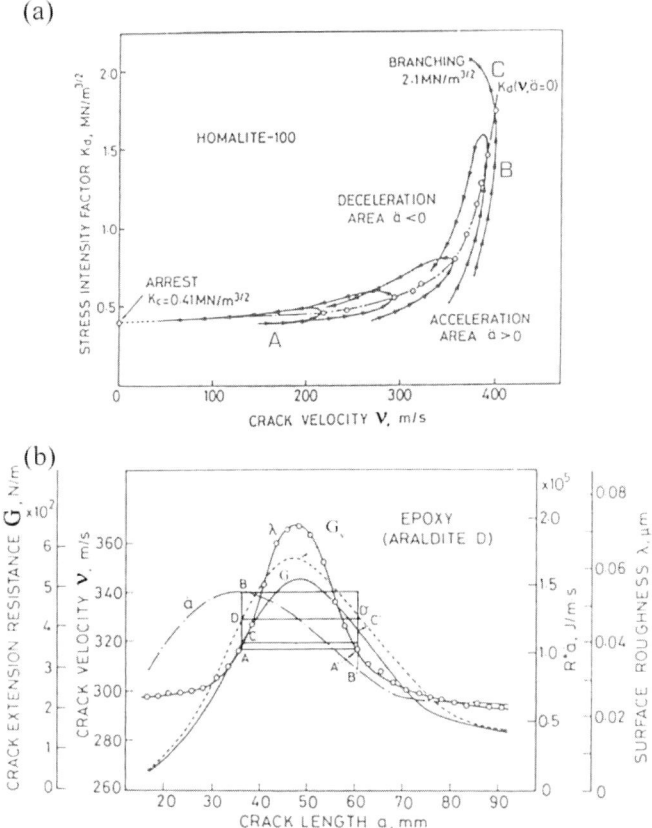

Fig. 7. (**a**) Dependence of stress intensity factor, K_d, fracture velocity, v, and fracture acceleration, $\ddot{a}$, in Homalite-100 on fracture length (after Arakawa & Takahasi 1991). (**b**) Surface roughness, λ, specific crack extension resistance, G, crack velocity, v, and the product of Gv for epoxy as a function of crack length (after Arakawa & Takahasi 1991).

crack lengths giving peak values of λ and v differ from each other (Fig. 7b). There also seems to exist a slight discrepancy between crack lengths giving peak values of λ and G: for the same λ values, indicated by points A and A$'$, the corresponding values of v (at points B and B$'$) or G (at points C and C$'$) slightly differed. However, a good agreement is shown between crack lengths giving the peak values of λ and Gv, and the values of Gv are almost equal at points D and D$'$, which correspond to points A and A$'$, respectively. These results support the suggestion of the dependence of λ on Gv. Hence, the implication is that, in the examined polymeric materials, λ behaviour (presumably along the traverse through the mirror–mist–hackle–branching, Fig. 1a) is correlated with Gv, K_d, G, and v, in a decreasing order. The results by Arakawa & Takahashi (1991) were confirmed by Arakawa *et al.* (2000) using different experimental conditions on polymethylmethacrylate (PMMA). These studies demonstrate the *association* of v with the formation of hackle in polymeric mate-

rials, showing, however, the subordinate *dependence* of this morphology on v. On the other hand, these results indicate the important influence that Gv and K_d have on the formation of hackles.

Continuum mechanics and hackle formation in silicon single crystal

We quote below Cramer *et al.* (2000), who explained the role of the term G (after Griffith) in fracture. An external load acting on a precracked body exerts a driving force on the crack. This driving force is equivalent to the mechanical energy release per unit crack advance, the energy release rate G. In an ideally brittle material crack extension takes place if the crack is supplied with a driving force larger than the specific energy 2γ required to create the fracture surfaces. This threshold driving force is called the critical energy release rate G_c. The driving force is generally a function of the geometry of the body, the external

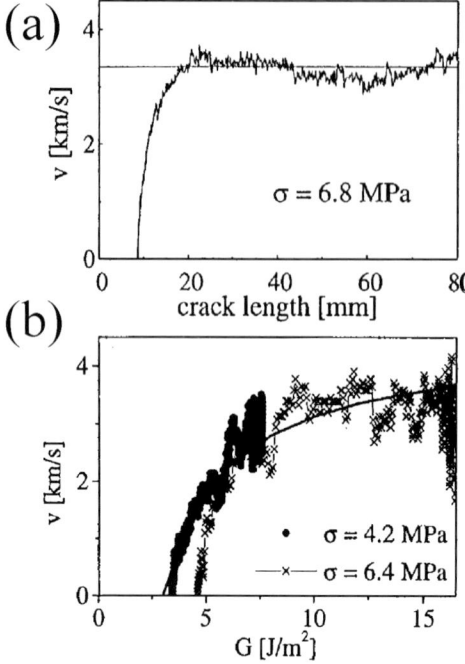

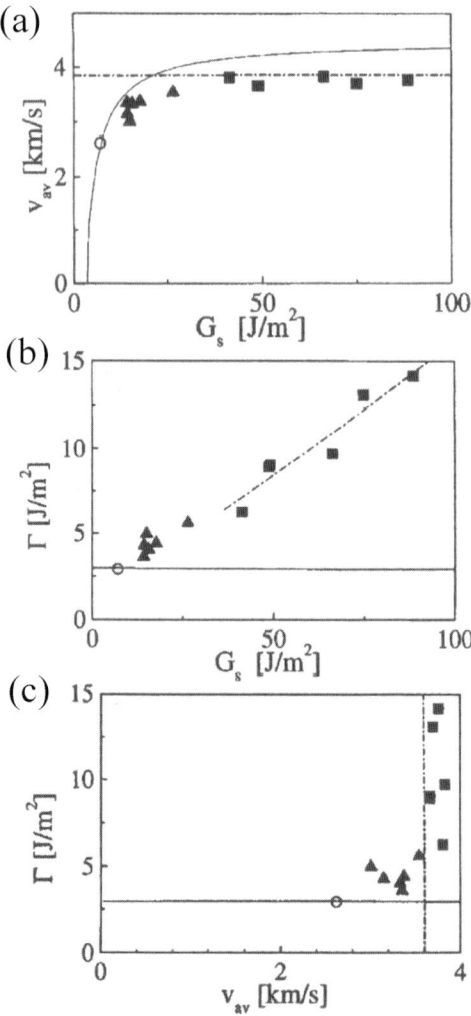

Fig. 8. (a) Dependence of measured instantaneous fracture verlocity on crack length. (b) Dependence of measured fracture velocity, *v*, on *G* is compared to the theoretical prediction from equation (3) for two specimens fractured at relatively low applied stress (after Cramer *et al.* 2000).

Fig. 9. (a) Dependence of the average fracture velocity, v_{av}, on the steady-state energy release rate, G_s. At the lowest G_s (open circle) the fracture surface is smooth. A faceted fracture surface is observed at higher G_s (triangles). The fracture surface is very rough at the highest G_s (squares). The solid line follows equation (3) using $\Gamma = 3$ Jm^{-2}. (b) Equation (3) is used to determine Γ (b) as a function of G_s, and (c) as a function of v_{av} (after Cramer *et al.* 2000).

forces, and the shape, size and orientation of the examined crack. Hence, the driving force may change and rise to levels well above G_c as the crack length *l* increases during propagation, raising the question, how is the surplus of supplied energy 'spent'?

According to continuum mechanical considerations, a straight crack is expected to attain increasingly higher velocities for increasing energy release rates (Cramer *et al.* 2000). The crack velocity asymptotically reaches an upper bound that is equal to the Rayleigh wave speed v_R. For a straight crack propagating at a velocity *v* below v_R, the dynamic fracture energy $\Gamma(v)$ can be approximated (Freund 1990) as:

$$\Gamma(v) = G(l, \sigma) (1 - v/v_R) \qquad (3)$$

where the energy release rate $G(l, \sigma)$ is the static, time-independent energy flux into the crack tip that represents the geometry of the specimen and the applied stress, σ, and, in the simplest case, $\Gamma(v) = 2\gamma$.

Cramer *et al.* (1999, 2000) conducted dynamic fracture experiments on silicon single crystal plates that were loaded to force a {110} cleavage crack in a <110> direction. The dependence of the crack velocity on the crack length is shown in Figure 8a for $\sigma_f = 6.8$ MPa. A short transient velocity overshoot

(predicted by Rabinovitch & Bahat 1979) is followed by propagation at almost constant velocity. The crack velocity closely follows the continuum mechanical solution $v(G)$ obtained from inversion of equation (3) for $\Gamma = 2\gamma = $ constant (110) (Fig. 8b). Cramer *et al.* (1999, 2000) plotted the average crack velocities from all their fracture experiments and found (Fig. 9a) that at the lowest steady energy release rates ($G = 7$ J m^{-2}) the measured velocities were in quantitative

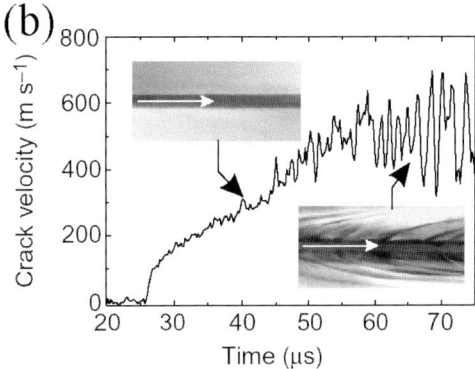

Fig. 10. (a) SEM micrograph, showing two smooth (110) cleavage planes (light) that are linked by a secondary shear fracture (dark) in the middle (after Cramer *et al.* 2000). **(b)** The velocity, v, of a crack in PMMA as a function of time. Strong oxcillations in v start at $v > v_{cbr} - 340$ m s^{-1}, resulting from microbranching instability. Insets show views of the two halves of the fractured plate: a single propagating crack for $v < v_{cbr}$ (upper left) and the main crack surrounded by subsurface microbranches for $v > v_{cbr}$ (lower right) (after Sharon & Fineberg 1999).

agreement with the continuum theory. However, at high G values, the crack velocity was lower than the continuum mechanical prediction. Apparently, the dynamic fracture energy Γ increases linearly with G (Fig. 9b) at high-energy release rates. The slope of the Γ (G) curve is given by $(v_R - v_t)/v_R$, which uniquely defines a terminal crack velocity $v_t = 3800$ m s^{-1} = 0.85 v_R. The fracture energy Γ increases at v_t due to the continued crearion of crack area (Fig. 9c).

The fractographic features vary with the changes in the average crack velocity, v_{av}, and in G. The fracture surface is smooth and mirror-like at the lowest fracture stress, corresponding to $G = 7$ J m^{-2} (circles in Fig. 9). At $G \leq 14$ J m^{-2} and above $v_{av} = 3000$ m s^{-1} = $(2/3)v_R$ (triangles in Fig. 9) Atomic Force Microscopy (AFM) reveals that the crack partially deviated from the initial (110) plane, and displays hills and valleys extending in the crack propagation direction. Facets appear on {111} planes next to the mirror with further increase in G. The size of the facets increases with increasing G. Instability, exhibited by an abrupt onset of rough hackle zones, starts at $G > 40$ J m^{-2} and at velocities close to v_t (squares in Fig. 9). The instabilities are manifested by the cracks on the {111} facets to 'overgrow' each other, manifested by the simultaneous propagation of two crack fronts and in secondary shear fracture of the material places between them (Fig. 10a).

Thus, there are three modes that show that the excess energy is spent in several ways. Initially, the excess energy contributes to the increase in crack velocity. More excess energy results in faceting. Finally, great excess energies lead to path instabilities, resulting in the propagation of multiple cracks. The path instabilities in the silicon plates bears similarities to the bifurcation instability found in PMMA by Sharon & Fineberg (1999) (Fig. 10b). The important difference, however, between the latter and the

result by Cramer *et al.* (2000) (Fig. 10a) is that the bifurcation in the isotropic amorphous PMMA occurs along various alternative fracture paths, whereas in the silicon crystal instability develops by tensile cleavage on particular planes, i.e. en echelon cracks, that are connected by shear fractures, often termed steps or bridges. Hence, under constraints of anisotropy en echelon segmentation may occur at fracture terminal velocities.

Dynamic fracture

Note that the earlier definition of dynamic fracture was based on crack velocities. The studies by Arakawa & Takahashi (1991, 2000) show that the characterization of dynamic fracture relate better to Gv, and, to a lesser extent, to K_d. However, the increases in Gv, K_d, G and v have similar trends, with some discrepancies that are not very large (Fig. 7b). Therefore, the earlier definition can remain as a reasonable approximation while these discrepancies are taken into account (note also that equation (3) is an approximation).

Formation of en echelon segmentation at terminal velocities

The results by Cramer *et al.* (2000) are quite intriguing, demonstrating that en echelon segmentation may be obtained at terminal velocities if mechanical constraints are imposed on the system. If this is so, what might be the additional constraints that would impose en echelon segmentation at high velocities? This problem requires additional studies in the future.

We thank the following institutions and colleagues for different kinds of support: one of us (P. Bankwitz) enjoyed the status of a 'Dozor Visiting Professor' during 1999 at the Ben Gurion University in Beer Sheva, for which he is most grateful, Dr R. Seltmann, London, is thanked for discussions on granites and their foliation, KAVEX (Plzen and Mrákotin) are thanked for allowing the work to be carried out in their quarries, and Dr P. Bosak and Prof. J. Ulrych, Geological Institute and Academy of Science, Prague, for assistance regarding the Czech stone industry. This paper greatly benefited from Prof. A. Rabinovitch, Prof. T. Engelder and Dr L. Savalli, who provided most useful reviews of various drafts.

References

ARAKAWA, K. & TAKAHASHI, K. 1991. Relationships between fracture parameters and fracture surface roughness of brittle polymers. *International Journal of Fracture,* **48**, 103–114.

ARAKAWA, K., MADA, T. & TAKAHASHI, K. 2000. Correlations among dynamic stress intensity factor, crack velocity and acceleration in brittle fracture. *International Journal of Fracture,* **105**, 311–320.

BAHAT, D. 1991. *Tectonofractography.* Springer, Berlin.

BAHAT, D. 1998. Quantitative tectonofractography – an appraisal. *In:* ROSSMANITH, H. P. (ed.) *Mechanics of Jointed and Faulted Rock, MJFR-3, Vienna, Austria,* Balkema, Rotterdam, 59–67.

BAHAT, D. & RABINOVITCH, A. 2000. New fractographic aspects of natural and artificial fractures in chalks, from the Upper Galilee, Israel, and experimental fracture in Perspex. *Journal of Structural Geology,* **22**, 1427–1435.

BAHAT, D., LEONARD, G. & RABINOVITCH, A. 1982. Analysis of symmetric fracture mirrors in glass bottles. *International Journal of Fracture,* **18**, 29–38.

BAHAT, D., BANKWITZ, P. & BANKWITZ, E. 2001. Joint formation in granite plutons: En echelon-hackle series on mirror fringes (Example: South Moldanubian Pluton, Czech Republic). *Zeitschrift der deutschen geologischen Gesellschaft,* **152**, 593–609.

BAHAT, D., BANKWITZ, P., BANKWITZ, E. & RABINOVITCH, E. 2002. Comparison of the new fracture areas created by the formation of en echelon and hackle fringes on joint surfaces. *Zeitschrift für Geologische Wissenschaften,* **30**, 1–12.

BAHAT, D., RABINOVITCH, A. & FRID, V. 2004. *Tensile Fracturing in Rocks.* Springer, Berlin.

BANKWITZ, P. 1966. Über Klüfte II. Die Bildung der Kluftfläche und eine Systematik ihrer Strukturen. *Geologie,* **15**, 896–941.

BANKWITZ, P. & BANKWITZ, E. 1984. Die Symmetrie von Kluftoberflächen und Ihre Nutzung für eine Paläospannungsanalyse. *Zeitschrift für Geologische Wissenschaften, Berlin,* **12**, 305–334.

BANKWITZ, P., BAHAT, D. & BANKWITZ, E. 2000. Joints in granite – knowledge 80 years after Hans Cloos [in German]. *Zeitschrift für Geologische Wissenschaften,* **28**, 87–110.

BARTON, N. R. & CHOUBEY, V. 1977. The shear strength of rock joints in theory and practice. *Rock Mechanics,* **10**, 1–54.

BENES, K. 1971. *Structure of Plutonic Bodies in the Bohemian Massif.* Upper Mantle Project (UMP), Final Geology Report. Praha, 111–119.

CRAMER, T., WANNER, A. & GUMBSCH, P. 1999. Dynamic fracture of glass and single-crystalline silicon. *Zeitschrift für Metallkunde,* **90**, 675–686.

CRAMER, T., WANNER, A. & GUMBSCH, P. 2000. Energy dissipation and path instabilities in dynamic fracture of silicon single crystals. *Physical Review Letters,* **85**, 788–791.

DALLMEYER, R . D., FRANKE, W. & WEBER, K. (eds) 1995. *Pre-Permian Geology of Central and Western Europe.* Springer, New York.

FREUND, L. B. 1990. *Dynamic Fracture Mechanics.* Cambridge University Press, New York.

GRIFFITH, A. A. 1920. The phenomena of rupture and flow in solids. *Philosophical Transactions of the Royal Society, London,* **A221**, 163–198.

HAHN, G. T. & ROSENFIELD, A. R. 1965. Local yielding and extension of a crack under plane stress. *Acta Metallica,* **13**, 293–306.

KERKHOF, F. 1975. Bruchmechanische Analyse von Schadensfällen an Gläsern. *Glastechnische Berichte,* **48**, 112–124.

KULANDER, B. R. & DEAN, S. 1985. Hackle plume geometry and joint propagation dynamics. *In:* STEPHANSSON, O. (ed.) *Fundamentals of Rock Joints. Proceedings of International Symposium on Fundamentals of Rock Joints.* Centek Publishers, Bjrkliden, Sweden, 85–94.

KULANDER, B. R., BARTON, C. C. & DEAN, S. C. 1979. The Application of Fractography to Core and Outcrop Fracture Investigations. Report to US DOE. Morgantown Energy Technology Center, **METC SP-79/3**.

LAWN, B. 1993. *Fracture in Brittle Solids,* Cambridge University Press, New York.

MATTE, P. H., MALUSKI, H., RAJLICH, P. & FRANKE, W. 1990. Terrane boundaries in the Bohemian Massif: Result of large-scale Variscan shearing. *Tectonophysics,* **177**, 151–170.

MECHOLSKY, J. J. JR. 1991. Quantitative fractography: an assessment. *In:* FRECHETTE, V. D. & VARNER, J.R. (eds) *Fractography of Glasses and Ceramics II: Ceramic Transactions, Vol. 17.* The American Ceramic Society, Westerville, OH, 413–451.

MÜLLER, G. & DAHM, T. 2000. Fracture morphology of tensile cracks and rupture velocity. *Journal of Geophysical Research,* **105**, 723–738.

RABINOVITCH, A. & BAHAT, D. 1979. Catastrophe theory: a technique for crack propagation analysis. *Journal of Applied Physics,* **50**, 231–234.

RABINOVITCH, A., BELIZOVSKY, G. & BAHAT, D. 2000*a*. Origin of mist and hackle patterns in brittle fracture. *Physical Review,* **B62**, 14968–14974.

RABINOVITCH, A., ZLOTNIKOV, R. & BAHAT, D. 2000*b*. Flaw length distribution measurement in brittle materials. *Journal of Applied Physics,* **87**, 7720–7725.

SHARON, E. & FINEBERG, J. 1999. Confirming the continuum theory of dynamic brittle fracture for fast cracks. *Nature,* **397**, 333–335.

SUK, M. (ed.) 1984. *Geological History of the Territory of the Czech Socialist Republic.* Geological Survey, Prague.

TSCHEGG, E. K. 1983. Mode 3 and mode 1 fatigue crack propagation behaviour under tortional loading. *Journal of Materials Science,* **18**, 1604–1614.

WIEDERHORN, S. M. & BOLZ, L. H. 1970. Stress corrosion and static fatigue of glass. *Journal of the American Ceramics Society,* **53**, 543–548.

WIEDERHORN, S. M., JOHNSON, H., DINESS, A. M. & HEUER, A. H. 1974. Fracture of glass in vacuum. *Journal of the American Ceramics Society,* **57**, 336–341.

Arrest and aperture variation of hydrofractures in layered reservoirs

SONJA L. BRENNER* & AGUST GUDMUNDSSON*

*Department of Earth Science, University of Bergen, Allégaten 41, 5007 Bergen, Norway
*Current address: Geoscience Centre, University of Göttingen, Department of Structural
Geology and Geodynamics, Goldschmidtstr. 3, 37077 Göttingen, Germany (e-mail:
Sonja.Brenner@geo.uni-goettingen.de)*

Abstract: Hydrofractures are extension fractures generated by internal fluid overpressure (net or driving pressure): they include dykes, veins and many joints. The growth of a hydrofracture depends primarily on the mechanical properties of the host rock and the overpressure of the hydrofracture. Field observations show that in heterogeneous and anisotropic rocks, many hydrofractures change their apertures on passing through layers with different mechanical properties. Alternatively, hydrofractures may become arrested at contacts (and other discontinuities) between layers. We present boundary-element models on hydrofracture arrest and aperture variation that focus on the effects of abrupt changes in Young's modulus in layered reservoirs. The results show that, for internal fluid pressure as the only loading, high tensile stresses concentrate in the stiff layers. When approaching a stiff layer, the hydrofracture tip becomes sharp and narrow, and would normally continue its propagation through that layer. By contrast, soft layers suppress the tensile stresses associated with the hydrofracture tip and blunt the tip itself. Without a nearby weak, subvertical discontinuity that could open up, the hydrofracture tends to become arrested on meeting with a soft layer. When fluid overpressure is the only loading, the aperture of a hydrofracture is normally larger in soft layers than that in stiff layers, which may lead to flow channelling in a layered reservoir.

Hydrofractures are defined as fractures that are generated partly or entirely by internal fluid overpressure; they are normally extension fractures (Gudmundsson *et al.* 2002). The fracture-generating fluid may be any crustal fluid that generates an open fracture, including groundwater, geothermal water, oil, gas and magma. For many hydrofractures, such as dykes, sills and mineral-filled veins, the fracture-generating fluid solidifies in the fracture subsequent to its formation. By contrast, joints are normally open fractures. Man-made hydrofractures in petroleum engineering, used to increase the permeability in reservoirs, are referred to as hydraulic fractures. In this chapter, we compare results on hydraulic-fracture propagation (Valko & Economides 1995; Charlez 1997; Yew 1997; Economides & Nolte 2000) with our models and observations of natural hydrofractures.

We suggest that the growth of a hydrofracture depends primarily on the mechanical properties of the host rock and the fluid overpressure in the hydrofracture at its time of emplacement. One of the most important mechanical properties to affect hydrofracture propagation is the host rock stiffness, that is its Young's modulus, *E*. Following the tradition in engineering rock mechanics, we refer to layers with high Young's moduli as stiff and those with low Young's moduli as soft. We define fluid overpressure (net pressure or driving pressure) as the fluid pressure exceeding the stress normal to the fracture, which for extension fractures (mode I cracks) is the maximum principal tensile stress (minimum principal compressive stress), σ_3 (cf. Gudmundsson *et al.* 2002).

Traditionally, most models of rock fractures assume the mechanical properties of the host rock to be homogeneous and isotropic to make the problem mathematically tractable. Although these models provide insights into the basic physics of the initiation, propagation and shape of hydrofractures, their applicability to a fluid reservoir consisting of heterogeneous and anisotropic rocks is limited. Indeed, our previous results (Gudmundsson & Brenner 2001) indicate that the heterogeneity and anisotropy of the mechanical properties in a layered reservoir greatly influence the arrest of hydrofractures. These results are supported by the new numerical models presented in this chapter.

In petroleum engineering experiments the aim is that the injected hydraulic fracture propagates only along the target layer (the reservoir) in order to increase the reservoir permeability. Thus, a hydraulic fracture should be confined to the target layer and be arrested at its contacts with the layers above and below. Natural hydrofractures that are confined to a single or a few layers contribute significantly less to the overall permeability of a reservoir than do fractures that propagate through many layers. This follows because only interconnected fracture systems reach the percolation threshold (Stauffer & Aharony 1994). Also, vertically confined fractures often form systems that are only interconnected

From: COSGROVE, J. W. & ENGELDER, T. (eds) 2004. *The Initiation, Propagation, and Arrest of Joints and Other Fractures.* Geological Society, London, Special Publications, **231**, 117–128. 0305-8719/04/$15 © The Geological Society of London 2004.

horizontally. By contrast, non-strata-bound (uncon-fined) veins may form extensive connected fracture systems that can transport fluids from remote sources (Gillespie *et al.* 1999). The conditions for arrest or, alternatively, propagation of natural hydrofractures are, therefore, important for understanding fluid transport in petroleum, groundwater and geothermal exploration.

For a fracture that remains open (is not filled with minerals or rock) the permeability depends much on its aperture. In particular, because of potential chan-nelling of the fluid flow along the widest parts of a fracture (Tsang & Neretnieks 1998), its aperture variation is of great importance. Hydrofracture aper-ture depends, among other parameters, on the mechanical properties of the host rocks. Thus, in a layered reservoir, even when the fluid overpressure is constant, the layering is likely to affect the size of the aperture.

In this chapter we summarize field observations on the arrest and aperture of hydrofractures to form the basis for testing the results of numerical models. We use boundary-element models to study the con-ditions for hydrofracture arrest, focusing on the effects of variation in stiffness (Young's modulus) between layers of the host rock and hydrofracture aperture variation. The effects of discontinuities (e.g. Cook & Gordon 1964; Daneshy 1978; Cooke & Underwood 2001) and stress barriers (e.g. Valko & Economides 1995; Charlez 1997; Yew 1997; Smith & Shlyapobersky 2000) have received more atten-tion than have changes in Young's moduli, which are emphasized here. We also discuss the implications of these results for permeability of, and flow chan-nelling in, fluid-filled layered reservoirs.

Hydrofracture arrest

Field observations

In relatively homogeneous and isotropic rocks, ver-tical hydrofractures often propagate to the free surface or if they became arrested then their tips taper away. In heterogeneous and anisotropic rocks, by contrast, many, and perhaps most, hydrofractures become arrested at discontinuities at various crustal depths. The term discontinuity denotes any signifi-cant mechanical break in the host rock, often of neg-ligible tensile strength (Priest 1993); its tensile strength may, however, reach or exceed that of the host rock when the discontinuity has an infill such as breccia or secondary minerals. Discontinuities at contacts between rock layers with different mechan-ical properties are of particular importance in fracture arrest (Gudmundsson & Brenner 2001). Fractures restricted to single layers are referred to as

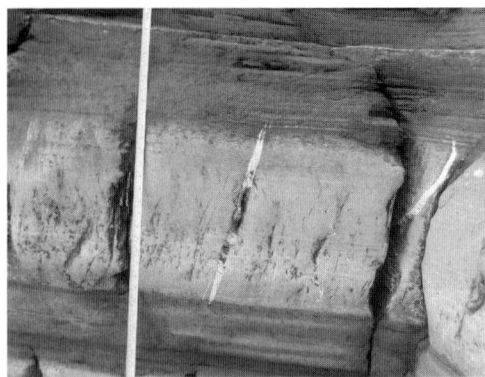

Fig. 1. Calcite veins restricted to a limestone layer, interbedded with shale, at Kilve, on the Somerset coast of the Bristol Channel, UK. These mineral veins were presumably arrested when the tips tried to propagate from the stiff limestone layer into the much softer shale layers. View ESE, the measuring steel tape is 0.7 m long.

strata- or layer-bound, whereas the term non-strata-bound is used to describe fractures where layering does not lead to fracture arrest (Odling *et al.* 1999; Gillespie *et al.* 2001).

Hydraulic fracture experiments in petroleum engineering indicate that vertically propagating tips of hydraulic fractures are commonly arrested at con-tacts between layers, particularly those contacts that separate layers with strong mechanical and stress contrasts (Valko & Economides 1995; Yew 1997; Charlez 1997; Economides & Nolte 2000). Simi-larly, our field observations show that many hydro-fractures become arrested at contacts between mechanically different rock layers (Marinoni & Gudmundsson 2000; Gudmundsson & Brenner 2001; Gudmundsson 2002).

In well-exposed outcrops in layered carbonate rocks at the Somerset coast near Kilve in SW England (cf. Peacock & Sanderson 1999), hydro-fracture arrest is commonly observed in steep sea cliffs (Figs 1 and 2). Most of the hydrofractures rep-resented as calcite veins are confined to limestone layers and terminate abruptly, some with blunt ends, where their tips attempt to enter shale layers (Fig. 1). Many joints, some of them presumably hydrofrac-tures, also end abruptly on meeting with contacts between mechanically dissimilar layers (Fig. 2). Our field observations show that there is no slip at bedding contacts associated with the arrested tips.

In igneous rocks, many dykes are arrested at con-tacts between lava flows and pyroclastic layers. This applies, for example, to many dyke tips in Tenerife (Canary Islands) and Iceland (Marinoni & Gud-mundsson 2000; Gudmundsson & Brenner 2001; Gudmundsson 2002). Often, dykes end abruptly with blunt, rounded tips at layer contacts that, never-

Fig. 2. Open joints restricted to limestone layers (between shale layers) at Kilve. Like the mineral veins at the same locality (Fig. 1), the joints tend to become arrested when their tips propagate into the softer shale layers. View south, the person provides a scale.

theless, show no evidence of slip (Gudmundsson 2003). Many arrested dyke tips have also been observed at bedding contacts in sedimentary rocks (Baer 1991). In metamorphic rocks with mechanical layering, there are commonly arrested joints and veins, for example at contacts between gneiss and amphibolite bands (Brenner & Gudmundsson 2002). These and other field observations indicate that the arrest of hydrofractures at contacts is a very common feature in mechanically layered rocks.

Numerical models

Analytical studies of theoretical tip tensile stresses of hydrofractures commonly use the mathematical crack or the elliptic hole as models. Closed-form solutions, reviewed in Sneddon & Lowengrub (1969), Valko & Economides (1995) and Maugis (2000), indicate that for homogeneous, isotropic rocks crack-tip tensile stresses are normally very high or, for mathematical cracks, infinite (Sneddon & Lowengrub 1969). Therefore, a continuous and buoyant hydrofracture in such rocks should normally continue its vertical propagation to the surface (e.g. Gudmundsson & Brenner 2001). To study the behaviour of hydrofractures in heterogeneous rocks, in which arrested hydrofractures are common, analytical solutions become too complex and numerical models must be used.

In the numerical models we focus on the effects of abrupt changes in rock stiffness (Young's modulus), such as are known to occur in layered reservoirs, on hydrofracture propagation and arrest. For example,

limestone and sandstone have normally much higher Young's moduli than shale (Bell 2000). Abrupt changes in Young's modulus often coincide with discontinuities and stress barriers (layers with high fracture-normal compressive stresses) (Gudmundsson & Brenner 2001).

We made many numerical models on hydrofractures in layered rock masses using the boundary-element program BEASY (1991). The boundary-element method (BEM), based on linear elasticity theory, is described in detail by Brebbia & Dominguez (1989), whereas specific information about BEASY is also available at www.beasy.com. The advantage of the BEM, compared with the finite-element method, is that only the boundaries of zones with certain mechanical properties are discretized into elements for calculation. It follows that in the BEM the problem dimensions are reduced by one, and that more accurate solutions for boundary problems (e.g. surface stresses) are obtained. By adding numerous internal calculation points to the BEM model very accurate results on the stress concentrations can be obtained, for example around fracture tips.

Because hydrofracture propagation is normally slow compared with the velocity of seismic waves (Valko & Economides 1995), static Young's moduli are appropriate for analysis of hydrofracture problems. The static moduli are normally 2–10 times lower than the dynamic moduli. *In situ* static moduli are as much as 1.5–5 times lower than static values from laboratory measurements (Heuze 1980), mainly because *in situ* fractures (absent from laboratory samples) lower the Young's moduli (Priest

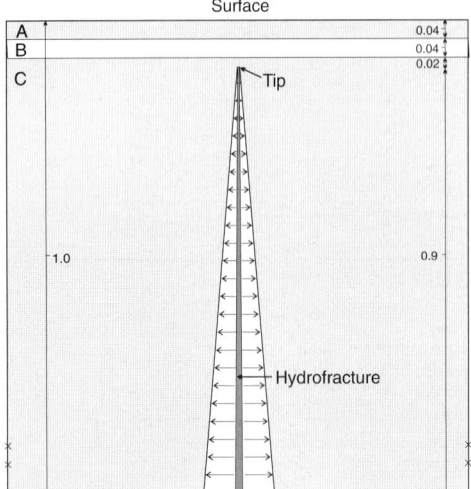

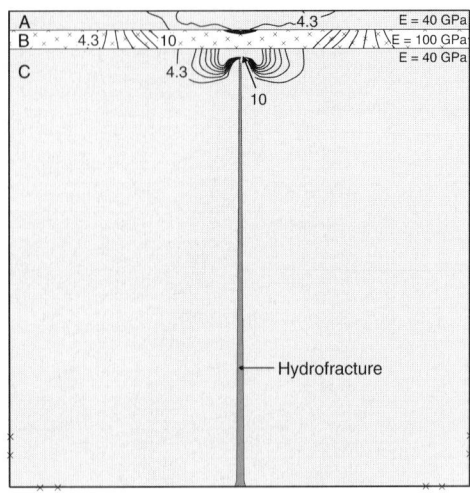

Fig. 3. Basic boundary-element configuration used for the models in Figures 4 and 5. Each model has unit dimensions, a uniform Poisson's ratio of 0.25 and is fastened in its lower corners. The fluid overpressure in the hydrofracture varies linearly from 10 MPa from the fracture bottom (centre) to 0 MPa at the tip, as indicated by horizontal arrows. The fluid overpressure is the only loading. Layer C at the bottom, hosting the hydrofracture, is 0.92 units thick and has a Young's modulus of 40 GPa. Both of the thin top layers are 0.04 units thick. In all the model runs the uppermost layer A has the same Young's modulus as layer C, whereas the stiffness of layer B varies.

Fig. 4. Tip of a hydrofracture located in the rather stiff layer C approaches a very stiff layer B with a Young's modulus E of 100 GPa near the free surface. The contours show the minimum principal compressive stress (maximum principal tensile stress), σ_3, in MPa (truncated at 1 and 10 in all the models). The stiff layer B concentrates tensile stresses that are much higher than those in the softer top layer A.

1993). In the models below we use static Young's moduli as estimated for *in situ* values.

All models are of unit length and height so that they are scale-independent. In nature, however, the hydrofracture could be of the order of tens or hundreds of metres high. To avoid rigid-body translation and rotation, the models are fastened in their lower corners. A Poisson's ratio of 0.25 is used for all the layers in all the models. We do this to emphasize the effects of Young's modulus on the hydrofractures, and also because variation in Poisson's ratio between rock layers is normally much smaller than the variation in Young's modulus (Bell 2000). Poisson's ratio can lead to a variation in horizontal stress between rocks of the order of several MPa, so the effects are much smaller than those induced by changes of Young's modulus. In accordance with our field observations, slip at layer contacts is not allowed in the models.

Each model (Figs 3–8) has a hydrofracture subject to a fluid overpressure that varies linearly from 10 MPa at the fracture centre (the bottom, as only the upper half of each hydrofracture is modelled) to 0 MPa at the fluid front. This overpressure variation is presumably appropriate for many hydrofractures,

but similar results are obtained with constant overpressure (Gudmundsson *et al.* 2002). The fluid overpressure is the only loading (applied stress) in all the models. The stress field around the hydrofracture tip is calculated for each model with the appropriate specific boundary conditions. Tip propagation can be modelled in a semi-static way with subsequent models simulating time-steps.

In the first set of models (Figs 3–5) the fluid front coincides with the fracture tip. Here, the focus is on the stress field around a hydrofracture tip approaching either a stiff or a soft layer where the fracture tip is very near to the free surface (at a distance of 0.1 units). In the second set of models (Figs 6–8) the hydrofracture propagates ahead of the fluid front along a weak vertical discontinuity and reaches a succession of layers with very different stiffnesses. Here the focus is on the shape of the hydrofracture tip, as well as the fracture-tip tensile stresses, and thus the probability of tip propagation through the layer above.

In the first model (Fig. 4) a hydrofracture in a moderately stiff layer (C) with a Young's modulus of 40 GPa approaches two thin layers of different stiffnesses. The lower layer (B) is very stiff, with a Young's modulus of 100 GPa. The surface layer A is much softer; its Young's modulus is 40 GPa (the same as that of layer C). Stiffness of 100 GPa is a common value for crystalline rocks, such as gneiss, or stiff basaltic lava flows. The layer with 40 GPa

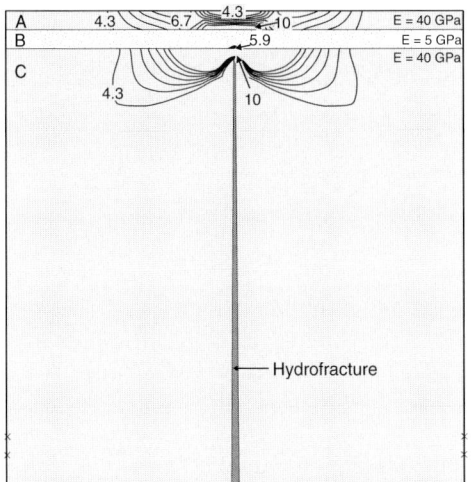

Fig. 5. Same model as in Figure 4, except that layer B is now soft (Young's modulus $E=5$ GPa). Very little tensile stress is transferred into this soft layer. Furthermore, the tensile stress concentration in the topmost layer A (with a moderate stiffness) is lower than in Figure 4.

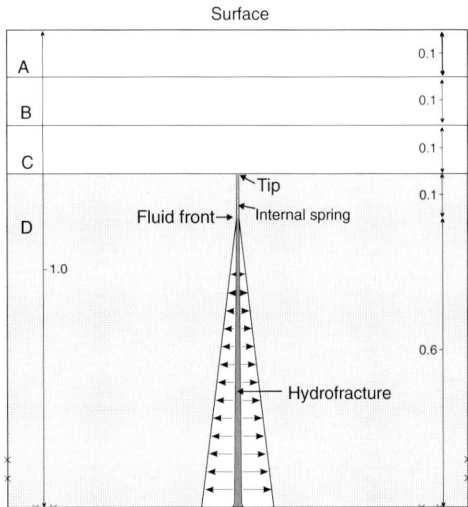

Fig. 6. Basic boundary-element configuration used for the models in Figures 7 and 8. Again, the models have unit dimensions, a uniform Poisson's ratio of 0.25 and are fastened in the lower corners. The fluid overpressure in the hydrofracture, indicated by horizontal arrows, is the only loading; here it has a linear variation from 10 MPa at the fracture bottom (centre) to 0 MPa at the fluid front, 0.4 units below the free surface. Between the fluid front and the tip there is an internal spring with a stiffness of 6 MPa m^{-1}, representing a discontinuity with a very soft but elastic infill. The spring was used to allow the hydrofracture tip to propagate upwards, meaning that the fracture tip can propagate ahead of the fluid front. The thickness of the lowermost layer D is 0.7 units and its Young's modulus of 10 GPa, is moderately stiff. The three layers above have identical thickness of 0.1 units, but Young's modulus varies between model runs.

Young's modulus could be either the same rock type, but highly fractured and with a reduced stiffness (Priest 1993), or a moderately stiff rock such as many sedimentary rocks, e.g. sandstone (Bell 2000). Associated with the hydrofracture tip there is tensile stress that concentrates in the stiff layer B next to the tip. This stress, however, falls off abruptly in the softer topmost layer A.

The results become very different when the lower layer (B) is soft ($E=5$ GPa) and the surface layer (A) is moderately stiff ($E=40$ GPa; Fig. 5). Here the lower layer (B) receives little tensile stress. The tensile stress concentration in the moderately stiff layer C is greater than in the first model (Fig. 4). In the stiff surface layer A there are comparatively high tensile stresses, but still lower than those in the stiff layer B of the first model (Fig. 4).

In the next models (Figs 6–8) we show the opening of the hydrofracture, and the deformation of the host rock in addition to the tensile stress concentration. Particular attention is paid here to the shape of the hydrofracture tip; therefore, the opening of the hydrofracture is exaggerated. The lower part of each model (layer D in Figs 6–8), in which the hydrofracture is located, has a uniform Young's modulus of 10 GPa. This value is similar to the lower range of values of many common rock types (Jumikis 1979; Bell 2000), particularly when the lowering *in situ* effects are taken into account (Heuze 1980).

The tip of a hydrofracture commonly propagates ahead of the fluid front. This is indicated by hydraulic

fracture experiments where there is normally an unwetted zone between the fluid front and the fracture tip (Advani *et al.* 1997; Yew 1997; Garagash & Detournay 2000). In the following models (Figs 6–8) the hydrofracture tip is allowed to propagate so that it reaches the top layers along a vertical internal spring with a stiffness of 6 MPa m^{-1} inside the layer hosting the hydrofracture. This spring represents a pre-existing vertical discontinuity (a joint or a fault) with a soft elastic infill such as a gouge. For comparison, weathered mudstone can have a Young's modulus as low as 3 MPa (Bell 2000).

The three top layers, A–C, used here have greater thicknesses than those in the first set of models (Figs 3–5), so that the fracture tips in Figures 7 and 8 are at 0.3 units below the free surface. The stiffnesses of the three layers were changed between model runs.

In the first model of this set (Fig. 7), the bottom layer C and the surface layer A are very stiff, with a Young's modulus of 100 GPa, whereas layer B is

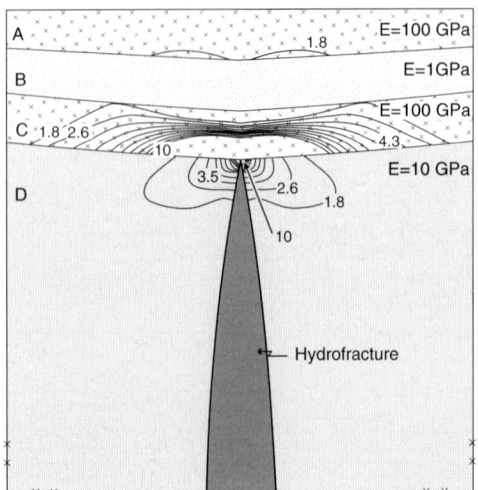

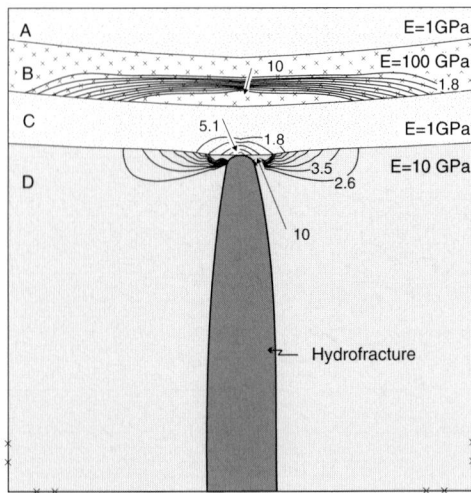

Fig. 7. Tip of a hydrofracture, located in the moderately stiff layer D, meets with the bottom of a very stiff layer C ($E = 100$ GPa). Layer B is very soft ($E = 1$ GPa), layer A has the same high stiffness as layer C. The wedge-shaped, sharp tip of the hydrofracture generates high tensile stresses in layer C and would normally be able to propagate through that layer.

Fig. 8. Same model as in Figure 7, except that here layers C and A are very soft ($E = 1$ GPa), whereas layer B is very stiff ($E = 100$ GPa). The tip of the hydrofracture is wide and blunt under these conditions. There is high tensile stress concentration in the stiff layer (here B) that could induce new fractures. But in the soft layer C next to the hydrofracture hardly any tensile stress is concentrated, so that it may act as a barrier to the vertical propagation of the hydrofracture.

very soft, its Young's modulus being only 1 GPa. The tip of the hydrofracture, at the contact with the stiff layer C, is sharp and narrow. The tensile stresses associated with the tip concentrate in the stiff layer C and exceed common *in situ* tensile strengths of rocks, 0.5–6 MPa (Haimson & Rummel 1982; Schultz 1995; Amadei & Stephansson 1997), so that fracture opening in layer C is expected. In the soft layer B, by contrast, there is virtually no tensile stress, whereas in the stiff topmost layer A there is again some tensile stress. It is likely that, for the given loading conditions, the hydrofracture would be able to propagate through the stiff layer C but might be arrested by the soft layer B.

Very different results are obtained when the lowermost of the three layers (C) is very soft ($E = 1$ GPa), in which case the hydrofracture tip first reaches a soft layer (C in Fig. 8). Layer B is now taken to be very stiff ($E = 100$ GPa), while the top layer A is very soft and with the same Young's modulus as layer C, 1 GPa. In this model, the soft layer C takes up little, but the stiff layer B much more, tensile stress. The hydrofracture tip at the contact with the soft layer C becomes rounded and relatively blunt. The soft layer C obviously suppresses the tensile stress around the hydrofracture tip and blunts the tip itself. Without a nearby weak subvertical discontinuity that could open up, it would be unlikely that the hydrofracture could propagate through the soft layer. Under these loading conditions, a soft layer would therefore favour hydrofracture arrest.

Aperture variation

Field observations

The variation in aperture of a single isolated hydrofracture in a horizontal section in a roughly homogeneous rock layer is often similar to a flat ellipse. This aperture variation has been observed in some dykes (Delaney & Pollard 1981) and mineral veins (Berg 2000), as well as in other extension fractures that may initially have been hydrofractures (Gjesdal 2001). Also, many tension fractures and normal faults in the Holocene rift zone of Iceland show a roughly elliptical variation in aperture (Gudmundsson 1987).

In vertical sections through layered rocks the aperture of a fracture may change between layers of different mechanical properties. Although described for a few dykes (Gudmundsson 1984), this type of aperture variation has rarely been reported for other hydrofractures. One reason for this may be that the difference in aperture, particularly for small-scale hydrofractures such as veins, is too small to be noticed. Another reason is that many soft layers are to some extent ductile and have no tensile strength; therefore, they cannot sustain tensile stresses. Hydrofractures, such as veins, dykes or joints, that enter such ductile layers are thus likely to trigger failure in shear rather than in extension. This is

Fig. 9. Partly inclined calcite vein with variation in aperture at Kilve, UK. In the grey limestone layer the vein is a vertical extension fracture. In the shale layer the vein is inclined and much thinner than in the subvertical parts above and below. View east, the limestone layer in the upper half of the picture is 25 cm thick.

because for a rock with zero tensile strength, the tensile stress cannot become negative, and the Mohr failure envelope would only be met on the positive side of the shear-stress axis forming a shear fracture. Thus, if hydrofractures propagate through very weak or ductile layers they are likely to follow inclined shear fractures rather than to go vertically through the layer as extension fractures. An inclined hydrofracture is thus normally a shear fracture and thus not perpendicular to the horizontal minimum principal compressive stress σ_3, as a vertical extension fractures would be, but rather subject to a higher normal stress, and thus becomes thin (Fig. 9).

The variation in aperture of calcite veins and open joints dissecting limestone and calcareous shale layers was studied in the coastal outcrops near Kilve, SW England (Fig. 10). Measurements indicate that the apertures tend to be greater in the relatively stiff limestone layers than in the soft shale layers. This can be explained as follows. Some veins die out quickly inside the soft shale layers and, being close to their tips, may be thinner for this reason. Other veins are inclined and therefore thinner inside the shale layers because they are not perpendicular to σ_3. Another reason for this aperture distribution might be that the hydrofractures propagated through a layered reservoir that was at that time subject to tension. Then the aperture tends to become greater in the stiff layers than in the soft layers because the stiff layers build up greater relative tensile stresses during tension than the soft layers (Gudmundsson & Brenner 2001). By contrast, in a reservoir that is not subject to tension so that the fluid overpressure of the hydrofractures is the only loading, the aperture variation is different, as is explored in the numerical models below.

Fig. 10. Open joints at the same locality as in Figures 1, 2 and 9 (Kilve, UK) dissecting layers of limestone and shale. Their apertures tend to be greater in limestone layers than in shale layers because these joints formed during tension (see text for discussion). View south, see the person for scale.

Numerical models

Analytical models of the aperture variation of a hydrofracture with a constant overpressure in a homogeneous, isotropic rock show an elliptical opening displacement profile (Sneddon & Lowengrub 1969). Generally, for many other overpressure distributions, the opening displacement profiles vary smoothly (Valko & Economides 1995). The aperture variation of hydrofractures in layered rocks may be much greater, depending on the mechanical contrast between the different layers. We made several boundary-element models using the program BEASY (1991) to explore the influence of layers with different stiffnesses. We also considered the effects of different overpressure distributions in the modelled hydrofracture.

The models (Figs 11–13) are all of unit height and with a Poisson's ratio of 0.25. We used 10 layers of equal thicknesses (0.09 units). The lowermost layer (J) is very stiff, with a Young's modulus E of 100

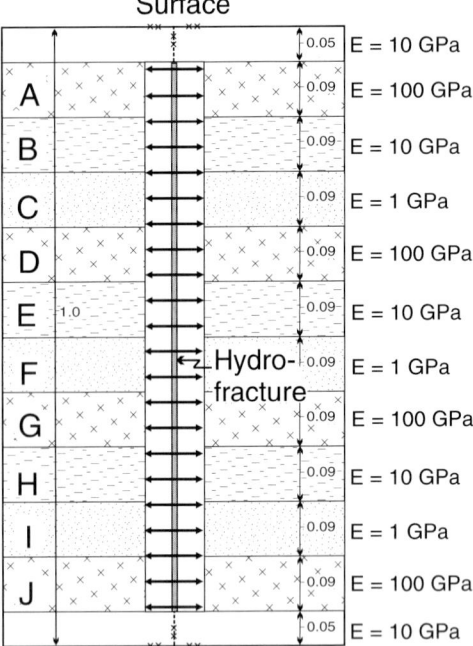

Fig. 11. Basic boundary-element configuration used for the models in Figures 12 and 13. The fluid overpressure in the hydrofracture, indicated by horizontal arrows, is the only loading; its distribution was changed between model runs. The models have unit heights (vertical dimensions), and all the 10 layers, A–J, have the same thickness (0.09 units) and Poisson's ratio (0.25). The thin layers at the top and bottom of each model are used to fasten the model (as indicated by crosses) and to confine the hydrofracture. Young's modulus of the lowermost layer under consideration J is very stiff ($E = 100$ GPa), layer I is, by contrast, very soft ($E = 1$ GPa), and layer H is moderately stiff ($E = 10$ GPa). This succession of three layers (stiff, soft and moderate) is repeated to the top, so that layer A has a Young's modulus of 100 GPa.

GPa, and is overlain by a very soft layer (I) with a Young's modulus of 1 GPa. The third layer in ascending order (H) is moderately stiff; its Young's modulus is 10 GPa. This three-layer sequence is repeated up to the surface, so that the topmost layer (A) has a Young's modulus of 100 GPa. The hydrofracture is confined and thus cannot propagate into the top and bottom layers, in which the models are also fastened. Fluid overpressure is applied along the entire height, or dip dimension, of the hydrofracture. In the first model (Fig. 12) there is a constant fluid overpressure of 6 MPa, whereas in the second model (Fig. 13) the fluid overpressure varies linearly from 10 MPa at the fracture bottom to 0 MPa at the fracture top.

Using a constant overpressure (Fig. 12) results in a wide aperture in the soft layers C, F and I. The aperture is greatly different between the soft (C, F, I) and

the moderate (B, E, H) layers, but less so between the moderate (B, E, H) and the stiff (A, D, G) layers. The linear overpressure distribution gives rise to a greater variation in aperture (Fig. 13). The largest aperture occurs where the soft layer I coincides with a high fluid overpressure near the fracture bottom. Generally, the aperture decreases towards the top of the fracture as the applied fluid overpressure decreases. But there remains a strong aperture dependence on the Young's modulus: the aperture in the soft layers is much larger than in the stiffer layers, similar to that in the models with a constant overpressure (Fig. 12), but opposite to the aperture variation of hydrofractures in layered rocks subject to tension.

Discussion and conclusions

Only an interconnected fracture system can contribute to the permeability in the host reservoir. It follows that, if most fractures in a reservoir become arrested, the reservoir may develop only a few or no vertically interconnected systems. Lateral interconnectivity may lead to the formation of horizontal units with high permeability, such as in many confined aquifers, but for the overall permeability in a reservoir the vertical interconnectivity is of special importance. In reservoirs without vertically interconnected fracture systems, the overall permeability depends entirely on the non-fracture porosity of the matrix, which may be very low.

We propose that permeability in fluid-filled heterogeneous (e.g. layered) reservoirs – for water, magma or oil – is generated and maintained by two main mechanisms. One mechanism, the formation of shear fractures (faults) through the linking up of small fractures, has been studied in detail (e.g. Cox & Scholz 1988; Martel & Pollard 1989; Cartwright *et al.* 1995; Acocella *et al.* 2000; Mansfield & Cartwright 2001), whereas the other, the propagation of fluid-driven extension fractures, i.e. hydrofractures, has received less attention and has been explored here.

Mechanical layering may determine whether developing hydrofractures become restricted to single layers or not (are strata-bound or non-strata-bound) and, therefore, if a vertically interconnected fracture system forms. Our numerical models indicate that, provided the only loading is their internal fluid pressure, hydrofractures are more likely to propagate through stiff layers than soft layers. Similarly, there are generally more fractures in stiff than in soft layers (Nelson 1985; Aguilera 1995). The results may be different, however, in reservoirs that contain stress barriers or weak discontinuities such as contacts that may open up and form T-shaped fractures (Valko & Economides 1995; Gudmundsson *et al.* 2002).

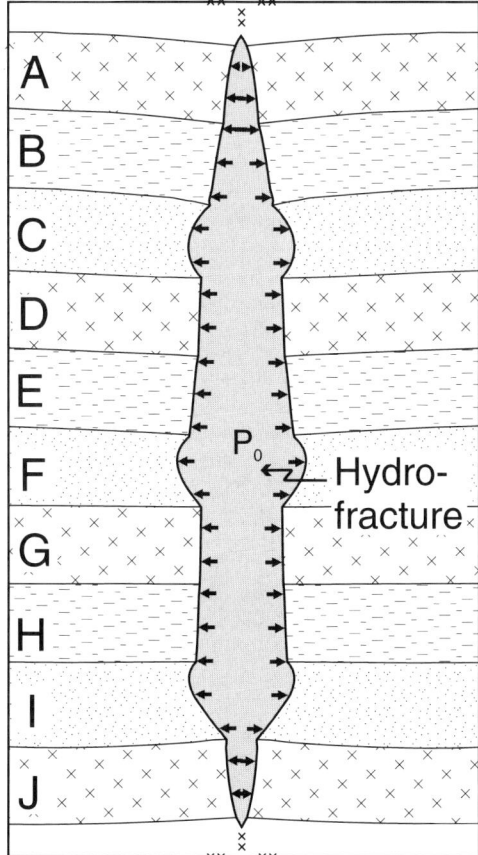

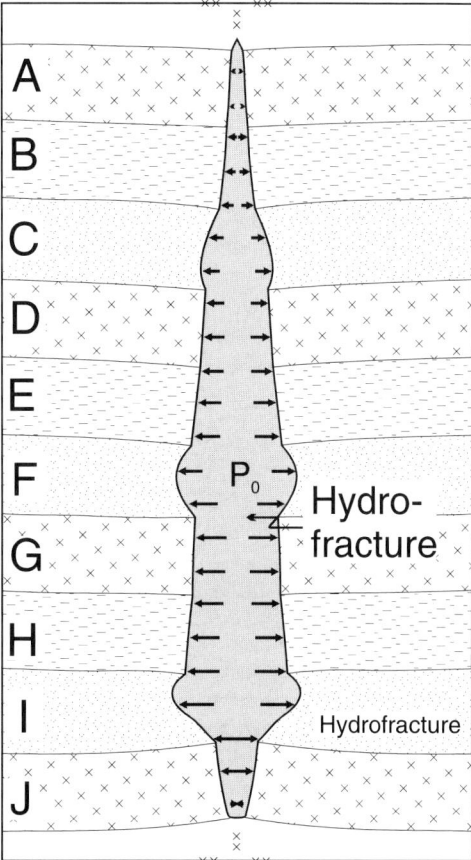

Fig. 12. Constant fluid overpressure of 6 MPa is applied along the whole height (dip dimension) of the fracture. The aperture is greatest where the fracture dissects the very soft layers C, I and F. This would normally encourage channelling of horizontal fluid flow.

Fig. 13. Same model as in Figure 12, except that the fluid overpressure has a linear variation from 10 MPa at the bottom of layer J to 0 MPa at the top of layer A. The aperture is greatest in the lower part of the hydrofracture. Again, the aperture in the soft layers C, F and I is greater than in the other layers.

Our field observations (Figs 1–2) (Gudmundsson *et al.* 2002, 2003) and numerical models (Figs 3–8) suggest that hydrofractures can become arrested at any depth if there is a strong contrast in the mechanical properties of adjacent rock layers. The observation that hydraulic fractures in petroleum engineering are commonly arrested at depths of several kilometres supports this conclusion (Valko & Economides 1995; Yew 1997; Economides & Nolte 2000), although strata-bound fractures may be more common at shallow than deep crustal levels (Odling *et al.* 1999).

In all our numerical models fluid overpressure is the only loading. Then, the stiff layers lead to hydrofracture tips being sharp and with high tensile stresses, thereby making propagation easier. By contrast, the soft layers tend to blunt the hydrofracture tips and suppress tensile stresses, thereby favouring

hydrofracture arrest. In nature, however, there is normally either an external compression or tension in addition to the fluid overpressure of the hydrofracture. For example, earlier hydrofractures normally generate compressive stresses in the adjacent rocks for some time after their emplacement. Therefore, for real reservoirs, external loading (in addition to internal fluid overpressure) needs to be taken into account when analysing hydrofracture propagation. In the present models the external loading is taken into account for those layers where the hydrofracture is located, as its fluid overpressure is the total fluid pressure minus the stress normal to the fracture (therefore lower under compression, higher under tension). However, for the layers above the hydrofractures (A and B in Figs 3–5, A–C in Figs 6–8) remote loading is not taken into account. If such a loading is considered, stiff layers would concentrate these stresses. Under tension, stiff layers would therefore concentrate tensile stresses and favour fracture propagation even further. However, under compression, stiff layers would concentrate compressive stresses, and possibly become stress barriers and tend to arrest propagating hydrofractures (cf. Gudmundsson & Brenner 2001).

Remote loading also influences the aperture variation of the resulting hydrofractures. Under compression, stiff layers concentrate compressive stresses and hydrofractures would open still less inside these layers, channelling fluid flow to the soft layers. Under tension, as mentioned earlier, stiff layers concentrate tensile stresses and lead to hydrofractures being wider than inside the soft layers. This is probably the reason why, in the field study area at Kilve, these joints dissect many layers (Fig. 10), and some dissecting vertical mineral veins have greater apertures in the stiff limestone layers than in the soft shale layers.

In the models presented in this chapter properties of layer contacts are not dealt with, but we have done so in other papers (Gudmundsson et al. 2002, 2003). The results show that weak horizontal contacts tend to open up as the tip of a vertically propagating hydrofracture approaches the contact. These results are in agreement with well-known data from fracture mechanics where crack arrest due to opening of or contacts is referred to as the Cook–Gordon debounding effect (Atkins & Mai 1985). This effect has been applied to many subjects of geology, for example the formation of T-shaped fractures in petroleum geology (Valko & Economides 1995) and to fracture step-over in structural geology (Cooke & Underwood 2001).

Hydrofractures formed at a later stage in the evolution of a fluid reservoir may propagate through contacts, which by then have become sealed and are thus no longer able to arrest fracture propagation. The stiffnesses of layers may also increase with mechanical compaction and diagenetic reactions, such as mineral transitions, precipitation (cementation) and pressure solution. These processes affect primarily originally soft rocks like shale and mudstone so that their stiffnesses gradually increase and approach that of the originally stiffer adjacent rock layers. Thus, during the evolution of a layered reservoir, when the initially very soft layers gradually become stiffer, they arrest fewer and fewer hydrofractures. Therefore, the conditions for hydrofracture arrest or propagation can change significantly over time. It follows that the permeability is a time-dependent parameter. Thus, in fluid reservoirs, especially hydrocarbon reservoirs, the permeability during migration is important for exploration (occurrence), whereas the present-day permeability is of interest for exploitation (production).

Soft layers like clay or gypsum often have extremely low permeabilities and are seals to the migration of petroleum or groundwater in reservoirs. The sealing is due to three factors. One is the typically low residual porosity following the compaction of fine-grained rocks, or, for gypsum, an originally low depositional porosity. A second factor is that the fracture-related porosity is low in soft rocks. This is partly because few hydrofractures are able to propagate through the soft clay and gypsum layers, and many (or most) become arrested at contacts with these layers. Partly, however, this factor is due to fewer fractures being initiated in soft than in stiff rocks. The third factor is that the permeability of individual fractures is directly related to their apertures. Many open fractures that occur inside soft layers are inclined (shear or hybrid) fractures that tend to be thinner than similar-sized simultaneous tension fractures (Fig. 9). If the reservoir is subject to tension, in addition to the internal fluid overpressure of hydrofractures, the fracture aperture will be smaller inside the soft layers (Fig. 10).

Sealing rocks in reservoirs are commonly soft so that fluids tend to accumulate beneath them and form fluid sills, such as water or petroleum sills (analogous to magma sills or small chambers). When the fluid excess pressure in such a reservoir reaches the *in situ* tensile strength of the host rock a hydrofracture initiates. Often, upward fluid transport, especially hydrocarbon migration in weak sedimentary rocks, is thought to occur through hydrofracturing (Iliffe *et al.* 1999; Nunn & Meulbroek 2002). Because fluid overpressure occurs mostly beneath or inside soft sealing layers, it follows that hydrofractures form more often in such soft layers than in stiff layers. As our models indicate (Figs 3–8), during the propagation of a hydrofracture, however, it is more likely to be arrested by a soft layer than by a stiff layer.

The conclusion that the apertures of hydrofractures tend to be greater in soft layers than in stiff layers (Figs 11–13), at least if fluid overpressure is the only loading, implies that there will be preferential flow (flow channelling) inside the soft layers. But this factor can only contribute to a higher overall permeability of the reservoir if the fracture system is well connected in all directions. Therefore, arrest or propagation and aperture variation both play important roles in the permeability development in a fluid reservoir.

We thank the referees N. Odling and G. Lash for helpful comments. This work was supported by a PhD grant (to A. Gudmundsson) for S. L. Brenner from Statoil, a grant from the European Commission (contract EVR1–CT-1999–40002) and several grants from the Research Council of Norway.

References

ACOCELLA, V., GUDMUNDSSON, A. & FUNICIELLO, R. 2000. Interaction and linkage of extension fractures and normal faults: examples from the rift zone of Iceland. *Journal of Structural Geology*, **22**, 1233-1246.

ADVANI, S. H., LEE, T. S., DEAN, R. H., PAK, C. K. & AVASTHI, J.M. 1997. Consequences of fluid lag in three-dimensional hydraulic fractures. *International Journal of Numerical and Analytical Methods in Geomechanics*, **21**, 229–240.

AGUILERA, R. 1995. *Naturally Fractured Reservoirs.* PennWell, Tulsa, OK.

AMADEI, B. & STEPHANSSON, O. 1997. *Rock Stress and its Measurement.* Chapman & Hall, London.

ATKINS, A. G. & MAI, Y. W. 1985. *Elastic and Plastic Fracture.* Horwood, Chichester.

BAER, G. 1991. Mechanisms of dike propagation in layered rocks and in massive, porous sedimentary rocks. *Journal of Geophysical Research,* **96**, 11911–11929.

BEASY. 1991. *The Boundary-element Analysis System User Guide.* Computational Mechanics, Boston, MA.

BELL, F. G. 2000. *Engineering Properties of Soils and Rocks,* 4th edn. Blackwell, Oxford.

BERG, S. S. 2000. *Structural analysis of fracture zones in relation to groundwater potential of fractured rocks* (In Norwegian.) MSc thesis, Bergen, Norway.

BREBBIA, C. A. & DOMINGUEZ, J. 1989. *Boundary Elements: an Introductory Course,* 2nd edn. Computational Mechanics, Boston, MA.

BRENNER, S. L. & GUDMUNDSSON, A. 2002. Permeability development during hydrofracture propagation in layered reservoirs. *Norges geologiske undersøkelse Bulletin,* **439**, 71–77.

CARTWRIGHT, J. A., TRUDGILL, B. D. & MANSFIELD, C. S. 1995. Fault growth by segment linkage – An explanation for scatter in maximum displacement and trace length data from the Canyonlands Grabens of SE Utah. *Journal of Structural Geology,* **17**, 1319–1326.

CHARLEZ, P. A. 1997. *Rock Mechanics, Volume 2: Petroleum Applications.* Editions Technip, Paris.

COOK, J. & GORDON, J. E. 1964. A mechanism for the control of crack growth in all-brittle systems. *Proceedings of the Royal Society of London,* **A282**, 508–520.

COOKE, M. L. & UNDERWOOD, C. A. 2001. Fracture termination and step-over at bedding interfaces due to frictional slip and interface debounding. *Journal of Structural Geology,* **23**, 223–238.

COX, S. J. D. & SCHOLZ, C. H. 1988. On the formation and growth of faults: an experimental study. *Journal of Structural Geology,* **10**, 413–430.

DANESHY, A. A. 1978. Hydraulic fracture propagation in layered formations. *AIME Society of Petroleum Engineers Journal,* **18**, 33–41.

DELANEY, P. T. & POLLARD, D. D. 1981. *Deformation of Host Rocks and Flow of Magma During Growth of Minette Dikes and Breccia-bearing Intrusions near Ship-Rock, New Mexico.* US Geological Survey, Professional Papers, **1202**.

ECONOMIDES, M. J. & NOLTE, K. G. (eds). 2000. *Reservoir Stimulation,* 3rd edn. Wiley, New York.

GARAGASH, D. & DETOURNAY, E. 2000. The tip region of a fluid-driven fracture in an elastic medium. *MSME Journal of Applied Mechanics,* **67**, 183–192.

GILLESPIE, P. A., JOHNSTON, D. J., LORIGA, M. A., MCCAFFREY, K. J. W., WALSH, J. J. & WATTERSON, J. 1999. Influence of layering on vein systematics in line samples. *In:* MCCAFFREY, K. J. W., LONERGAN, L. & WILKINSON, J. J. (eds) *Fractures, Fluid Flow and Mineralization.* Geological Society, London, Special Publications, **155**, 601–611.

GILLESPIE, P. A., WALSH, J. J., WATTERSON, J., BONSON, C. G. & MANZOCCHI, T. 2001. Scaling relationships of joint and vein arrays from The Burren, Co. Clare, Ireland. *Journal of Structural Geology,* **23**, 183–201.

GJESDAL, O. 2001. *Changes in hydraulic conductivity during the evolution of fracture systems.* (In Norwegian.) MSc thesis, Bergen, Norway.

GUDMUNDSSON, A. 1984. *A study of dykes, fissures and faults in selected areas of Iceland.* PhD Thesis, University of London.

GUDMUNDSSON, A. 1987. Geometry, formation and development of tectonic fractures on the Reykjanes Peninsula, southwest Iceland. *Tectonophysics,* **139**, 295–308.

GUDMUNDSSON, A. 2002. Emplacement and arrest of sheets and dykes in central volcanoes. *Journal of Volcanology and Geothermal Research,* **116**, 279–298.

GUDMUNDSSON, A. 2003. Surface stresses associated with arrested dykes in rift zones. *Bulletin of Volcanology,* **65**, 606–619.

GUDMUNDSSON, A. & BRENNER, S. L. 2001. How hydrofractures become arrested. *Terra Nova,* **13**, 456–462.

GUDMUNDSSON, A., FJELDSKAAR, I. & BRENNER, S. L. 2002. Propagation pathways and fluid transport of hydrofractures in jointed and layered rocks in geothermal fields. *Journal of Volcanology and Geothermal Research,* **116**, 257–278.

GUDMUNDSSON, A., GJESDAL, O., BRENNER, S. L. & FJELDSKAAR, I. 2003. Effects of linking up of discontinuities on fracture growth and groundwater transport. *Hydrogeology Journal,* **11**, 84-99

HAIMSON, B. C. & RUMMEL, F. 1982. Hydrofracturing stress measurements in the Iceland research drilling project drill hole at Reydarfjordur, Iceland. *Journal of Geophysical Research,* **87**, 6631–6649.

HEUZE 1980. Scale effects in the determination of rock mass strength and deformability. *Rock Mechanics,* **12**, 167–192.

ILIFFE, J. E., ROBERTSON, A. G., WARD, G. H. F., WYNN, C., PEAD, S. D. M. & CAMERON, N. 1999. The importance of fluid pressures and migration to the hydrocarbon prospectivity of the Faeroe–Shetland While Zone. *In:* FLEET, A. J. & BOLDY, S. A. R. (eds) *Petroleum Geology of Northwest Europe,* Volume 1. Geological Society, London, London, 601–611.

JUMIKIS, A. R. 1979. *Rock Mechanics.* Trans Tech Publications, Clausthal.

MANSFIELD, C. & CARTWRIGHT, J. 2001. Fault growth by linkage: observations and implications from analogue models. *Journal of Structural Geology,* **23**, 745–763.

MARINONI, L. B. & GUDMUNDSSON, A. 2000. Dykes, faults and palaeostresses in the Teno and Anaga massifs of

Tenerife (Canary Islands). *Journal of Volcanology and Geothermal Research*, **103**, 83–103.

MARTEL, S. J. & POLLARD, D. D. 1989. Mechanics of slip and fracture along small faults and simple strike-slip-fault zones in granitic rock. *Journal of Geophysical Research*, **94**, 9417–9428.

MAUGIS, D. 2000. *Contact, Adhesion and Rupture of Elastic Solids.* Springer, Berlin.

NELSON, R. A. 1985. *Geologic Analysis of Naturally Fractured Reservoirs.* Gulf Publishing, Houston, TX.

NUNN, J. A. & MEULBROEK, P. 2002. Kilometer-scale upward migration of hydrocarbons in geopressured sediments by buoyancy-driven propagation of methane filled fractures. *AAPG Bulletin*, **86**, 907–918.

ODLING, N. E., GILLESPIE, P., BOURGINE, B., CASTAING, C., CHILES, J. P., CHRISTENSEN, N. P., FILLION, E., GENTER, A., OLSEN, C., THRANE, L., TRICE, R., AARSETH, E., WALSH, J. J. & WATTERSON, J. 1999. Variations in fracture system geometry and their implications for fluid flow in fractured hydrocarbon reservoirs. *Petroleum Geoscience*, **5**, 373–384.

PEACOCK, D. C. P. & SANDERSON, D. J. 1999. Deformation history and basin-controlling faults in the Mesozoic sedimentary rocks of the Somerset coast. *Proceedings of the Geologists Association*, **110**, 41–52.

PRIEST, S. D. 1993. *Discontinuity Analysis for Rock Engineering.* Chapman & Hall, London.

SCHULTZ, R. A. 1995. Limits on strength and deformation properties of jointed basaltic rock masses. *Rock Mechanics and Rock Engineering*, **28**, 1–15.

SMITH, M. B. & SHLYAPOBERSKY, J. W. 2000. Basics of hydraulic fracturing. *In*: ECONOMIDES, M. J. & NOLTE, K. G. (eds) *Reservoir Stimulation*, 3rd edn. Wiley, New York, 5-1–5-28.

SNEDDON, I. N. & LOWENGRUB, M. 1969. *Crack Problems in the Classical Theory of Elasticity.* Wiley, New York.

STAUFFER, D. & AHARONY, A. 1994. *Introduction to Percolation Theory*, 2nd edn. Taylor & Francis, London.

TSANG, C. F. & NERETNIEKS, I. 1998. Flow channeling in heterogeneous fractured rocks. *Reviews of Geophysics*, **36**, 275–298.

VALKO, P. & ECONOMIDES, M. J. 1995. *Hydraulic Fracture Mechanics.* Wiley, New York.

YEW, C. H. 1997. *Mechanics of Hydraulic Fracturing.* Gulf Publishing, Houston, TX.

Preferential jointing of Upper Devonian black shale, Appalachian Plateau, USA: evidence supporting hydrocarbon generation as a joint-driving mechanism

GARY LASH[1], STACI LOEWY[2] & TERRY ENGELDER[3]

[1]*Department of Geosciences, State University of New York – College at Fredonia, Fredonia, NY 14063, USA*

[2]*Department of Geological Sciences, University of Texas, Austin, TX 78712, USA*

[3]*Department of Geosciences, Pennsylvania State University, University Park, PA 16802, USA*

Abstract: The Catskill Delta Complex of western New York State contains fractured Upper Devonian black shales throughout a 300 km-transect from the more distal, somewhat shallower, deposits of the western region of the state eastward to more proximal and more deeply buried deposits. Each black shale unit grades upward into organically lean grey shale and abruptly overlies another grey shale unit. Within each black shale–grey shale sequence, ENE-trending vertical joints, interpreted to be hydraulic fractures, are best developed (i.e. more closely and uniformly spaced) in the organic-rich shale. Moreover, the density of ENE joints diminishes up-section through each black shale unit, as does the total organic carbon (TOC) content. While ENE joints are less well developed outside the black shale intervals, joints that formed during the Alleghanian orogeny (NW-trending) are found throughout the Upper Devonian shale sequence. Both sets are best developed in black shales in the distal delta sequence, whereas in more proximal deposits the Alleghanian joint sets are best developed in grey shales. Moreover, the density of ENE joints within each stratigraphic level of the black shale exceeds that of Alleghanian joints at the same level, except in the deepest black shale where Alleghanian joints are locally best developed at the top of the black shale interval. The preferential jointing of black shale units in the Appalachian Plateau reflects an extended hydrocarbon generation history. In the distal delta, hydrocarbon generation began when black shale was close to or at maximum burial depth (*c.* 2.3 km) during the Alleghanian orogeny with the propagation of a NW joint set and continued through post-Alleghanian uplift of the Appalachian Plateau when the ENE joints propagated. In the proximal delta deposits ENE joints propagated before the onset of Alleghanian deformation suggesting that the base of the Upper Devonian section was buried to thermal maturity by progradation of the Catskill Delta Complex before the advent of Alleghanian sedimentation.

Joints can enhance the bulk permeability of hydrocarbon source rocks, particularly black shales, because their aperture is significantly larger than matrix pore throat diameters (Tissot & Welte 1984). If joints remain confined within source rocks they may serve as a reservoir within the source rock. Those joints that have propagated to the boundaries of the source rock are efficient drains that can enhance secondary migration of hydrocarbons. Yet, even if individual joints do not propagate across the entire bed or unit, a network of smaller joints that become interconnected during growth can serve as an effective drain. Hence, an understanding of the orientation and density of joints (i.e. the joint pattern) and timing of joint propagation in Devonian black shales of North America is important to the natural gas industry in predicting whether Devonian source rocks are also reservoir rocks. In this chapter we examine the connection of joint development to burial history and organic carbon content in Devonian black shales of the northern Appalachian Basin. It is these organic-rich shales that serve as reservoir rocks within the more central portions of the basin.

We studied joint development in Devonian black shales of the Catskill Delta Complex along a 300 km-transect across the Southern Tier of New York State from the more proximal and deeply buried deposits of the Sonyea and Genesee groups in the vicinity of the Finger Lakes District to the more distal and somewhat shallower strata of the Canadaway and West Falls groups of the Lake Erie District (Figs 1 and 2). Our transect trends obliquely across very low-amplitude folds (<30 m) of the outer Appalachian Plateau in the Finger Lakes District to the unfolded foreland of the Lake Erie District. In sampling for joint development in the more distal portions of the Catskill Delta Complex, we hoped to generate a control data set that could be used to further our understanding of joint development in the more deeply buried black shales of the folded Appalachian Plateau.

From: COSGROVE, J. W. & ENGELDER, T. (eds) 2004. *The Initiation, Propagation, and Arrest of Joints and Other Fractures.* Geological Society, London, Special Publications, **231**, 129–151. 0305-8719/04/$15 © The Geological Society of London 2004.

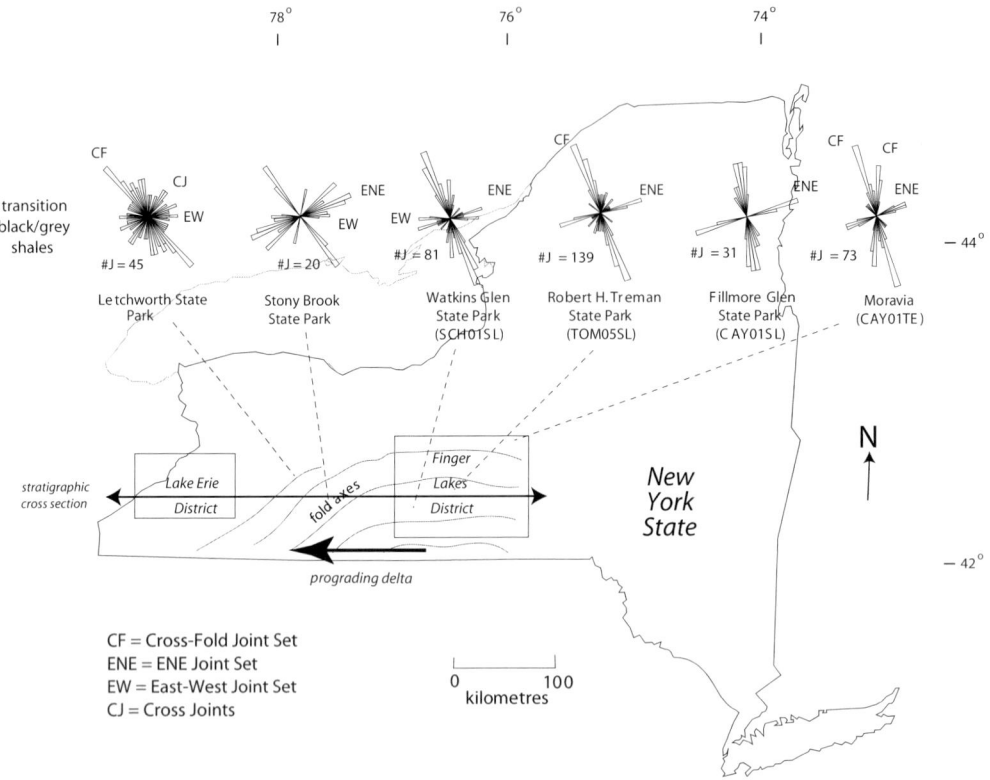

Fig. 1. Location map of the study areas in the Lake Erie and Finger Lakes districts. Approximate direction of progradation of the Catskill Delta location of the stratigraphic cross-section in Figure 4 is shown. Rose diagrams for joints in shales at six locations are superimposed on Alleghanian fold axes. Cross-fold and fold-parallel joints change orientation from east to west to remain roughly orthogonal to Alleghanian folds. Note ENE and E–W joints remain consistent in orientation across the Finger Lakes District.

Our work was stimulated by industry reports from both the Appalachian (e.g. Kubik 1993) and Michigan basins (e.g. Decker *et al.* 1992) that describe a strong relationship between production from Devonian shales and a penetrative fracture permeability that has transformed source rock into reservoir rock. Gas production from Devonian black shales of the Michigan Basin is, in part, a consequence of the desorption of methane from the surface of residual organic material (kerogen and bitumen) and clay minerals (e.g. illite: Schettler & Parmely 1990; Manger & Curtis 1991). However, well logs show that production is principally dependent on a natural joint permeability (Manger & Curtis 1991; Apotria *et al.* 1994). Similarly, there is compelling evidence that organic-rich Devonian shales of the Appalachian Basin have, on average, higher joint densities than interlayered lean grey shales (Soeder 1986; Jochen & Hopkins 1993; Kubik 1993). The robust correlation between joint development and organic carbon content is well defined in Devonian core recovered from the

Appalachian Basin as part of the Eastern Gas Shales Project (EGSP; Fig. 3).

Geological setting and stratigraphic framework

Our field area includes a vast stretch of the Catskill Delta Complex extending more than 300 km across the Southern Tier of New York State (Figs 1 and 4). The Catskill Delta Complex thickens towards its source area in the Acadian Highlands of New England (Fig. 4). By the end of the Acadian orogeny (i.e. post-Pocono Group time) the burial depth of the Geneseo black shale in the Finger Lakes District was somewhere between 1.6 and 2.3 km (Lindberg 1985). At the same time, the Dunkirk Shale of the Canadaway Group in the Lake Erie District had roughly 0.6–0.7 km of overburden.

The Upper Devonian sequence of western New York grades upwards from a base of marine shales and scattered turbidite siltstones into shallow-

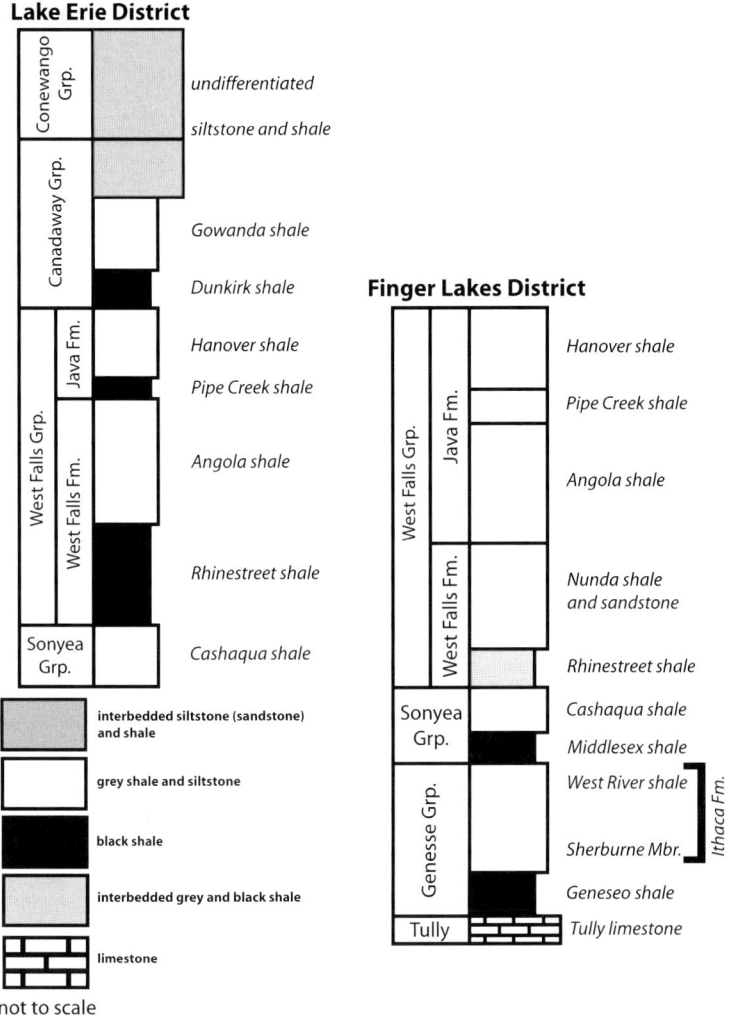

Fig. 2. Generalized Upper Devonian stratigraphic columns of the Lake Erie and Finger Lakes districts.

marine or brackish-water deposits (Baird & Lash 1990), thus recording progradation of the Catskill Delta across the Acadian foreland basin (Faill 1985; Ettensohn 1992). Marine deposits of the Catskill Delta Complex in the northern Appalachian Plateau are arranged in several cycles, each one defined by a basal unit of uniformly laminated fissile black shale that passes upwards through a transition zone of alternating black and grey shale beds into strata dominated by poorly bedded (poorly fissile) grey shale and occasional turbidite siltstone and thin black shale beds (Fig. 2). We documented joint development in four of these black shale cycles within the Genesee, Sonyea, West Falls and Canadaway groups (Figs 2 and 4). The basal black shale unit of each cycle has been interpreted as a

record of rapid cratonward movement of the Acadian fold and thrust load followed by deposition of coarser grey shale and occasional silt turbidites (Ettensohn 1985, 1992). Each phase of thrust-sheet imbrication was accompanied by rapid subsidence of the basin and deposition of clastic-starved, organic-rich black shales. Overlying shales and siltstones reflect tectonic relaxation, establishment of terrestrial drainage systems and delta progradation (Ettensohn 1985, 1992). However, Ettensohn's tectonostratigraphic explanation for the cyclic deposition of black shales in the Appalachian Basin has been challenged by models that involve eustatic oscillations and/or fluctuations in productivity of marine organic matter (Johnson *et al.* 1985; Werne *et al.* 2002).

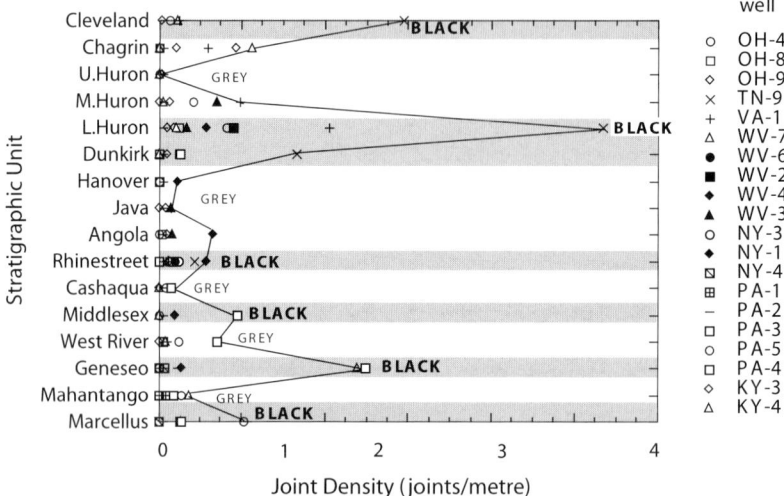

Fig. 3. Plot of joint density (number of joints per unit length of core) within various Devonian formations of the Appalachian Basin (black shale units are shaded). Data are from shale core drilled as part of the Eastern Gas Shales Project (EGSP) and published in a report from Cliffs Minerals, Inc. (1982). The key lists the various EGSP wells. The plot suggests that some black shale sections contain more joints per unit length of core than grey shale. Some unit names apply to the greater Appalachian Basin whereas other unit names are restricted to the New York State portion of the basin. Hence, there is not a one-for-one correlation between the unit names on this plot and stratigraphic terminology employed in the text. The Lower Huron of Ohio and Kentucky is correlated with the Dunkirk Shale of western New York and NW Pennsylvania (de Witt *et al.* 1993). Thus, the single shaded block defining the Dunkirk and Lower Huron black shales does not imply a single very thick unit.

Organic geochemistry of the Catskill Delta Complex

Previous work: the Finger Lakes District

The most complete body of data on the distribution of organic carbon within Devonian shale of the Finger Lakes District of New York State comes from drill cuttings on file with the New York State Geological Survey (Claypool *et al.* 1980). The Catskill Delta Complex in this area of the Appalachian Plateau is dominated by grey shale and siltstone with black shale comprising less than 10% of the section. On average, black shales of the Catskill Delta Complex in the Finger Lakes District contain three–four times the total organic carbon (TOC) of the grey shales (Fig. 5). The organic content of the grey shale serves as a background level against which TOC of the black shale may be compared. The highest TOC in the Catskill Delta Complex of the Finger Lakes District is found in the Middle Devonian Marcellus Formation, which contains almost 10% TOC (Claypool *et al.* 1980). By comparison, the Norwood Member of the Upper Devonian Antrim Formation in the Michigan Basin comprises more than 15% TOC (Loewy 1995).

Thermal maturation, mostly burial-related, of black shale in the Finger Lakes District resulted in vitrinite reflectance ($\%R_o$) values of between 1.5 and 2% (Weary *et al.*, 2000). $\%R_o$ of the younger black shales of the Lake Erie District is of the order of 0.5–0.6 (Weary *et al.* 2000, and measurements of this study). Any of three explanations may account for this difference in thermal maturity. First, it may reflect the rapid thickening by sedimentation or tectonics of the Appalachian Basin from the Lake Erie District eastward to the Finger Lakes District at the end of the Alleghanian orogeny (Johnsson 1986). Second, it may be a response to an elevated geothermal flux associated with emplacement of Cretaceous-age ultramafic intrusions in central New York State (Kay *et al.* 1983). Finally, the marked increase in thermal maturity eastward from the Lake Erie District may reflect a regional fluid flow of heated brines from the hinterland (Oliver 1986).

New work: the Lake Erie District

To supplement the Claypool *et al.* (1980) data we measured TOC in shale units of the Lake Erie District from the Cashaqua Shale up-section into the Gowanda Shale (Fig. 2). The Cashaqua Shale, approximately 30 m of light-grey organically lean shale ($0.32\% < TOC < 0.77\%$) and sparse thin–thick

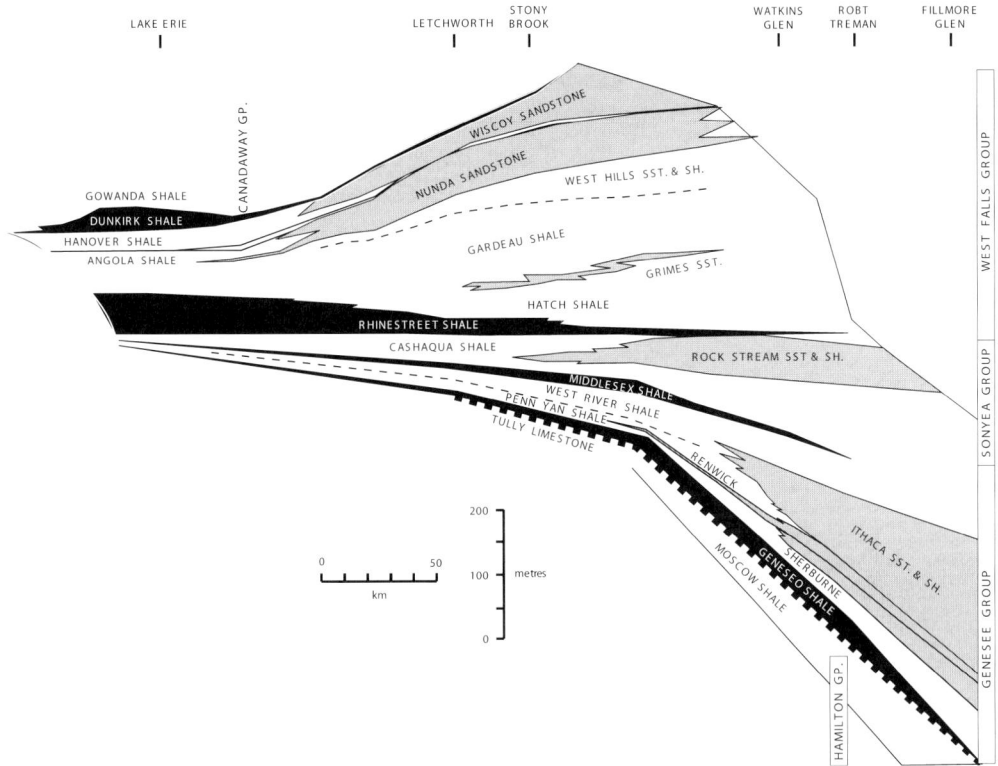

Fig. 4. East–west cross-section through the Catskill Delta Complex showing four phases of black shale deposition defined by the Geneseo, Middlesex, Rhinestreet and Dunkirk shales (adapted from Woodrow *et al.* 1988). Because of regional changes in stratigraphy and stratigraphic nomenclature, names of some rock units in the cross-section do not correspond to names of rock units in Figure 2. Figure 1 shows the location and orientation of the cross-section.

beds of siltstone and black shale, is abruptly overlain by the Rhinestreet Shale, 60–80 m of black shale containing horizons of very large (>2 m diameter) septarial carbonate concretions. Geochemical analysis of the Rhinestreet Shale reveals it to be organic-rich (1.8% < TOC < 8.01%). The Rhinestreet intertongues with increasingly greater proportions of grey shale and siltstone, and thickens to several hundred metres in the Finger Lakes District (Roen 1984; Evans *et al.* 1989; de Witt *et al.* 1993). The Rhinestreet Shale is gradationally overlain by the organically lean (0.18% < TOC < 0.98%) Angola Shale in the Lake Erie District, which comprises about 65 m of grey shale and sporadic thin beds of siltstone and black shale. Overlying the Angola Shale is the carbon-rich (4.85% < TOC < 7.37%) Pipe Creek Shale, which thickens from 0.6 m along the Lake Erie shoreline near Silver Creek, New York, to more than 5 m east of Hamburg, New York (Fig. 6). The Pipe Creek is overlain by the Hanover Shale, roughly 30 m of organically lean (0.09% < TOC < 0.93%) grey shale and occasional turbidite siltstone and thin black shale beds. The poorly bedded character of the grey

silty shale is testimony to its highly bioturbated condition (Baird & Lash 1990). The Hanover Shale is abruptly overlain by the Dunkirk Shale, approximately 15 m of laminated fissile greyish-black and black shale, sparse thin siltstone beds and large (1.5 m maximum diameter) septarial carbonate concretions.

The organic carbon content of the Dunkirk Shale varies at two levels. First, TOC diminishes up-section from 4.63% at the base to 2.74% at the top of the unit along the Canadaway Creek section and the Lake Erie shoreline in the vicinity of Dunkirk, New York (Figs 6 and 7). Roughly 70 km to the east (Cazenovia Creek section), however, TOC of the Dunkirk Shale diminishes from 2.2% at its base to 1.1% at the top of the unit (Figs 6 and 7). Rock-Eval parameters provide information regarding the type of organic material within a source rock as well as its level of thermal maturation (Peters 1986). Comparison of the S2 Rock-Eval parameter (a measure of the hydrocarbon generative potential of a source rock) with TOC suggests that organic matter in the Dunkirk is dominantly oil-prone Type II kerogen of marine origin (Langford & Blanc-Valleron 1990),

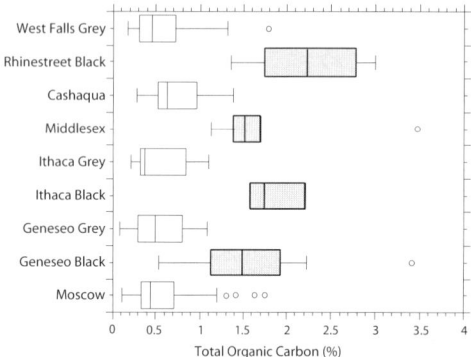

Fig. 5. Percentage of total organic carbon (TOC) averaged for various shales of the Finger Lakes District of the Appalachian Basin. TOC data were taken from cutting samples as reported in Claypool *et al.* (1980).

testimony to the distal location of the Dunkirk Basin. Rock-Eval production index (0.121–0.163) and T_{max} (439–444 °C) parameters place the Dunkirk Shale of the Lake Erie District within the thermal zone of oil generation (Tissot & Welte 1984; Peters 1986; Espitalie 1986). Vitrinite reflectance of the Dunkirk Shale (0.62%) also places this unit within the oil-generating window (Tissot & Welte 1984; Espitalie 1986). Abundant solid bitumen observed in Dunkirk Shale samples (J. Ruffin written comm. 2001) indicates that these rocks generated hydrocarbons (e.g. Momper 1978; Comer & Hinch 1987).

The Dunkirk Shale grades upward through a sequence of interbedded black and grey shale and thin turbidite siltstones into the Gowanda Shale, roughly 70 m of bioturbated and non-bioturbated grey and lesser black shale and bundles of very thin- to medium-bedded siltstone. Black shale beds within the Dunkirk–Gowanda transition zone are laminated and range from several millimetres to more than 55 cm thick. TOC values of these black shale beds (1.07% < TOC < 2.34%) compare favourably with TOC reported from the top of the Dunkirk Shale. Grey shale interbedded with carbonaceous beds of the transition zone and poorly bedded grey shale in the upper two-thirds of the Gowanda Shale is organically lean (0.52% < TOC < 0.83%). The Gowanda Shale is overlain by more than 560 m of shale and very thin- to thick-bedded siltstone that records the gradual infilling of the Acadian foreland basin by the prograding Catskill Delta in this region of the Appalachian Basin (Baird & Lash 1990). Work reported herein focuses on the Hanover–Dunkirk–Gowanda sequence (Fig. 2). These data collected from the Lake Erie District are then compared with data sampled in the more proximal, deeper portion of the Catskill Delta sequence in the Finger Lakes District.

Joint development

The concept of joint development was well entrenched in the literature nearly 100 years ago when Sheldon (1912) reported from the Appalachian Plateau that a particular joint set was 'best developed in shale beds'. Sheldon's usage of the word 'developed' (i.e. development) reflected *joint density*, the number of joints per unit length of scanline and the inverse of joint spacing. In Figure 3 joint density is the number of joints per unit length of core (recalculated as joints m^{-1}) where the direction of coring defines the scanline orientation (i.e. subvertical in this case). Analysis of joint density data derived from EGSP cores reveals that while joint density is greatest in some black shale units, jointing is not uniformly developed among organic-rich shale units throughout the Appalachian Basin. If hydrocarbon production from black shales is dependent on joint density (i.e. development) and interconnectivity, then we conclude from Figure 3 that some stratigraphic levels are better targets for exploration than others.

Joint development was quantified by Wu & Pollard (1995), who suggested that a two-dimensional (2D) analysis of joint density is a more robust measure of joint development than that obtained by 1D scanlines. In their analysis of cumulative joint length per unit area of outcrop, poorly developed joint sets are those with joint lengths less than or roughly equal to orthogonal spacing. Well-developed joint sets are those whose component joint lengths are much greater than spacing. Regardless of lithology, most outcrops of Devonian shales on the Appalachian Plateau carry at least one well-developed regional joint set according to the definition of Wu & Pollard (1995). Still, development of a particular joint set as measured by spacing data may not be uniform in a temporal or spatial sense as indicated by data from the EGSP cores (Fig. 3).

Joint development in the Hanover–Dunkirk–Gowanda sequence, Lake Erie District

The control sample in our study of the relationship of TOC and joint development in Devonian shales is the Hanover–Dunkirk–Gowanda sequence of the Lake Erie District, which lies on the North American craton beyond the influence of Alleghanian folding above a detachment on Silurian salt (Fig. 1). Nevertheless, finite-strain analysis of Devonian rocks of far-western New York indicates that the imprint of the Alleghanian orogeny does reach into the craton through the Lake Erie District (Engelder 1979) and much farther to the west (Craddock & van der Pluijm 1989).

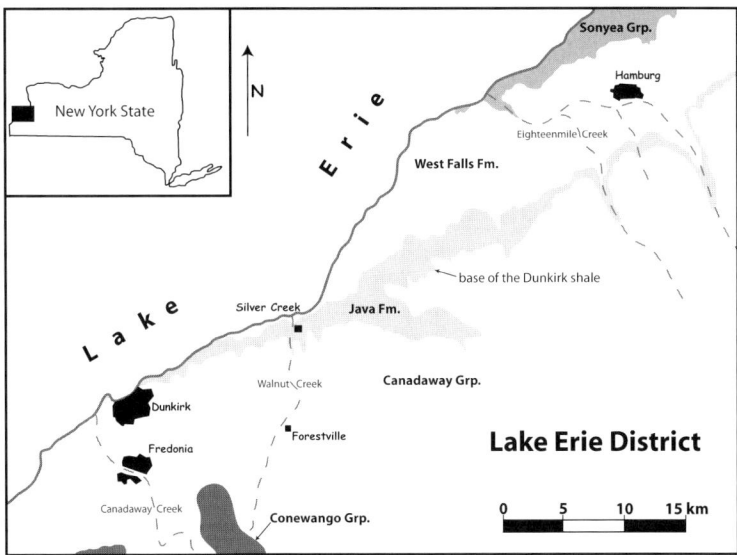

Fig. 6. Map of the Lake Erie shoreline region of the Lake Erie District (refer to Fig. 1) showing the location of Canadaway Creek, and the outcrop distribution of the Hanover Shale (Java Formation) and the Dunkirk and Gowanda shales (Canadaway Group) and other units.

Orientation data. Rocks of the Hanover–Dunkirk–Gowanda sequence typically carry two or three of five regional joint sets recognized in the Lake Erie District. NW (*c.* 310°), NNW (*c.* 352°) and ENE (*c.* 072°) sets are dominant over a WNW (*c.* 275°) set and a NE (*c.* 050°) set that appears to be the most recent (Fig. 8). Almost all joints studied are near vertical and none show evidence of slip, an observation that contrasts with reactivated joints observed deeper within the Catskill Delta Complex (e.g. Engelder *et al.* 2001). Of particular interest is the ENE set, which shows a very strong affinity for the Dunkirk black shale as well as black shale beds within the Dunkirk–Gowanda transition zone. Preferential jointing of organic-rich rocks in the Lake Erie District accords well with Sheldon's (1912) early observation of joint development in black shale further to the east and deeper in the sedimentary pile. ENE joints typically are planar and very continuous (locally >50 m, extending beyond the limits of exposure). Their observed heights (occasionally in excess of 4 m, the height of the exposure) are sometimes an order of magnitude greater than their spacing (Fig. 9A). The large height–spacing ratio indicates that ENE joints propagated through the mechanically isotropic Dunkirk Shale unimpeded by bedding interfaces. The planarity and continuity of these joints, as well as their straight overlapping geometries, suggest that ENE joints formed under conditions of relatively high (for mode I cracks) differential stress (Olson & Pollard 1989) perhaps as natural hydraulic fractures (Fischer

et al. 1995). ENE joints appear to have propagated upward from the Dunkirk Shale into the Gowanda grey shale, yet very few ENE joints extend more than a few centimetres from the bottom of the Dunkirk into the Hanover grey shale. Thus, the base of the black shale unit is a sharp mechanical boundary (e.g. Gross 1993).

NW joints, too, are very planar and continuous in outcrop, extending more than 40 m, beyond the limits of outcrop. They differ from ENE-trending joints by being more a bit more pervasive throughout the Lake Erie District Upper Devonian shale section, but, like ENE joints, NW joints are more closely spaced in black shale. Locally within the Dunkirk Shale, NW joints attain heights of 2–3 m, and few NW joints extend from the base of the Dunkirk Shale into the Hanover Shale (Fig. 9B).

Joint development (spacing and density). We used simple scanline techniques to assess the relative uniformity of development of one or more joint sets. Our data are corrected for the orientation of the scanline by employing Terzaghi's (1965) geometrical formula and are then plotted in the form of box-and-whisker diagrams (e.g. Fig. 10). Spacing values of a particular joint set at a specific outcrop are plotted horizontally. The box encloses the interquartile range of the data-set population; a vertical line drawn through the box defines the median value of the data population. The interquartile range is bounded on the left by the 25th percentile (lower quartile) and on the right by the 75th percentile

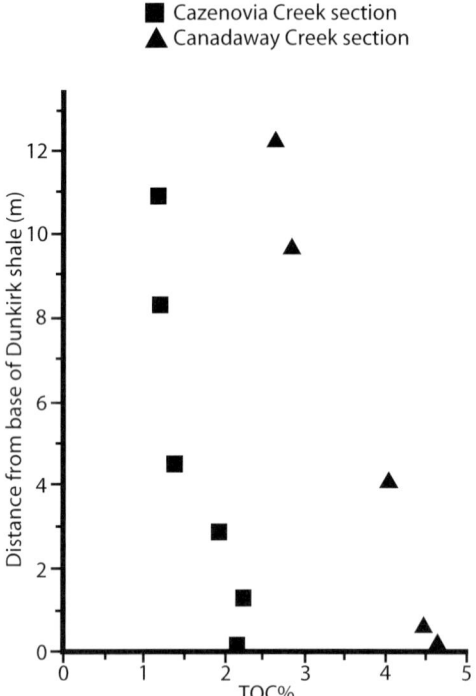

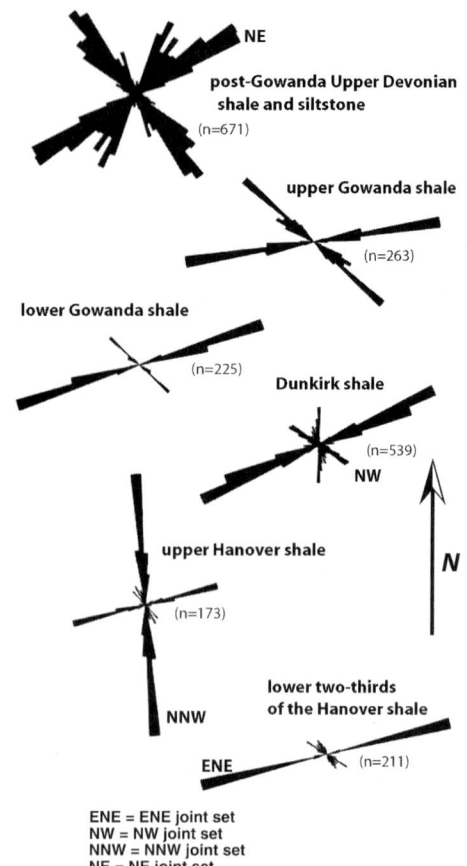

Fig. 7. Plot of total organic carbon (TOC) v. distance from the base of the Dunkirk Shale along the Canadaway Creek and Cazenovia Creek sections, Lake Erie District (refer to Fig. 6 for locations).

Fig. 8. Rose diagrams of joint orientations through the Hanover–Dunkirk–Gowanda sequence and overlying Devonian deposits of the Lake Erie District.

(upper quartile); the 'whiskers' mark the extremes of the sample range. We conclude that spacing data from two samples (i.e. spacing values collected from the same joint set at two sampling stations) is statistically related if the 25th percentile line of one plot (sample station) does not exceed the median value of the other (Walpole *et al.* 2002). If several samples of a specific joint set are statistically related, we infer that a joint set is similarly developed (i.e. similar spacing characteristics) among the sampling stations. One joint set is said to be better developed than another set at the same sampling station if the median value of the former lay to the left of (i.e. is less than) the 25th percentile value of the other joint set.

Scanline analysis of the Hanover–Dunkirk–Gowanda sequence along the Canadaway Creek section and the Lake Erie shoreline in the western part of the Lake Erie District (Fig. 6) reveals that the degree of development of ENE joints, as measured by orthogonal spacing, decreases up-section from the base of the unit (Fig. 10). Specifically, ENE joints are more evenly and closely spaced (better developed) in the Dunkirk black shale than they are in the overlying Gowanda grey shale. Moreover, the

density of ENE joints is highest in the lower part of the Dunkirk Shale, the most organically rich rocks (compare Figs 7 and 10). ENE joint set scanline data collected from three exposures in the lower half of the Dunkirk Shale show little variation and have median values of less than 75 cm (Fig. 10). However, 25th percentile spacing values of the upper half of the Dunkirk Shale exceed median spacing values of ENE joints in the lower part of the unit (Fig. 10) suggesting that ENE joints in the lower half of the Dunkirk Shale are the better developed. ENE joints are present in the Gowanda Shale, although spacing increases markedly upwards through the grey shale (Fig. 10). The great variation in joint spacing/density within the Dunkirk–Gowanda transition zone (Fig. 10) reflects the preferential ENE fracturing of black shale interbedded with grey shale. Although some widely spaced joints escape the carbon-rich layers into encapsulating grey shale, most ENE joints are confined to black shale beds. Finally, ENE jointing

Fig. 9. (**A**) ENE (at high angle to the plane of the photograph) and NW (roughly parallel to the plane of the photograph) joints in the Dunkirk Shale, Lake Erie shoreline. (**B**) Large NW joints at the contact (white bar) of the Dunkirk Shale and underlying Hanover grey shale along Eighteenmile Creek (see Fig. 6 for location). Note the close spacing of these tall joints. White bar = 0.5 m. (**C**) NNW joints in the Hanover Shale (below white bar) terminating at the base of the Dunkirk Shale, Lake Erie shoreline.

in poorly bedded grey shale higher in the Gowanda Shale demonstrates a diminished degree of development, as revealed by a low joint density (Fig. 10).

Qualitative field observations of NW joints suggest a more subdued link to TOC in the Dunkirk Shale. Two trends that are recognized in the NW joint spacing data lend support to this hypothesis. First, the 25th percentile values from three scanlines across NW joints in the Gowanda grey shale far exceed the median spacing of NW joints in the three stratigraphically lowest scanlines of the Dunkirk Shale (Fig. 11). Second, the 25th percentiles of NW joints in the grey shale exceed median spacing of ENE joints carried by rocks at the same exposures (e.g. compare locations 2B34CC and 3B8CC in Figs 10 and 11). Similarly, the 25th percentile values of NW joints in the Dunkirk Shale exceed the median spacing of ENE joints at the same sampling stations (e.g. compare locations BD1CC, MdkCC and WC28CC in Figs 10 and 11). We conclude that, although both ENE and NW joint sets are best developed in black shale, ENE joints are the better developed.

The occurrence of joints in the Hanover Shale differs in several respects from jointing patterns in the Gowanda and Dunkirk shales. ENE joints are present at intervals throughout the Hanover, especially in the lower two-thirds of the unit, yet they are infrequent (very widely spaced) or absent near the top of the unit (Fig. 8). Scanlines completed in the Hanover Shale failed to yield enough data to evaluate; still, observations of ENE joints in these organically lean shales reveal that spacing typically exceeds 2 m. NW joints are not well represented in the Hanover Shale; indeed, they are less common in the upper half of the unit than are ENE joints (Fig. 8). The most intriguing joints carried by the Hanover Shale are those of the NNW set, which are found only in the upper half of this grey shale sequence (Fig. 8). NNW joints observed from bluffs along the Lake Erie shoreline are continuous for well over 100 m and display spacings of the order of 1–5 m,

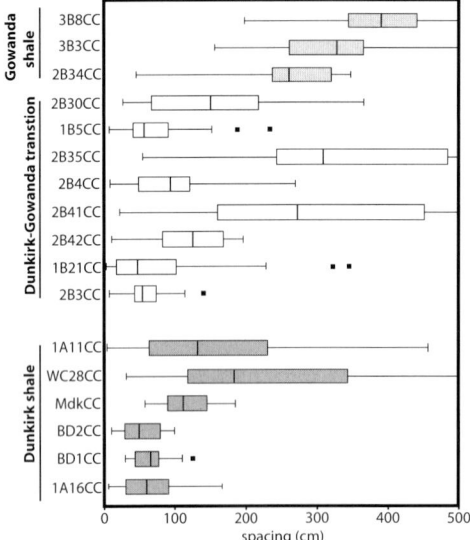

Fig. 10. Box-and-whisker diagrams representing joint-spacing distribution for ENE joints from the base of the Dunkirk Shale into the Gowanda Shale. Scanline data collected from Canadaway Creek and Lake Erie shoreline (as indicated by 'CC' in station locations), Lake Erie District. Stations are arranged in stratigraphic order.

depending on proximity to the contact with the Dunkirk Shale. Moreover, few of these joints extend more than several tens of centimetres into the Dunkirk (Fig. 9C).

Field evidence in the form of abutting relations suggests that NNW joints are older than both the NW and ENE joints. Moreover, ENE joints commonly abut (i.e. a T-shaped pattern or curving perpendicular geometry) NW and NNW joints, suggesting that they were open discontinuities during propagation of ENE joints (Dyer 1988). The classic interpretation for such abutting is that the ENE joints are the younger, a view held by Engelder (1982) and Hancock & Engelder (1989). Nevertheless, 20% mutually cross-cutting joint interactions between NW and ENE joints and the rare NW joint that abuts an ENE joint reveal the difficulties that attend relative age determination of jointing in black shale.

Joint development in the deeper portion of the Catskill Delta Complex, Finger Lakes District

Having established that the ENE joint set is best developed in the Dunkirk black shale but later than the NW joint set, we now move east and deeper into the stratigraphic pile of the Catskill Delta Complex (Fig. 12). In so doing we also move from rocks that

were unaffected by Alleghanian folding to a section that was folded and experienced as much as 10% layer-parallel shortening. Furthermore, we progress much deeper into the Catskill Delta Complex where black shales may have entered the oil window by the close of the Acadian orogeny in the Early Carboniferous. We focus on joint development in the Middlesex Shale of the Sonyea Group and the Geneseo Shale of the Genesee Group (Figs 2 and 4).

Orientation data. Similar to previous work in the Catskill Delta Complex by Parker (1942), Nickelsen & Hough (1967) and Engelder & Geiser (1980), we identified multiple joint sets in both black and grey shales based on the clustering of orientation data. Three major joint sets are found in grey shales, the most prominent being a cross-fold joint set that varies in strike with the trend of Alleghanian folds from NNW–SSE to N–S (CF in Fig. 1). Some outcrops carry multiple cross-fold joint sets, one set in siltstone beds and another in interlayered grey shale (Younes & Engelder 1999). Two cross-fold joint sets are frequently observed in thick, weakly bedded shales, in some cases mutually cross-cutting; elsewhere a later set abuts an earlier set. There remains disagreement as to how to distinguish between these cross-fold joint sets and their mode of origin (e.g. Nickelsen & Hough, 1967; Engelder & Geiser 1980; Bahat & Engelder 1984; Helgeson and Aydin 1991; Evans 1994). We correlate the cross-fold joints carried by black shale in the Finger Lakes District with the NW joint set in the Dunkirk Shale of the Lake Erie District. Both sets reflect the transport direction during Alleghanian deformation of the Appalachian Plateau and, thus, are manifestations of the orientation of the maximum horizontal stress, S_H, in an Alleghanian stress field. Although NW joints in the Lake Erie District are best developed in the black shale, cross-fold joints in the Finger Lakes District are better developed in grey shale and siltstones.

A second joint set, particularly well developed in the thicker, weakly bedded shales of the Hamilton, Genesee and Sonyea groups, strikes consistently near 070°. This is set III of Parker (1942) and the strike set of Sheldon (1912). Some cross-joints and curvy cross-joints have the same orientation as the 070° joints, but joints of the 070° or ENE set are large and planar, cross-cut cross-fold joints, and are best developed in the deeper portion of the Catskill Delta Complex where shale is more common (ENE in Fig. 1). These joints do not vary in orientation with changes in the trend of the Alleghanian folds like cross-fold joints; rather, ENE joints transect folds obliquely (Fig. 1).

ENE and cross-fold joints mutually cross-cut in black shale precluding use of an abutting criterion to infer the relative ages of the two joint sets. In fact,

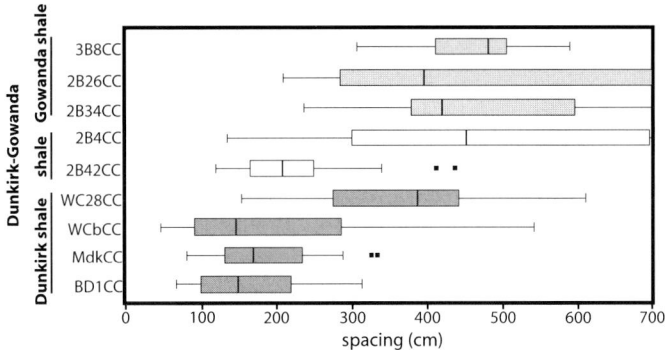

Fig. 11. Box-and-whisker diagrams representing joint-spacing distribution for NW joints from the base of the Dunkirk Shale into the Gowanda Shale. Scanline data collected from Canadaway Creek and Lake Erie shoreline (as indicated by 'CC' in station locations), Lake Erie District. Stations are arranged in stratigraphic order.

occasional ENE joints are displaced by slip on Alleghanian cross-fold joints, indicating that the ENE joints in the Finger Lakes District are pre-Alleghanian (Engelder *et al.* 2001). For these reasons, the ENE joints of the Finger Lakes District do not correlate in time with the ENE joints reported from the Lake Erie District despite having nearly the same orientation.

The thick, weakly-bedded black shales of the Catskill Delta Complex in the Finger Lakes District carry the same cross-fold joints, cross-joints (i.e. neotectonic) and ENE joints as the grey shales, but the thinner-bedded black shales of the Rhinestreet Shale and the Sherburne Member of the Ithaca Formation carry an additional joint set that is most common in these thin (<1 m) organic-rich beds (EW in Fig. 1). This set, which strikes approximately 085°, is rare in the grey shales and siltstones as well as in the thick, weakly bedded Geneseo and Middlesex black shale units. Like the ENE joint set, the EW (E–W) set does not vary in orientation with the trend of the local Alleghanian folds (Fig. 1).

Finally, outcrops of the Catskill Delta Complex carry non-systematic cross-joints (CJ in Fig. 1) that extend between systematic joints to form an outcrop pattern resembling the rungs on a ladder (e.g. Gross 1993). Some cross-joints have a sigmoidal shape in plan view. Engelder & Gross (1993) refer to the sigmoidal joints as 'curvy cross-joints' that propagated as neotectonic joints in the contemporary stress field. Cross-joints, both straight and curvy, appear best developed in siltstone beds within formations of the West Falls Group and younger strata. Many of these joints strike parallel to Appalachian Basin fold axes and orthogonal to bounding cross-fold joints. The strike of cross-fold joints in siltstone beds ranges from 045° in the west of the Finger Lakes District to 070° in eastern outcrops. Farther to the east, the fold-parallel set strikes E–W (Parker 1942).

Our observations indicate that the fold-parallel joint set propagated as late-stage cross-joints whose orientations were controlled by either the contemporary tectonic stress field or bounding cross-fold joints.

Joint development (spacing and density). We have scanline data collected from more than two dozen high-quality outcrops in black shale (Loewy 1995). Results of these scanlines are best captured in the data from Boyd Point (STE01AY) in the Middlesex Shale of the Sonyea Group, and Squaw Point (YAT03AY) and Fillmore Glen (CAY-01-SL) in the Geneseo Shale of the Genesee Group (Fig. 12). The transition from the organic-rich Middlesex Shale to the Cashaqua grey shale at Boyd Point carries the ENE joint set and two cross-fold joint sets (326° and 008°) with the anticlockwise set of this pair the better developed set (Fig. 13). Two trends in the joint spacing data are evident. First, the 25th percentile for the ENE joint set in grey shale at the top of the Cashaqua Shale (scanline #5, Fig. 13) equals or exceeds the median spacing of the ENE joint set for two scanlines lower in the section where black shale is found (scanlines #1, #2 and #3, Fig. 13). Second, 25th percentiles of cross-fold joint sets in three scanlines exceed the median spacing of the ENE joint sets in all scanlines in both the black Middlesex and grey Cashaqua shales (Fig. 13). Both the cross-fold and ENE joints at Boyd Point illustrate relatively consistent spacing characteristics within their respective set, although each set is defined by distinctly different spacing populations. Hence, we conclude that the ENE set is better developed than the cross-fold joints. Moreover, the density of ENE joints gradually decreases up-section into the grey shale, similar to our observations of the transition from black to grey shale in the Lake Erie District.

The transition from the Geneseo black shale to the

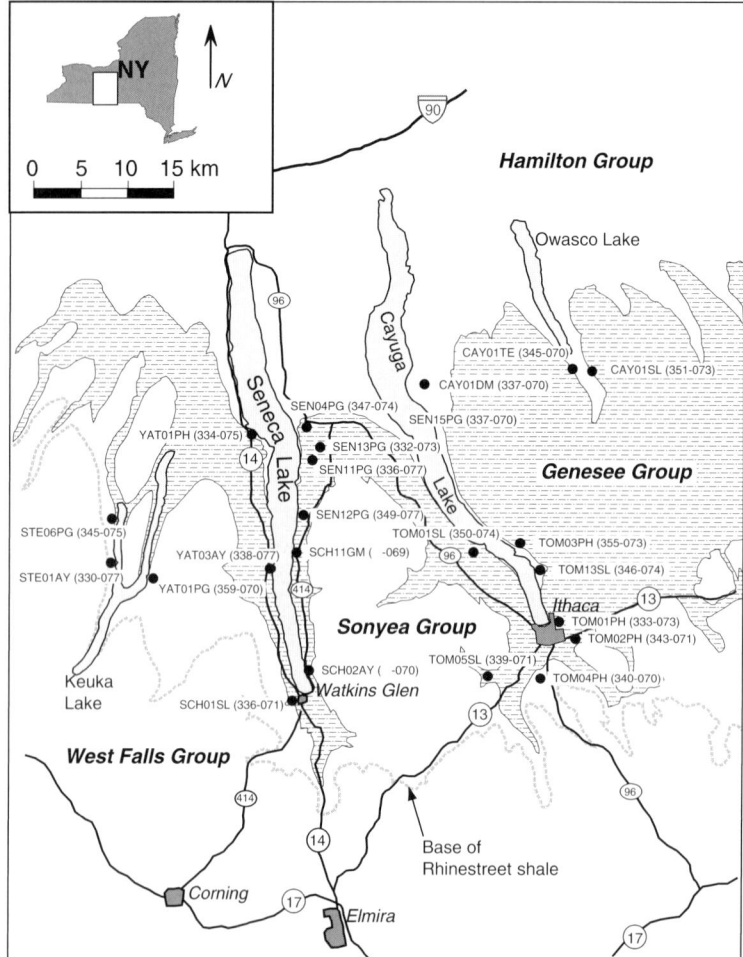

Fig. 12. Map of the Senaca–Cayuga lakes region of the Finger Lakes District showing the outcrop distribution of the Hamilton, Genesee, Sonyea and West Falls groups. The base of the Rhinestreet Shale marks the boundary between the West Falls and Sonyea groups. The Geneseo black shale crops out just above the base of the Genesee Group and the Middlesex black shale crops out just above the top of the Genesee Group (see stratigraphic section in Fig. 2). Median joint orientations for ENE and cross-fold joint sets are indicated for each outcrop.

grey shale of the Sherburne Member (Ithaca Formation) is exposed at Squaw Point (Fig. 12). One cross-fold joint set (i.e. 338°) and the ENE joint set occur throughout the section. A second cross-fold set is too poorly developed to be of use in our analysis. Again, two trends in the joint spacing data can be seen (Fig. 14). First, 25th percentile values for the three scanlines through the ENE joint set in the grey shale of the Sherburne Member equals or exceeds the median spacing of ENE joints in two scanlines through the Geneseo black shale and a scanline through the transition (Fig. 14). Second, the spacing of cross-fold joints in the black Geneseo and grey shale of the Sherburne Member is more variable, as indicated by the width of the interquartile range and

range of median values. At Squaw Point cross-fold and ENE joints show dissimilar spacing characteristics. Here again, we conclude that ENE joints, most densely distributed in the black shale, constitute the better developed set.

The highest quality exposure through the Geneseo black shale and into the grey shale of the Sherburne Member is found at Fillmore Glen, an outcrop to the east of Squaw Point (Fig. 12) that was buried deeper than all sample stations to the west. As in previous examples, the ENE joint set is best developed at the base of the Geneseo and gradually decreases in density upwards into the overlying grey shale (Fig. 15). However, in the grey shale the cross-fold joint set is consistently better developed than is the ENE

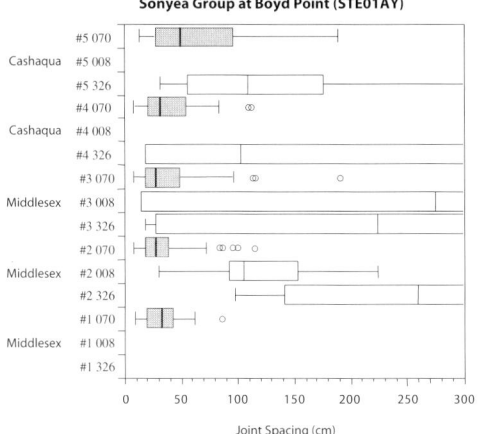

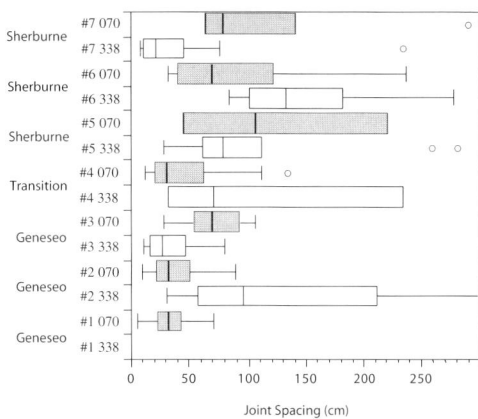

Fig. 13. Box-and-whisker diagrams representing the joint-spacing distribution for two joint sets (i.e. one cross-fold set and the 070° set) across the transition from the Middlesex Shale to the Cashaqua Shale at Boyd Point outcrop on Keuka Lake. Refer to Figure 12 for outcrop location STE01AY.

Fig. 14. Box-and-whisker diagrams representing the joint-spacing distribution for two joint sets (i.e. one cross-fold set and the 070° set) across the transition from the Geneseo Shale to the Sherburne Member of the Ithaca Formation at the Squaw Point outcrop on the west side of Seneca Lake. Refer to Figure 12 for outcrop location YAT03AY.

set (Fig. 15). The distinguishing feature of the Fillmore Glen section is that cross-fold joints are unusually well developed at the top of the black shale with spacing values less than those observed in any joint set at any other location in the Catskill Delta Complex. It is the focusing of cross-fold joint development at the top of this most deeply buried black shale section that provides a major clue to our understanding of the role of hydrocarbon generation as a joint-driving mechanism.

Data from the Boyd Point, Squaw Point and Fillmore Glen exposures, as well as other outcrops shown in Figure 12, enable us to distinguish joint development in black and grey shale. First, the ENE joint set is better developed in black shales than in grey shales. Second, in most black shale exposures that carry the ENE and cross-fold joint sets, the former is the better developed. Third, throughout the region shown in Figures 1 and 12, the orientation of the ENE joint set remains relatively consistent, unaffected by changes in the trend of local folds. Finally, the orientations of the cross-fold joint sets vary to maintain an orientation roughly normal to Alleghanian folds axes. Coupling these observations with our observations from the Lake Erie District, we conclude that conditions in the black shales most favoured development of the ENE joint set regardless of structural position and depth of burial within the northern portion of the Appalachian Basin. However, the ENE joints do not constitute a single set because they predate joints of the Alleghanian orogeny in the deeper, proximal portion of the Catskill Delta Complex and post-date joints of the

Alleghanian orogeny in the shallower more distal region of the basin.

Discussion

In the following discussion we search for a self-consistent interpretation of joint development in black shale across the northern portion of the Appalachian Basin. We begin with the premise that cross-fold joints in the Finger Lakes District are Alleghanian in age (Engelder & Geiser 1980), an assertion that has not been challenged to date (Younes & Engelder 1999). The same premise applies to the Lake Erie District where the NW joint set is approximately coaxial with a very modest Alleghanian strain (e.g. Engelder 1979; Craddock & van der Pliujm 1989). Despite the strength of evidence pointing to an Alleghanian age for the cross-fold and NW joints, we are left with the dilemma that the best-developed joint set in Devonian black shale throughout the Catskill Delta Complex, the ENE set, appears to have no tectonic affinity for Alleghanian structures and is not even contemporaneous throughout the delta complex. Indeed, the ENE set seems to be uniformly oriented across the transition from the folded Appalachian Plateau of the Finger Lakes District to the unfolded rocks of the Lake Erie District. In addition, a joint set of the same orientation is the best developed set in Devonian black shales of the Michigan Basin (Apotria et al. 1994). While it might seem appropriate to correlate the ENE joint sets in the

Genesee Group at Fillmore Glen (CAY01SL)

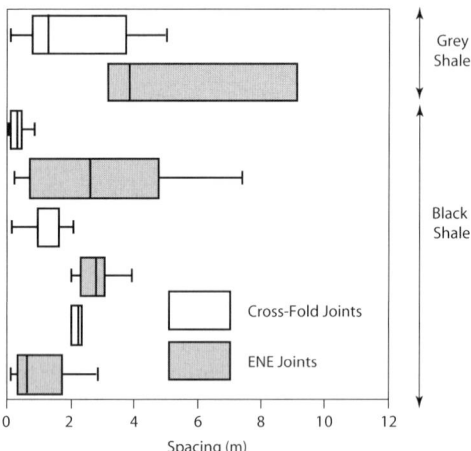

Fig. 15. Box-and-whisker diagrams representing the joint-spacing distribution for one cross-fold set and the 070° set across the transition from the Geneseo black shale into the grey shale of the Sherburne Member of the Ithaca Formation at the Fillmore Glen outcrop. Refer to Figure 12 for outcrop location CAY01SL.

Lake Erie and Finger Lake Districts based on orientation, evidence from abutting relationships makes it clear that these joints do not correlate in age.

Joint-driving mechanism

In order to more fully understand the role of TOC in jointing of the Devonian shale of the Appalachian Plateau, we need to understand the nature of joint-driving mechanisms in the carbon-rich shale. Analysis of joint-surface ornamentation facilitates the identification of joint-driving mechanisms. For example, cross-fold joints of the Finger Lakes District have a cyclic propagation pattern indicative of hydraulic fracturing (Lacazette & Engelder 1992). But, with the exception of the rare arrest line and plumose structure, all observed joint surfaces within Upper Devonian black shales are devoid of any surface morphology that might provide information regarding the origin of the joints. This is largely a consequence of the rock being so fine grained that the rupture path is not disrupted by small-scale heterogeneities. It is these small-scale heterogeneities that give rise to surface morphology in coarser-grained clastic rocks (Scott *et al.* 1992).

Still, there is evidence that both ENE and NW joints in black shales in both the Lake Erie and Finger Lakes Districts are natural hydraulic fractures. For example, the locally large height–spacing ratio of the ENE joints is at odds with spacing as a consequence

of stress-reduction shadows in joints developed under extension of beds (Gross *et al.* 1995). Rather, the close spacing of the studied joints is more consistent with their formation as hydraulic fractures than tensile joints produced by joint-normal loading or stretching (Laderia & Price 1981; Fischer *et al.* 1995; Engelder & Fischer 1996). The NW joint set of the Lake Erie District also displays the same large height–spacing ratio as do cross-fold joints at the top of the Geneseo black shale in the Fillmore Glen section of the Finger Lakes District. Equally compelling evidence comes from abutting relations of NW and ENE joints and carbonate concretions within black shale throughout the delta sequence. Joints of both sets consistently retain their planar nature as they approach and eventually make contact with concretions. Most joints terminate at concretion surfaces, although some penetrate or even cleave small concretions (Fig. 16). These relations suggest that NW and ENE joints propagated under a fluid-driven mechanism characterized by high crack-tip stress (McConaughy & Engelder 1999).

The degree of development of the ENE joints in post-Gowanda units in the Lake Erie District (as revealed by low joint density) is markedly lower than that in the underlying shale-dominated section. This may mean that the fluid-driven mechanism was not as effective in these coarser-grained rocks, which could have effectively drained the section thereby preventing a build-up of higher fluid pressure. Individual joints are curvilinear, almost always less than 0.5 m high and seldom extend more than 10 m horizontally. Moreover, siltstone beds thicker than 3 cm virtually never carry systematic ENE joints. Two joint sets, a NW and younger NE (*c.* 050°) set, are recognized in these rocks; both display a greater level of circular variance than that seen in any systematic joint set recognized in the Gowanda Shale and older units (see Fig. 8). The NE set is interpreted to have formed at relatively shallow depths under the influence of the contemporary stress field in this region of the Appalachian Plateau ($S_H = N58°E \pm 8°$; Plumb & Cox 1987). Where NE-trending joints can be observed in the Hanover–Dunkirk–Gowanda sequence, abutting relationships clearly indicate that these joints are younger than ENE joints.

The transition in jointing style from the shale-dominated to siltstone-rich section in the Lake Erie District resembles the same transition from the deeper shale section of the Sonyea and Genesee groups to the overlying sandstones in the Finger Lakes District (Engelder & Oertel 1985). Circular variance of Alleghanian cross-fold joints of the Finger Lakes District increases upward from the deeper, overpressured section (as indicated by undercompaction) to the overlying sandier section (Engelder & Oertel 1985). ENE joints, best developed in black shale sections throughout the

Fig. 16. (**A**) Small carbonate concretion in the Dunkirk Shale cleaved by an ENE joint (Walnut Creek). (**B**) Carbonate concretion penetrated (indicated by vertical scar on the concretion surface on the immediate right of the scale) by a NW joint (perpendicular to the plane of the photograph). Note that an ENE joint (subparallel to plane of the image) shows no obvious deviation around the concretion.

Appalachian Plateau, disappear upwards into siltier and sandier units in both the Finger Lakes and Lake Erie districts. There seems to be little doubt that the hydraulic-fracture-driving mechanism of the ENE joints was tied directly to the black shales. The mechanism responsible for driving the Alleghanian cross-fold joints into the coarser-grained section of the Devonian sequence is the same. However, the striking difference in orientation of the ENE and cross-fold joints is testimony to the asynchronicity of these fracturing events. Moreover, the fact that joints of both sets are hydraulic fractures suggests that asynchronous jointing was a function not only of the stratigraphic host but also of the specific timing and mechanism by which overpressure was generated in the first place.

Mechanisms for generating overpressure

The generation of abnormal pressures in a sedimentary sequence commonly reflects the interplay of two or more mechanisms (Magara 1981; Gaarenstroom *et al.* 1993; Sweeney *et al.* 1995; Swarbrick & Osborne 1998). It is possible that accumulation of the increasingly coarse-grained deposits that overlie the Gowanda Shale resulted in disequilibrium compaction of deeper shales, thereby inhibiting expulsion of pore water from these low-permeability rocks (e.g. Swarbrick & Osborne,1998). Chapman (1980) noted that disequilibrium compaction is most

common to regressive sequences, and Burrus *et al.* (1993) pointed out that deeper shales within the Mahakan Delta Complex, Indonesia, are overpressured whereas shallow sand-rich deposits are hydrostatically pressured. In fact, undercompaction and, hence, palaeodisequilibrium compaction has been documented in shales of the Finger Lakes District (Engelder & Oertel 1985). Nevertheless, disequilibrium compaction alone is insufficient to generate fluid pressures capable of driving natural hydraulic fractures (Hart *et al.* 1995; Kooi 1997).

The strong correlation between the degree of development of ENE joints and TOC in thermally mature Upper Devonian black shale of the northern Appalachian Basin suggests a genetic link. Numerous authors have cited thermal maturation of organic material as a mechanism capable of generating abnormal fluid pressures in source rocks (Snarsky 1962; Meissner 1978; Momper 1978; Spencer 1987; Stainforth 1984; Buhrig 1989; Gaarenstroom *et al.* 1993; Leonard 1993; Holm 1998, among others). We suggest that the active generation of hydrocarbons boosted formation pressures high enough to induce propagation of hydraulic fractures within the impermeable, organic-rich Dunkirk Shale of the Lake Erie District and the Middlesex and Geneseo black shales deeper in the section and farther to the east in the Catskill Delta Complex. Organically lean strata of the Gowanda and Hanover shales of the Lake Erie District, as well as their counterparts in the Finger Lakes District, are not nearly as well jointed as vertically adjacent black shale and could not have developed high formation pressures by internal hydrocarbon generation. Instead, episodic expulsion of highly pressured fluids (water, oil, gas) from mature source rocks, or perhaps vertically propagating water and/or methane-filled hydraulic fractures (e.g. Nunn & Meulbrock 2002), may have been crucial to the hydraulic fracturing of the Gowanda grey shale. The same may be said for the presence of ENE joints in the immediate post-Middlesex and Geneseo grey shale in the more proximal portion of the Catskill Delta Complex

Alleghanian joints (i.e. the NW or cross-fold joints) are best developed in the black shale in the Lake Erie District but pervade the section in the Finger Lakes District. Although the generation of hydrocarbons was responsible for the production of overpressures in the distal, organic-rich, portion of the delta complex during the Allghanian orogeny, the level of pressure or volume of fluid was insufficient to fracture the entire *Upper Devonian* sedimentary pile. Therefore, another mechanism may have been active in the folded region of the complex to drive a more pervasive hydraulic fracturing event. The strongly oriented character of ENE joints, and NW and cross-fold joints, along our transect suggests that hydraulic fracturing occurred under an anisotropic horizontal stress field associated with tectonics in the upper crust. However, the presence of an anisotropic horizontal stress field does not mean that the differential stress was high enough to cause lateral compaction, which can lead to a marked increase in pore pressure in a sequence of rocks encompassing zones of under-compacted low permeability strata (Hubbert & Rubey 1959; Pickering & Indelicato 1985; Grauls 1998). Layer-parallel shortening strain of the folded region of the Appalachian Plateau evinces a large component of lateral compaction during the Alleghanian orogeny in the Finger Lakes District; however, layer-parallel strain in the Lake Erie District was minimal (Engelder 1979). We believe that while the orientation of NW hydraulic fractures in black shale of the Lake Erie District was probably controlled by an Alleghanian stress field, the hard overpressures required to hydraulically fracture the rocks were generated during a period of hydrocarbon maturation without the benefit of lateral compaction. For the same reason, the presence of ENE joints in both districts, unrelated to documented folding of the western New York Appalachian Plateau, reflects a well-organized stress field but one that did not generate elevated pore pressure as a consequence of significant lateral strain. However, further to the east and down-section in the Catskill Delta Complex, tectonic compaction associated with the Alleghanian orogeny may have played a very important role in boosting the pore pressure to induce hydraulic fractures throughout the non-source rock portion of the section.

Relative timing of joint propagation: cross-cutting v. abutting joint relationships

The one significant difference between ENE joints in the Lake Erie District and those of the Finger Lakes District is the nature of their relationships with cross-fold or NW joints (i.e. Alleghanian jointing). In brief, only 20% of the interactions between NW and ENE joints in the Lake Erie District are mutually cross-cutting. The situation is very different in the Finger Lakes District, where more than 90% of all joint interactions are mutually cross-cutting. The high frequency of mutually cross-cutting joints in the Finger Lakes District is a function of the greater overburden at the time of joint propagation. Isopach maps of the Appalachian Basin show that Devonian rocks of the Finger Lakes District were more deeply buried than Devonian shales of the Lake Erie District (Colton 1970), a relationship confirmed by the analyses of vitrinite reflectance of black shales of both districts described earlier. In addition, Engelder *et al.* (2001) described rare occurrences of slippage of ENE joints along cross-fold joints that they believed to be contemporaneous with the main phase of layer-parallel shortening during the Alleghanian orogeny necessitating that ENE joints were in place by this time. This scenario requires that the thermal generation of hydrocarbons in Devonian strata of the Finger Lakes District started before major Alleghanian layer-parallel shortening. Evidence from the Lake Erie District, notably the preferential development of NW joints in the black shale discussed earlier, suggests that thermal generation of hydrocarbons in this region of the Appalachian Plateau occurred during the Alleghanian orogeny. However, because of its deeper burial, we entertain the possibility that Upper Devonian black shale of the Finger Lakes District reached the hydrocarbon window prior to the onset of the Alleghanian orogeny. Below we discuss joint development in the Lake Erie District by distinguishing its burial history from that recognized in the Finger Lakes District. We conclude that despite the fact that all other evidence points to a common mechanism of origin for ENE joints throughout the Catskill Delta Complex, the actual history of joint development along our transect varied markedly because of differences in thermal and tectonic history.

Relative timing of maturation of source rocks in the Catskill Delta Complex

NNW and NW joints in the shale-dominated Lake Erie District sequence record an anticlockwise rotation of the remote Alleghanian stress field, an interpretation consistent with observations made elsewhere in the western New York Appalachian Plateau (Zhao & Jacobi 1997). NW joints are pervasive throughout much of the Hanover–Dunkirk–Gowanda sequence and underlying shales, yet there are two noteworthy observations regarding their occurrence: (1) NW joints are uncommon (locally absent) at the top of the Hanover Shale and generally infrequent throughout this unit; and (2) NW joints display a preference for the Dunkirk black shale, especially the organic-rich lower part of the unit. The latter point suggests that early hydrocarbon generation may have worked in tandem with disequilibrium compaction to elevate pore pressure to the fracture gradient inducing formation of NW-trending hydraulic fractures in the Dunkirk Shale during the Alleghanian orogeny. Plentiful solid bitumen in Dunkirk Shale samples provides evidence for the generation and expulsion of oil at some point in the thermal history of these rocks (Momper 1978; Comer & Hinch 1987).

The high height–spacing ratios of NW joints at the bottom of the Dunkirk Shale suggest that hydraulic fracturing occurred episodically (e.g. Roberts & Nunn 1995). Elevation of pore pressure to the fracture gradient resulted in hydraulic fracturing followed by rapid dissipation of fluid pressure and closure of the joint. When pore pressure again rose new joints formed well within stress-reduction shadows of early formed joints, and/or older joints were re-opened and lengthened (Roberts & Nunn 1995; Holm 1998). Buoyancy pressure created by overpressured water and newly formed hydrocarbons (e.g. Pickering & Indelicato 1985; Zieglar 1992; Roberts & Nunn 1995) kept NW joints from propagating downwards from the base of the Dunkirk Shale and may have fostered joint propagation into overlying organically lean rocks (e.g. Nunn & Meulbrock 2002). Similarly, highly pressured fluids generated in the underlying organic-rich Pipe Creek Shale may have opened NW joints deeper in the Hanover Shale.

The remote stress field had undergone a major change in orientation by the time ENE joints formed in the Lake Erie District. This is one of the clearest facts but one of the most difficult to explain. There is no independent evidence for an Alleghanian stress field defined by a maximum horizontal stress in the ENE direction. However, the chronological age of ENE joints in the Lake Erie District suggests that they may have formed during Late Permian–Early Jurassic erosion-related rebound of the Appalachian foreland basin and related relaxation of horizontal stress (Blackmer et al. 1994). The strongly oriented character of ENE joints may reflect the Early Cretaceous change in the remote stress system from one dominated by rift-related dynamics to one of compression caused by seafloor spreading of the North Atlantic Ocean (Miller & Duddy 1989).

We used the EASY%R_o kinetic model of vitrinite reflectance (Sweeney & Burnham 1990) to model the thermal history of the black shale in the Catskill Delta Complex (Figs 17 and 18). Thermal modelling based on the EASY%R_o algorithm requires knowledge of: (1) the age(s) of the unit(s) of interest (the base of the Dunkirk and Geneseo shales for our models); (2) at least a partial thickness of the local stratigraphic sequence; and (3) the measured vitrinite reflectance of the unit(s) of interest (vitrinite reflectance of the base of the Dunkirk Shale = 0.62% and the Geneseo Shale = 1.74%). Our model assumes a geothermal gradient of 30 °C km^{-1} and a 20 °C seabed temperature (e.g. Gerlach & Cercone 1993). For the Lake Erie District we estimate that the Hanover Shale along the Lake Erie shoreline was overlain by approximately 660 m of Devonian strata (including the Dunkirk Shale), and that the age of the Hanover–Dunkirk contact (essentially the Frasnian–Famennian boundary) is 376.5 Ma (Tucker et al. 1998). For the Finger Lakes District, we estimate that the Tully Limestone between Fillmore Glen and Squaw Point was overlain by approximately 2120 m of Devonian strata (including the Geneseo Shale; Lindberg 1985) and that the age of the Tully–Geneseo contact is 383.5 Ma. The Devonian–Carboniferous boundary is 362 Ma (Tucker et al. 1998). We further assume that all post-Devonian strata had accumulated by the end of the Carboniferous (i.e. essentially no net sediment accumulation during Permian time; Gerlach & Cercone 1993). Finally, we adopt the Appalachian Basin unroofing history detailed by Blackmer et al. (1994) in which post-Alleghanian uplift due to flexural rebound of the foreland basin occurred from Late Permian to Early/Middle Jurassic time followed by a period of little or no unroofing that persisted until the Late Oligocene. Rapid uplift occurred from the Miocene to the present.

For the Lake Erie District, EASY%R_o modelling indicates that by the end of the Devonian, following accumulation of 660 m of sediment over the Hanover Shale, the vitrinite reflectance of the Dunkirk shale was 0.38%, a value well shy of the top of the oil window (Fig. 17). Thermal maturation was more advanced by the end of the Devonian in the Finger Lakes Distict, where a minimum estimate for overburden (i.e. 1642 m) brings the vitrinite reflectance of the Geneseo to 0.39%; the addition of 150 m of Pocono Group deposits brings R_o to 0.43%. This leaves the Finger Lakes District thermally immature, as was the case for its counterpart to the west. A less conservative extrapolation of Devonian stratigraphic thickness from Pennsylvania using Lindberg's (1985) compilation yields a post-Geneseo Devonian thickness of 2120 m (Fig. 18). Assuming a Devonian section this thick, the estimated temperature of the Geneseo was 83.6 °C and R_o = 0.45% placing the Geneseo close to the top of the oil window. The addition of 170 m of late Acadian strata (Pocono Group) during Early Carboniferous time brings the Geneseo inside the oil

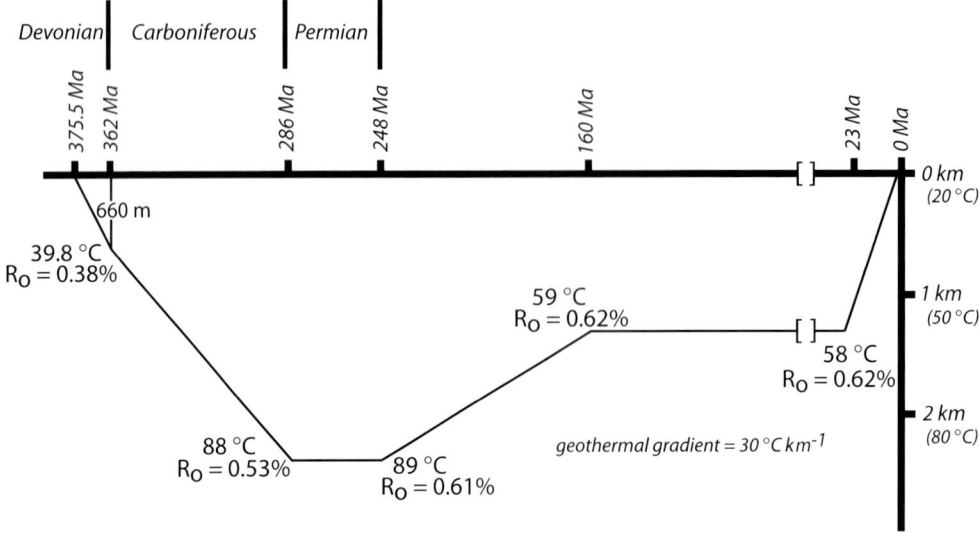

Fig. 17. EASY%R_0 thermal model for the burial history of the Dunkirk Shale, Lake Erie District. Refer to text for discussion.

window with a temperature of 88.7 °C and $R_0 = 0.52\%$ (Fig. 18).

By the end of the Carboniferous the depth of burial of the base of the Dunkirk Shale was a bit short of 2.3 km and the calculated vitrinite reflectance was 0.53% (burial temperature approximately 88 °C), meaning that the Dunkirk Shale had just entered the oil window (Fig. 17) (Peters 1986; Tissot *et al.* 1987; Hunt 1996). Assuming virtually no net sediment accumulation during Permian time, the vitrinite reflectance of the Dunkirk Shale at the end of the Palaeozoic would have been 0.61%, farther into the oil window (Fig. 17). Therefore, according to our model, the thermal maturity of the Dunkirk Shale at the end of the Carboniferous (362 Ma) equalled that of the Geneseo at the end of Pocono deposition (*c.* 355 Ma).

Earliest joint propagation by hydrocarbon-generated hydraulic fracturing in the deeper, more proximal portions of the Catskill Delta Complex could have taken place in an Acadian (Early Carboniferous) stress field and, indeed, the earliest joints strike ENE, an orientation consistent with Acadian tectonics to the east in the Acadian Highlands of New England (Engelder *et al.* 2001). Earliest joint propagation by hydraulic fracturing in the distal Lake Erie District was delayed until the effects of Alleghanian sedimentation had pushed the Dunkirk black shale into the oil window. We see the first stages of hydro-carbon-induced hydraulic fracturing (i.e. NW jointing) in the Lake Erie District when a second set of joints has started to propagate in the Finger Lakes District (i.e. the cross-fold joints). This second set cross-cuts the original ENE joint set because depth

of burial maintained a compressive effective stress on the original ENE joint set.

Uplift of the Dunkirk Shale of, perhaps, 1 km from the end of the Permian–Early/Middle Jurassic (*c.* 160 Ma) time would have resulted in minor further maturation to a reflectance of 0.62%, the measured value. Erosion of only 750 m during this time interval would have raised the vitrinite reflectance only to 0.63%. Regardless of the amount of uplift, no further maturation of the Dunkirk Shale would have occurred after Jurassic time. While the generation of new joint sets was completed by the end of the Alleghanian orogeny in the Finger Lakes District, it continued in the Lake Erie District during the Permian–Early/Middle Jurassic.

Mechanisms for overpressure generation during the Alleghanian orogeny

Our thermal models, given their constraints, suggest that the Dunkirk Shale reached the oil window during the Permian (perhaps near the end of Carboniferous time). At the same time, hydraulic fractures were propagating through much of the Devonian section of the Finger Lakes District, especially in that part of the section below the Rhinestreet Shale. Our observations in the Lake Erie District suggest that thermal maturation of organic shale alone is not sufficient to generate pervasive overpressures throughout a section containing significant thicknesses of grey shale and siltstone. Thus, an additional mechanism must be invoked to explain the more pervasive development of cross-fold joints in the Finger Lakes

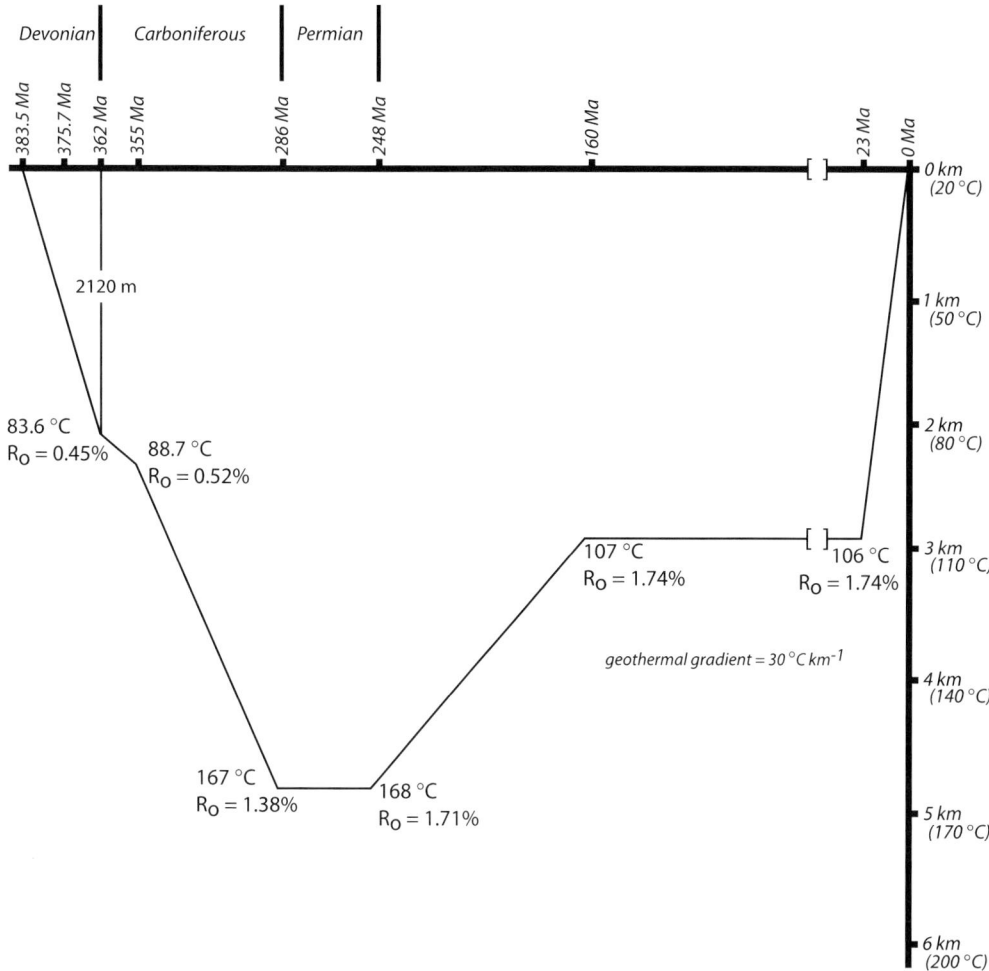

Fig. 18. EASY%R_o thermal model for the burial history of the Geneseo Shale, Finger Lakes District. Refer to text for discussion.

District. Lateral compaction of the Devonian section could have been the mechanism that pervasively overpressured the sub-Rhinestreet sequence. At the same time maturation continued in the black shales as witnessed by the density of cross-fold joints at the top of the Geneseo Shale at Fillmore Glen (note the modelled increase in the level of thermal maturation in these rocks during the Permian; Fig. 18). A hydrocarbon column produced by continued maturation of organic matter could have further elevated the level of overpressures of the upper part of a pressure compartment characterized by well-connected fractures. Indeed, modelled temperatures of the Geneseo black shale (>160 °C; Fig. 18) suggest that these rocks may have been affected by the cracking of oil and bitumen to lighter hydrocarbons such as methane, a particularly efficient mode of boosting formation

pressures to lithostatic levels (Barker 1990; Swarbrick & Osborne 1998).

Further maturation of the impermeable Dunkirk Shale and associated hydrocarbon generation during the Permian in the absence of increasing overburden (i.e. no change in the vertical component of stress) would have reduced local differential stress (σ_d) by increasing the least horizontal component of stress as a result of elevated pore pressure, a poroelastic response (e.g. Engelder & Leftwich 1997). Diminished σ_d coupled with a reduction in effective stress caused by the build-up of pore pressure resulted in the propagation of the NW-trending hydraulic fractures in the black shale. Our thermal model indicates that the Dunkirk Shale could have continued to mature, albeit minimally, during post-Alleghanian uplift, which may explain the high degree of development of ENE joints in the black shale. That is, at about the time the Dunkirk had reached its peak thermal maturity (and hydrocarbon generation), unroofing of the Appalachian Plateau brought the overpressured black shale closer to the surface, resulting in thermoelastic contraction. Moreover, relaxation of the regional horizontal compressive stress combined with the overpressured nature of the impermeable black shale provided the mechanism for the preferential development of effective tensile stress within the black shale relative to grey shale (Hanover and Gowanda shales). Propagation of ENE joints in the Lake Erie District, thus, may reflect development of effective tension primarily as a consequence of post-Alleghanian uplift rather than active hydrocarbon generation.

The strong correlation of ENE joint development and TOC links formation of these joints to organic content. The documented decrease in degree of development of ENE joints from the base of the Dunkirk Shale upwards follows dwindling TOC and may chronicle a reduction in pore pressure generated by conversion of organic matter to hydrocarbons. Hydrocarbon buoyancy prevented ENE joints at the base of the Dunkirk Shale from propagating downwards into the lean Hanover Shale. Thus, overpressured fluids were transmitted vertically towards the normally pressured part of the sediment pile, the post-Gowanda Shale strata, inducing modest ENE hydraulic fracturing of the lean Gowanda grey shale.

Locally, high height–spacing ratios of ENE joints within the Dunkirk Shale suggest that jointing was episodic, each episode being followed by closure of at least part of the fracture network as pore pressure leaked off. Formation pressure would again build as a consequence of continued hydrocarbon generation until another episode of hydraulic fracturing reopened and lengthened early formed joints and formed new ones. All of this forces one to question how it was that overpressures generated during formation of ENE joints did not bleed off through older NW joints. Although there are no easy answers to this question, it is conceivable that the spacing of NW joints and their lack of interconnectivity failed to provide the necessary permeability capable of keeping pace with the generation of hydrocarbons in the Dunkirk Shale during propagation of ENE joints. Moreover, mutually cross-cutting NW and ENE joints indicate that the apertures of some of the former joints had been diminished enough during formation of the latter to enable the transmission of ENE joint-tip stress across the NW joints.

Conclusion

ENE joints (and to a lesser extent NW joints of the Lake Erie District) of the Appalachian Plateau

display an especially strong affinity for Devonian black shale units suggesting that their propagation was linked in some way to the generation of hydrocarbons. Crucial to understanding the complex nature of jointing in Devonian black shale deposits of the Appalachian Plateau is sorting out the timing of the ENE joints. Field evidence cited here suggests that ENE jointing of organic-rich rocks of the Lake Erie District occurred late in its deformation history, during post-Alleghanian uplift of the Appalachian Plateau. To the east, though, in the Finger Lakes District of New York State, ENE joints in older black shale deposits appear to have formed before all other joint sets, perhaps near the close of the Acadian orogeny. Our thermal modelling suggests that while the Dunkirk Shale remained thermally immature at the end of Devonian time, the deeper Geneseo Shale of the Finger Lakes District was closely approaching the oil window. Further burial of the Geneseo beneath late Acadian strata during Early Carboniferous time carried the Geneseo inside the oil window. It was at this time that ENE hydraulic fractures, consistent with the orientation of Acadian tectonics to the east, formed in black shales of the Finger Lakers District. The earliest phase of hydrocarbon-generated hydraulic fracturing in the Lake Erie District – the NW joint set – did not occur until the Dunkirk Shale was buried to the oil window during the Alleghanian orogeny. Continued thermal maturation of the Finger Lakes District section, perhaps enhanced by the thermal cracking of oil to methane, coupled with lateral compaction resulted in the pervasive development of cross-fold joints. Lateral compaction was minimal in the Lake Erie District, yet NW joints propagated upwards from the thermally mature organic-rich Dunkirk Shale. ENE joints in the Lake Erie District are interpreted to have formed when unroofing of the Appalachian Plateau brought the overpressured Dunkirk Shale closer to the surface, resulting in thermoelastic contraction and relaxation of the regional least horizontal compressive stress. The complex jointing history of the Devonian black shales of the Appalachian Plateau decribed here demonstrates the critical role that variations in the timing of thermal maturity may play in the basin-wide fracturing history of shale deposits, and reveals how the structural history of potential hydrocarbon source and reservoir rocks can very over relatively small distances. Further, it provides a coincidence of nature in which seemingly contemporaneous joints formed at different times and under very different conditions

G. Lash acknowledges the help of former and current students R. Blood, P. Bembia, P. Case, P. Dwyer, A. Sorricelli and J. Greene, and financial support provided by the State University of New York, College at Fredonia. This work was also supported by funds from Penn State's Seal

Evaluation Consortium (SEC). In particular, we thank C. Kaiser, B. Cai, and T. Apotria for their encouragement during this project when all three worked for Shell Western E&P. Discussions with A. Younes and D. McConaughy were important in developing ideas for this chapter, and we thank the two of them for reviews of earlier drafts. We thank R. Jolly and P. Connolly for their thorough reviews of the manuscript.

References

APOTRIA, T. G., KAISER, C. J. & CAIN, A. B. 1994. Fracturing and stress history on the Devonian Antrim Shale, *In:* NELSON, P. P., & LAUBACH, S. E. (eds), *Michigan Basin. Rock Mechanics, Models and Measurements, Challenges from Industry.* A. A. Balkema, Brookfield, 809–816.

BAHAT, D. & ENGELDER, T. 1984. Surface morphology on joints of the Appalachian Plateau, New York and Pennsylvania. *Tectonophysics*, **104**, 299–313.

BAIRD, G. C. & LASH, G. G. 1990. Devonian strata and environments: Chautauqua County region: New York State. *In*: *62nd Annual Meeting Guidebook*. New York State Geological Association, A1–A46.

BARKER, C. E., 1990, Calculated volume and pressure changes during the thermal cracking of oil to gas in reservoirs. *AAPG Bulletin*, **74**, 1254–1261.

BLACKMER, G. C., OMAR, G. I. & GOLD, D. P. 1994. Post-Alleghanian unroofing history of the Appalachian Basin, Pennsylvania, from apatite fission track analysis and thermal models. *Tectonics*, **13**, 1259–1276.

BUHRIG, C. 1989. Geopressured Jurassic reservoirs in the Viking Graben: modeling and geological significance. *Marine and Petroleum Geology*, **6**, 31–48.

BURRUS, J., OSADETZ, K., GAULIER, J. M., BROSSE, E., DOLIGEZ, B., CHOPPIN DE JANVRY, G., BARLIER, J. & VISSER, K. 1993, Source rock permeability and petroleum expulsion efficiency: modelling examples from the Mahakam delta, the Williston Basin and the Paris Basin. *In*: PARKER, J. R. (ed.), *Petroleum Geology of Northwest Europe: Proceedings of the 4th Conference.* The Geological Society, London, 1317–1332.

CHAPMAN, R. E. 1980, Mechanical versus thermal cause of abnormally high pore pressures in shales. *AAPG Bulletin*, **64**, 2179–2183.

CLAYPOOL, G. E., HOSTERMAN, J. W., MALONE, D. R. and STONE, A.W. 1980. *Organic Carbon Content of Devonian Shale Sequence Sampled in Drill Cuttings from Twenty Wells in Southwestern New York.* US Geological Survey, Open-File Report, **80–810**.

CLIFFS MINERALS, INC. 1982. *Analysis of the Devonian Shales in the Appalachian Basin.* US DOE contract **DE-AS21-80MC14693**. Final Report. Cliffs Minerals Inc., Washington, Springfield Clearing House.

COLTON, G. W, 1970. The Appalachian Basin – its depositional sequences and their geologic relationships. *In*: FISHER, G. W., PETTIJOHN, F. J., REED, J. C. & WEAVER, K. N. (eds) *Studies of Appalachian Geology: Central and Southern*. Wiley, New York, 5–47.

COMER, J. B. & HINCH, H. H. 1987. Recognizing and quan-tifying expulsion of oil from the Woodford Formation and age-equivalent rocks in Oklahoma and Arkansas. *AAPG Bulletin*, **71**, 844–858.

CRADDOCK, J. P & VAN DER PLUIJM, B. A. 1989. Late Paleozoic deformation of the cratonic carbonate cover of eastern North America. *Geology*, **17**, 416–419.

DE WITT, W., JR, ROEN, J. B. & WALLACE, L. G. 1993. Stratigraphy of Devonian black shales and associated rocks in the Appalachian Basin. *US Geological Survey Bulletin*, **1909**, B1–B47.

DECKER, D., COATES, J.-M. P. & WICKS, D. 1992. *Stratigraphy, Gas Occurrence, Formation Evaluation and Fracture Characterization of the Antrim Shale, Michigan Basin.* Gas Research Institute Topical Report, **GRI 5091-213-2305**.

DYER, R. 1988. Using joint interactions to estimate paleo-stress ratios. *Journal of Structural Geology*, **10**, 685–699.

ENGELDER, T. 1979. The nature of deformation within the outer limits of the central Appalachian foreland fold and thrust belt in New York State. *Tectonophysics*, **55**, 289–310.

ENGELDER, T.1982. Is there a genetic relationship between selected regional joints and contemporary stress within the lithosphere of North America? *Tectonics*, **1**, 161–177.

ENGELDER, T. & FISCHER, M. P. 1996. Loading configurations and driving mechanisms for joints based on the Griffith energy-balance concept. *Tectonophysics*, **256**, 253–277.

ENGELDER, T. & GEISER, P. A. 1980. On the use of regional joint sets as trajectories of paleostress fields during the development of the Appalachian Plateau, New York. *Journal of Geophysical Research*, **94**, 6319–6341.

ENGELDER, T. & GROSS, M. R. 1993. Curving joints and the lithospheric stress field in eastern North America. *Geology*, **21**, 817–820.

ENGELDER, T. & LEFTWICH, J. T., JR. 1997. A pore-pressure limit in overpressured South Texas oil and gas fields. *In*: SURDAM, R. C. (ed.) *Seals, Traps and the Petroleum System*. American Association of Petroleum Geologists, Memoirs, **67**, 255–267.

ENGELDER, T. & OERTEL, G. 1985. The correlation between undercompaction and tectonic jointing within the Devonian Catskill Delta. *Geology*, **13**, 863–866.

ENGELDER, T., HAITH, B. F & YOUNES, A. 2001. Horizontal slip along Alleghanian joints of the Appalachian plateau: evidence showing that mild penetrative strain does little to change the pristine appearance of early joints. *Tectonophysics*, **336**, 31–41.

ESPITALIE, J. 1986. Use of T_{max} as a maturation index for different types of organic matter. Comparison with vitrinite reflectance, *In*: BURRUS, J. (ed.) *Thermal Modelling in Sedimentary Basins*. Editions Technip, Paris, 475–496.

ETTENSOHN, F. R. 1985. The Catskill Delta complex and the Acadian Orogeny: A model. *In*: WOODROW, D. L. & SEVON, W. D. (eds) *The Catskill Delta*. Geological Society of America, Special Papers, **201**, 39–49.

ETTENSOHN, F. R. 1992. Controls on the origin of the Devonian–Mississippian oil and gas shales, east-central United States. *Fuel*, **71**, 1487–1492.

EVANS, K. F., ENGELDER, T. & PLUMB, R. A. 1989. Appalachian stress study 1. A detailed description of in situ stress variations in Devonian shales of the Appalachian Plateau. *Journal of Geophysical Research*, **94**, 7129–7154.

EVANS, M. A. 1994. Joints and decollement zones in the Middle Devonian shale: evidence for multiple deformation event in the central Appalachian Plateau. *Geological Society of America Bulletin*, **106**, 447–460.

FAILL, R. T. 1985. The Acadian orogeny and the Catskill Delta. *In*: WOODROW, D. L. & SEVON, W. D. (eds) *The Catskill Delta*. Geological Society of America Special Papers, **201**, 15–37.

FISCHER, M., GROSS, M. R., ENGELDER, T. & GREENFIELD, R. J. 1995. Finite element analysis of the stress distribution around a pressurized crack in a layered elastic medium: implications for the spacing of fluid-driven joints in bedded sedimentary rock. *Tectonophysics*, **247**, 49–64.

GAARENSTROOM, L., TROMP, R. A. J., DE JONG, M. C. & BRANDENBURG, A. M. 1993. Overpressures in the Central North Sea: implications for trap integrity and drilling safety. *In*: PARKER, J. R. (ed.), *Petroleum Geology of Northwest Europe: Proceedings of the 4th Conference*. Geological Society, London, 1305–1313.

GERLACH, J. B. & CERCONE, K. R. 1993. Former Carboniferous overburden in the northern Appalachian Basin: a reconstruction based on vitrinite reflectance. *Organic Geochemistry*, **20**, 223–232.

GRAULS, D. 1998. Overpressure assessment using a minimum principal stress approach, *In*: MITCHELL, A. & GRAULS, D. (eds) *Overpressures in Petroleum Exploration*. Memoir **22**. Elf EP-Editions, Pau, 137–147.

GROSS, M. R, 1993. The origin and spacing of cross joints: Examples from the Monterey Formation, Santa Barbara coastline, California. *Journal of Structural Geology*, **15**, 737–751.

GROSS, M. R., FISCHER, M. P., ENGELDER, T. & GREENFIELD, R. 1995. Factors controlling joint spacing in interbedded sedimentary rocks; integrating numerical models with field observations from the Monterey Formation. U.S.A., *In*: AMEEN, M. S (ed.) *Fractography; Fracture Topography as a Tool in Fracture Mechanics and Stress Analysis*. Geological Society, London, Special Publications, **92**, 215–233.

HANCOCK, P. L. & ENGELDER, T. 1989. Neotectonic joints: *Geological Society of America Bulletin*, **101**, 1197–1208.

HART, B. S., FLEMINGS, P. B. & DESHPANDE, A. 1995. Porosity and pressure: Role of compaction disequilibrium in the development of geopresures in a Gulf Coast Pleistocene basin. *Geology*, **23**, 45–48.

HELGESON, D. E. & AYDIN, A. 1991. Characteristics of joint propagation across layer interfaces in sedimentary rocks. *Journal of Structural Geology*, **13**, 897–911.

HOLM, G. M., 1998, Distribution and origin of overpressures in the Central Graben of the North Sea. *In*: LAW, B. E ., ULMISHEK, G. F. & SLAVIN, V. I. (eds) *Abnormal Pressures in Hydrocarbon Environments*. Association of Petroleum Geologists, Memoirs, **70**, 123–144.

HUBBERT, M. K. & RUBEY, W. W. 1959. Role of fluid pressure in mechanics of overthrust faulting. Part 1: mechanics of fluid-filled porous solids and its application to overthrust faulting. *Geological Society of America Bulletin*, **70**, 115–166.

HUNT, J. M. 1996. *Petroleum Geochemistry and Geology*, 2nd edn. W.H. Freeman, New York.

JOCHEN, J. E. & HOPKINS, C. W. 1993. Development of a layered reservoir description. *Gas Shales Technology Review*, **8**, 79–93.

JOHNSON, J. G., KLAPPER, G. & SANDBERG, C. A. 1985. Devonian eustatic fluctuations in Euramerica. *Geological Society of America Bulletin*, **96**, 567–587.

JOHNSSON, M. J. 1986. Distribution of maximum burial temperatures across northern Appalachian Basin and implications for Carboniferous sedimentation patterns. *Geology*, **14**, 384–387.

KAY, S. M., SNEDDEN, W. T., FOSTER, B. P. & KAY, R. W. 1983. Upper mantle and crustal fragments in the Ithaca Kimberlites. *Journal of Geology*, **91**, 277–290.

KOOI, H. 1997. Insufficiency of compaction disequilibrium as the sole cause of high pore fluid pressures in pre-Cenozoic sediments. *Basin Research*, **9**, 227–241.

KUBIK, W. 1993. Natural fracturing style and control on Devonian shale gas production, Pike County, KY. *Gas Shales Technology Review*, **8**, 1–25.

LACAZETTE, A. & ENGELDER, T. 1992. Fluid-driven cyclic propagation of a joint in the Ithaca siltstone, Appalachian Basin. *In*: EVANS, B., WONG, T.-F. (eds) *Fault Mechanics and Transport Properties of Rocks*. Academic Press, London, 297–324.

LADEIRA, F. L. & PRICE, N. J. 1981. Relationship between fracture spacing and bed thickness. *Journal of Structural Geology*, **3**, 179–183.

LANGFORD, F. F. & BLANC-VALLERON, M.-M., 1990. Interpreting Rock-Eval pyrolysis data using graphs of pyrolizable hydrocarbons vs. total organic carbon. *AAPG Bulletin*, **74**, 799–804.

LEONARD, R. C. 1993. Distribution of sub-surface pressure in the Norwegian Central Graben and applications for exploration. *In*: PARKER, J. R. (ed.) *Petroleum Geology of Northwest Europe: Proceedings of the 4th Conference*. Geological Society, London, 1295–1303.

LINDBERG, F. A. (ed.) 1985. *Northern Appalachian Region: COSUNA Project*, American Association of Petroleum Geologists, Tulsa, OK.

LOEWY, S. 1995. *The post-Alleghanian tectonic history of the Appalachian Basin based on joint patterns in Devonian black shales*. MS thesis, Pennsylvania State University, University Park.

MAGARA, K. 1981. Mechanisms of natural fracturing in a sedimentary basin. *AAPG Bulletin*, **65**, 123–132.

MANGER, K. C. & CURTIS, J. B. 1991. Geological influences on location and production of Antrim shale gas. *Devonian Gas Shales Technology Review*, **7**, 5–16.

McCONAUGHY, D. T & ENGELDER, T. 1999. Joint interaction with embedded concretions: joint loading configurations inferred from propagation paths. *Journal of Structural Geology*, **21**, 1637–1652.

MEISSNER, F. F. 1978. Petroleum geology of the Bakken Formation, Wlliston Basin, North Dakota and Montana. *In*: *Williston Basin Symposium*. Montana Geological Society, Billings, MT, 207–227.

MILLER, D. S. & DUDDY, I. R. 1989. Early Cretaceous uplift and erosion of the northern Appalachian Basin, New York, based on apatite track analysis. *Earth and Planetary Science Letters*, **93**, 35–49.

MOMPER, J. A. 1978. Oil migration limitations suggested by geological and geochemical considerations. *American*

Association of Petroleum Geologists Continuing Education Course Notes Series, **8**, B1–B60.

NICKELSEN, R. P. & HOUGH, V D. 1967. Jointing in the Appalachian Plateau of Pennsylvania. *Geological Society of America Bulletin*, **78**, 609–630.

NUNN, J. A, & MUELBROCK, P. 2002. Kilometer-scale upward migration of hydrocarbons in geopressured sediments by buoyancy-driven propagation of methane-filled fractures. *AAPG Bulletin*, **86**, 907–918.

OLIVER, J. 1986. Fluids expelled tectonically from orogenic belts: their role in hydrocarbon migration and other geologic phenomena. *Geology*, **14**, 99–102.

OLSON, J. & POLLARD, D. D. 1989. Inferring paleostresses from natural fracture patterns: a new method. *Geology*, **17**, 345–348.

PARKER, J. M. 1942. Regional systematic jointing in slightly deformed sedimentary rocks. *Geological Society of America Bulletin*, **53**, 381–408.

PETERS, K. E. 1986. Guidelines for evaluating petroleum source rock using programmed pyrolysis. *AAPG Bulletin*, **70**, 318–329.

PICKERING, L. A. & INDELICATO, G. J. 1985. Abnormal formation pressure: a review. *Mountain Geologist*, **22**, 78–89.

PLUMB, R. A. & COX, J. W. 1987. Stress directions in eastern North America determined to 4.5 km from borehole elongation measurements. *Journal of Geophysical Research*, **92**, 4805–4816.

ROBERTS, S. J. & NUNN, J. A. 1995. Episodic fluid expulsion from geopressured sediments. *Marine and Petroleum Geology*, **12**, 195–204.

ROEN, J. B. 1984. Geology of the Devonian black shales of the Appalachian Basin. *Organic Geochemistry*, **5**, 241–254.

SCHETTLER, P. D. & PARMELY, C. R. 1990. The measurement of gas desorption isotherms for Devonian shale. *Devonian Gas Shales Technology Review*, **7**, 4–9.

SCOTT, P. A., ENGELDER, T. & MECHOLSKY, J. J. 1992. The correlation between fracture toughness anisotropy and surface morphology of the siltstones in the Ithaca Formation, Appalachian Basin. *In*: EVANS, B. & WONG, T.-F. (eds) *Fault Mechanics and Transport Properties of Rocks*. Academic Press, London, 341–370.

SHELDON, P. 1912. Some observations and experiments on joint planes. *Journal of Geology*, **20**, 53–70.

SNARSKY, A. N. 1962. Die primare migration des erdols. *Freiberger Forschungsch*, **C123**, 63–73.

SOEDER, D. J. 1986. Porosity and permeability of Eastern Devonian gas shale. *In*: HOLDITCH, S. A. (chairperson) *Proceedings, SPE Unconventional Gas Technology Symposium*, Society of Petroleum Engineering, Richardson, TX, 75–84.

SPENCER, C. W. 1987. Hydrocarbon generation as a mechanism for overpressuring in Rocky Mountain Region. *AAPG Bulletin*, **71**, 368–388.

STAINFORTH, J. G. 1984. Gippsland hydrocarbons – a perspective from the basin edge. *APEA Journal*, **24**, 91–100.

SWARBRICK, R. E. & OSBORNE, M. J. 1998. Mechanisms that generate abnormal pressures: an overview. *In*: LAW, B. E., ULMISHEK, G. F. & SLAVIN, V. I (eds) *Abnormal Pressures in Hydrocarbon Environments*. American Association of Petroleum Geologists, Memoirs, **70**, 13–34.

SWEENEY, J. J. & BURNHAM, A. K. 1990. Evaluation of a simple model of vitrinite reflectance based on chemical kinetics. *AAPG Bulletin*, **74**, 1559–1570.

SWEENEY, J. J., BRAUN, R. L., BURNHAM, A. K., TALUKDAR, S. & VALLEJOS, C. 1995. Chemical kinetic model of hydrocarbon generation, expulsion, and destruction applied to the Maracaibo Basin, Venezuela. *AAPG Bulletin*, **79**, 1515–1532.

TERZAGHI, R. D. 1965. Sources of error in joint surveys. *Geotechnique*, **15**, 287–303.

TISSOT, B. P. & WELTE, D. H. 1984. *Petroleum Formation and Occurrence*, 2nd edn. Springer, New York.

TISSOT, B. P., PELET, R. & UNGERER, P. 1987. Thermal history of sedimentary basins, maturation indices, and kinetics of oil and gas generation. *AAPG Bulletin*, **71**, 1445–1466.

TUCKER, R. D., BRADLEY, D. C., VER STRAETEN, C. A., HARRIS, A. G., EBERT, J. R. & McCUTCHEON, S. R. 1998. New U–Pb zircon ages and the duration and division of Devonian time. *Earth and Planetary Science Letters*, **158**, 175–186.

WALPOLE, R. E., MYERS, R. H., MYERS, S. L., and YE, K. 2002. *Probability and Statistics for Engineers and Scientists*, 7th edn. Prentice Hall, Englewood Cliffs, NJ.

WEARY, D. J., RYDER, R. T. & NYAHAY, R. 2000. *Thermal maturity patterns (CAI and %Ro) in the Ordovician and Devonian rocks of the Appalachian basin in New York State*. US Geological Survey Open-file Report, **00-496**.

WERNE, J. P., SAGEMAN, B. B., LYONS, T. W. & HOLLANDER, D. J. 2002. An integrated assessment of a 'type euxinic' deposit: evidence for multiple controls on black shale deposition in the Middle Devonian Oatka Creek Formation. *American Journal of Science*, **302**, 110–143.

WOODROW, D. L, DENNISON, J. M., ETTENSOHN, F.R., SEVON, W. T. & KIRCHGASSER, W. T. 1988. Middle and Upper Devonian stratigraphy and paleontology of the central and southern Appalachians and eastern Midcontinent, U.S.A. *In*: McMILLAN, N.J., EMBRY, A.F. & GLAN, D. J. (eds) *Devonian of the World; Proceedings of the Second International Symposium on the Devonian System; Volume I, Regional Syntheses*. Canadian Society of Petroleum Geologists, Memoirs, **14**, 277–301.

WU, H. & POLLARD, D. D., 1995, An experimental study of the relationship between joint spacing and layer thickness. *Journal of Structural Geology*, **17**, 887–905.

YOUNES, A. & ENGELDER, T. 1999. Fringe cracks: key structures for the interpretation of progressive Alleghanian deformation of the Appalachian Plateau. *Geological Society of America Bulletin*, **111**, 219–239.

ZHAO, M. & JACOBI, R. D. 1997. Formation of regional cross-fold joints in the northern Appalachian Plateau. *Journal of Structural Geology*, **19**, 817–834.

ZIEGLAR, D. L. 1992. Hydrocarbon columns, buoyancy pressures, and seal efficiency: comparisons of oil and gas accumulations in California and the Rocky Mountain area. *AAPG Bulletin*, **76**, 501–508.

Using differential geometry to characterize and analyse the morphology of joints

D. D. POLLARD, S. BERGBAUER & I. MYNATT

*Department of Geological and Environmental Sciences, Stanford University, Stanford, CA
94305-2115, U.S.A. (e-mail: dpollard@pangea.stanford.edu)*

Abstract: Concepts of differential geometry are reviewed, and it is demonstrated through examples
that the main joint surface, rib marks and hackle of a joint may be described using parametric representations such that the first and second fundamental forms fully characterize these surfaces. Other
useful quantities are the unit normal vector, the principal normal curvatures, and the Gaussian and
the mean curvature. Sufficiently close to any point on a surface the shape is planar, parabolic, elliptical or hyperbolic. The surface of a joint in chert and another in siltstone were scanned and the
resulting data analysed. Although the main joint surface of the chert sample is approximately
planar, it is composed of low-amplitude undulations with elliptical and hyperbolic forms. The unit
normal vector does not vary by more than about 3.4° over this surface, which is consistent with the
threshold angle for the initiation of hackle based on laboratory experiments. An individual hackle is
found to be approximately helicoidal in shape, but only in the breakdown zone. Rib marks on the
siltstone sample have distinct and similar morphologies, with a concave base and convex peak.
Field and laboratory campaigns designed to test hypotheses about the geometry of joints should use
the principles and tools of differential geometry.

The principal objective of this chapter is to demonstrate how the morphology of joint surfaces may be
quantified using modern scanning technology and
then analysed using the fundamental concepts of differential geometry. From such an investigation we
expect to learn how to quantitatively characterize the
forms of joint surfaces and thereby advance beyond
the qualitative descriptive names employed by geologists to date. Also, we expect to develop rigorous
techniques for testing hypotheses about the shapes
of joint surfaces. Finally, although this chapter is
about fracture geometry and not about fracture
mechanics, we anticipate gaining some insight about
how to address the mechanics of joint formation in
future investigations.

Many of the morphological features of joints that
are discussed by geologists today were carefully
described and beautifully illustrated by J. B.
Woodworth more than 100 years ago from his field
study of pelitic rocks of the Mystic River region near
Somerville, Massachusetts, USA. These features
include what Woodworth referred to as feather fractures, fringe, border planes, and cross-fractures (Fig.
1) (Woodworth 1896). The block diagram of
Kulander *et al.* (1979), published 83 years later,
emphasizes the three-dimensional nature of these
and some additional features, and compares the
descriptive classification from Hodgson (1961*a*) to a
genetic classification (Fig. 2). Despite differences in
nomenclature, these two illustrative figures and the
body of literature accumulated over the past century
(Pollard & Aydin 1988) demonstrate that these geological structures exhibit systematic morphological

features that serve to characterize joints, and to distinguish them from other tabular structures such as
faults, solution surfaces and deformation bands.

Examples of joint origins have been identified at
geometric irregularities, cavities or material heterogeneities that serve as points of stress concentration
(Hodgson 1961*a*; Wise 1964; Kulander *et al.* 1979;
Bahat & Engelder 1984; Pollard & Aydin 1988;
Wienberger 2001). Hackles that radiate from an
origin, or fan out from the axis of a plumose structure (feather fracture), have proved useful in interpreting the kinematics of joint propagation (Parker
1942; Hodgson 1961*b*; Bankwitz 1965, 1966; Bahat
& Engelder 1984;Kulander & Dean 1985; DeGraff
& Aydin 1987; Pollard & Aydin 1988; Helgeson &
Aydin 1991; Bai & Pollard 2000*b*). Rib marks (arrest
or hesitation lines) appearing on some joint surfaces
as curvilinear ridges or troughs that are more or less
orthogonal to hackle serve a similar purpose
(Bankwitz 1965; Kulander *et al.* 1979; Engelder
1987). Fine hackle that form a component of the
plumose structure may merge distally into the intersection lines between adjacent echelon segments
(border planes, twist-hackle, partial fractures). The
surfaces of these segments twist out of the plane of
the main (parent) joint to form the joint (twist-
hackle) fringe (Hodgson 1961*a*; Bankwitz 1965,
1966; Pollard *et al.* 1982; Mandl 1987; Pollard &
Aydin 1988; Cooke & Pollard 1996; Treagus & Lisle
1997; Bahat *et al.* 2001). Individual echelon
segment surfaces in the fringe region may display
plumose structures indicating local propagation
directions that diverge from the plume axes as

From: COSGROVE, J. W. & ENGELDER, T. (eds) 2004. *The Initiation, Propagation, and Arrest of Joints and Other Fractures.*
Geological Society, London, Special Publications, **231**, 153–182. 0305-8719/04/$15 © The Geological Society of London
2004.

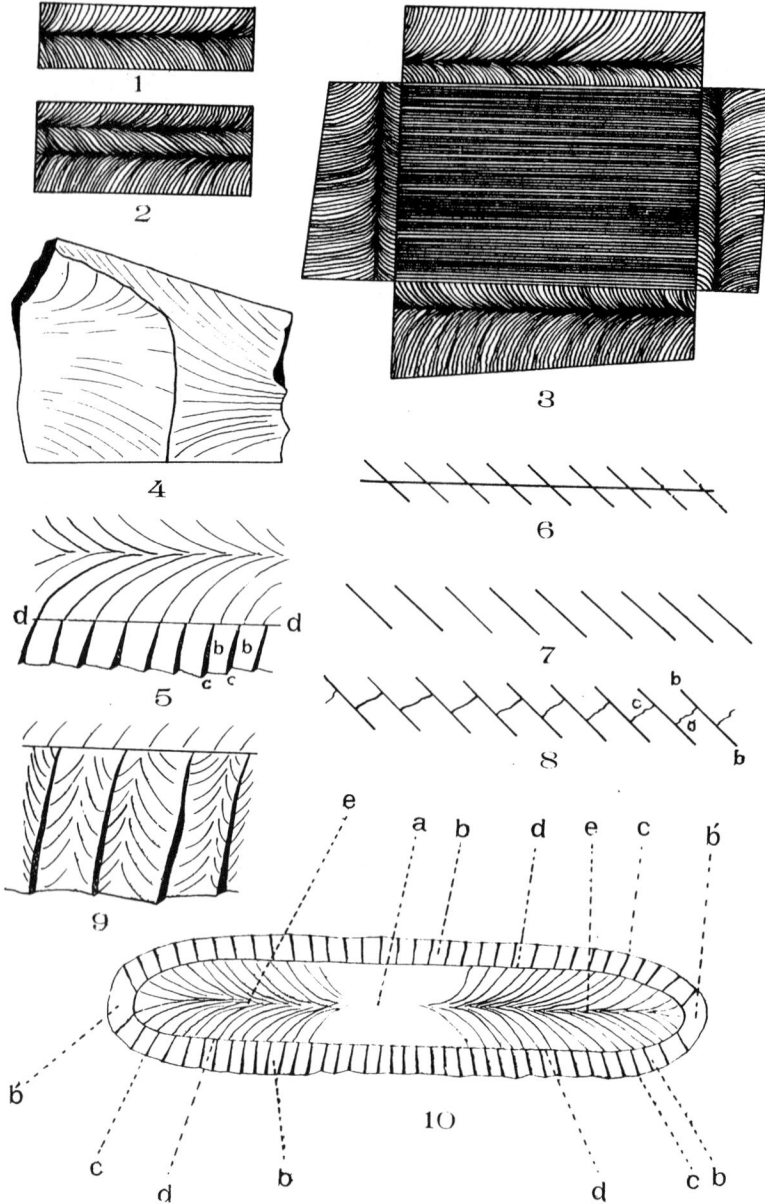

Fig. 1. Reproduction of plate 1 Woodworth (1896). The following captions are quoted with minor clarifications and omissions from this paper. (**1**) Diagram of feather fracture as exhibited on joint plane cutting a stratum whose upper and lower surfaces form the edge of the joint plane. (**2**) Diagram of side of a joint block with twinned feather fracture and two axial planes. (**3**) Diagram of the four sides of a joint block drawn as if lying in one plane. The lines covering the interior square or surface of the joint block represent shading only. (**4**) Intersection of two joint planes, slightly departing from a common plane. (**5**) Joint plane with fringe, showing relation of feather fracture to fracture system in the fringe: *d–d*, edge of the joint plane; *b*, border-plane of the fringe; *c*, cross-fracture. (**6**) Diagram showing direction of main joint plane and border planes of the fringe. (**7**) Plan of border planes prior to formation of cross-fractures in joints. (**8**) Diagram showing relation between border planes and cross-fractures: *b–b*, border plane; *c*, cross-fracture. (**9**) Distribution of feather fractures on border planes. (**10**) Ideal scheme of elliptical joint in stratified rock: *a*, centre of figure of the fractured surface; *b*, border plane of the fringe; *b'*, distal border plane, often extending from upper to lower limits of the joint plane; *c*, cross-fracture; *d*, edge of joint plane, and inner margin of the fringe; *e*, axis of feather fracture, parallel with the stratification.

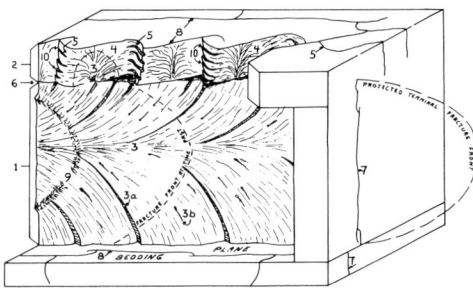

Fig. 2. Reproduction of figure 77 from the report by Kulander *et al.* (1979). The descriptive classification of Hodgson includes: 1 main joint face; 2, fringe; 3, plumose structure; 4, F-joints; 5, C-fractures; 6, shoulder; 7, trace of main joint face; 9, conchoidal ridge (Hodgson 1961*a*). The genetic classification of Kulander *et al.* (1979) includes: 1, main fracture face; 2, twist hackle fringe; 3, fracture plume; 3a, twist hackle; 3b, inclusion hackle; 4, twist hackle face; 5, twist hackle step; 6, plumose–coarse twist hackle boundary; 7, tendential penetration perpendicular to bedding; 8, tendential extent parallel to bedding; 9, arrest or Wallner line; 10, second order twist hackle.

created surfaces, but commonly include lesser components of displacement parallel to the surfaces. Thus, the displacement discontinuity that accompanies formation of a joint results in an opening that is generally much greater than any shearing of the two surfaces. Furthermore, the displacement discontinuities during joint formation are very small compared to a characteristic length measured between two distal points on the tipline. For joints that are reactivated by frictional sliding to become faulted joints, shearing may exceed opening (Segall & Pollard 1983; Martel *et al.* 1988; Martel & Pollard 1989; Martel & Boger 1998).

The geometric and kinematic attributes of joints mentioned above place them, for the most part, in the context of linear elastic fracture mechanics (Lawn & Wilshaw 1975; Kanninen & Popelar 1985; Lawn 1993; Anderson 1995), and this discipline provides many of the principles and methods that have been used to interpret field observations of joints (Engelder & Geiser 1980; Engelder 1985; Helgeson & Aydin 1991; Gross 1993, 1995; Renshaw & Pollard 1994*b*; Fischer *et al.* 1995; Wu & Pollard 1995; Bai & Pollard 2000*a, b*; Cooke *et al.* 2000). In particular, the form and structure of joint surfaces are a direct expression of the propagation path taken by the tipline as it moves through a rock mass and these paths may be interpreted using the three modes of fracture mechanics (Fig. 3). Focusing on the joint surfaces very near the tipline, mode I refers to an opening-displacement discontinuity; mode II refers to a shearing-displacement discontinuity directed perpendicular to the tipline; and mode III refers to a tearing-displacement discontinuity directed parallel to the tipline. For mixed-mode cases applicable to the continuous propagation of a joint it is understood that opening is the dominant mode, whereas shearing and/or tearing are the perturbing modes.

Straight paths (planar joint surfaces) are associated with pure mode I because the near-tip stress field is symmetric and the energy release for an increment of propagation is greatest along this path (Erdogan & Sih 1963; Gell & Smith 1967; Hussain *et al.* 1974; Sih 1974; Cotterell & Rice 1980; Pollard *et al.* 1982; Pollard & Aydin 1988). Using similar stress or energy arguments, tilted or curved paths (rib marks, tail or wing cracks, sigmoidal and hook-shaped segment traces, and tilted fringe segments) are associated with mixed modes I + II (Ingraffea 1981; Olson & Pollard 1989, 1991; Thomas & Pollard 1993; Renshaw & Pollard 1994*a*). Twisted paths (border planes, twist hackle, echelon segments) are associated with mixed modes I + III (Sommer 1969; Ueda *et al.* 1983; Mandl 1987; Yates & Miller 1989; Cooke & Pollard 1996; Treagus & Lisle 1997; Lazarus *et al.* 2001*a, b*). A general mixture of modes I + II + III results in combinations of these characteristic structures. Experiments have

segments grow laterally and overlap with their neighbours along curved paths (twist-hackle steps). The intervening bridge of rock between adjacent echelon segments may be broken by later (cross-) fractures or solution seams (Segall & Pollard 1983; Nicholson & Pollard 1985; Martel *et al.* 1988; Martel & Pollard 1989; Cruikshank *et al.* 1991*a*; Willemse *et al.* 1997; Willemse & Pollard 1998), or continued lateral propagation of the segments may result in intersecting or subparallel geometries observed in the bedding plane (Pollard *et al.* 1982; Nicholson & Pollard 1985; DeGraff & Aydin 1987; Pollard & Aydin 1988; Olson & Pollard 1989; Cruikshank *et al.* 1991*b*).

Joints are fractures that usually form within a brittle body of rock at some depth in the Earth's crust. As such, they share geometric and kinematic attributes that are common to most opening fractures (Pollard 1987). Veins and dykes share many of the morphological features of joints, because they too are dominantly opening fractures, but fillings of hydrothermal minerals and igneous rock distinguish these structures from joints. Each joint is composed of two continuous surfaces that are bounded by a continuous curve referred to as the tipline or fracture tip. The two surfaces are more or less mirror images of one another and they may be decorated with some or all of the structures mentioned in the preceding paragraphs. Particles of rock that were originally adjacent and bonded together are separated during formation of the joint by relative displacements that are predominantly perpendicular to the newly

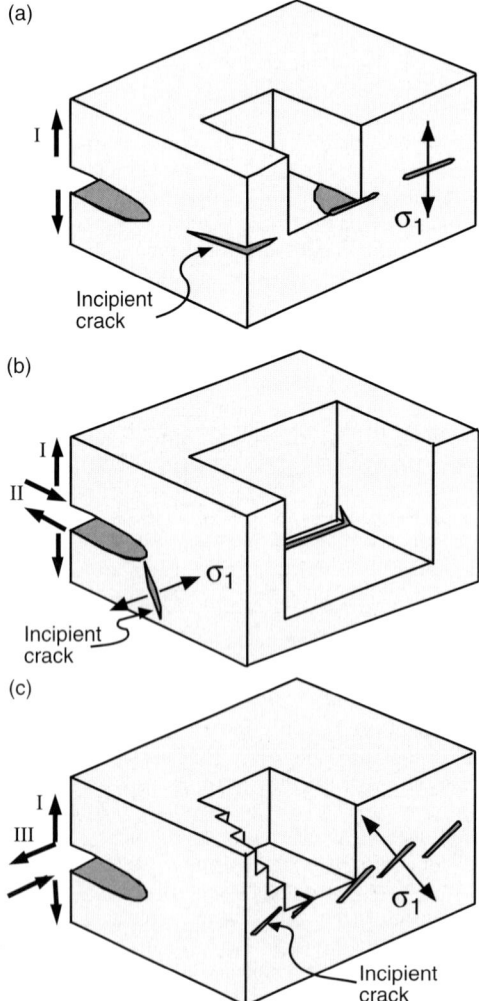

Incipient
crack

(b)

Incipient
crack

(c)

Incipient
crack

Fig. 3. Schematic illustration of the region near the tipline of a fracture, such as a joint, subject to a dominant mode I (opening) deformation (Pollard & Aydin 1988). Three cases of discontinuous growth (from an arrested tipline) are identified with fundamentally different fracture paths that can be distinguished based on the loading of a putative increment of the fracture in its own plane. (**a**) For pure mode I there is no shear stress resolved on the increment and the fracture path is coincident with this increment, but an arrest line marks the position of the tipline. (**b**) For a perturbing mode II deformation a shear stress is resolved on the increment acting in the direction normal to the tipline and the fracture path tilts sharply about an axis parallel to the tipline. (**c**) For a perturbing mode III deformation a shear stress is resolved on the increment acting in the direction parallel to the tipline and the fracture breaks down into segments with paths that twist about an axis perpendicular to the tipline.

produced secondary fractures around the perimeters of preformed cracks or notches subjected to mixed-mode loading in laboratory specimens and these fractures share many of the geometric features found on joints (Brace & Bombolakis 1963; Adams & Sines 1978; Cox & Scholz 1988*a*, *b*; Germanovich *et al.* 1994, 1996).

The surface of a joint often is idealized on a map or diagram as planar, and a single attitude (strike and dip) serves to characterize the orientation of this plane. Nonetheless, some of the most interesting features of natural joints, as illustrated in Figures 1 and 2, are the curves, kinks and twists of their surfaces. Furthermore, joint surfaces are sometimes characterized as undecorated, yet curved hackle, plumose structure and rib marks form intriguing curves on these surfaces. Perhaps because ready explanations can be found for some of these non-planar surfaces and curved lineations within the context of elastic fracture mechanics (Pollard & Aydin 1988), and because these explanations facilitate the geological interpretation of the structures, relatively little effort has been expended to characterize their detailed geometry quantitatively. In the literature of structural geology we are presented with abundant photographs and drawings, and a plethora of names, but little numerical data that would serve to document the actual forms of these surfaces or lineations. Here we suggest that differential geometry provides the mathematical basis for the quantitative description of joints.

Differential geometry includes the analytical study of points, curves and surfaces in three-dimensional space using vectors and the methods of calculus. For in-depth treatments of differential geometry that provide a rigorous mathematical basis, the reader is referred to textbooks on the subject (Struik 1961; Lipschultz 1969; Stoker 1969) or to the website http://mathworld.wolfram.com. Here we review some of the elementary concepts of differential geometry that are helpful to describe surfaces, including the parametric representation of surfaces; the normal and normal curvature vectors; the first and second fundamental form coefficients; the surface area and normal curvature; and the mean and Gaussian curvatures. Where appropriate these concepts are related to the morphology of joints. We also provide examples that demonstrate how to analyse the digitized surface of a joint to compute the relevant quantities from differential geometry that serve to characterize the joint morphology quantitatively.

Geologists have made limited use of differential geometry despite the obvious need to describe surfaces and the incentives for doing this quantitatively. Methods have been proposed to use triangulated or gridded data (e.g. from reflection seismic surveys) to calculate surface curvature and other attributes of surface geometry (Samson & Mallet 1997; Stewart

& Podolski 1998), and to filter surface data for subsequent analysis using differential geometry (Stewart & Wynn 2000; Bergbauer *et al.* 2003). Applications to geological structures include the prediction of strains in folds (Lisle 1994, 2000), the description of folded surfaces (Lisle 1992; Lisle & Robinson 1995), and the relationships between folding and fracturing.

Readers who are familiar with the basic concepts of differential geometry may omit the next section of this chapter on 'The analytical description of surfaces', as it reviews these topics. The section on 'An analytical model for fracture surfaces' introduces the helicoid as a geometric model for some fracture surfaces and describes the various quantities from differential geometry that are relevant to this model. The final section on 'The quantitative analysis of joint surfaces' takes up the analysis of sample joint surfaces using the tools of differential geometry and digitized data to quantitatively compare the actual surfaces to idealized model geometries.

The analytical description of geological surfaces

Vector functions of two real variables may be used to describe geological surfaces such as the surfaces of a joint (Figs 1 and 2). The analytical description of curved surfaces is approached from the intuitive notion of a set of points arranged in some continuous fashion in three-dimensional space. Sufficiently close to any particular point, the neighbouring points are distributed such that they resemble a plane. This leads to the definition of a curved surface in terms of a set of position vectors, **s**, defined by a continuous vector function of two scalar variables (u, v), called the parameters of the surface (Fig. 4a), such that $\mathbf{s} = \mathbf{f}(u, v)$. As the two parameters vary, the position vectors trace out the curved surface in three-dimensional space (Fig. 4b). The two Cartesian coordinate axes (Ou, Ov) define the parameter plane on which the two parameters are u and v. The three Cartesian coordinate axes (Ox, Oy, Oz) and the base vectors $(\mathbf{e}_x, \mathbf{e}_y, \mathbf{e}_z)$ comprise the system for the curved surface.

Parametric representation of a surface

Position vectors, $\mathbf{s}(u, v)$, locate every point on a curved surface and these are determined by the vector function of the parameters u and v (Lipschultz 1969, p. 128):

$$\mathbf{s} = \mathbf{f}(u,v) = f_x(u,v)\mathbf{e}_x + f_y(u,v)\mathbf{e}_y + f_z(u,v)\mathbf{e}_z. \quad (1)$$

The three scalar functions $[f_x(u,v), f_y(u,v), f_z(u,v)]$ are the components of the vector function, $\mathbf{f}(u, v)$, with

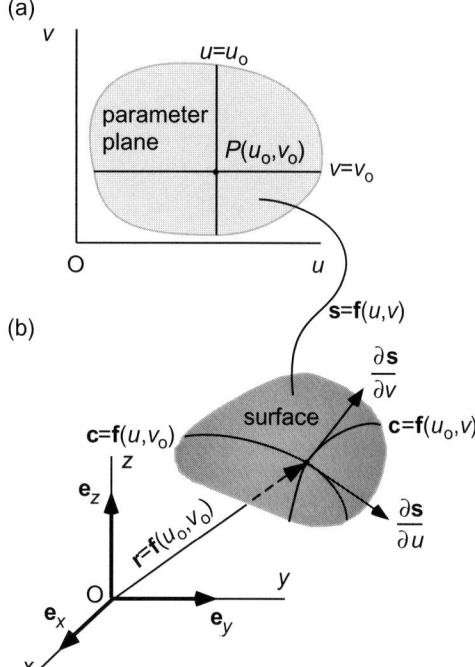

Fig. 4. Parametric representation of a surface. (**a**) The parameter plane with an arbitrary point $P(u_o, v_o)$ at the intersection of the two coordinate lines, $u = u_o$ and $v = v_o$. (**b**) Three-dimensional space with base vectors $(\mathbf{e}_x, \mathbf{e}_y$, and $\mathbf{e}_z)$ and position vectors for a point, $\mathbf{r}(u_o, v_o)$, the u-parameter curve $\mathbf{c}(u, v_o)$, the v-parameter curve $\mathbf{c}(u_o, v)$ and the surface, $\mathbf{s}(u, v)$. Partial derivatives are the tangent vectors to the u- and v-parameter curves.

respect to the base vectors $(\mathbf{e}_x, \mathbf{e}_y, \mathbf{e}_z)$. These functions, along with the base vectors, determine the position vectors, **s**, for all points on the curved surface. The components of the vector function in equation (1) may be complicated functions of u and v, subject only to constraints that ensure the functions are continuous and can be differentiated, and that the surface has a well-defined tangent plane at each point (Lipschultz 1969, p. 150). The vector equation is called the parametric representation of the surface.

An individual point in the parameter plane (Fig. 4a), such as $P(u_o, v_o)$, maps onto the surface (Fig. 4b) using the position vector $\mathbf{r} = \mathbf{f}(u_o, v_o)$. Similarly, the coordinate lines $u = u_o$ and $v = v_o$ in the parameter plane map onto the curves $\mathbf{c} = \mathbf{f}(u_o, v)$ and $\mathbf{c} = \mathbf{f}(u, v_o)$ on the surface. The two sets of orthogonal lines, $u =$ constant and $v =$ constant, cover the parameter plane. The curve $\mathbf{c} = \mathbf{f}(u, v_o)$ is referred to as a u-parameter curve on the surface: as the value of v_o is changed, the set of u-parameter curves is defined. The curve $\mathbf{c} = \mathbf{f}(u_o, v)$ is referred to as a v-parameter curve on the surface: as the value of u_0 is changed, the set of v-parameter curves is defined. In this way the two sets of coordinate lines, $v = $ constant and $u = $ constant, that cover the parameter plane map to the two sets of curves that cover the three-dimensional surface.

Because the parametric representation of a surface, $\mathbf{s} = \mathbf{f}(u, v)$, describes a vector function of two variable parameters, there is a partial derivative associated with each parameter. To calculate the partial derivative $\partial\mathbf{s}/\partial u$, for example, one takes the derivative with respect to u of each component of the vector function while holding v constant, and then uses these derivatives as the components of a new vector function. Thus, the partial derivatives of $\mathbf{s} = \mathbf{f}(u, v)$ with respect to the two parameters are (Lipschultz 1969, p. 126):

$$\frac{\partial\mathbf{s}}{\partial u} = \frac{\partial\mathbf{f}(u,v)}{\partial u} = \frac{\partial f_x}{\partial u}\mathbf{e}_x + \frac{\partial f_y}{\partial u}\mathbf{e}_y + \frac{\partial f_z}{\partial u}\mathbf{e}_z, \; v = \text{constant} \quad (2)$$

$$\frac{\partial\mathbf{s}}{\partial v} = \frac{\partial\mathbf{f}(u,v)}{\partial v} = \frac{\partial f_x}{\partial v}\mathbf{e}_x + \frac{\partial f_y}{\partial v}\mathbf{e}_y + \frac{\partial f_z}{\partial v}\mathbf{e}_z, \; u = \text{constant.} \quad (3)$$

The partial derivative, $\partial\mathbf{s}/\partial u$, represents vectors that are tangent to the u-parameter curves and point in the direction of increasing u (Fig. 4b). Similarly, $\partial\mathbf{s}/\partial v$ represents vectors that are tangent to the v-parameter curves and point in the direction of increasing v. The two partial derivatives of the vector function for a curved surface are used to define the parametric representation of planes that are tangent to this surface.

The tangent plane and unit normal vector to a surface

Now we are in a position to understand the concept of an arbitrary curve lying on a particular surface, such as the hackle or rib marks on a joint surface (Figs 1 and 2). This concept is important because the curvature at a point on a surface may vary with direction, and these directions are defined in terms of curves passing through the point and lying on the curved surface. Consider an arbitrary curve (Fig. 5a) in the parameter plane (u, v) that is defined by the functions $u = u(t)$, $v = v(t)$ and passes through the point $P(u_0, v_0)$. The two coordinate lines, $u = u_0$ and $v = v_0$, are parallel to the axes and also pass through this point. The parametric representation of the surface, $\mathbf{s} = \mathbf{f}(u, v)$, along with the u-parameter curve, $\mathbf{c}(u, v_0)$, and the v-parameter curve, $\mathbf{c}(u_0, v)$, are shown in Figure 5b. The arbitrary curve in the parameter plane is a function of the two 'surface' parameters, u and v, and these are, in turn, functions of the one 'curve' parameter, t. Thus, the parametric representation of the arbitrary curve is given by the vector function $\mathbf{c} = \mathbf{f}[u(t), v(t)]$.

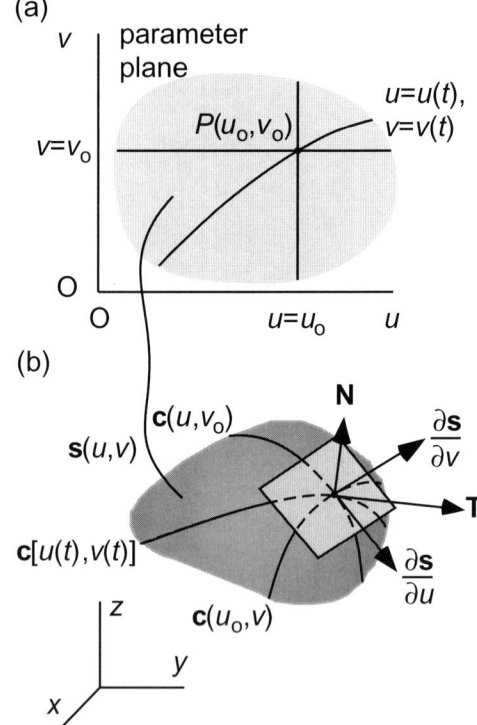

Fig. 5. Parametric representation of a surface with an arbitrary curve, $\mathbf{c}[u(t), v(t)]$. (**a**) Parameter plane with coordinate lines and arbitrary curve, $u = u(t)$, $v = v(t)$, through the point $P(u_0, v_0)$. (**b**) Three-dimensional space with mapping of these curves onto the surface. Partial derivatives are the tangent vectors to the u- and v-parameter curves; $\mathbf{N}$ is the unit normal vector; $\mathbf{T}$ is the tangent vector to the arbitrary curve. All tangent vectors lie in the tangent plane.

The tangent vector, $\mathbf{T}$, to the arbitrary curve is given by the derivative of the vector function $\mathbf{c}[u(t), v(t)]$ with respect to the parameter, t. Because $\mathbf{c}$ is a vector function of two variable parameters that are in turn functions of a single variable parameter, the derivative is evaluated using the chain rule as (Lipschultz 1969, p. 158):

$$\mathbf{T} = \frac{d\mathbf{c}[u(t), v(t)]}{dt} = \frac{d\mathbf{c}(u, v_0)}{du}\frac{du}{dt} +$$

$$\frac{d\mathbf{c}(u_0, v)}{dv}\frac{dv}{dt} = \frac{\partial\mathbf{s}}{\partial u}\frac{du}{dt} + \frac{\partial\mathbf{s}}{\partial v}\frac{dv}{dt}. \quad (4)$$

In the last step we used the fact that the partial derivatives, $\partial\mathbf{s}/\partial u$ and $\partial\mathbf{s}/\partial v$, are the tangent vectors to the u- and v-parameter curves, respectively (Fig. 5b). Because the tangent vector $\mathbf{T}$ is linearly dependent on these two partial derivatives, it also lies in the tangent plane. In this way the tangent vector to the surface

$\mathbf{s}(u, v)$ at an arbitrary point in any arbitrary direction is related to the partial derivatives of the parametric representation of the curved surface at that point.

The orientation of the tangent plane is uniquely determined by either of the two unit normal vectors to that plane (Fig. 5b), and the choice between these two oppositely directed vectors is determined by a right-hand rule. The unit normal vector, $\mathbf{N}$, to the tangent plane makes a right-handed orthogonal system with the two tangent vectors, $\partial \mathbf{s}/\partial u$ and $\partial \mathbf{s}/\partial v$, and therefore is defined in terms of the vector product as (Lipschultz 1969, p. 158):

$$\mathbf{N} = \frac{\dfrac{\partial \mathbf{s}}{\partial u} \times \dfrac{\partial \mathbf{s}}{\partial v}}{\left| \dfrac{\partial \mathbf{s}}{\partial u} \times \dfrac{\partial \mathbf{s}}{\partial v} \right|}. \tag{5}$$

Using equation (5) the unit normal at any point on a surface can be calculated from its parametric representation $\mathbf{s}(u, v)$.

The first and second fundamental forms of a surface

The shape of continuous curved surfaces is uniquely described at an arbitrary point in terms of two differential quantities called the first and second fundamental forms (Lipschultz 1969, Chap. 9). The first fundamental form, I, of a curved surface, $\mathbf{s} = \mathbf{f}(u, v)$, is a measure of the differential arc length, $d\mathbf{c}$, of curves lying on the surface and oriented in all possible directions at the arbitrary point. To define this property of a curved surface consider the two points, $P(u, v)$ and $P(u + du, v + dv)$, that lie along an arbitrary line, $u = u(t)$ and $v = v(t)$, in the parameter plane and are separated by an arbitrarily small distance (Fig. 6a). These points map onto the curved surface as the points $\mathbf{r}(u, v)$ and $\mathbf{r}(u + du, v + dv)$ along the curve $\mathbf{c}[u(t), v(t)]$. The tangent vector to this arbitrary curve is defined using equation (4) as $\mathbf{T} = d\mathbf{c}/dt$, so the differential vector quantity $d\mathbf{c}$ is (Lipschultz 1969, p. 171):

$$d\mathbf{c} = \mathbf{T}dt = \frac{\partial \mathbf{s}}{\partial u}du + \frac{\partial \mathbf{s}}{\partial v}dv. \tag{6}$$

Because this vector is parallel to $\mathbf{T}$, it lies in the tangent plane and we refer to it as the differential tangent vector. As shown in Figure 6b, this vector is not exactly parallel to the secant line that passes through the two points $\mathbf{r}(u, v)$ and $\mathbf{r}(u + du, v + dv)$ on the curve. However, the tangent line and the secant line become parallel as the distance between the two points goes to zero, so the tangent line is the best-fitting line to the curve at the point in question. The differential tangent vector, $d\mathbf{c}$, is a measure of the differential arc length of a curve lying in the

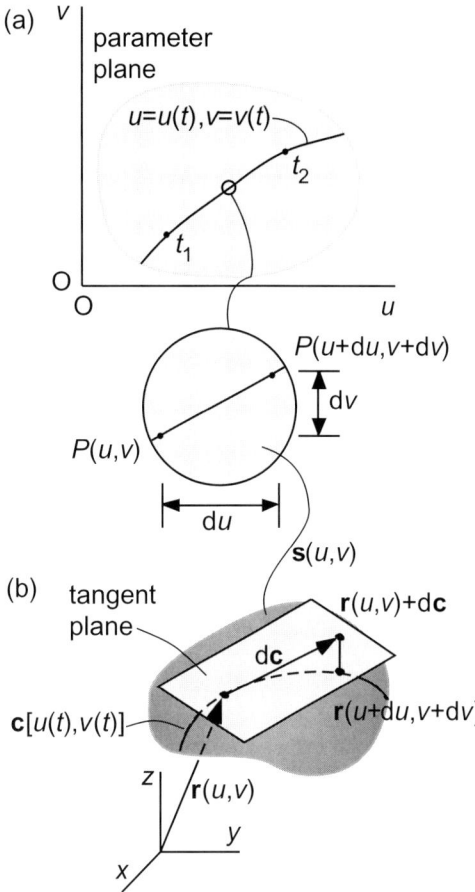

Fig. 6. Parametric representation of a surface with an arbitrary curve. (**a**) Parameter plane with curve and limiting points t_1 and t_2 for determination of the length of arc. Inset: differential arc length. (**b**) Three-dimensional space with surface and arbitrary curve including the tangent plane and the differential tangent vector $d\mathbf{c}$.

surface, $\mathbf{s}(u, v)$, and oriented in a direction determined by the differential parameters du and dv.

The first fundamental form, I, is a differential quantity defined as the scalar product of the differential tangent vector with itself (Lipschultz 1969, p. 171). This scalar product is a measure of arc length on the surface and it is expanded as follows:

$$I = d\mathbf{c} \cdot d\mathbf{c} = \left(\frac{\partial \mathbf{s}}{\partial u} \cdot \frac{\partial \mathbf{s}}{\partial u} \right)du^2 + 2\left(\frac{\partial \mathbf{s}}{\partial u} \cdot \frac{\partial \mathbf{s}}{\partial v} \right)du\,dv +$$
$$\left(\frac{\partial \mathbf{s}}{\partial v} \cdot \frac{\partial \mathbf{s}}{\partial v} \right)dv^2. \tag{7}$$

Note that the first fundamental form characterizes lengths in all possible directions at a particular point,

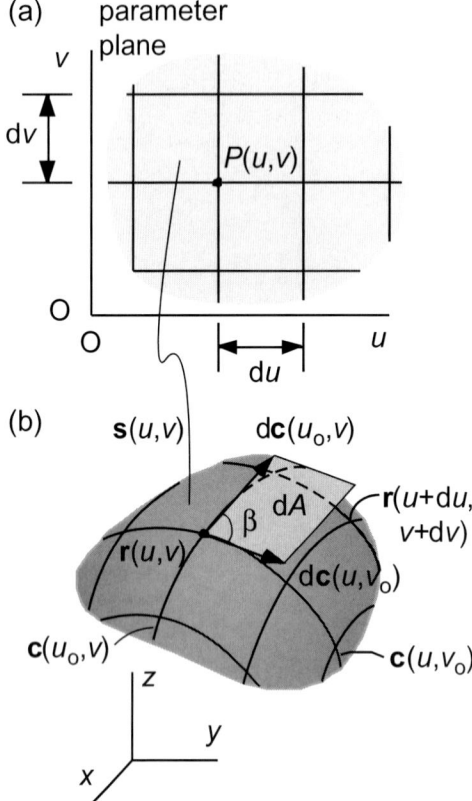

Fig. 7. Parametric representation of a surface with a grid of *u*- and *v*-parameter curves. (**a**) Parameter plane with coordinate lines separated by differential distances d*u* and d*v*. (**b**) Three-dimensional surface with adjacent *u*- and *v*-parameter curves making an angle β and defining a differential area, d*A*.

and that the differential parameters d*u* and d*v* are used to define the direction (Fig. 6a). Because of their role in defining I and in the calculation of useful quantities, such as the arc length and area on a curved surface, the coefficients in equation (7) are denoted with special symbols and called the coefficients of the first fundamental form. These coefficients depend on the choice of parameters used to represent the surface, but the first fundamental form itself is invariant with respect to this choice (Lipschultz 1969, p. 172). Because they are scalar products of the tangent vectors to the *u*- and *v*-parameter curves, the coefficients can be interpreted geometrically as:

$$E = \left|\frac{\partial \mathbf{s}}{\partial u}\right|^2, F = \left|\frac{\partial \mathbf{s}}{\partial u}\right|\left|\frac{\partial \mathbf{s}}{\partial v}\right|\cos\beta, G = \left|\frac{\partial \mathbf{s}}{\partial v}\right|^2. \quad (8)$$

Here β is the smaller angle between the two tangent vectors to the *u*- and *v*-parameter curves at a particu-

lar point on the surface (Fig. 7b). *E* and *G* are, respectively, the squares of the magnitudes of these vectors. Furthermore, the *u*- and *v*-parameter curves are orthogonal (β = 90°) if and only if *F* = 0 (Lipschultz 1969, p. 173).

The coefficients of the first fundamental form are useful for calculating the area of a surface, given its parametric representation. Consider adjacent members of the two families of coordinate lines on the parameter plane that are separated by small differential distances d*u* and d*v* (Fig. 7a). These lines partition the parameter plane into a rectangular grid that is, in turn, mapped onto the surface as a curvilinear grid (Fig. 7b). The differential tangent vectors at the point **r**(*u*,*v*) in the directions of the *u*- and *v*-parameter curves are found from equation (6):

$$d\mathbf{c}(u, v_o) = \frac{\partial \mathbf{s}}{\partial u}du, \, d\mathbf{c}(u_o, v) = \frac{\partial \mathbf{s}}{\partial u}dv. \quad (9)$$

These vectors form two sides of a small parallelogram and the differential area, d*A*, of this planar figure is used to approximate the area of the curved surface between the adjacent parameter curves. Recall that the area of a parallelogram with sides *a* and *b* and included angle β is *A* = *ab*sinβ, so the differential area is d*A* = |d**c**(*u*, *v*ₒ)|| d**c**(*u*, *v*ₒ)|sinβ. Substituting from (9), and noting that the differential area is in the form of the absolute value of a cross product, we have:

$$dA = |d\mathbf{c}(u, v_o) \times d\mathbf{c}(u_o, v)| = \left|\frac{\partial \mathbf{s}}{\partial u} \times \frac{\partial \mathbf{s}}{\partial v}\right| du\, dv =$$

$$\left[\left(\frac{\partial \mathbf{s}}{\partial u} \times \frac{\partial \mathbf{s}}{\partial v}\right) \cdot \left(\frac{\partial \mathbf{s}}{\partial u} \times \frac{\partial \mathbf{s}}{\partial v}\right)\right]^{1/2} du\, dv = [EG - F^2]^{1/2} du\, dv.$$
$$(10)$$

The second line uses an identity relating the scalar and vector products to convert the equation to one that involves the coefficients of the first fundamental form (Lipschultz 1969, p. 10). The area of a curved surface is found by integrating the differential area over the surface

$$A = \iint (EG - F^2)^{1/2} du\, dv. \quad (11)$$

The surface area is a function of the coefficients of the first fundamental form and, in turn, of the parameters *u* and *v*.

The second fundamental form provides a measure of the shape of the surface. To understand how this is accomplished we focus on a very small part of the parameter plane, so lengths along the coordinate axes are measured using the differential quantities d*u* and d*v* (Fig. 8a). Consider the arbitrary curve *u* = *u*(*t*), *v* = *v*(*t*) in the parameter plane that maps to the curve **c**[*u*(*t*), *v*(*t*)] on the curved surface, **s** = **f**(*u*, *v*). At an arbitrary point along this curve the differential

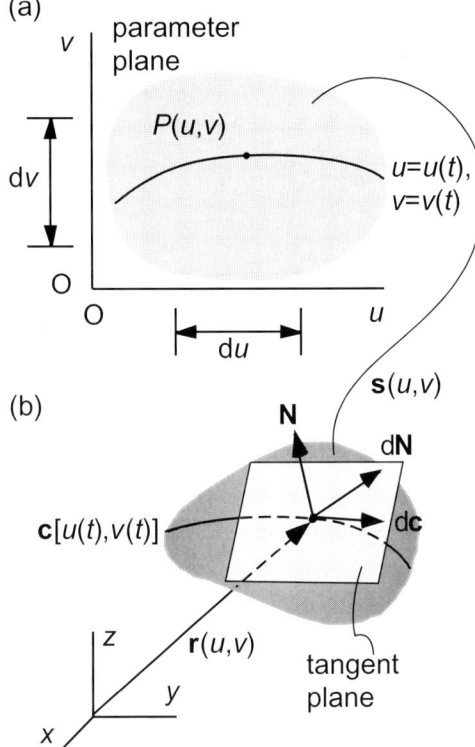

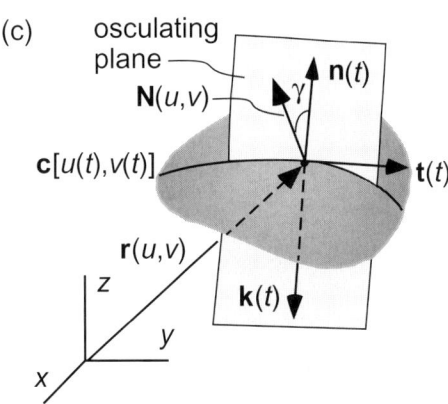

Fig. 8. Parametric representation of a surface used to define the normal curvature. (**a**) Parameter plane with arbitrary curve, $u = u(t)$, $v = v(t)$, viewed only over differential distances du and dv. (**b**) Three-dimensional space with surface, tangent plane, unit normal vector, **N**, differential tangent vector, dc, and differential normal vector, dN. (**c**) Same surface but with osculating plane, and both the unit principal normal vector, $n(t)$, and curvature vector, $k(t)$, to the arbitrary curve.

tangent vector, dc, is defined by equation (6) and lies in the tangent plane to the surface at this point (Fig. 8b). The unit vector, **N**, is normal to the surface at this point and is defined by equation (5). This vector is normal to the tangent plane and is a function of the two parameters u and v, such that its differential is (Lipschultz 1969, p. 175):

$$dN = \frac{\partial N}{\partial u}du + \frac{\partial N}{\partial v}dv. \qquad (12)$$

This vector is a measure of the change in orientation of **N** with position on the surface. Although dc and dN both lie in the tangent plane (Fig. 8b), these vectors are not necessarily parallel to one another. Thus, the shape of the surface in the particular direction specified by dc is characterized by the component of dN on an axis that is parallel to dc, in other words by the scalar product of these two vectors.

The second fundamental form, II, is a differential quantity defined as the negative of the scalar product of the differential normal vector and the differential tangent vector (Lipschultz 1969, p. 175):

$$II = -dN \cdot dc = -\left(\frac{\partial N}{\partial u} \cdot \frac{\partial s}{\partial u}\right)du^2 - \left(\frac{\partial N}{\partial u} \cdot \frac{\partial s}{\partial v} + \frac{\partial N}{\partial v} \cdot \frac{\partial s}{\partial u}\right)$$

$$du \, dv - \left(\frac{\partial N}{\partial v} \cdot \frac{\partial s}{\partial v}\right)dv^2. \qquad (13)$$

The quantities L, M and N are functions of the two parameters, u and v, and are called the coefficients of the second fundamental form of the surface. Do not confuse the scalar quantity N, and the vector quantity **N**. These coefficients depend upon the choice of parameters used to represent the surface, but the second fundamental form itself is invariant with respect to this choice. In this sense II is a property of the surface and is fundamental to characterizing the shape of the surface. The second fundamental form characterizes the changing shape of the surface in all directions at a particular point, and the differential parameters du and dv define the direction. The coefficients of the second fundamental form may be rewritten in the following way that is useful for computations (Lipschultz 1969, p. 176):

$$L = N \cdot \frac{\partial^2 s}{\partial u^2}, \; M = N \cdot \frac{\partial^2 s}{\partial u \partial v}, \; N = N \cdot \frac{\partial^2 s}{\partial v^2}. \qquad (14)$$

Using these equations and equation (5) for the unit normal vector, the coefficients may be calculated for any position on the surface $s = f(u, v)$.

The coefficients of the second fundamental form determine the local shape of a surface in the vicinity of a particular point according to the following classification (Lipschultz 1969, p. 177):

$$LN - M^2 \begin{cases} >0, \text{ elliptic point} \\ =0, \text{ parabolic point} \\ <0, \text{ hyperbolic point} \end{cases}$$
$$L = M = N = 0, \text{ planar point.} \qquad (15)$$

For the elliptic point the local surface lies entirely on one side of the tangent plane to that point (Fig. 9a). Planes that are parallel to the tangent plane and intersect the local surface cut out elliptical curves. For the parabolic point (Fig. 9b) not all of the coefficients are zero, but the combination $LN - M^2$ is zero, and the local surface is cylindrical. Planes that are parallel to the tangent plane intersect the local surface in two straight and parallel lines, or in one line if the surface passes from one side to the other of the tangent plane. For the hyperbolic point (Fig. 9c) the local surface lies on both sides of the tangent plane to that point and intersects the tangent plane along two straight and non-parallel lines where the surface passes from one side to the other of the tangent plane. For the special case where all the coefficients of the second fundamental form are zero, the local surface is planar.

The normal curvature of a surface

The shape of a curved line is characterized, in part, using the curvature vector, $\mathbf{k}$, and the scalar curvature, $\kappa = |\mathbf{k}|$ (Lipschultz 1969, p. 62). The curvature vector is not generally a unit vector, but it is orthogonal to the unit tangent vector and is directed away from the curve on its concave side (Fig. 8c). An intuitive understanding of the scalar curvature at a point on a curve may be obtained by considering the circle that passes through this point, lies on the concave side of the curve and has the closest possible contact with the curve. In other words it is the 'best-fitting' circle to the curve at that point. The radius of this circle is equal to the radius of curvature of the curve at the point, and this in turn is equal to the reciprocal of the scalar curvature (Lipschultz 1969, p. 1).

For a curve on a surface two analogous measures of shape are the normal curvature vector, $\mathbf{k}_n$, and the normal curvature, κ_n. Both of these quantities are defined by considering an arbitrary curve $u = u(t)$, $v = v(t)$ in the parameter plane (Fig. 8a), which maps to the curve $\mathbf{c}[u(t), v(t)]$ on the surface (Fig. 8c). At a point on this curve the curvature vector $\mathbf{k}$ is a function of the parameter t; it lies in the osculating plane of the curve, and it extends away from the concave side of the curve. The osculating plane of a curve is the plane that contains both the unit tangent vector, $\mathbf{t}(t)$, and the principal unit normal vector, $\mathbf{n}(t)$, to the curve (Fig. 8c). In contrast, the unit normal vector to the surface, $\mathbf{N}$, is perpendicular to the surface; it is a function of the parameters u and v, and it may not lie in the osculating plane to the curve. The normal curvature

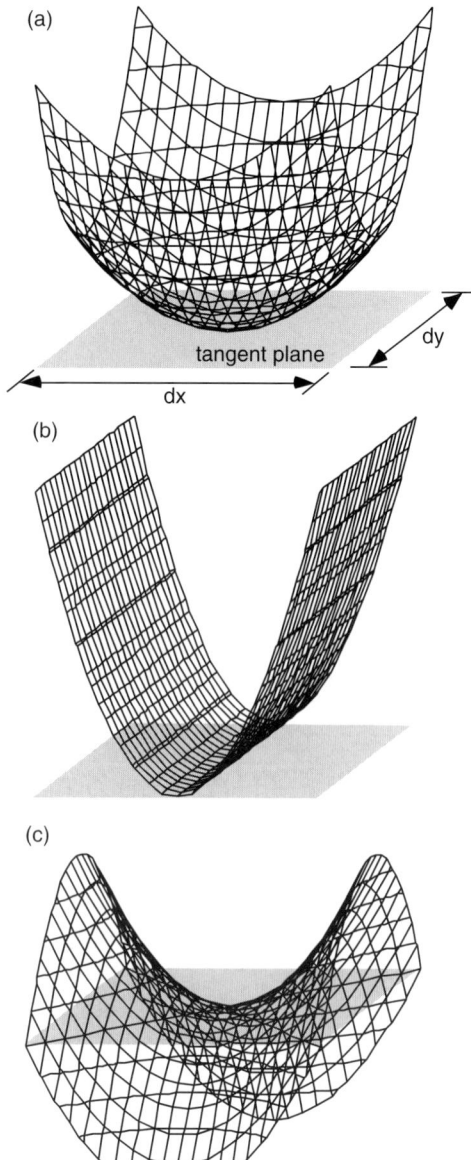

Fig. 9. Wire-frame diagrams of possible surface shapes near a point with the tangent plane at that point: (**a**) elliptical point; (**b**) parabolic point; and (**c**) hyperbolic point.

vector, $\mathbf{k}_n$, and the normal curvature, κ_n, are defined in terms of $\mathbf{k}(t)$ and $\mathbf{N}(u, v)$ as (Lipschultz 1969, p. 174):

$$\mathbf{k}_n = (\mathbf{k} \cdot \mathbf{N})\, \mathbf{N}, \quad \kappa_n = \mathbf{k} \cdot \mathbf{N}. \qquad (16)$$

Geometrically, we understand that $\mathbf{k}_n$ is the vector projection of the curvature vector, $\mathbf{k}$, onto the line parallel to the unit normal vector, $\mathbf{N}$. The normal curvature vector may have the same, or the opposite, direction as $\mathbf{N}$.

The principal unit normal vector, $\mathbf{n}(t)$, to the curve $\mathbf{c}[u(t), v(t)]$ lies in the osculating plane perpendicular to $\mathbf{t}(t)$, but its direction is chosen for consistency along the curve (Fig. 7). If we choose the direction of $\mathbf{n}(t)$ such that the angle, γ, between $\mathbf{n}$ and $\mathbf{N}$ is in the range $0 \le \gamma \le \pi/2$ (Fig. 8c), then the curvature and normal curvature are related as (Lipschultz 1969, p. 180):

$$\kappa_n = \kappa \cos \gamma. \qquad (17)$$

In other words, the normal curvature for the arbitrary curve $\mathbf{c}[u(t), v(t)]$ on the surface $\mathbf{s} = \mathbf{f}(u, v)$ is equal to the curvature of this curve times the cosine of the angle between $\mathbf{n}(t)$ and $\mathbf{N}(u, v)$. If the osculating plane of the curve contains the unit normal vector for the surface, then $\mathbf{n}$ and $\mathbf{N}$ are parallel, and $\kappa_n = \kappa$. In this case the curve is the intersection of the plane containing $\mathbf{N}$ and the surface. On the other hand, if the osculating plane of the curve is parallel to the tangent plane for the surface, then $\kappa_n = 0$. In this case the surface is planar near the point in question and the curve lies in that plane.

It is important to understand that the normal curvature at a point on a surface depends on the direction of the designated curve through that point and lying in the surface. That is, if one chooses two differently directed curved lines through the same point on the surface, the respective values of κ_n may be different. The curvature, κ, at a point on a curve is a unique number that is a property of the curve. In contrast, the normal curvature, κ_n, at a point on a curved surface varies systematically with the direction of the curves through that point. Because the normal curvature is a property of the surface at any point it can be written as a function of the fundamental forms (Lipschultz 1969, p. 180):

$$\kappa_n = \frac{L du^2 + 2M du\, dv + N dv^2}{E du^2 + 2F du\, dv + G dv^2} = \frac{\mathrm{II}}{\mathrm{I}}. \qquad (18)$$

The differentials, du and dv, determine the orientation of the curve, and the normal curvature is simply the ratio of the second to the first fundamental form.

The normal curvature at a point varies in a smooth and systematic manner with the direction of the tangent line through the point of interest from a maximum value, κ_1, to a minimum value, κ_2 (Fig. 10a). This variation with direction is of the same form (Fig. 10b) for all surfaces with continuous partial derivatives of the order of 2 or greater such that (Lipschultz 1969, p. 196):

$$\kappa_n = \kappa_1 \cos^2 \alpha = \kappa_2 \sin^2 \alpha. \qquad (19)$$

This relationship is known as Euler's theorem and the angle α is measured in the tangent plane from the direction of the tangent line corresponding to the curvature κ_n, to the tangent line corresponding to the curvature κ_1. The two extreme values, κ_1 and κ_2, are called the principal normal curvatures. The directions of the tangent lines associated with the extreme values of normal curvature are called the principal directions of the normal curvature and they are orthogonal (Fig. 10a).

Referring collectively to the values of the principal curvatures as κ_0, they are found by solving a quadratic equation, with coefficients that depend on the coefficients of the fundamental forms (Lipschultz 1969, p. 183):

$$(EG - F^2)\kappa_0^2 - (EN + GL - 2FM)\kappa_0 + (LM - M^2) = 0. \qquad (20)$$

The two roots of equation (20) are always real numbers, but two relevant cases are distinguished. If the two roots are unequal they are κ_1 and κ_2. If two roots are equal then the normal curvature is constant for all directions of the tangent lines through the point in question. In the second case the point is referred to as an umbilical point and the local surface either is planar ($\kappa_0 = 0$) or spherical ($\kappa_0 = $ constant). The orientations of the orthogonal tangent lines that correspond to the principal curvatures are given by the direction numbers (du, dv) for values of these differential quantities that satisfy the following equation (Lipschultz 1969, p. 185):

$$(EM - LF)(du)^2 + (EN - LG)du\, dv + (FN - MG) (dv)^2 = 0. \qquad (21)$$

For a non-umbilical point the solution to equation (21) provides the direction numbers for the lines in the parameter plane that map to curves with the principal curvatures.

The sum and product of the principal normal curvatures serve to define two useful parameters (Lipschultz 1969, p. 184) that characterize the local shape of a curved surface near a point:

$$\kappa_m = \frac{(EN - 2FM + GL)}{2(EG - F^2)} = \tfrac{1}{2}(\kappa_1 + \kappa_2),$$

$$\kappa_g = \frac{(LN - M^2)}{(EG - F^2)} = \kappa_1 \kappa_2. \qquad (22)$$

The mean principal normal curvature is κ_m and κ_g is called the Gaussian curvature. The Gaussian and mean curvatures of a surface at a point are independent of the parametric representation of the surface and therefore can be used to categorize the local shape of the surface (Roberts 2001; Bergbauer & Pollard 2003). For $\kappa_g > 0$ the point on the surface is elliptical (Fig. 9a); it is a dome if $\kappa_m > 0$ and it is a basin if $\kappa_m < 0$ (Fig. 10b). For $\kappa_g = 0$ the surface is

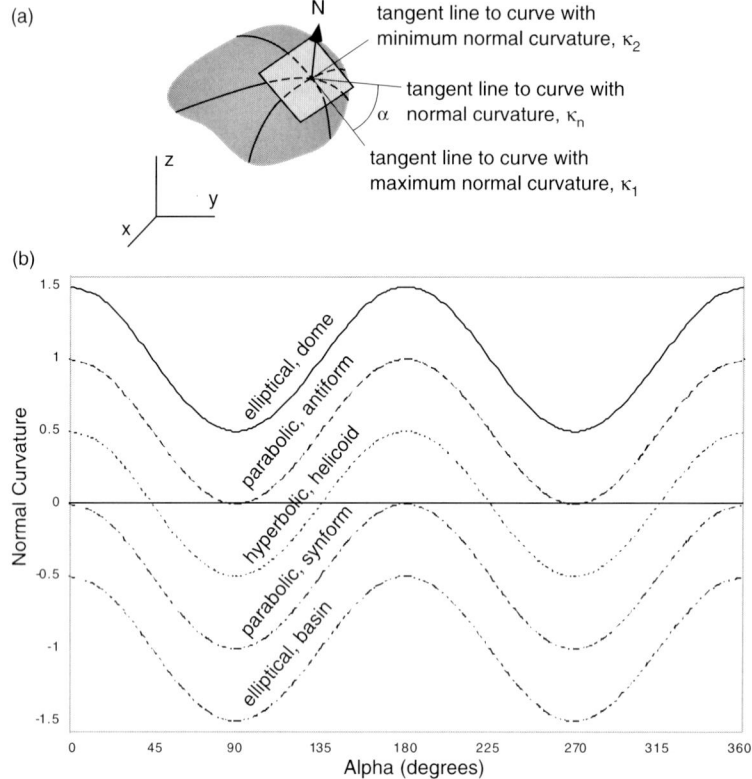

Fig. 10. Variation in the normal curvature at a point with the direction of the tangent line through that point. (**a**) Three-dimensional space with surface, tangent plane and unit normal vector. The angle α is measured in the tangent plane between the tangent line for the curve with maximum curvature, κ_1, and the tangent line with normal curvature, κ_n. (**b**) Illustration of Euler's theorem (19) showing the distribution of normal curvature with angle α. Different curves represent the different possible surface shapes.

parabolic (Fig. 9b); it is antiformal if $\kappa_m > 0$, planar if $\kappa_m = 0$ and synformal if $\kappa_m < 0$ (Fig. 10b). For $\kappa_g < 0$ the surface is hyperbolic (Fig. 9c); it is helicoidal if $\kappa_m = 0$ (Fig. 10b).

An analytical model for fracture surfaces

Based on observations and mapping of opening fractures, including basaltic dykes and joints, Pollard *et al.* (1982) proposed that the surfaces of these fractures can be idealized as helicoids. They observed that traces of some echelon fractures are approximately straight when viewed in cross-sectional exposures that apparently are perpendicular to the propagation direction (e.g. Fig. 1, parts 6–8). However, exposures at different levels (serial cross-sections) reveal different orientations, such that the surfaces appear to twist about an axis that is parallel to the propagation direction (Fig. 1, parts 5 and 9). They suggested that a straight line that is

perpendicular to the propagation direction could generate these fracture surfaces. If the generating line is rotated about an axis that is parallel to the propagation direction as it is translated in that direction, a twisted surface is produced that is a helicoid. In this section we define the analytical description of helicoids and the quantities that serve to characterize fracture surfaces with this shape.

The parametric representation of helicoids is based on equation (1) where $v=$ constant and $u=$ constant are coordinate lines in the parameter plane that map onto the u- and v-parameter curves on the helicoidal surface defined by the position vectors (Weisstein 2002):

$$\mathbf{s}(u, v) = (u \cos v)\mathbf{e}_x + (u \sin v)\mathbf{e}_y + (cv)\mathbf{e}_z. \quad (23)$$

The u-parameter curves on the surface are straight lines that intersect and are perpendicular to the z-axis. Each v-parameter curve is a helix that intersects the x-axis and curves around the z-axis. The tangent vectors to the u- and v-parameter curves are found using (23) in equations (2) and (3) such that:

$$\frac{\partial \mathbf{s}}{\partial u} = (\cos v)\mathbf{e}_x + (\sin v)\mathbf{e}_y, \quad \frac{\partial \mathbf{s}}{\partial v} = -(u \sin v)\mathbf{e}_x + (u \cos v)\mathbf{e}_y + (c)\mathbf{e}_z. \quad (24)$$

Using equation (24) in (5) the unit normal vector at any point on the helicoid is:

$$\mathbf{N}(u, v) = (1/\sqrt{c^2 + u^2})[(c \sin v)\mathbf{e}_x - (c \cos v)\mathbf{e}_y + (u)\mathbf{e}_z]. \quad (25)$$

The unit normal completely determines the orientation of the surface at any point specified by the parameters (u, v) and for a given value of the constant c.

Geometric interpretation of fracture surfaces using the helicoid

The helicoid may be visualized as shown in Figure 11a, where the ranges of the parameters are taken as $-1 \leq u \leq 1$ and $0 \leq v \leq 2\pi$. This image bears little resemblance to the surfaces illustrated by Woodworth (1896) (Fig. 1, parts 5, 9 and 10). In fact the patch of this surface that we use to model a fracture surface is viewed in Figure 11b. To appreciate the geometric meaning of the parameters and constants associated with this patch of the helicoid consider the following special cases. For $u = 0$ the surface is coincident with the z-axis such that:

$$\mathbf{s}(0, v) = (cv)\mathbf{e}_z, \mathbf{N}(0, v) = (\sin v)\mathbf{e}_x - (\cos v)\mathbf{e}_y. \quad (26)$$

Position on this part of the surface is determined by the value of v scaled by the constant c. The unit normal is independent of c and varies, for example, from $-\mathbf{e}_y$ for $v = 0$ to $+\mathbf{e}_x$ for $v = \pi/2$ as the local surface twists from the (x, z)-plane to the (y, z)-plane.

We define the twist of the surface as the angle between the reference unit normal, $\mathbf{N}(0, 0)$, and the unit normal at the point in question, $\mathbf{N}(0, v)$. Considering the scalar product of these two vectors we have:

$$\mathbf{N}(0, 0) \cdot \mathbf{N}(0, v) = [-(1)\mathbf{e}_y] \cdot [(\sin v)\mathbf{e}_x - (\cos v)\mathbf{e}_y] = (\cos v). \quad (27)$$

Because the scalar product of any two unit vectors is equal to the cosine of the angle between them, the twist angle is equal to the parameter v. From equation (26) the component of $\mathbf{s}$ in the z-direction is $s_z = cv$, but $s_z = z$, so $dz/dv = c$. Considering the reciprocal of this derivative, we define the spatial rate of twist of the surface as:

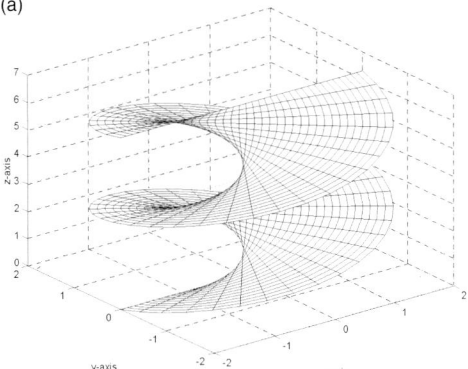

(a)

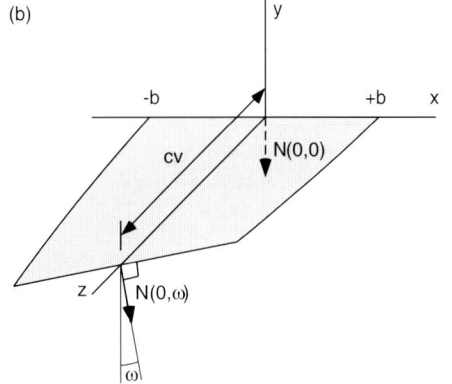

(b)

Fig. 11. Visualizations of the helicoids. (**a**) The standard form illustrated in textbooks with ranges of the parameters $-1 \leq u \leq 1$ and $0 \leq v \leq 2\pi$. (**b**) The form of the helicoids taken as a model for a fracture surface with ranges of the parameters $-b \leq u \leq +b$ and $0 \leq v \leq \omega$.

$$\frac{dv}{dv} = \frac{1}{c} = \text{constant}. \quad (28)$$

Because we use the unit normal at the point $\mathbf{s}(0, 0)$ as a reference orientation and define the derivative in (28) for positive dz, the patch of the surface that will be considered as an analytical description of a fracture surface covers the range $0 \leq v \leq \omega$ on the parameter plane (Fig. 11b) Thus, ω is the maximum twist angle at the distal edge of the fracture surface.

For $v = 0$ the surface is coincident with the x-axis such that:

$$\mathbf{s}(u, 0) = (u)\mathbf{e}_x, \mathbf{N}(u, 0) = (1/\sqrt{c^2 + u^2})[-(c)\mathbf{e}_y + (u)\mathbf{e}_z]. \quad (29)$$

Position on this part of the surface is given by the value of u. For the analytical description of a helicoidal fracture surface we take the range $-b \leq u \leq +b$, so the width of the fracture is $2b$ and the fracture midline is coincident with the z-axis (Fig. 11b). We take the x-axis as the intersection between a planar fracture surface that lies in the (x, z)-plane, where $z \leq 0$, and a helicoidal fracture surface that twists about the positive z-axis. Thus, the x-axis is equivalent to the line of breakdown from a single parent (main) fracture to multiple echelon fractures (feather fractures, twist hackle) in the fringe (Fig. 1, part 5, line dd; Fig. 2, item 6).

For $u = 0$ and $v = 0$ the unit normal for the helicoidal fracture surface is $\mathbf{N}(0, 0) = -\mathbf{e}_y$, which is in the y-coordinate direction and is parallel to the unit normal for the planar fracture surface. However, for all other points along the x-axis, $0 < |u| \leq b$ and $v = 0$, there is a component of the unit normal in the z-coordinate direction that is proportional to $u/(c^2 + u^2)^{1/2}$. Thus, the planar and helicoidal fracture surfaces may be continuous with one another at the line of intersection (x-axis), but there is a discontinuity in the orientation of the two surfaces except at the midline of the helicoidal surface ($u = 0, v = 0$).

For $c = 0$ the rate of twist is infinite and the surface is restricted to the (x, y)-plane with a unit normal in the z-direction:

$$\mathbf{s}(u, v) = (u \cos v)\mathbf{e}_x + (u \sin v)\mathbf{e}_y, \ \mathbf{N}(u, v) = \mathbf{e}_z. \ (30)$$

Positions on this surface are determined by both parameters u and v. For $|u| \leq b$ and $v \leq \pi$ the surface is a circular disk of radius b. This limiting case is not of any relevance to the physical phenomenon of fracture and will not be considered further.

For $c \to \infty$ the rate of twist goes toward zero. To examine this limit we take the surface length, $c\omega$, as finite and less than some multiple of the half width, b. Commonly, this multiple is less than 10, but taking the generous value of 100 we have $cv \leq c\omega < 100b$ so $v < 100b/c$. Then, as $c \to \infty$, we find $\sin v \to 0$ and $\cos v \to 1$ so:

$$\mathbf{s}(u, v) \cong (u)\mathbf{e}_x + (cv)\mathbf{e}_z, \ \mathbf{N}(u, v) \cong -\mathbf{e}_y. \ (31)$$

The helicoidal surface collapses onto the (x, z)-plane with a unit normal in the negative y-direction. Positions on this rectangular surface are determined by both parameters u and v. For this special case there is no discontinuity in the orientation of the unit normal.

Fundamental forms for the helicoidal fracture surface

The helicoidal surface is characterized by the coefficients of the fundamental forms that are calculated from derivatives of the position vector for the surface $\mathbf{s}(u, v)$. Using the first derivatives and we find (Weisstein 2002):

$$E = \frac{\partial \mathbf{s}}{\partial u} \cdot \frac{\partial \mathbf{s}}{\partial u} = 1, \ F = \frac{\partial \mathbf{s}}{\partial u} \cdot \frac{\partial \mathbf{s}}{\partial v} = 0, \ G = \frac{\partial \mathbf{s}}{\partial v} \cdot \frac{\partial \mathbf{s}}{\partial v}$$

$$= c^2 = u^2. \ (32)$$

The coefficients of the second fundamental form are calculated using the second derivatives of $\mathbf{s}(u, v)$ and these are found from (24) to be:

$$\frac{\partial^2 \mathbf{s}}{\partial u^2} = 0, \ \frac{\partial^2 \mathbf{s}}{\partial u \, \partial v} = -(\sin v)\mathbf{e}_x + (\cos v)\mathbf{e}_y, \ \frac{\partial^2 \mathbf{s}}{\partial u^2}$$

$$= -(u \cos v)\mathbf{e}_x - (u \sin v)\mathbf{e}_y. \ (33)$$

Using equations (25) and (33) in (14) we find (Weisstein 2002):

$$L = \mathbf{N} \cdot \frac{\partial^2 \mathbf{s}}{\partial u^2} = 0, \ M = \mathbf{N} \cdot \frac{\partial^2 \mathbf{s}}{\partial u \, \partial v} = \frac{-c}{\sqrt{c^2 + u^2}},$$

$$N = \mathbf{N} \cdot \frac{\partial^2 \mathbf{s}}{\partial v^2} = 0.$$

$$(34)$$

The two fundamental forms, as defined in equations (7) and (13), are written for the helicoidal surface as:

$$I = \left(\frac{du}{dt}\right)^2 + (c^2 + u^2)\left(\frac{dv}{dt}\right)^2, \ II = \frac{-2c}{\sqrt{c^2 + u^2}}$$

$$\left(\frac{du}{dt}\frac{dv}{dt}\right). \ (35)$$

The first fundamental form is used to characterize lengths and areas on the surface, and the second fundamental form is used to characterize the shape of the surface.

The coefficients of the first fundamental form are scalar products of the tangent vectors and, in general, a scalar product of two vectors, $\mathbf{U}$ and $\mathbf{V}$, is related to the angle between them, θ, as $\mathbf{U} \cdot \mathbf{V} = |\mathbf{U}||\mathbf{V}|\cos\theta$. We note that E is the squared magnitude of the tangent vector to the u-parameter curves and this vector is a unit vector. Furthermore, because this vector lies in the (x, y)-plane and is not a function of u, the u-parameter curve is a straight line. Similarly, G is the squared magnitude of the tangent vector to the v-parameter curves and this vector has a magnitude to $(c^2 + u^2)^{1/2}$. We note that the parametric representation of a circular helix of radius u_o and pitch c, and the tangent vector to this curve, are (Lipschultz 1969, p. 63):

$$\mathbf{c}(t) = (u_o \cos t)\mathbf{e}_x + (u_o \sin t)\mathbf{e}_y + (ct)\mathbf{e}_z \ (36)$$

$$\frac{d\mathbf{c}}{dt} = -(u_o \sin t)\mathbf{e}_x + (u_o \cos t)\mathbf{e}_y + (c)\mathbf{e}_z. \ (37)$$

Here t is the single arbitrary parameter for the curve. Comparing these equations to equations (23) and (24) it is clear that each v-parameter curve is a helix with radius u_0 and pitch c. Because neither tangent vector is zero, yet from $F = 0$, it must be the case that $\cos\theta = 0$ and $\theta = \pi/2$, so the u- and v-parameter curves are orthogonal everywhere on the helicoidal surface. The set of straight lines and the set of helixes cover the helicoidal surface with an orthogonal network.

The area of the patch of a helicoid that we take as a fracture surface is found using equation (32) in with the limits $-b \le u \le +b$ and $0 \le v \le \omega$. From symmetry this is equivalent to twice the area using the range $0 \le u \le +b$ such that (Weisstein 2002):

$$A = 2 \int_0^\omega \left[\int_0^b \sqrt{c^2 + u^2}\, du \right] dv. \quad (38)$$

The inner integral is evaluated using (Selby 1975, p. 424, #156):

$$\int \sqrt{a^2 + x^2}\, dx = \tfrac{1}{2}\left[x\sqrt{a^2 + x^2} + a^2 \ln\left(x + \sqrt{a^2 + x^2}\right) \right]. \quad (39)$$

The area of the helicoidal fracture surface is:

$$A = \omega\left[b\sqrt{b^2 + c^2} + c^2 \ln\left(\frac{b + \sqrt{b^2 + c^2}}{c} \right) \right]. \quad (40)$$

In the limit as the twist angle goes to zero, $\omega \to 0$, the area goes to that of a rectangular plane, $A \to A_0$. For unit half-width ($b = 1$) and unit length ($\omega c = 1$), we have $A_0 = 2$. The variation of normalized area, A/A_0, with twist angle, ω, is illustrated in Figure 12a for the range $0 \le \omega \le \pi/4$, thought to be representative of most echelon fractures. Note that the area increases slowly and non-linearly with twist angle over this range: increasing by about 9.5% for an angle of 45°.

Using the dimensionless ratio b/c in equation (40) the area takes the form (Pollard et al. 1982, equation 6):

$$A = \omega c^2$$
$$\left[\frac{b}{c}\sqrt{\left(\frac{b}{c}\right)^2 + 1} + \ln\left(\frac{b}{c} + \sqrt{\left(\frac{b}{c}\right)^2 + 1} \right) \right]. \quad (41)$$

These authors used equation (41) to point out that the surface area of n helicoidal fractures, each of half-width b and length ωc, is less than the surface area of a single helicoidal fracture of half-width nb and length ωc. Taking the n fractures as a model for the partial fractures (hackle) in the fringe region of a joint the non-intuitive result is that the surface area decreases as the number of fractures increases (Fig. 12b). On this figure the parameter for each curve is the twist angle, ω. For a twist angle of 1° ($\omega = \pi/180$) the relative surface area of 10 fractures is 99.5% of that for the single fracture, only marginally less. However, for

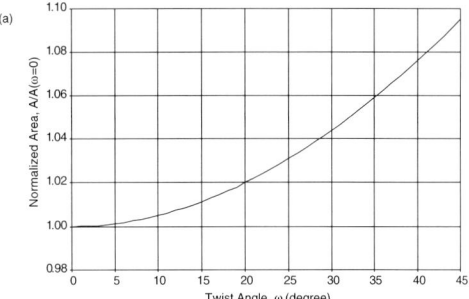

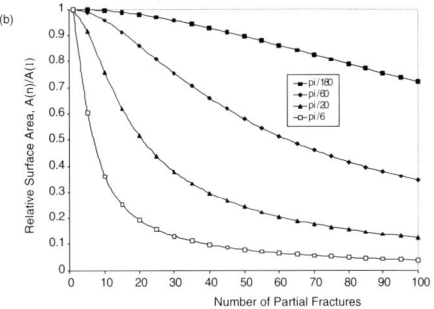

Fig. 12. Surface area of the helicoidal fractures. (a) Surface area, normalized by the surface area for $\omega = 0$ (rectangular plane), plotted v. twist angle, ω, using equation (40). (b) Total surface area of n helicoidal partial fractures, each of half-width b, normalized by the surface area of one helicoidal fracture of half-width nb, plotted v. number of partial fractures.

a twist angle of 30° ($\omega = \pi/6$) the relative surface area of 10 fractures is 36.1% of that for the single fracture, dramatically less. Because the energy required to form a fracture scales with its surface area, this result demonstrates that the breakdown of joints into hackle with helicoidal shapes is consistent with a condition of lesser energy expended during propagation.

The local shape of the surface is determined by the coefficients of the second fundamental form, as described in equation (15). In particular for the helicoid we have:

$$LN - M^2 = \frac{-c^2}{c^2 + u^2} < 0. \quad (42)$$

Because this combination of coefficients is less than zero, every point on the helicoid is a hyperbolic point (see Fig. 9c).

Curvature of the helicoidal fracture surface

The general equation relates the normal curvature of a surface to the ratio of the fundamental forms. For the helicoid we use equation (35) to find:

$$\kappa_n = \frac{\dfrac{-2c}{\sqrt{c^2 + u^2}}\left(\dfrac{du}{dt}\dfrac{dv}{dt}\right)}{\left(\dfrac{du}{dt}\right)^2 + (c^2 + u^2)\left(\dfrac{dv}{dt}\right)^2}. \tag{43a}$$

For a coordinate line $v = $ constant that maps onto the helicoid as a u-parameter curve with $u = t$, we have $du/dt = 1$, $dv/dt = 0$, so $\kappa_n = 0$ as expected for these straight lines. For a coordinate line $u = $ constant that maps onto the helicoid as a v-parameter curve (a helix) with $v = t$, we have $du/dt = 0$, $dv/dt = 1$, so $\kappa_n = 0$. This is, perhaps, not expected as a helix does not have zero curvature.

The non-intuitive result mentioned in the preceding paragraph is resolved by considering the alternative definition of the normal curvature provided by equation (16). The normal curvature is the scalar product of the unit curvature vector for the helix, equation (36), and the unit normal vector for the helicoid, equation (25). Normalizing the tangent vector for the helix we find the unit curvature vector for the helix:

$$\mathbf{k}(t) = \frac{d\mathbf{t}}{dt} / \left|\frac{d\mathbf{c}}{dt}\right| = [1/(c^2 + u_o^2)][-(u_o\cos t)\mathbf{e}_x$$
$$-(u_o\sin t)\mathbf{e}_y]. \tag{43b}$$

This vector is not zero but it is directed in the (x, y)-plane toward the z-axis. On the other hand, the unit normal for the helicoids along a v-parameter curve where $u = u_o$ is:

$$\mathbf{N}(u_o, v) = (1/\sqrt{u_o^2 + c^2})[(c\sin v)\mathbf{e}_x$$
$$-(c\cos v)\mathbf{e}_y - (u_o)\mathbf{e}_z]. \tag{44}$$

This vector is not zero, but the scalar product $\mathbf{k}(t) \cdot \mathbf{N}(u_o, v) = 0$, so the two vectors are orthogonal. The curvature vector for the helix does not resolve any component onto the line normal to the helicoidal surface and therefore the normal curvature, κ_n, along v-parameter curves is zero.

The mean and Gaussian curvatures for the helicoidal surface are found using the fundamental form coefficients in equation (22):

$$\kappa_m = 0, \kappa_g = \frac{-c^2}{(c^2 + u^2)^2}. \tag{45}$$

The helicoid is a so-called minimal surface, which requires that the mean normal curvature, κ_m, be exactly zero, and it is the only ruled minimal surface (Weisstein 2002). This property is appealing for a fracture surface and may be an outcome of the process of propagation in which the energy necessary to create a new surface is minimized. A ruled surface can be generated by moving a straight line in space, in this case perpendicular to the line itself while twisting the line about the direction of movement. The fact that many echelon fractures have approximately straight traces in cross-section suggests that they are ruled surfaces. Also note that κ_g is a function of c and u, but not of v. Therefore, the Gaussian curvature is constant on any v-parameter curve (helix) on the surface. For the special case $u = 0$, along the z-axis, the Gaussian curvature is $\kappa_g = -(1/c)^2$, the negative of the square of the spatial rate of twist.

The values of the principal curvatures satisfy equation (20), which for the helicoid reduces to:

$$(c^2 + u^2)\kappa_0^2 - [c^2/(c^2 + u^2)] = 0. \tag{46}$$

Solving for the principal curvatures we have:

$$\kappa_1, \kappa_2 = \pm c/(c^2 + u^2). \tag{47}$$

Note that the principal curvatures are equal in magnitude and opposite in sign, a fact that is consistent with the mean curvature being zero. For a given rate of twist the principal curvatures decrease in magnitude with distance, u, from the z-axis. Along the z-axis, where $u = 0$, the principal curvatures are $\kappa_1 = +1/c$ and $\kappa_2 = -1/c$, that is they are equal in magnitude to the spatial rate of twist.

The principal directions are found using equation (21), and for the helicoid this reduces to:

$$(-c/\sqrt{c^2 + u^2})(du)^2 - (-c/\sqrt{c^2 + u^2})$$
$$(c^2 + u^2)(dv)^2 = 0. \tag{48}$$

Solving for the ratio $dv/du = \tan\theta$:

$$\tan\theta = 1/\sqrt{c^2 + u^2}; \theta_1, \theta_2 = \tan^{-1}[1/\sqrt{c^2 + u^2}]. \tag{49}$$

These two angles determine the orientations of the lines in the parameter plane (Fig. 9c) that map onto the helicoidal surface with tangent lines that correspond to the directions of the principal curvatures.

The quantitative analysis of joint surfaces

Numerous publications on joint-surface morphology describe qualitatively the geometric relationships of a main (parent) joint with twist hackle (border planes, echelon segments) and rib marks (arrest lines) on the basis of macroscopic observations (e.g. Woodworth 1896; Hodgson 1961*b*; Bankwitz 1965, 1966; Price 1966; Sommer 1969; Pollard *et al.* 1982; Pollard & Aydin 1988; Bahat 1991). Here we describe these geometries quantitatively using high-precision scanned joint surfaces and the concepts of differential geometry. From this quantitative investigation we expect to learn how to characterize the forms of joint surfaces and, perhaps, gain insight about the formation of hackle and rib marks. Hackle are interpreted to form because of changes in the orientation of the principal stresses (Pollard *et al.* 1982; Cooke & Pollard 1996). If the joint breaks down into hackle in a continuous fashion, i.e. during the same increment of growth, the hackle should curve smoothly out of the plane of the main joint to re-align with the principal stresses. For the sample under investigation we test the continuity of hackle with the main joint surface, and evaluate the geometric form of these surfaces. Rib marks are interpreted to form as a result of slowing or stopping of the joint propagation front (Engelder 1987). Here we examine the geometric form of these structures and their relationship to the main joint surface.

Description of the samples, data acquisition and data analysis

The first sample (sample 1) is a fine-grained chert from an unknown location. One side of this sample (Fig. 13a) exhibits the remnant of a main joint surface (*c.* $8 \times c.$ 4 cm), interpreted as forming at depth in the Earth. The main joint surface is truncated by rough fracture surfaces, interpreted as forming when the sample broke free of its surroundings during erosion. The main joint surface is nominally smooth and planar to the eye, but displays a faint plumose structure with a gently curved centreline that is approximately parallel to a breakdown zone along the distal edge of the main joint surface (Fig. 13b). This breakdown zone is marked by the transition from the main joint surface to a set of fine hackle. The edge of the breakdown zone adjacent to the main joint surface is sharply defined by a change in orientation of the surface. Individual lineations, which make up the plumose structure on the main joint surface, follow curved paths from the centreline towards the breakdown zone, becoming subparallel to the steps between the fine hackle. The width of the breakdown zone is 1–2 cm.

The fine hackle of the breakdown zone appear to gradually merge into the coarse hackle of the fringe (Fig. 13). The fringe itself is about 5×12 cm in aerial extent, and seven prominent hackle extend along the 5 cm-length of exposure, ranging in width from 0.5 to 1.5 cm. These hackle may have been longer than 5 cm, but are truncated by a fracture forming another side of the sample. Individual hackle are easily distinguished from the main joint surface because they have a different orientation; the hackle appear to twist continuously through the breakdown zone about an axis perpendicular to this zone. There is no sign of plumose structure on individual hackle, which appear to have very smooth surfaces.

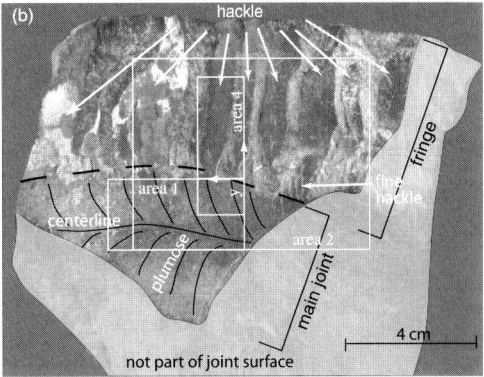

Fig. 13. Chert sample with joint remnant (sample 1). (**a**) Photograph. (**b**) Structural interpretation. The joint surface is composed of part of a main joint surface and several fringe hackle. Cross-fractures between individual hackle are interpreted as due to a later fracturing event. Locations of subareas (1), (2) and (4), used for analysis, are indicated by white rectangles.

Cross-fractures link each twist hackle to its immediate neighbours forming a crude 'stair-step' structure, but the angles between the hackle and cross-fractures are generally obtuse (100°–130°). The hackle have nominally smooth surfaces, but the cross-fracture surfaces are rough and irregular. We interpret the cross-fractures as a later phenomenon, perhaps accompanying the fracturing events that led to isolation of this hand sample from the surrounding rock mass. In cross-section, the individual hackle are approximately straight, and their orientations change by a few degrees in a consistent fashion along the cross-section. Most cross-fractures do not appear to overlap in cross-section, but the exposure is poor and this conclusion may not be definitive.

The second sample (sample 2) is a siltstone from the Monterey Formation collected near Davenport, California. One of several joint surfaces displayed on the sample was examined in detail (Fig. 14a). The main joint surface (*c.* $10 \times c.$ 5 cm) has no plumose

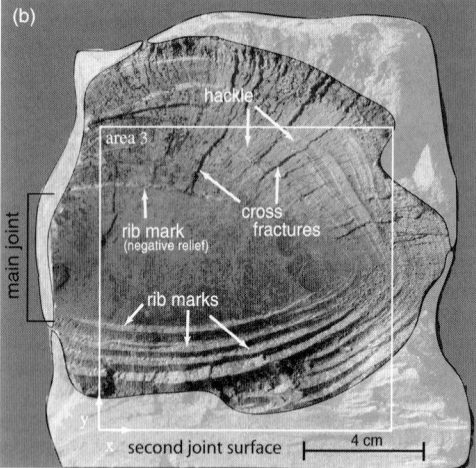

Fig. 14. Joint surface on Monterey Formation siltstone sample (sample 2). (**a**) Photograph. (**b**) Structural interpretation. A main joint surface surrounded on all remaining sides by rib marks. Upper rib marks interact with shallow-angle twist hackle. Location of subarea (3) indicated by white rectangle.

structure and has transitions into rib marks at its approximately elliptical edge. This transition is generally immediate and distinct, with an obvious change in orientation from the visually planar main joint surface to the oblique angle of the adjacent rib mark. However, at the narrow tip of the main joint both the height and width of the rib marks are greatly decreased, almost to the point of imperceptibility. Here only minor undulations define the individual rib marks, although some of these undulations are distinct enough to follow a particular rib mark continuously around the main joint surface.

The rib marks on the lower part of the joint are distinctly different in form from those on the upper part (Fig. 14a). The lower rib marks have approximately uniform shape and spacing, and are similar in size, with widths from approximately 3 to 8 mm and heights of *c.* 0.5–3 mm. They generally increase in size in the direction of propagation, with the rib mark of the least height and width lying next to the parent surface. All of the rib marks of this set stand out in positive relief from the joint surface. The upper set of rib marks is much less regular in spacing and shape. The upper rib mark closest to the main joint surface is similar in form and size (*c.* 4 mm width, *c.* 0.5 mm height) to the corresponding rib mark on the lower side, but stands in negative relief. Successive rib marks appear as minor steps both into and out of the joint surface, and are irregularly spaced. Unlike the lower set, the rib marks of the upper set do not follow the shape of the edge of the parent surface and diverge from its elliptical shape.

The upper side of the joint is marked by shallow hackle that are approximately perpendicular to the rib marks (Fig. 14b). The hackle are more easily discerned by the presence of their cross-fractures, which make small (<0.3 mm) sharp drops in the joint surface, than by the hackle surfaces themselves. The hackle appear to cut through, but not to alter, the shape of the rib marks other than by the small discontinuities of the cross-fractures. The cross-fractures are spaced from 1 to 7 mm and demarcate areas of high and low hackle density. The hackle die out near the right side of the joint surface, thereby not affecting the lower set of rib marks.

Both samples were scanned using a three-dimensional (3D) Cyberware® diode scanner. The output of a scan consists of the *x*-, *y*- and *z*-coordinates of the points on the sample surface relative to a reference frame. Average resolution in the *z*-direction is 0.050–0.200 mm. The data were sampled on several, differently oriented, partly overlapping rectangular grids to better capture changes in surface morphology. Sample spacing is approximately 0.125–0.35 mm. For the data analysis, the surface data were re-sampled over a single, rectangular grid in a new (*x*, *y*)-coordinate plane using linear data interpolation. Grid spacing of either 0.35 or 0.20 mm was used depending on the resolution desired.

For the purposes of the analyses performed in this study, the parametric representation of a surface given above equation (1) can be simplified. The parameter plane is instead the (*x*, *y*)-coordinate plane and the elevation of the surface above this plane, *z*, is a function of *x* and *y*. For this reason the surface can be described as (Lipschultz 1969, p. 128):

$$\mathbf{s} = \mathbf{f}(x, y) = x\mathbf{e}_x + y\mathbf{e}_y + f_z(x, y)\mathbf{e}_z. \qquad (50)$$

First and second partial derivatives, which are necessary for calculating the unit normal, the coefficients of the first fundamental form, the coefficients of the second fundamental form (14) and the normal curvature (18), are estimated from the gridded data using the finite-difference method. Data points are numbered in the coordinate directions such that $1 \leq i \leq n$ in the x-direction and $1 \leq j \leq m$ in the y-direction. The slopes (first partial derivatives of z) in x and y are calculated at each grid point using the previous and following data points, in other words the central difference method (equations 1–37 and 1–39 of Ames 1992):

$$\frac{\partial z}{\partial x}\Big|_{i,j} = \frac{z_{i+1,j} - z_{i-1,j}}{2\Delta x}, \frac{\partial z}{\partial y}\Big|_{i,j} = \frac{z_{i,j+1} - z_{i,j-1}}{2\Delta y}. \quad (51)$$

Here z is the elevation of the sample point from the (x, y)-coordinate plane, and Δx and Δy are the grid spacing in x and y, respectively. The change of slopes (second partial derivatives) are also calculated using the central difference method:

$$\frac{\partial^2 z}{\partial x^2}\Big|_{i,j} = \frac{z_{i+1,j} - 2z_{i,j} + z_{i-1,j}}{2\Delta x}, \frac{\partial^2 z}{\partial y^2}\Big|_{i,j} =$$

$$\frac{z_{i,j+1} - 2z_{i,j} + z_{i,j-1}}{2\Delta y}, \frac{\partial^2 z}{\partial x \partial y}\Big|_{i,j} =$$

$$\frac{z_{i-1,j-1} - z_{i-1,j+1} - z_{i+1,j-1} + z_{i+1,j+1}}{4\Delta x \Delta y}. \quad (52)$$

Differential geometry of a main joint surface

Given the common descriptions of joints in the geological literature (Figs 1 and 2) we would expect a main joint surface to be planar. A glance at the main joint surface of sample 1 (Fig. 13a) suggests that it is approximately planar, although the plumose structure decorates the surface with minor undulations, and incipient hackle form low escarpments near the breakdown zone. To analyse this surface we take the rectangular area (area 1, Fig. 13b), which covers the majority of the available main joint surface. A colour map of the topography of this area (Fig. 15a) shows that the z-coordinate (height) ranges from -0.25 to $+0.35$ mm. The topographic contours (white lines) suggest a succession of low valleys and ridges with a wavelength of about 30 mm in the y-coordinate direction. The ratio of the amplitude to wavelength is about 0.01, so the surface is better described as gently undulating than as planar.

The colour map and contours of the main joint surface (Fig. 15a) reveal a small-scale roughness with a wavelength about at the scale of the sampling grid (0.35 mm) that appears to be superimposed on the gentle undulations. Distinctly different colours are juxtaposed and locally the contours take sharp bends. To analyse these different contributions to the

shape of the surface a filtering technique was used that is based on 2D Fourier analysis, in which the surface is approximated by a summation of sine and cosine functions of varying amplitude and wavelength (Bracewell 2000). After transforming the surface into the Fourier domain, the filter removed frequencies greater than a given cut-off, and the surface was reconstructed (Fig. 15b) using only the lower frequencies (Bergbauer et al. 2001). For practical purposes this was carried out using a Fast Fourier Transform, which divides the frequency spectrum into discrete frequency bins.

The greatest discrete wavelengths contained in the original surface data (Fig. 15a) are 20 mm in the x-direction, and 35 mm in the y-direction. The least wavelength contained in either coordinate direction is 0.7 mm, which is equal to twice the grid spacing and corresponds to the Nyquist frequency. The filter removed all undulations with wavelengths smaller than 9 mm, to create the smoothed surface (Fig. 15b). Note that the low valley and ridge pattern of the original data is preserved, and that this filtering smoothes the contours. The difference in heights between the sampled data and the filtered data (Fig. 15c) reveals a systematic pattern of smaller-scale valleys and ridges that grow in amplitude away from the centreline towards the breakdown zone on the right-hand side of the image. These correspond to the most prominent incipient hackle on the main joint surface. Apparently, these incipient hackle are superimposed on the gentle undulations revealed in the smoothed data (Fig. 15b).

Locally, a surface is planar near any point if all the coefficients of the second fundamental form are zero (equation 15). To obtain a more accurate representation of the main joint surface to examine its planar nature, a second filtering of the data was performed with more of the original surface data retained. The grid spacing was reduced to 0.20 mm to increase resolution and only wavelengths smaller than 1.42 mm were removed (Fig. 15d). This cut-off value was chosen to retain as much original surface information as possible while eliminating those highest frequencies, interpreted as 'noise'. We infer that this 'noise' is composed primarily of dust, dirt and incidental damage not related to the formation of the surface.

In terms of curvatures, both principal curvatures and the mean and Gaussian curvatures (equation 22) are zero for planar surfaces. For the scanned data, no point will fulfill these criteria exactly as any minor undulation will lead to non-zero values. To correct for this a threshold was specified, under which any curvature was set to zero. This threshold was based on the manufacturer's specifications that the smallest possible variation from a plane observable by the scanner was a point 0.050 mm above the plane. A plane with this elevated point was created and

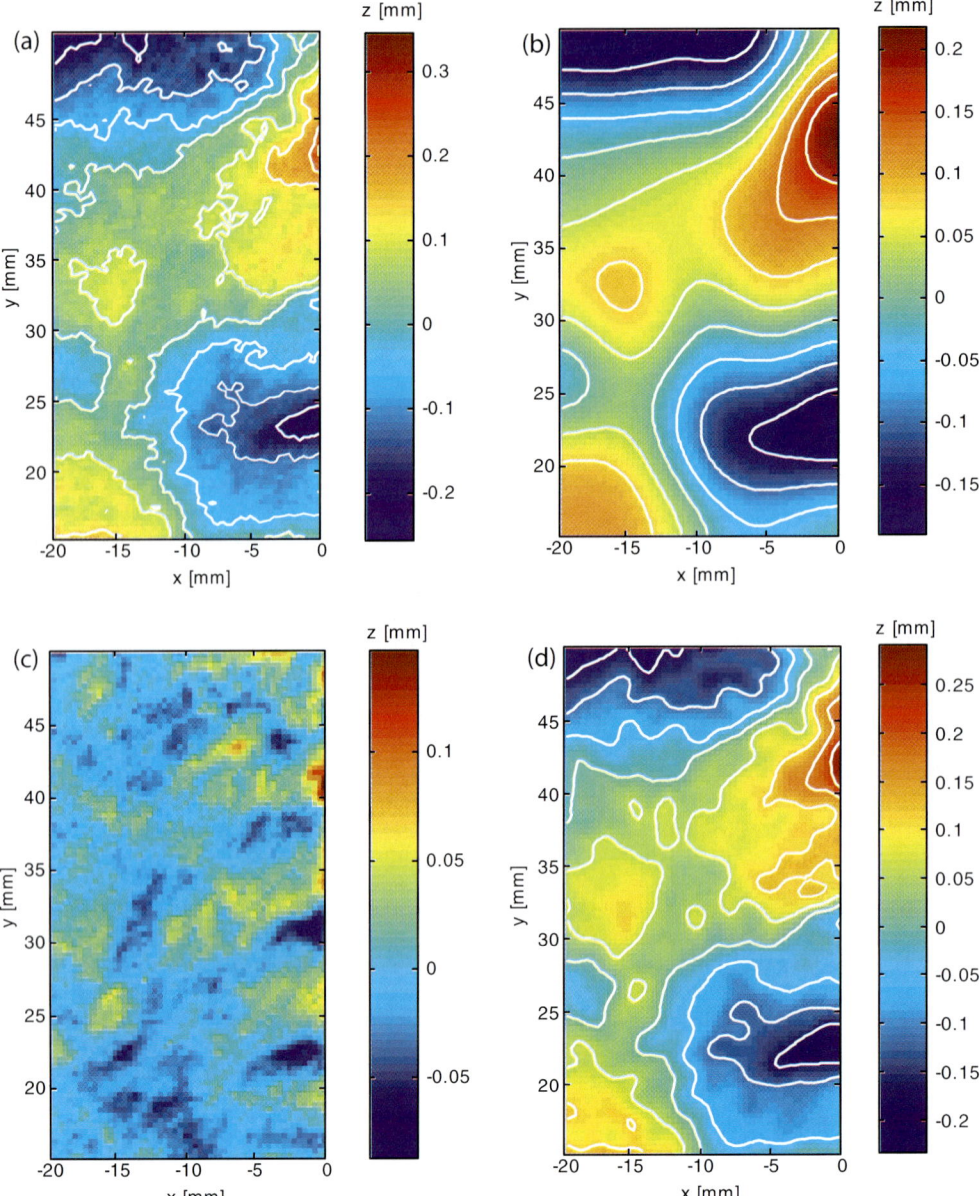

Fig. 15. Topography of area 1 (main joint surface of sample 1). (**a**) Topography after re-sampling of scanned data. (**b**) Topography after filtering all wavelengths smaller than 9 mm. (**c**) Surface undulations removed by filtering. Note incipient hackle as indicated by curved domains of alternating highs and lows on the right-hand side of this image. (**d**) Topography of area 1 after filtering all wavelengths smaller than 1.42 mm. Note that more of the original surface undulations are retained than in (b).

filtered using the 1.42 mm-wavelength specification, as was performed with the actual data. The filter smoothed the point to 0.0024 mm above its surroundings. This value was used as the smallest variation that represented actual surface data and not imprecision due to the scanning process. Any variation less than this value was assumed to be noise and indistinguishable from a planar surface. The plane curvature associated with this value was calculated using:

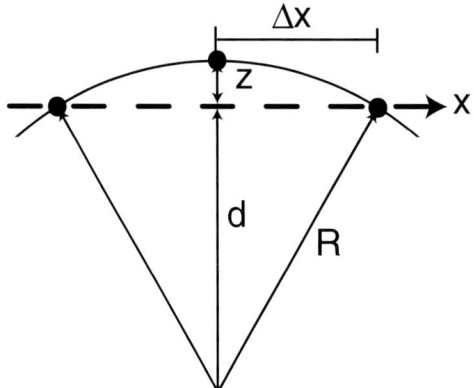

Fig. 16. Geometric relationships used to calculate the plane curvature, κ, from elevation data, z, along a line with a grid spacing Δx.

$$\kappa = 1/R = \frac{2z}{\Delta x^2 + z^2}. \qquad (53)$$

Here R is the radius of curvature, z is the smallest measurable height (0.0024 mm) and Δx is the grid spacing in the x-direction (0.20 mm) (Fig. 16). The curvature was also calculated using equation (18). Both give $\kappa = 0.12$ mm^{-1}. This value was used as the curvature threshold under which any curvature value was set to zero to define a statistically planar surface.

A curvature analysis of this filtered surface was then performed. For every filtered data point in area 1 the mean and Gaussian curvature were calculated, and these values were used to assign each point a particular shape curvature (Fig. 17a) (Roberts 2001; Bergbauer & Pollard 2003). A colour map (Fig. 17b) corresponding to the six different shape curvatures demonstrates that the surface in area 1 is predominantly planar with small areas of parabolic antiformal ridges and synformal valleys. Also locally seen are hyperbolic saddles, and elliptical basins and domes, but only within otherwise parabolic areas. For example, hyperbolic saddles occur where parabolic ridges and valleys intersect. The least planar areas correspond to incipient hackle along the right-hand side of the image.

The formation of hackle under mixed mode I–III loading apparently depends on a minimum angle of rotation of the greatest tensile stress about the axis of propagation (Sommer 1969). Sommer determined that this threshold angle for glass samples is about 3.3°. Using equation (5) we computed the unit normal vectors, **N**, for each data point on the filtered surface shown in Figure 15b for area 1 (Fig. 13a). A unit normal from near the centreline of the main joint surface at the point (−17 mm, 15 mm) was used as a reference orientation, and the angle between the normal at each data point and this reference was calculated and plotted on a colour map (Fig. 18). Across most of the surface of area 1 the unit normals vary by less than 3° from the reference orientation. The greatest angular differences occur at two locations near the zone of breakdown where incipient hackle are present, with a maximum value of approximately 3.4°. Apparently, the main joint surface was capable of undulating with variations in the orientation of the unit normal up to about 3° without breaking down into well-defined hackle.

Differential geometry of joint surfaces

Rib marks and hackle are common structures on joints, and many qualitative descriptions of their forms exist in the literature. To describe these structures more precisely we examine the shape curvatures of the two joint surfaces as a whole. Area 2 on sample 1 (Fig. 13b) and area 3 on sample 2 (Fig. 14b) were analysed. Area 2 includes most of the hackle and a significant portion of the main joint surface of sample 1. Area 3 includes the main joint and most of the rib marks of sample 2. To achieve consistency and allow comparison between samples, both areas were analysed using the same parameters. Sample point spacing was 0.20 mm and all wavelengths less than 1.42 mm were removed using filtering. The threshold value for zero curvature was specified as 0.12 mm^{-1} to define a statistically planar surface.

The results of this analysis are consistent with the above findings for area 1 (Fig. 17). Significant portions of both main joint surfaces are statistically planar with local sinuous valleys and ridges (Figs 19 and 20). The main joint of sample 1 appears to have fewer of these undulations than the main joint of sample 2. On both, where these parabolic valleys and ridges intersect, hyperbolic saddles are seen. Elliptical basins are seen within valleys, and domes on ridges, but basins and domes are not seen isolated in planar areas.

The hackle of both samples have similar shape characteristics. For sample 1 the coarse hackle display as yellow bands starting near the y-axis and running parallel to the x-axis (Fig. 19). They appear to be approximately planar, but with greater portions of their surface areas consisting of non-planar features than the main joint surface (i.e. less yellow area). This is generally consistent with the visual appearance of the two areas, with the hackle seeming to be not as uniformly planar as the main joint surface. Interestingly, the cross-fractures also show systematic curvatures, in contrast to the 'noisy' unrelated fracture in the lower-left corner. They appear as bands parallel to the hackle with a distinct repeating pattern of reds turning to blues (in the positive y-direction). The blues correspond to

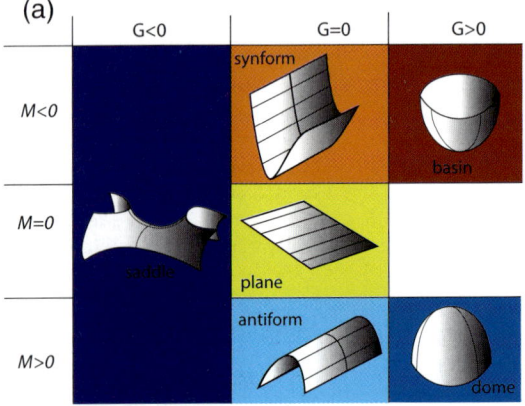

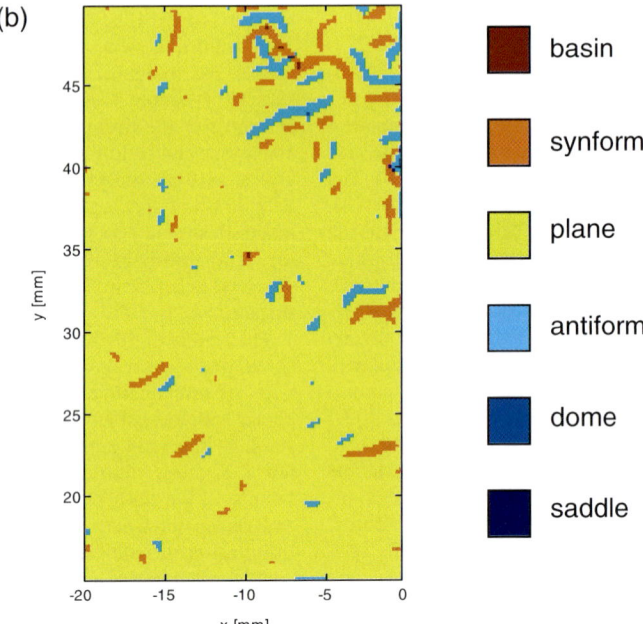

Fig. 17. Shape curvature. (**a**) A combination of mean (κ_m) and Gaussian (κ_g) curvatures, called shape curvature, can be used to infer the local shape of a surface. (**b**) Shape-curvature of area 1. Note the generally planar nature with some parabolic valleys and ridges.

ridges and domes, or convex features, while the reds correspond to concave valleys and basins. Saddles appear to be distributed fairly evenly in both areas.

To examine this characteristic in more detail a profile of the hackle was created along the $x = 25$ mm line from 40 mm $\geq y \geq -25$ mm (Fig. 19). The profile of the filtered z values as a function of y shows the 'stair-step' shape of the hackle and cross-fractures, and the gradual change in orientation of the hackle across the surface (Fig. 21a). The hackle

are the slopes ascending from left to right and the cross-fractures are the descending slopes. Figure 21b limits the profile from 20 mm $\geq y \geq -10$ mm to show two complete hackle and parts of two more with three cross-fractures joining them. The hackle and cross-fractures themselves appear to be approximately planar, while the transitions from hackle to cross-fractures are rounded peaks and the transitions from cross-fractures to hackle are rounded valleys.

To quantify the profile shape the change in

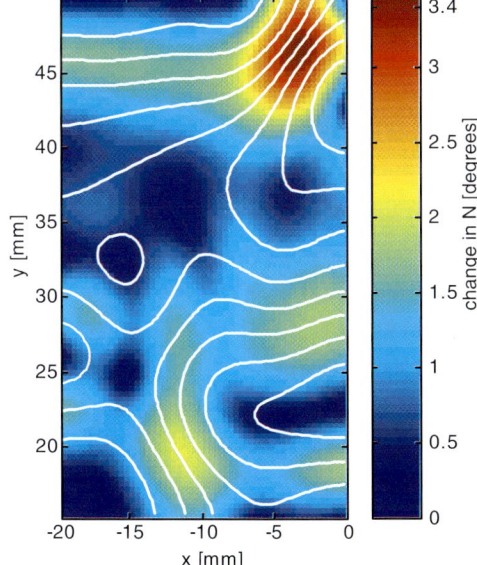

Fig. 18. Change in the orientation of the surface normal vector, **N**, with respect to a reference normal vector located at $x = -17$ mm, $y = 15$ mm. The main joint surface varies in orientation by up to 3.4°.

degrees of the normal vector, **N**, of the profile was calculated at each point. For a linear section, the angle of the normal vector does not change over distance. On a convex section moving from left to right the angle changes in a clockwise (positive) manner. Conversely, concave sections display negative changes in the orientation of the normal. A plot of the change in orientation of the normal vector for the section of the profile 20 mm $\geq y \geq -10$ mm (Fig. 21c) shows which points along the profile are linear, concave or convex. The peaks of the hackle to cross-fracture transitions correspond to positive peaks in the normal vector plot, with the most concave point (greatest change in **N** angle) occurring at the peak. The cross-fracture to hackle transitions similarly correspond to negative peaks in the normal vector plot. The hackle faces correspond to values around zero, in agreement with the above conclusion that they are statistically planar. Short segments near the middle of the cross-fractures are approximately linear. This can be seen in the plot of the shape curvature of sample 1 (Fig. 19). There are small planar areas of yellow between the blue (convex hackle to cross-fracture transition) and red (concave cross-fracture to hackle transition) bands demarcating the cross-fractures.

Many of these characteristics also are seen in the hackle and cross-fractures of sample 2 along the top of the figure (Fig. 20). The cross-fractures stand out as the blue over red bands trending away from the main joint surface toward the upper right. The blue over red pattern is consistent with that described for sample 1, implying the fracturing style may be similar. The hackle themselves show an approximately planar shape, as evidenced by the yellow areas between cross-fractures. The similarities between the samples are even more striking given the difference in the size and orientation of the hackle on each. The coarse hackle of sample 1 are easily distinguishable and differ in orientation from the main joint surface by as much as 25°. The hackle of sample 2 more closely resemble the fine hackle of sample 1, being harder to distinguish individually, and having closer spacing and only minor differences in orientation (<5°) from the main joint surface.

Rib marks have been described qualitatively as ranging in shape from cusped (concave on both sides) to asymmetrical, with a convex face in the direction of joint propagation and a concave face nearest the main joint surface (Price 1966; Engelder 1987; Bahat 1991). Here we look quantitatively at the shape of these features using differential geometry for the first time. The rib marks on sample 2 form a distinct, repeating pattern of red and blue bands running parallel to the main joint surface edge and orthogonal to the hackle cross-fractures (Fig. 20). The well-formed rib marks along the lower edge of the joint surface (-60 mm $< y < -45$ mm) show an alternating red–blue pattern in the negative y-direction that is repeated consistently. The rib marks along the upper edge of the joint surface (-10 mm $< y < 20$ mm) are less well formed and do not show as regular a pattern, in part due to their interaction with the hackle in that area. The exception is the rib mark immediately adjacent to the main joint surface. This rib mark is in negative relief and so the pattern is reversed with blue (convex) next to the main joint surface and then red (concave).

A profile of sample 2 was plotted from 20 mm $\geq y \geq -80$ mm (Fig. 20) showing the main joint surface, the rib mark in negative relief and six rib marks in positive relief from the set along the lower edge of the main joint (Fig. 22a). A section from -35 mm $\geq y \geq -60$ mm was examined in detail and a plot of the change in angle of the normal vectors made (Figs 22b, c). The main joint surface can be seen from -35 mm $\geq y \geq -40$ mm oscillating around 0, indicating it is approximately linear (Fig. 22c). At $y = -40$ mm the first rib mark begins as a sudden negative change in the orientation of the normal vector. The angle decreases for 0.4 mm (concave shape) and then begins increasing to intercept the origin at around -40.8 mm. At this point there is an inflection point on the rib mark face and the curve changes to convex, with increasing positive values. The convexity reaches a maximum near the peak of the rib mark and decreases down the face in the

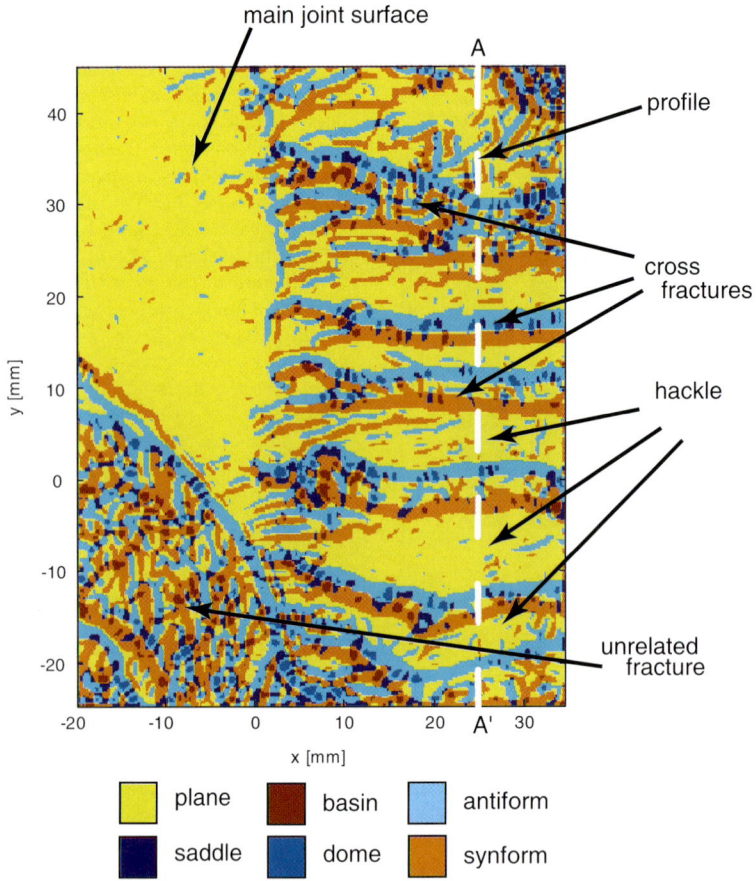

Fig. 19. Shape curvature plot of sample 1. Visible are the main joint surface, the hackle and the cross-fractures joining the hackle. The transect used to make a profile (Fig. 21) is indicated by the dashed white line, A–A'.

inferred direction of joint propagation to reach an approximately linear section. Beyond this the face increases in concavity to a maximum in a basin before the next rib mark. The pattern is then repeated over the next several rib marks. The rib mark shape we have quantitatively described here is neither cusped nor concave on one side and convex on the other. Instead, on the side facing the main joint surface we see a concave lower face with an inflection point followed by a convex peak. The peak continues convexly to the far side of the rib mark, where there is a second inflection point and a concave descent.

Differential geometry of twist hackle

The geometry of twist-hackle surfaces has been hypothesized to be similar to a helicoidal surface (Pollard *et al.* 1982). To test this hypothesis we ana-

lysed area 4 (Fig. 13b), which covers a small portion of the main joint surface along with a portion of the breakdown zone, most of one coarse hackle in the fringe region and the adjacent cross-fractures. An oblique rendering of area 4 (Fig. 23) depicts the local geometry of the main joint surface in the background, the breakdown zone, the hackle and the two cross-fractures. Note that this view foreshortens the x-axis but leaves the y-axis and z-axis approximately equal. It should be noted that this rendering of the surface does not capture the sharp discontinuity of the exposed surface where the cross-fractures meet the hackle, but it does provide an accurate image of the main joint, the hackle and the cross-fractures. Also, it should be noted that a small piece of the hackle surface has been plucked away from the breakdown zone where the ridge along $y = 18$ mm would have intersected and extended above the main joint surface.

The centreline of the hackle (Fig. 23) lies some-

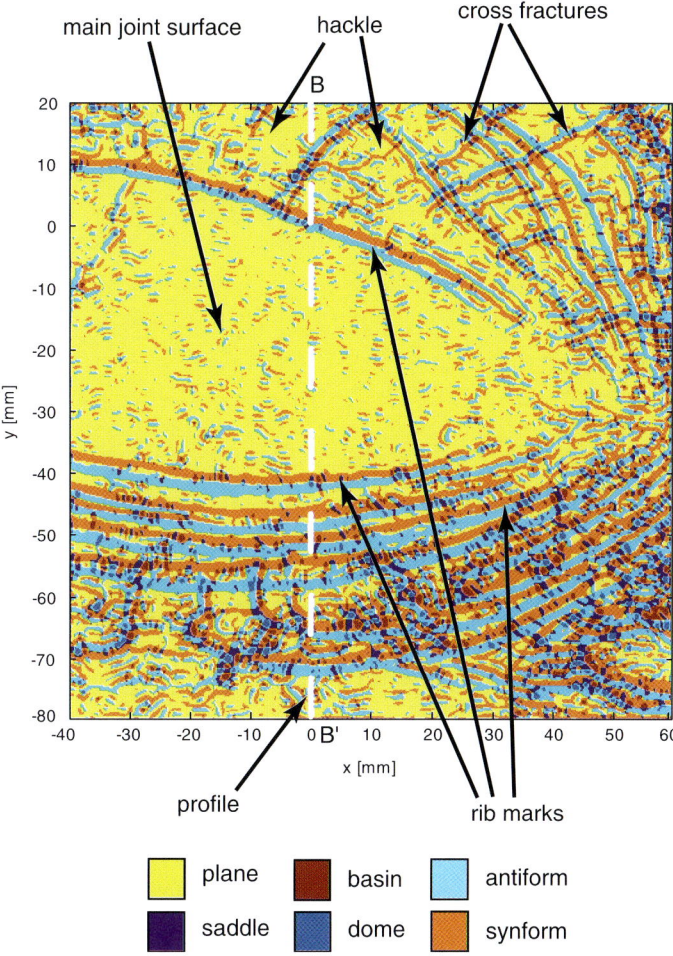

Fig. 20. Shape curvature plot of sample 2. Visible are the main joint surface, shallow-angle hackle, cross-fractures and rib marks. The transect used to make a profile (Fig. 22) is indicated by the dashed white line, B – B′.

where between the lines $y = 19.5$ and 20.5 mm, and the elevation of the surface there is approximately zero, equal to the local elevation of the main joint surface. The elevation of the hackle is greater than zero for lines between $y = 17$ and 19.5 mm (except for the plucked portion), and less than zero for lines between $y = 20.5$ and 23 mm. Furthermore, the ridge along $y = 17$ mm increases in elevation and the trough along $y = 23$ mm decreases in elevation away from the main joint. These observations are consistent with the hypothesis that the hackle approximates a helicoidal surface.

To further test the hypothesis that the hackle is helicoidal we recall that the centreline of a helicoid is defined as the line along which the parameter $u = 0$. Also, the unit normal (equation 26) on the centreline of a helicoid rotates smoothly about this line

with distance along the line (Fig. 11b). The difference between the orientation of this unit normal at the proximal edge of the helicoid and that at any position along the centreline was defined as the twist angle (equation 27), which is equal to the parameter v. Using data for the sample joint (Fig. 23) we calculated the change in orientation of the unit normal vector, $\mathbf{N}$, relative to a reference point on the main joint surface, along selected x-grid lines over the range -10 mm $\leq x \leq 37$ mm. These x-grid lines are approximately parallel to the centreline of the hackle, and the range in y-values is chosen to encompass the centreline.

The orientation of the unit normal relative to the reference value changes systematically with the x-coordinate for all of the selected x-grid lines. On the main joint surface ($x < 4$ mm) the change in angle is

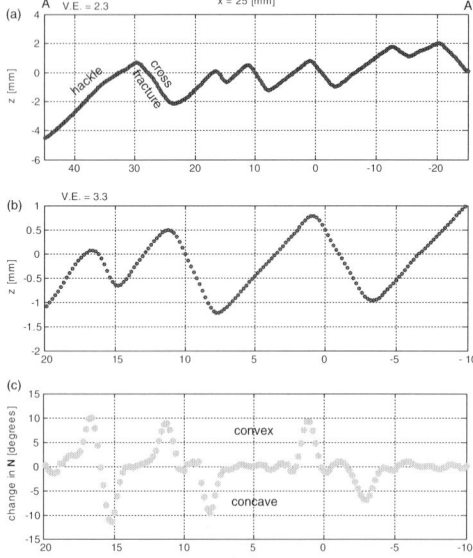

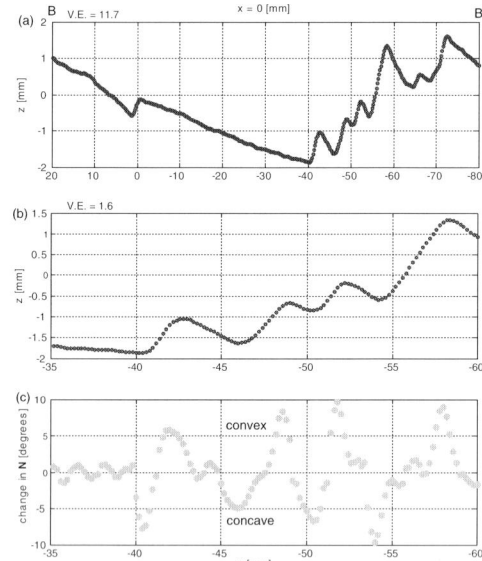

Fig. 21. Profiles made from transect of sample 1 (Fig. 19). (**a**) Profile from 40 mm $\geq y \geq -25$ mm showing hackle (ascending slopes from left to right) and cross-fractures (descending slopes from left to right). (**b**) Profile limited to 20 mm $\geq y \geq -10$ mm. (**c**) Plot of change in orientation of normal vector (**N**) for each point in (b). Values above zero are increasingly convex and values below are concave. Values near zero are approximately linear.

Fig. 22. Profiles made from transect of sample 2 (Fig. 20). (**a**) Profile from 20 mm $\geq y \geq -80$ mm showing rib mark in negative relief (*c.* $y = 0$ mm), the main joint surface (*c.* $0 \geq y \geq -40$ mm) and six rib marks in positive relief (*c.* -4 mm $0 \geq y \geq -80$ mm). (**b**) Profile limited to -35 mm $\geq y \geq -60$ mm. (**c**) Plot of change in orientation of normal vector (**N**) for each point in (b). Values above zero are increasingly convex and values below concave. Values near zero are approximately linear.

roughly constant and less than about 4°. Over the breakdown zone (4 mm $< x <$ 19 mm) the angle increases to between 18° and 27°. For the distal portion of the hackle (19 mm $< x <$ 37 mm) the angle oscillates about a more-or-less constant value. Data are plotted (Fig. 24a) for our interpretation of the centreline ($y = 20$ mm). The spatial rate of twist ($1/c$), as defined in equation (34), is the slope of the curve along the centreline in the breakdown zone. For the x-grid line interpreted as the centreline the slope is roughly constant over the range 7 mm $< x <$ 18 mm and is about 2° per mm (0.035 mm^{-1}). A constant rate of twist is consistent with the hypothesis that the hackle approximates a helicoidal surface.

Recall from equation (47) that the principal normal curvatures for helicoids are equal in magnitude and opposite in sign, so the mean normal curvature is identically zero (45). Furthermore, along the centreline of helicoids, the magnitude of the principal curvature is equal to the spatial rate of twist ($1/c$). The maximum principal curvature, κ_1, the minimum principal curvature, κ_2, and mean curvature, κ_m, along the x-grid line selected as the centreline ($y = 20$ mm) are plotted v. the x-coordinate (Fig. 24b). Note that all three curvatures are approximately con-

stant on the main joint surface: the minimum curvature is approximately zero and the maximum curvature is about $+0.5$ mm^{-1}. In the breakdown zone the minimum curvature decreases to negative values and the maximum curvature increases, so the mean value varies from approximately zero to small positive values. At the transition to the distal portion of the hackle the magnitudes of the principal curvatures sharply decrease and thereafter oscillate about zero. Although the surface of the hackle is not a perfect helicoid, the principal curvatures are generally opposite in sign and of similar magnitude.

Discussion

The quantitative investigation of the differential geometry of two joint surfaces has revealed a number of interesting similarities to concepts that have been developed during earlier field, laboratory and theoretical studies of joints, and other dominantly opening-mode fractures. For example, qualitative field observations have led many geologists to characterize the shape of main joint surfaces as planar (Woodworth 1896; Kulander *et al.* 1979), and

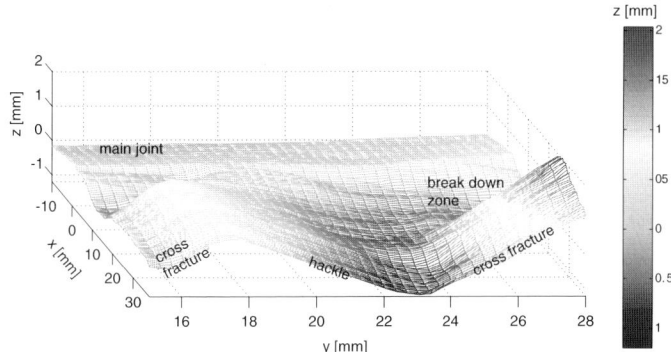

Fig. 23. Oblique view of the transitions from main joint to breakdown zone to fringe. An abrupt change in orientation of marks the transition from main joint to breakdown where the surface changes to finer hackle. The fine hackle gradually merge into the more prominent hackle of the fringe, which are themselves joined by cross-fractures.

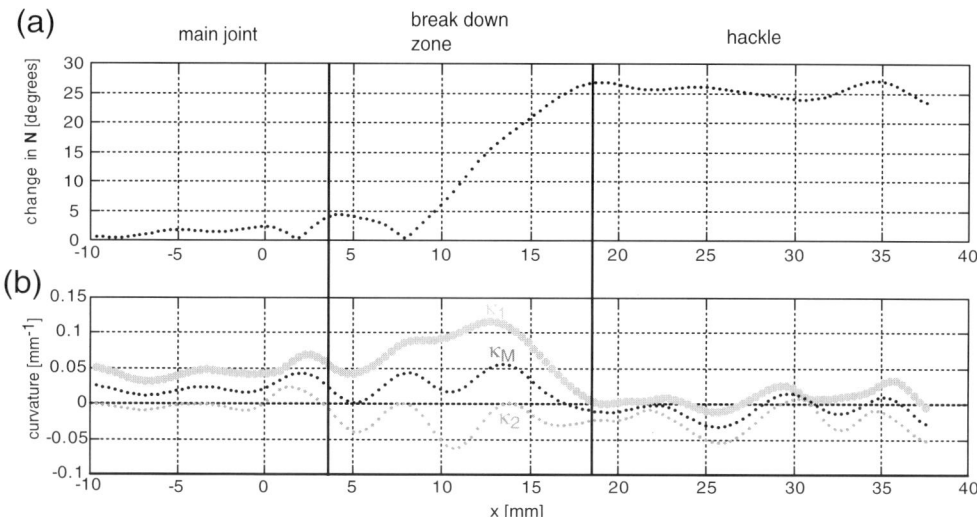

Fig. 24. Geometry along the centreline ($y = 20$ mm) of the hackle shown in Figure 23. (**a**) Change in orientation of the surface normal vector (**N**) along a grid line parallel to the x-axis near the centreline. Angular changes are less than $5°$ along the main joint. The majority of the twist occurs between 4 mm $< x <$ 19 mm in the breakdown zone, and the change in orientation is systematic and approximately linear. The finer hackle merge into the more prominent hackle of the fringe where the surface orientation changes are again less than $5°$. (**b**) Principal curvatures and mean curvatures for the same gridline as shown in (a).

the main joint surfaces of the samples analysed here are approximately planar. Also, qualitative observations of echelon dyke, vein and joint segments that are nominally straight in cross-section suggested that the 3D form of these fracture surfaces is a ruled surface such as a helicoid (Pollard *et al.* 1982), and the hackle analysed here approximates to a helicoid. Laboratory investigations of the propagation of an opening crack when subjected to mixed mode I–III loading (Sommer 1969) suggest a threshold angle of approximately $3°$ for the rotation of the greatest

tensile stress that must be exceeded for the crack to break down into multiple echelon segments (fracture lances) that twist about their centreline. The main joint surface analysed for this property undulates through angles up to this threshold without break down and appears to approach this angle where incipient hackle are observed. Finally, we recall that the calculation of surface area for an idealized fracture system was simplified by taking the rate of twist of echelon segments as constant (Pollard *et al.* 1982). Analysis of data from a prominent twist hackle demonstrates that the rate of twist of the hackle is approximately constant through the breakdown zone.

On the other hand, there are some noteworthy differences between widely held geometric conceptions of joints and the detailed shape of the actual joint surface investigated here. These differences are directly attributable to the unprecedented abundance and precision of data obtained through the scanning process, and to the novel application of differential geometry to analyse these data. Precise measurement of the sample joint has revealed that the main joint surface is not perfectly planar, but includes low-amplitude elliptical, parabolic and hyperbolic forms. The hackle do approximate a helicoidal shape through the breakdown zone, but the rate of twist then declines to near zero and the surface takes on an undulating form similar to the main joint surface with elliptical, parabolic and hyperbolic shapes. Even within the breakdown zone the shape of the sample surface is not precisely helicoidal. This could be a consequence of rock being removed from the surface, or may indicate that the path of joint propagation was perturbed from the ideal state. Such perturbations could be attributed to material heterogeneity of the rock or to mechanical interactions between this joint and neighbouring joints. The rib marks examined here have distinct and similar morphologies, with a concave base and convex peak. This is unlike shapes described in the literature, although it certainly does not preclude other possibilities. In any case the recognition of these differences motivates further study of these and other joint surfaces using these techniques.

It would seem imperative for geologists, who are interested in the study of joints and other fractures in rock, to learn enough differential geometry to carry out effective field and laboratory campaigns, and to design appropriate tests for their hypotheses about the geometry of joints and the mechanics of their formation. There are an abundance of qualitative terms (Figs 1 and 2) that serve a useful purpose in the everyday conversations of geologists about joints. However, this nomenclature and the associated classifications schemes are not sufficient to analyse the geometry of joints quantitatively or to compare field measurements with idealized geometric models. Furthermore, this nomenclature and classification is inadequate to compare field measurements with fractures produced in laboratory or numerical experiments. The features of a joint that can be represented by curves (e.g. plumose structure, arrest or hesitation lines and tiplines) may be characterized by fundamental quantities such as the curvature and torsion. Similarly, those features of a joint that can be represented by surfaces (e.g. the main joint, hackle and feather fractures), may be characterized by the first and second fundamental forms. These, in turn, enable one to calculate the unit normal vector, the principal curvatures, and the Gaussian and mean curvatures for any point on the surface. These quantities from differential geometry determine what is sensible to measure in the field or laboratory in order to characterize the features of a joint. Clearly, one needs to measure the joint surface in such a way as to admit the calculation of the fundamental quantities.

The authors wish to thank R. Lisle and B. J. Carter for thoughtful reviews. D. D. Pollard thanks J. Cosgrove and T. Engelder for organizing the meeting at Weston-super-mare on Mechanics of Jointing, which stimulated his re-examination of joint surface morphology in light of differential geometry. This material is based upon work supported by the National Science Foundation under Grant No. EAR-0125935-001.

References

ADAMS, M. & SINES, G. 1978. Crack extension from flaws in a brittle material subject to compression. *Tectonophysics*, **49**, 97–118.

AMES, W. F. 1992. *Numerical Methods for Partial Differential Equations*. Academic Press, Boston, MA.

ANDERSON, T. L. 1995. *Fracture Mechanics: Fundamentals and Applications*. CRC Press, Boca Raton, FL.

BAHAT, D. 1991. *Tectonofractography*. Springer, New York.

BAHAT, D. & ENGELDER, T. 1984. Surface morphology on cross-fold joints of the Appalachian Plateau, New York and Pennsylvania. *Tectonophysics*, **104**, 299–313.

BAHAT, D., BANKWITZ, P. & BANKWITZ, E. 2001. Changes of crack velocities at the transition from the parent joint through the en echelon fringe to a secondary mirror plane. *Journal of Structural Geology*, **23**, 1215–1221.

BAI, T. & POLLARD, D. D. 2000a. Fracture spacing in layered rocks: a new explanation based on the stress transition. *Journal of Structural Geology*, **22**, 43–57.

BAI, T. & POLLARD, D. D. 2000b. Closely spaced fractures in layered rocks: initiation mechanism and propagation kinematics. *Journal of Structural Geology*, **22**, 1409–1425.

BANKWITZ, P. 1965. Uber Klüfte I. Beobachtungen im Thüringischen Schiefergebirge. *Geologie*, **14**, 241–253.

BANKWITZ, P. 1966. Uber Klüfte II. Die Bildung der Kluftfläche und eine Systematik ihrer Strukturen. *Geologie*, **15**, 896–941.

BERGBAUER, S. & POLLARD, D. D. 2003. How to calculate normal curvatures of sampled geological surfaces. *Journal of Structural Geology*, **25**, 277–289.

BERGBAUER, S., MUKERJI, T. & HENNINGS, P. H. 2003. Improving curvature analyses of deformed horizons using scale-dependent filtering techniques. *AAPG Bulletin*, **87**, 1255–1272.

BERGBAUER, S., MUKERJI, T., POLLARD, D. D. & HENNINGS, P. H. 2004. Calculation of scale-dependent curvatures of geological surfaces. *Eos, Transactions of the American Geophysical Union*, **82**, F1231.

BRACE, W. F. & BOMBOLAKIS, E. G. 1963. A note on brittle crack growth in compression. *Journal of Geophysical Research*, **68**, 3709–3713.

BRACEWELL, R. N. 2000. *The Fourier Transform and its Applications*. McGraw-Hill, Boston, MA.

COOKE, M. & POLLARD, D. D. 1996. Fracture propagation paths under mixed mode loading within rectangular blocks of polymethyl methacrylate. *Journal of Geophysical Research*, **101**, 3387–3400.

COOKE, M., MOLLEMA, P., POLLARD, D. D. & AYDIN, A. 2000. Interlayer slip and joint localization in East Kaibab Monocline, Utah: field evidence and results from numerical modeling. *In*: COSGROVE, J. W. & AMEEN, M. S. (eds) *Forced Folds and Fractures*. Geological Society, London, Special Publications, **169**, 23–49.

COTTERELL, B. & RICE, J. R. 1980. Slightly curved or kinked cracks. *International Journal of Fracture*, **16**, 155–169.

COX, S. J. D. & SCHOLZ, C. H. 1988*a*. On the formation and growth of faults: an experimental study. *Journal of Structural Geology*, **10**, 413–430.

COX, S. J. D. & SCHOLZ, C. H. 1988*b*. Rupture initiation in shear fracture of rocks: an experimental study. *Journal of Geophysical Research*, **93**, 3307–3320.

CRUIKSHANK, K. M., ZHAO, G. & JOHNSON, A. M. 1991*a*. Analysis of minor fractures associated with joints and faulted joints. *Journal of Structural Geology*, **13**, 865–886.

CRUIKSHANK, K. M., ZHAO, G. & JOHNSON, A. M. 1991*b*. Duplex structures connecting fault segments in Entrada Sandstone. *Journal of Structural Geology*, **13**, 1185–1196.

DEGRAFF, J. M. & AYDIN, A. 1987. Surface morphology of columnar joints and its significance to mechanics and directions of joint growth. *Geological Society of America Bulletin*, **99**, 605–617.

ENGELDER, T. 1985. Loading paths to joint propagation during a tectonic cycle: an example from the Appalachian Plateau, U.S.A. *Journal of Structural Geology*, **7**, 459–476.

ENGELDER, T. 1987. Joints and shear fractures in rock. *In*: Atkinson, B. K. (eds) *Fracture Mechanics of Rock*. Academic Press, London, 27–69.

ENGELDER, T. & GEISER, P. 1980. On the use of regional joint sets as trajectories of paleostress fields during the development of the Appalachian Plateau, New York. *Journal of Geophysical Research*, **85**, 6319–6341.

ERDOGAN, F. & SIH, G. C. 1963. On the crack extension in plates under plane loading and transverse shear. *Journal of Basic Engineering – Transactions of American Society of Mechanical Engineers*, **85**, 519–527.

FISCHER, M. P., GROSS, M. R., ENGELDER, T. & GREENFIELD, R. J. 1995. Finite element analysis of the stress distribution around a pressurized crack in a layered elastic medium: implications for the spacing of fluid-driven joints in bedded sedimentary rock. *Tectonophysics*, **247**, 49–64.

GELL, M. & SMITH, E. 1967. The propagation of cracks through grain boundaries in polycrystalline 3% silicon–iron. *Acta Metallurgica*, **15**, 253–258.

GERMANOVICH, L. N., CARTER, B. J., INGRAFFEA, A. R., DUSKIN, A. V. & LEE, K. K. 1996. Mechanics of 3-D crack growth under compressive loads. *In*: AUBERTIN, M., HASSANI, F. & MITNI, H. (eds) *Rock Mechanics Tools and Techniques*. Balkema, Rotterdam, 1151–1159.

GERMANOVICH, L. N., SALGANIK, R. L., DYSKIN, A. V. & LEE, K. K. 1994. Mechanics of brittle fracture of rock with pre-existing cracks in compression. *Pure and Applied Physics*, **143**, 117–149.

GROSS, M. R. 1993. The origin and spacing of cross joints: examples from the Monterey Formation, Santa Barbara coastline, California. *Journal of Structural Geology*, **15**, 737–751.

GROSS, M. R. 1995. Fracture partitioning: failure mode as a function of lithology in the Monterey Formation of coastal California. *Geological Society of America Bulletin*, **107**, 779–792.

HELGESON, D. E. & AYDIN, A. 1991. Characteristics of joint propagation across layer interfaces in sedimentary rocks. *Journal of Structural Geology*, **13**, 897–911.

HODGSON, R. A. 1961*a*. Classification of structures on joint surfaces. *American Journal of Science*, **259**, 493–502.

HODGSON, R. A. 1961*b*. Regional study of jointing in Comb Ridge–Navajo Mountain area, Arizona and Utah. *AAPG Bulletin*, **45**, 1–38.

HUSSAIN, M. A., PU, S. L. & UNDERWOOD, J. 1974. Strain energy release rate for a crack under combined mode I and mode II. *In*: *Fracture Analysis*. ASTM Special Technical Publications, STP 560, 2–28.

INGRAFFEA, A. R. 1981. Mixed-mode fracture initiation in Indiana limestone and Westerly granite. *In*: EINSTEIN, H. (ed.) *Rock Mechanics from Research to Application: Proceedings of the 22nd US Symposium on Rock Mechanics*. Massachusetts Institute of Technology, Cambridge, MA, 186–191.

KANNINEN, M. F. & POPELAR, C. H. 1985. *Advanced Fracture Mechanics*. Oxford University Press, New York.

KULANDER, B. R., BARTON, C. C. & DEAN, S. L. 1979. *The Application of Fractography to Core and Outcrop Fracture Investigations*. Morgantown Energy Technology Center Report, **METC/SP-79/3**.

KULANDER, B. R. & DEAN, S. L. 1985. Hackle plume geometry and joint propagation dynamics. *In*: STEPHANSSON, O. (ed.) *Fundamentals of Rock Joints*. Centek, Luleå, Sweden, 85–94.

LAWN, B. R. 1993. *Fracture of Brittle Solids*. Cambridge University Press, New York.

LAWN, B. R. & WILSHAW, T. R. 1975. *Fracture of Brittle Solids*. Cambridge University Press, Cambridge.

LAZARUS, V., LEBLOND, J.-B. & MOUCHRIF, S.-E. 2001*a*. Crack front rotation and segmentation in mixed mode I + III or I + II + III. Part I: Calculation of stress intensity factors. *Journal of the Mechanics and Physics of Solids*, **49**, 1399–1420.

LAZARUS, V., LEBLOND, J.-B. & MOUCHRIF, S.-E. 2001*b*. Crack front rotation and segmentation in mixed mode I + III or I + II + III. Part II: comparison with experiments. *Journal of the Mechanics and Physics of Solids*, **49**, 1421–1443.

LIPSCHULTZ, M. M. 1969. *Theory and Problems of Differential Geometry*. McGraw-Hill, New York.

LISLE, R. J. 1992. Constant bed-length folding: three-dimensional geometrical implications. *Journal of Structural Geology*, **14**, 245–252.

LISLE, R. J. 1994. Detection of zones of abnormal strains in structures using Gaussian curvature analysis. *AAPG Bulletin*, **78**, 1811–1819.

LISLE, R. J. 2000. Predicting patterns of strain from three-dimensional fold geometries: neutral surface folds and forced folds. *In*: COSGROVE, J. W. & AMEEN, M. S. (eds) *Forced Folds and Fractures*. Geological Society, Special Publications, London, **169**, 213–221.

LISLE, R. J. & ROBINSON, J. M. 1995. The Mohr circle for curvature and its application to fold description. *Journal of Structural Geology*, **17**, 739–750.

MANDL, G. 1987. Discontinuous fault zones. *Journal of Structural Geology*, **9**, 105 – 110.

MARTEL, S. J. & BOGER, W. A. 1998. Geometry and mechanics of secondary fracturing around small three-dimensional faults in granitic rock. *Journal of Geophysical Research*, **103**, 21299–21314.

MARTEL, S. J. & POLLARD, D. D. 1989. Mechanics of slip and fracture along small faults and simple strike-slip fault zones in granitic rock. *Journal of Geophysical Research*, **94**, 9417–9428.

MARTEL, S. J., POLLARD, D. D. & SEGALL, P. 1988. Development of simple strike-slip fault zones, Mount Abbot Quadrangle, Sierra Nevada, California. *Geological Society of America Bulletin*, **100**, 1451–1465.

NICHOLSON, R. & POLLARD, D. D. 1985. Dilation and linkage of echelon cracks. *Journal of Structural Geology*, **7**, 583–590.

OLSON, J. & POLLARD, D. D. 1989. Inferring paleostresses from natural fracture patterns: A new method. *Geology*, **17**, 345–348.

OLSON, J. & POLLARD, D. D. 1991. The initiation and growth of en echelon veins. *Journal of Structural Geology*, **13**, 595–608.

PARKER, J. M. 1942. Regional systematic jointing in slightly deformed sedimentary rocks. *Geological Society of America Bulletin*, **53**, 381–408.

POLLARD, D. D. 1987. Elementary fracture mechanics applied to the structural interpretation of dykes. *In*: HALLS, H. C. & FAHRIG, W. F. (eds) *Mafic Dyke Swarms*. Geological Association of Canada, Special Papers, **34**, 5–24.

POLLARD, D. D. & AYDIN, A. 1988. Progress in understanding jointing over the past century. *Geological Society of America Bulletin*, **100**, 1181–1204.

POLLARD, D. D., SEGALL, P. & DELANEY, P. T. 1982. Formation and interpretation of dilatant echelon cracks. *Geological Society of America Bulletin*, **93**, 1291–1303.

PRICE, N. J. 1966. *Fault and Joint Development in Brittle and Semi-brittle Rock*. Pergamon, Oxford.

RENSHAW, C. E. & POLLARD, D. D. 1994*a*. Are large differential stresses required for straight fracture propagation paths? *Journal of Structural Geology*, **16**, 817–822.

RENSHAW, C. E. & POLLARD, D. D. 1994*b*. Numerical simulation of fracture set formation: a fracture mechanics model consistent with experimental observations. *Journal of Geophysical Research*, **99**, 9359–9372.

ROBERTS, A. 2001. Curvature attributes and their application to 3D interpreted horizons. *First Break*, **19**, 85–100.

SAMSON, P. & MALLET, J. L. 1997. Curvature analysis of triangulated surfaces in structural geology. *Mathematical Geology*, **29**, 391–412.

SEGALL, P. & POLLARD, D. D. 1983. Nucleation and growth of strike slip faults in granite. *Journal of Geophysical Research*, **88**, 555–568.

SELBY, S. M. 1975. *Standard Mathematical Tables*. CRC Press, Cleveland, Ohio, 756.

SIH, G. C. 1974. Strain-energy-density factor applied to mixed mode crack problems. *International Journal of Fracture*, **10**, 305–321.

SOMMER, E. 1969. Formation of fracture 'lances' in glass. *Engineering Fracture Mechanics*, **1**, 539–546.

STEWART, S. A. & PODOLSKI, R. 1998. Curvature analysis of gridded surfaces. *In*: COWARD, M. P., DALTABAN, T. S. & JOHNSON, H. (eds) *Structural Geology in Reservoir Characterization*. Geological Society, London, Special Publications, **127**, 133–147.

STEWART, S. A. & WYNN, T. J. 2000. Mapping spatial variation in rock properties in relationship to scale-dependent structure using spectral curvature. *Geology*, **28**, 691–694.

STOKER, J. J. 1969. *Differential Geometry*. Wiley-Interscience, New York.

STRUIK, D. J. 1961. *Lectures on Classical Differential Geometry*. Addison-Wesley, Reading, MA.

THOMAS, A. L. & POLLARD, D. D. 1993. The geometry of echelon fractures in rock: implications from laboratory and numerical experiments. *Journal of Structural Geology*, **15**, 323–334.

TREAGUS, S. H. & LISLE, R. J. 1997. Do principal surfaces of stress and strain exist? *Journal of Structural Geology*, **19**, 997–1010.

UEDA, Y., IKEDA, K., YAO, T. & AOKI, M. 1983. Characteristics of brittle fracture under general combined modes including those under bi-axial tensile loads. *Engineering Fracture Mechanics*, **18**, 1131–1158.

WEISSTEIN, E. 2002. *Eric Weisstein's World of Mathematics 2002*. Wolfram Research, Champaign, IL.

WIENBERGER, R. 2001. Joint nucleation in layered rocks with non-uniform distribution of cavities. *Journal of Structural Geology*, **23**, 1241–1254.

WILLEMSE, E. J. M. & POLLARD, D. D. 1998. On the orientation and patterns of wing cracks and solution surfaces at the tips of a sliding flaw or fault. *Journal of Geophysical Research*, **103**, 2427–2438.

WILLEMSE, E. J. M., PEACOCK, D. C. P. & AYDIN, A. 1997. Nucleation and growth of strike-slip faults in limestones from Somerset, U.K. *Journal of Structural Geology*, **19**, 1461–1477.

WISE, D. U. 1964. Microjointing in basement, middle Rocky Mountains of Montana and Wyoming. *Geological Society of America Bulletin*, **75**, 287–306.

WOODWORTH, J. B. 1896. On the fracture system of joints, with remarks on certain great fractures. *Boston Society of Natural History, Proceedings*, **27**, 163–183.

WU, H. & POLLARD, D. D. 1995. An experimental study of the relationship between joint spacing and layer thickness. *Journal of Structural Geology*, **17**, 887–905.

YATES, J. R. & MILLER, K. J. 1989. Mixed mode (I + III) fatigue thresholds in a forging steel. *Fatigue and Fracture of Engineering Materials and Structures*, **12**, 259–270.

The relationship of tilt and twist of fringe cracks in granite plutons

P. BANKWITZ & E. BANKWITZ

Gutenbergstraße 60, 14467 Potsdam, Germany (e-mail: epbank@web.de)

Abstract: Joint fractography in European plutons frequently shows large fringe-tilt angles connected to small fringe-crack twist angles (type A). In contrast to this first type, fringes of joints that tilt at small angles out of the parent joint plane are often associated with high twist angles of en echelon fringe cracks (type B). The interaction of tilt and twist angles gives evidence for the mode (I, II and III) acting at the advancing crack front during fracture propagation and the formation of fringes. The different fringe types depend on the varying influence of mode II or mode III, which establishes the degree of tilt or twist, in addition to opening-mode I that governs the crack propagation. Fringe types A and B are not randomly distributed. Within several plutons the first joints are characterized by a frequency of type A, and in other plutons by the dominance of type B. A third group of plutons is characterized by low tilt and low angle of twist for early joint fringes. The range of tilt/twist ratios of the earliest joints decreases with increasing depth of pluton emplacement and joint formation. The trend of the ratio approaches a value of between 0.5 and 1.0 at greater depth (*c.* 15 km). The ratio seems to be suitable for a prognosis of the possible depth of first joint formation.

The importance of phenomenological studies, now and in the future, is made obvious by our research of granite fractures. Fractographic features that comprise the topography of a fracture surface are common on joints, but were predominantly investigated in sedimentary rocks. The bedding of sedimentary rocks often influences the propagation of fractures to such a degree that, in many cases, they cannot propagate through in an unconstrained process. An unrestricted fracturing can only occur within an isotropic or quasi-isotropic rock such as a massive sandstone cliff (Cruikshank & Aydin 1995; Kulander pers. comm. 2002), or particular plutons. This study demonstrates observations made of varying joint morphology in several granites (Fig. 1). It is probable that the prototype of a rock fracture can grow in an undeformed plutonic rock, and good exposures can represent such preserved earliest joints. Regardless of some degree of faint magmatic foliation or layering, several plutons (e.g. the South Bohemian Pluton in the Czech Republic or parts of the Erzgebirge plutons in Germany) can be considered as quasi-isotropic rocks in relation to the size and the undisturbed formation of joints (length of 10–50 m and more than 100 m at exposures that are large enough).

Statement of the problem

This study contributes to our understanding of the shape that a growing opening-mode fracture will have, and the propagation path the joint follows, if an unconstrained fracture process takes place in a natural geological quasi-isotropic rock. Such conditions can be assumed in plutons at the moment when earliest joints were initiated. Therefore, one purpose

of the investigation was to determine the joint sets within the granite and the age relations of joints to establish the joint succession. Another purpose was to determine if the individual joints formed during a single fracture process or if portions of a joint surface developed after the change in direction of the maximum compression. This investigation also aims to ascertain the high-angle or low-angle tilt and the twist of fringe cracks, and to consider what factors influenced their occurrence if previously existing unhealed joints or free surface boundaries are absent at time of fracture. Finally, one objective was to find out whether special joint morphology is associated with specific stages of the pluton history or with different depths of joint formation.

With regard to the joints of the Appalachian Platform, the abrupt fringe planes seen after a kink were the result of a temporal change in the orientation of the remote stress field, indicating an arrest of the initial joint propagation (Engelder & Geiser 1980). The smooth curve at the end of a main joint, and its gradual tilt into the following fringe zone, represents a spatial change in the orientation of the local stresses, caused by the interaction of the advancing crack tip with previous joints or with inclusions (Kulander & Dean 1985; Olson & Pollard 1989; McConaughy & Engelder 1999). An inclusion can comprise small-scale material possessing elastic behaviour in contrast to the matrix, and is generally a more singular phenomenon. Within the granites discussed in this chapter both of the above reasons can be excluded for the formation of many first joints. The early joints initiated within quasi-isotropic granites and mostly without visible inhomogeneities.

Considering the unconstrained formation of earliest steeply dipping joints, one restriction arises from the common temporal sequence of fractures in

From: COSGROVE, J. W. & ENGELDER, T. (eds) 2004. *The Initiation, Propagation, and Arrest of Joints and Other Fractures.* Geological Society, London, Special Publications, **231**, 183–208. 0305-8719/04/$15 © The Geological Society of London 2004.

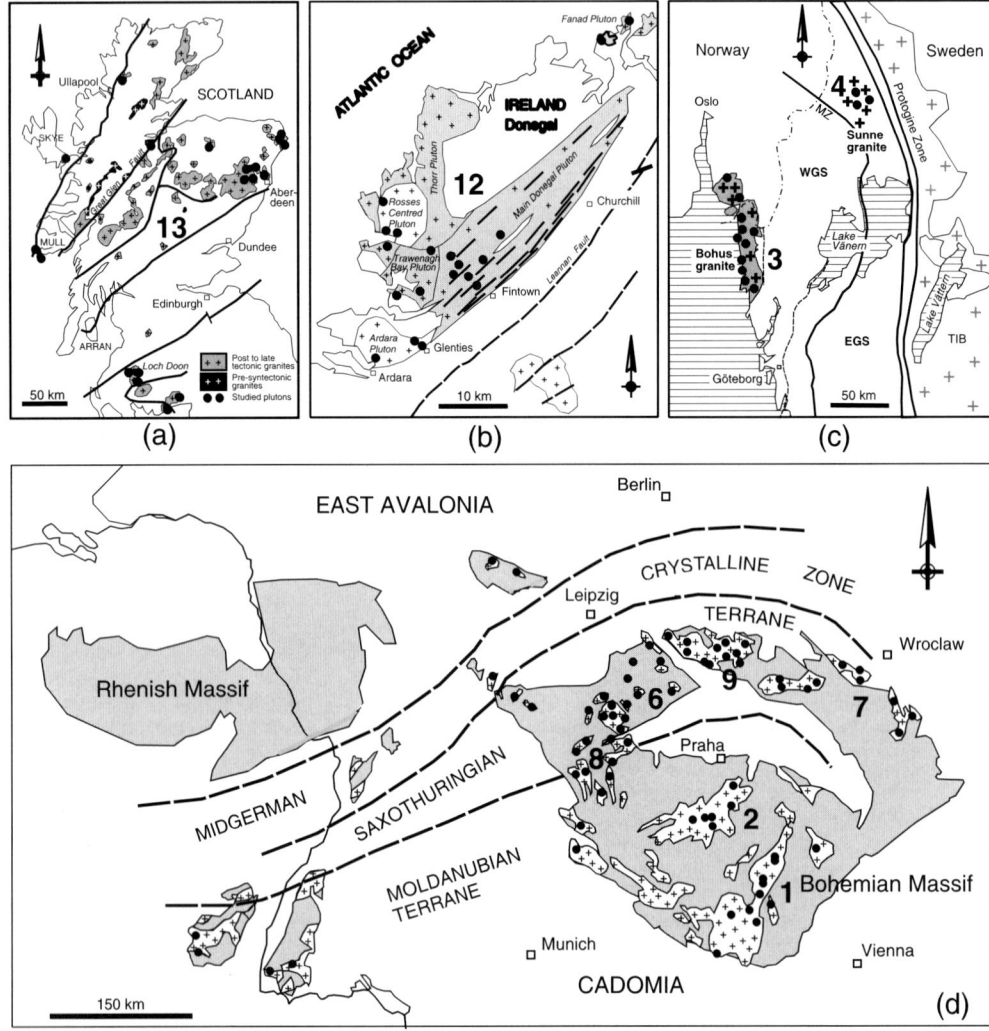

Fig. 1. Granites in Europe. Dots indicate areas in plutons where jointing was studied by the authors at many outcrops. Numbers 1–12: the main areas of joint studies. Denotation of the plutons within the text (list of granites). Simplified geological maps: (**a**) after Harris (in Craig 1991); (**b**) after Pitcher & Berger (1972); (**c**) after Freden (1994); (**d**) after Walter (1992). In (**c**) MZ, Mylonite Zone; WGS, Western Gneiss Segment; EGS, Eastern Gneiss Segment; TIB, Transscandinavian Igneous Belt. (**a**) and (**b**): Caledonian granites; (**c**) and 9 in (**d**): Precambrian granites; (**d**) Variscan granites. Granite localites in Denmark (5), Elba Island (10), the Urals (11) and Cornwall (14) are not shown.

plutons. Within the greater part of the studied plutons the first fractures are widely spaced subhorizontal joints or obliquely dipping joints, mostly at an angle of c. $0°–30°$ up to $40°$ ('lager' joints according to Cloos 1921). They occur in small numbers and, in some cases, could have influenced propagating vertical joints when approaching pre-existing subhorizontal joints. Pre-existing cracks interrupt a homogeneous stress field, and the wall of an unhealed fracture may prevent the transmission of the advancing crack-tip stress field (Engelder *et al.*

1993). As a consequence, the younger joint terminates at the previous joint. The principal stresses rotate locally in the vicinity of the wall of the pre-existing joints and therefore the propagating younger joint changes in orientation (Pollard & Segall 1987; Dyer 1988) forming tilted fringes. Trend changes in tensile/compressive stresses at the advancing crack tip are responsible for the development of twist hackles or tilt.

The causes of why the local stresses at the crack tip change direction and magnitude in the absence of pre-existing joints and other inhomogeneities are not finally known in each case. Factors influencing the propagation of early joints in granites were considered in an attempt to determine with which temporal moment of the pluton history the joint formation correlates, and at which position of the pluton within the crust the joints developed. If the joints were formed at the depth of pluton emplacement at an early stage of cooling (Bankwitz et al. 2004) the cooling and crystallization of the pluton can change the magnitude of stresses.

Peculiarities of the fringe and the fringe cracks in regard to unconstrained fracture propagation are one topic of this study. In many plutons high-angle tilted fringes are combined with low-angle twist hackles (en echelon segments) and, vice versa, fringe cracks with high-twist angle occur at low curved fringes. It was our aim in this chapter to prove if this combination is a common rule or occurs only randomly.

Method of investigation

Natural joints in plutons (Fig. 1) were considered with regard to their geometry, size, spacing, surface morphology and relative timing of formation by means of fractography and other known geological data. Measurements of angle and planes, and analysis of fractographic features, were the tools of the study. Fractographic surface pattern are well known from sedimentary rocks, and have been described and interpreted by many authors (e.g. Woodworth 1895; Bankwitz 1965; Kulander et al. 1979, 1990; Pollard & Aydin 1988; Bahat 1991, 1997; Kulander & Dean 1995; Younes & Engelder 1999).

We studied the earliest steeply dipping fractures that frequently represent the dominant systems of joints in undeformed plutons (e.g. in the South Bohemian Pluton of the Bohemian Massif). The relative ages of parent joints (sequence of joints) were ascertained for each pluton using the rotation of younger joints to become either parallel or perpendicular to the older joint (Dyer 1988), or their abutting relationships. We also noted their size (from several metres to nearly 100 m) and the remarkable distance (several metres to >30 m) between them. It was found that the shape and the fractographic features of granite joints differ widely, even of those granites that were similar in composition and dimension, and where the environmental conditions at the time of fracture were probably similar. The difference in joint morphology may be the consequence of the moment of fracturing in the history of the pluton, and with that are dependent on various driving mechanisms and stress conditions.

The joint face and the entire shape were considered, not just traces of fracture planes on a free surface. Younes & Engelder (1999) published tilt and twist angles of joint fringes and their cracks of up to 32° (with a few exceptions) in clastic rocks. Many granite joints propagated continuously into high-angle (30° to >60°) tilted large fringes several metres in length (Fig. 2a), or formed fringe cracks with large twist angles (Fig. 2d). In many cases the advancing crack front of the main joint plane curved gradually into the fringe, thus indicating a single fracture event. It was often found that fringe cracks had already initiated on the main joint face.

At an early stage, during the incipient formation of the joints, the granite supports an unconstrained propagation of fractures. Only such joints leave a record of undisturbed fracture propagation without forced arrest along older fractures or inhomogeneities. In contrast to granites, volcanic rocks consisting of several flows behave more like layered sedimentary rocks when fracturing (Fig. 3).

To be sure that the shape and the morphology of joint surfaces discussed in this chapter are common phenomena in granites, investigations of many plutons were necessary. On the basis of the comparison of jointing in a sufficient number of granites, it was possible to determine characteristic joint features. Our investigation of joint morphology included measurements of strike, dip and bending of main joint planes, fringes and associated en echelon segments, and related tilt and twist angles between them. In addition, the size and frequency of all joints and their echelon fringe cracks were documented. More than 10 000 individual joint surfaces and their fractographical features were studied, including the chronological succession of sets.

Plutons studied with regard to their joint features

Plutons hosting joints are often of different ages and have developed due to a variety of different crustal processes. In this study fractographic features were recognized on joints in all investigated granites. Several plutons, such as the Precambrian Bohus and Sunne granites in Sweden (Fig. 1c), and the Bornholm granites (Denmark; Bankwitz & Bankwitz 1994), remain for the most part undeformed after the emplacement, independent of their age. Others intruded prominent shear zones and were strongly deformed, such as the Caledonian granites in Scotland and Ireland (Fig. 1a, b). In our experience, the Caledonian granites are generally less suitable for studying the complete shape of early unconstrained formed joints, and their tilt and twist angles. In part they intruded into shear zones and underwent a strong blastesis (e.g. the Fanad Granite: Fig. 1c) by shearing under intracrustal conditions. In addition, the greater part of the exposures of Caledonian granites are small considering the large size of the granite joints, so that

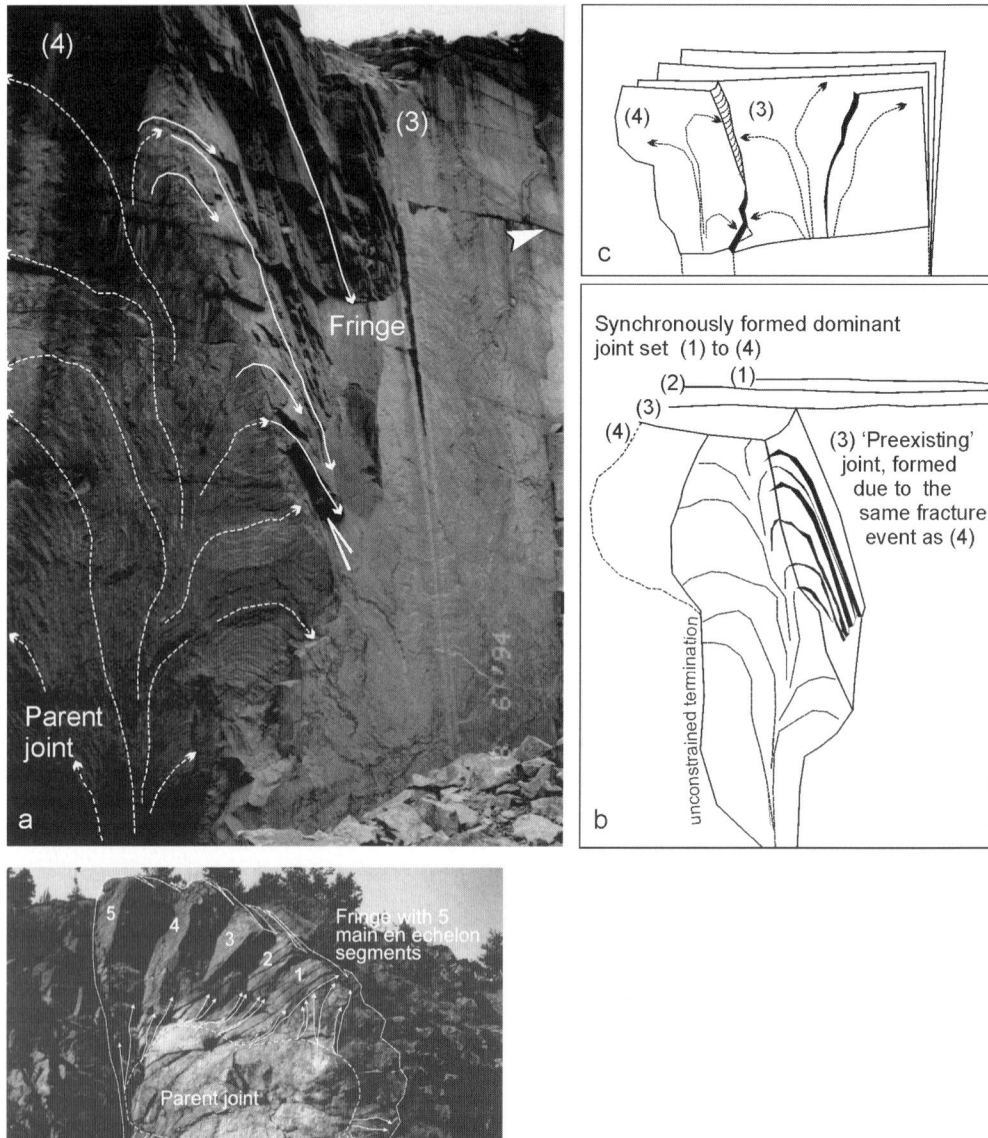

Fig. 2. Flossenbürg Granite, Oberpfalz, Germany. (**a**) One of the earliest 010°-joints: The parent joint (left part of the photograph) is covered with an upward-growing plume. The joint bends at the right end into a fringe tilted at about 65°. In part gradual en echelon fringe cracks initiated on the main plane, but mostly at the boundary, finally arranged subparallel to the fringe orientation. Very small twist angles may exist at the lower end of the fringe (see <). Subhorizontal traces of late post-uplift sheet fractures (large arrow) cross-cut the older vertical parent joint and the whole face. The height of the face is *c*. 18 m. (**b**) and (**c**) The joint is part of a subparallel joint bundle (1–4) more or less contemporaneously formed. The joint bundle is a fan-shaped structure, closely spaced near the bottom and diverging upwards. At the deepest part of the photograph (a) joint 4 and joint 3 approach another nearly in the same plane and form a composite single face (c). (**d**) The parent joint (bright) has an asymmetrical fringe, in its upper part represented by five gradual en echelon segments (1–5) with varying twist angles. The height of the face is *c*. 16 m. The joint initiated anywhere in the central plane and propagated in each direction, but dominantly to the top.

Fig. 3. Tongue-like parent joints (**a**) with large twist-hackle fringes (**b**) in the layered volcanic rock at Duved, Norway. The height of the face (left) is 4 m. The individual tongues change the lateral propagation direction from the right to the left and reverse, restricted by the layering. Gradual (arrows) and abrupt en echelon segments were formed at the same rim of a parent joint. The vertically propagated en echelon fractures suggest a far-stress field influence. (**c**) Fringe cracks in the same Proterozoic volcanic rock at Duved, Norway (Strömberg *et al.* 1984).

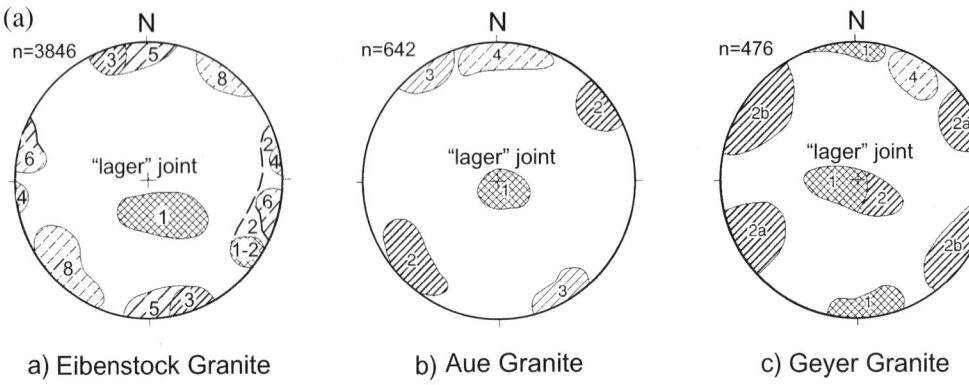

a) Eibenstock Granite   b) Aue Granite   c) Geyer Granite

Fig. 4. (**a**) Synoptic diagrams of lower-hemisphere stereographic projection of joints within three Variscan Erzgebirge granites (Germany). Numbers (1–8) of the maxima indicate the relative sequence of joint formation: 1, earliest formed joint set; 8, latest joints (Bankwitz & Bankwitz 1995). The same number in different maxima indicates changes in the interaction of the joint (e.g. 1 and 1–2). (**b**) Schematical illustration of the mapped joint sequence in the Eibenstock Pluton (Bankwitz & Bankwitz 1995) north of the national boundary between Germany (*D*) and Czech Republic (*CZ*). Jointing started with flat to oblique dipping (18°–35°) 'lager' joints. The other sets were steeply dipping, with the exception of set 9. The first subvertical joints (2) followed the shape of the exposed pluton. Numbers 1–10, temporal and spatial varying development of joint sets during three stages. The relative timing of the various joint sets give evidence for an area of preferred fracture formation that shifted in time, e.g. from phase 3 to 4 and 5, and from 6 to 7 and 8. Stage 3 comprises the development of sheet fractures and, finally, a second generation of smaller N–S and E–W-trending joints that terminate at the sheet fractures 'lager' joints (Fig. 8b). Dot, Blauenthal quarry (location of the photographs in Fig. 8). Joint sets 2 and 4 with dense greenish Uranium-micas at the fracture surface).

only parts of joint surfaces can be observed. Some larger quarries and cliffs (Aberdeen and Peterhead in Scotland, and Donegal in Ireland) reveal joints with mostly low surface morphology regardless of some few high-angle tilted fringes.

In many plutons multiple joint sets exist. The determination of the relative joint sequence is an important element for this study. However, the timing of the formation of small joints (about 1 m) subparallel to much larger fractures (tens of metres) is problematic. Both can be formed contemporaneously, in which smaller joints are prevented from further propagation by their large neighbours, as Segall & Pollard (1983) found in granitic rocks of the Sierra Nevada; on the other hand, however, the small joints can be formed later. In each case this needs to be established by their interaction with other joint sets. This applies not only to small joints, others can also present a second generation with the same orientation as an older set. Based on detailed joint mapping in the Eibenstock Pluton, two steeply dipping N–S and two E–W joint sets formed at different times could be recognized (Fig. 4b; sets 4 and 5, and set 10) by their peripheric twist-hackle deviation (Bankwitz & Bankwitz 1995). Here, the younger joints of these sets abut onto late formed, subhorizontal sheet fractures. They are distinctly smaller than the earlier joints, which are cross-cut by the sheet fractures, indicating that these vertical joints formed prior to the

present erosional surface. Within several granites, for example the Sunne Granite, a similar joint sequence with repeated formation and subparallel joint orientation exists.

The plutons are listed according to their suitability to this problem. Reference is also made to additional granites to demonstrate the wide range of observations. Most of the list numbers are given in Figure 1 (where each dot stands for a number of exposures).

First group: granites with finely developed joint surface features and that are largely well exposed

(1) South Bohemian Pluton (SBP), Czech Republic and Austria (330–320 Ma; Mrákotín (e.g. Mrákotín and Boršov quarry), Eisgarn and Weinsberg granites). Located in the SE part of the Bohemian Massif (CR). The pluton (length in the NNE direction of about 150 km) consists of several late to post-tectonical Variscan granites (geochronology by Scharbert 1998; Breiter & Koller 1999) that are related to the Variscan convergence of the Bohemian and Saxothuringian terranes. Gravity anomalies reflect the roots of the deep intruded bodies at c. 15 km depth (Breiter 2001). The palaeodepth of the granite at the present erosion level was determined by fluid inclusions at 7.4 km in the northern part (Boršov

(b)

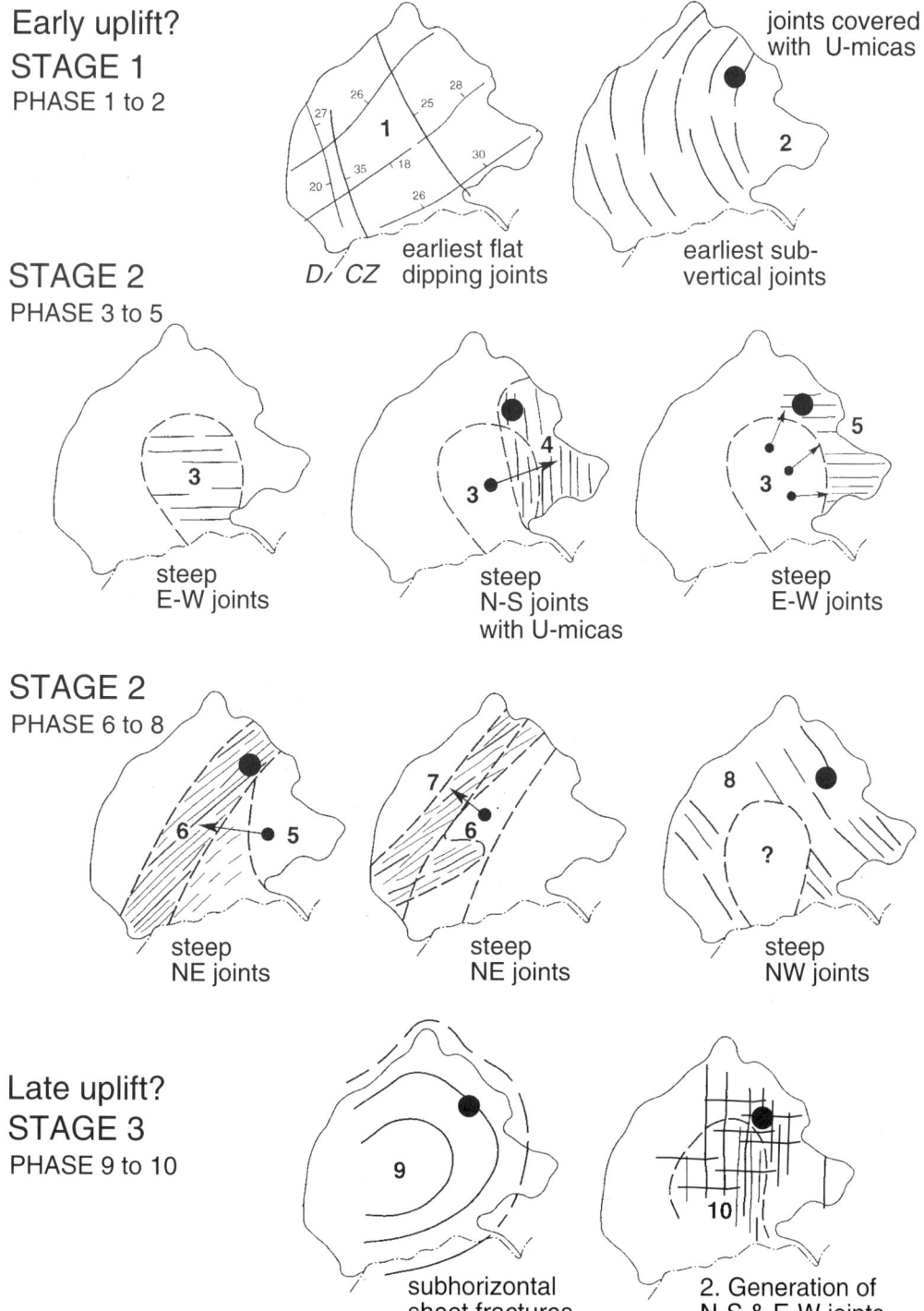

Early uplift?
STAGE 1
PHASE 1 to 2

earliest flat
D/ CZ dipping joints

joints covered
with U-micas

earliest sub-
vertical joints

STAGE 2
PHASE 3 to 5

steep
E-W joints

steep
N-S joints
with U-micas

steep
E-W joints

STAGE 2
PHASE 6 to 8

steep
NE joints

steep
NE joints

steep
NW joints

Late uplift?
STAGE 3
PHASE 9 to 10

subhorizontal
sheet fractures

2. Generation of
N-S & E-W joints

quarry, Mrákotín Granite, Czech Republic) and at 14.3 km in the southern Austrian Weinsberg Granite (Friepeß quarry). Evidence for the minimum age of early fractures comes from muscovites within thin microgranites (1 cm) on exposed fringe crack surfaces 324.9 ± 6.7 Ma in age (Bankwitz *et al.* 2004).

The samples were taken from the en echelon fringe cracks of two 025°-trending early joints: one from the vertically propagated fringe of the 'Trefoil' joint (Fig. 5b), and the other sample from a preferred lateral propagated fringe that is developed almost around the joint. Both samples were used for the determination of the palaeodepths of the exposed granite found today, and for K/Ar-age dating of muscovites within small microgranites at the fracture surfaces (Boršov quarry). Thus, the formation of the 025°-joints, including their fringes, could be post-dated at 324.9 Ma. The muscovite K/Ar-age of samples from a N–S joint surface in the Austrian part of the pluton (Friepeß quarry) was given as 318.9 Ma. The intrusion age of the pluton is 330–318 Ma, and the cooling age ranges between 328–325 and 320 Ma (Mrákotín Granite) and 313–308 Ma (Weinsberg Granite; Scharbert 1998; Breiter & Koller 1999). The Borsov joint was formed prior to or during injection of the microgranitic vein, that is 3 Ma after the start of cooling. Taking into consideration the fact that first cooling occurs at the boundary of the pluton, we can assume that granite cooling began later in the axial region of the pluton where Boršov is located. This means that the Boršov joints were formed less than 3 Ma after the beginning of cooling within the pluton. These joints are characterized by exquisite fractographic features.

(2) Central Bohemian Pluton (CBP) Czech Republic (340 Ma). The pluton consists of several Variscan granites (e.g. Hudčice and Solopysky quarry) that started to generate earlier than that in the SBP, with crustal conditions similar to those given above (Holub *et al.* 1997). Vertically to obliquely dipping joints, in part with subhorizontal fringes, are conspicuous at several places, as well as huge plumes with very broad branches (up to 5 cm-wide ridges and grooves). Other fractographic features are well developed.

(3) Bohus Granite, Sweden (950 Ma). Located along the west coast of South Sweden between Göteborg and Oslo, the granite intruded during late- or post-Grenvillian–Svekofenian orogenic time (Zheng 1996). The emplacement has been suggested to have been at a mid-crustal level of above 15 km (Eliasson 1992) related to shear-zone movement. The granite massif forms the reefs along the Skagerrak Sea, and is exposed at many intensely fractured cliffs and quarries, offering decametre to several hundreds of metres of extended joints. Some fracture sets represent impressive surface morphology including high-angle tilted fringes, in part with oblique dip.

(4) Sunne Granite, Sweden (1650 Ma). The granite is part of the Eastern Gneiss Segment (Freden 1994) in mid Sweden. Large exposures exist along roadcuts (length of several hundreds of metres; height of up to >20 m). The joints have similar morphology to granite (3).

(5) Bornholm Island granites, Denmark (1400 Ma; e.g. Rönne, Vang, Hammer, Svaneke, Almendingen granites; Larsen 1980). Reefs along the shoreline of the Baltic Sea, cliffs, and large and deep quarries with early formed joints exist. Younger, uplift-related vertical joints cross-cut the older joint system (Bankwitz & Bankwitz 1994). All of the common fractographic features exist in these granites.

(6) Erzgebirge granites, Germany (325–318 Ma; e.g. Eibenstock, Kirchberg, Bergen, Aue, Geyer and Niederbobritzsch plutons). Post-tectonical Variscan granites in SE Germany that are related to the Variscan convergence (Förster *et al.* 1999). The palaeodepth of the granites (today at the erosion level) was determined by fluid-inclusions at <3 km (Thomas & Klemm 1997). These highly evolved intrusions are, in part, of subvolcanic type and enriched in ore deposits (Erzgebirge: Ore Mountains). Fractographic features are common, but predominant with low joint morphology.

(7) Granites of the West-Sudetes, Poland (330–310 Ma; e.g. Riesengebirge/Krkonosze, Strzegom, Strzelin granites). Late syn- to post-tectonic Variscan granites in SW Poland that are related to the Variscan convergence (Mierzejewski 2001). They are the proper 'Cloos granites', where Cloos 80 years ago established for the first time the 'Granite Tectonics'. In part, magmatic and tectonic foliation is weakly developed. The joint morphology is well developed; however, the morphology in the N–S and E–W sets is very different to that recognized in the SBP (1).

(8) Karlovy Vary Granite and Bavarian granites (Fichtelgebirge granites, 320–290 Ma; Oberpfalz granites, 325, 305 and 290 Ma, e.g. Flossenbürg Granite, 312 Ma). Located in the northern Czech Republic and in the SE part of Germany (Bavaria), these plutons are of Variscan age (Siebel 1998). The fracture morphology usable for this study is only partly exposed (Bankwitz *et al.* 2000).

(9) West- and East-Lusatia Granite Massifs, Germany (540–530 Ma). Partly deformed Cadomian (up to Lower Cambrian) granites that were generated at the end of the Cadomian orogeny (Schust

Fig. 5. N-S-trending early joints with a curved boundary of the main plane against the fringes. Curved rims and undulations on the joint face mark the temporary crack front and indicate an unrestricted fracture process resulting as a quasi-circular fracture. Twist hackles (plumes) propagated from the main joint onto the fringe without interruption, partly producing fringe cracks. Such features indicate continuous fracturing. (**a**) Sunne Granite, mid-Sweden, Tossebergsklättan north of Sunne, Värmland. Road cut consisting of a single joint (total length of 25 m; total height of *c*. 12 m). The overall plumose structure becomes visible within wet stripes on the face. Arrows indicate the local propagation direction; arrows 1–4 mark places where steps of the first fringe run directly further to the second fringe. (**b**) South Bohemian Pluton, Boršov quarry. Features as in (**a**). Detail of the 'trefoil' joint face (total length of 28 m; total height of 15 m) with an excellent plumose structure. (**c**) Locality as in (**a**); the structures of the joint surface are similar to the features of cracks in glass, independent of the difference in size. The ellipse marks the site of an pegmatitic injection. From that inhomogeneity the 'augen' fracture initiated.

2000; Eidam & Krauss 2001; Tichomirova 2001). Magmatic and, at many locations, tectonical foliation exists (Eidam & Krauss 2001; Lobst 2001). Several superimposed joint systems indicate repeated events of fracturing and deformation.

(10) Mte. Capanne, Elba Island (6 Ma). This is the youngest granite massif in western Europe (Bussy 1991). The morphology of some joints reflect fracturing under a high stress magnitude that branches early after initiation close to the point of origin into en echelon segments with a large overlap. In some places vertical joint surfaces contained by dense hydrothermal tourmaline crystals with preferred vertical orientation. The temperature of tourmaline crystallization was determined to be 400 °C within a fluid with low fO_2 (oxygen fugacity), indicating a closed system (Y. Fuchs pers. comm. 2002). Good exposures for joint investigations are rare.

Second group: sheared or strongly weathered granites, in part with inadequate exposures

(11) Many granites of the Southern Urals (e.g. Dzabyk Granite, 276 Ma, Sanarka Pluton). Low to strongly deformed syntectonic granite massifs of the Variscan accretion wedge in the South Uralian hinterland (Fershtater *et al.* 1997). Here, in many plutons, magmatic and strong tectonic layering exists, along with foliation; several cleavage plane systems can also be recognized. Most of the granites intruded into shear zones of the accretion wedge and were deformed by transpressive strike-slip movements. Impressive joint-surface plumes with a rough morphology occur in the latest undeformed 'stock' granites.

(12) The granites of Donegal in the Caledonides of NW Ireland (e.g. Ardara, Ross, Thorr, Fanad and Main Donegal granites). Pitcher & Berger (1972) and Hutton & Alsopp (1996) investigated these Caledonian age granites, some of which have magmatic and quite strong syntectonic foliation; the Main Donegal Granite and the Fanad Granite are related to the Gweebarra and the Mossfield shear zone, which belongs together with the Leannan Fault and other wrench faults to the system representing the prolongation of the Great Glen Fault of Scotland. Ardara and Ross are diapirs, others are cauldrons. All of these are considered to be shallow intruded granites. More or less rough joint surface structures with low morphology were recognizable, however, in some parts high-angle tilted fringes were visible.

(13) The younger granites of the Scottish Highlands (456–397 Ma; Loch Doon, Ross of Mull, Aberdeen, *Peterhead, Great Glen, Borrolan).* Craig (1991) mapped these Caledonian granites (from Peterhead to Fionnphort) that host regular, relatively planar joints with only a smooth fracture morphology.

(14) The granite chain of Cornwall (c. 290 Ma; as the Dartmoor, Bodmin, Austell and Land's End granites). Post-tectonical Variscan collisional granite formation, often kaolinized; the exposures are not suitable for joint studies in the presented sense of this chapter.

Fractography of granite joints

In granites all fractographic features known from other rocks are developed, corresponding with the fact that the same fractographic features can develop in any brittle material. The length of joints range between a few metres and nearly 100 m, therefore shape and surface morphology of granite joints are best exposed in large quarries (e.g. SBP), or in cliffs or roadcuts of decametre height and length in mountaineous areas (e.g. Sunne Granite, Sweden, Fig. 5).

According to our field observation the petrological composition and the lithological grain pattern of the granites can be ignored for the purpose of this study, because in most plutons the grain size and fabric are small features relative to the size of the joints. However, the precise modulation of joint surface features is best developed in fine-grained rock.

Some joints in granites reflect an undisturbed fracture progress from the point of origin to the natural termination of the fracture, including the adjacent unfractured rock. Unconstrained fractures form as quasi-circular or quasi-elliptical surfaces that are seldom truely planar, the low bending of the joint surface is sometimes beyond measurability. Some joint surfaces have multiple curved boundaries between the parent joint and the fringe (Figs 2, 3, 5 and 6c) that demonstrate individual propagating crack-tip sections. Only constrained fracture propagation results in all the other fracture types that are discussed in literature.

Fine or strong plumose structures (Hodgson 1961), consisting of ridges and grooves of mm- to cm-width marking the propagation path, and locally frequent rib marks or undulations in the form of tilted panels or rounded forms (Kulander & Dean 1995) occur on joint faces of magmatic rocks (Figs 3 and 5). Near the periphery the fracture often deviates from the main plane and forms well-developed tilted fringes with twist hackles (Kulander *et al.* 1979). They propagate in the form of gradual en echelon segments, termed by Pollard *et al.* (1982) as dilatant echelon cracks. The fringe is most commonly divided into planar en echelon fringe cracks,

Fig. 6. Gradual en echelon fringe cracks with low twist angles. (**a**) Upper part of a much larger granite joint that was propagating to the top, indicated by the increasing main step. The steps represent the temporary rim of large en echelon cracks that initiated near the centre of the joint like playing cards. One of them formed at the top fringe crack. Height of *c*. 5 m. Sunne Granite, Värmland, Sweden. (**b**) The lateral (left) fringe cracks were initiated at the parent joint (arrow). This N–S-trending joint (entire length >20 m) terminated at the earlier formed E–W joint that forms an open rupture at the face. Height of *c*. 6 m. Bohus Granite, Sweden, part of the Kungsklyftan near Fjällbacka. (**c**)–(**e**) Parent joints and their fringes with gradual twist hackles. Length of the fringe sections was about 70 cm. (**c**) Sunne Granite near Tosseberg, Sweden. (**d**) Fringe in Ordovician shale (Steinach Thuringia, Germany) compared to granite joints (**c, e**). (**e**) South Bohemian Pluton, Boršov quarry: fringe (right) with two rows of twist hackles (1) and (2). Length of the fringe section is 1.5 m.

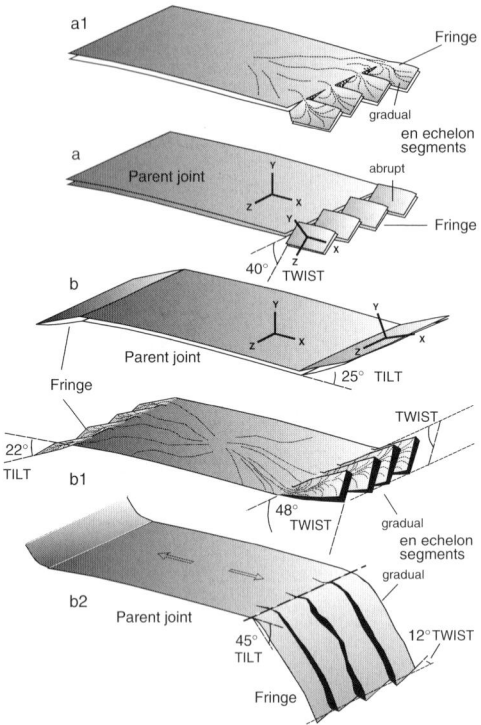

Fig. 7. Sections of joints and their fringes presented schematically. Basic features are shown in (**a**) and (**b**) modified after Engelder *et al.* (1993). (**a**) Joint with untilted fringe, occupied by abrupt en echelon fringe cracks. (**a1**) The untilted fringe with gradual en echelon segments (common in granites). (**b**) The fracture terminates with a tilted fringe, due to the rotation of the local stress field (*x, y, z*). (**b1**) Example of a parent joint initiated at the centre and propagated into a low-angle (22°) tilted fringe and gradual en echelon segments with large twist angles (48°). (**b2**) This fracture ends with a high-angle (45°) tilted fringe that is covered by low-angle (12°) twisted en echelon segments. The fractures in (b1) and (b2) are the result of mixed modes I + II + III acting at the advancing crack-tip. Type b1 occurs on a decametre scale in the Central and the South Bohemian Pluton, and in the Strzegom Granite of the Sudetes; type b2 occurs within the Bohus, Flossenbürg and Mte. Capanne granites.

whereas the hooked type (Engelder *et al.* 1993) in the studied granites is rare. Figure 7 shows gradual and abrupt fringe cracks, and various degrees of fringe tilt and fringe-crack twist. The figure only illustrates sections of a joint given in the photographs of Figure 6, and not the complete rock fracture. Many gradual fringe cracks occur in various directions around the parent joint, following the curved rim of the parent joint (Fig. 5); however, these fringe cracks are often concentrated at two opposite edges of the joint. Within granites the low-

grade curvature of the main planes and the deflection of their fringes reflect the local change in the principal stresses at the advancing crack tip and the response of the fracture to grow perpendicular to the direction of least stress (Lawn & Wilshaw 1975).

It is common knowledge that fractographic features are produced when the joint deviates locally from its mean plane of propagation. Deviations are attributed to tensile crack-tip stresses that are affected by various factors, mainly by the amount of strain energy released, and changes in propagation rate and in far-field stresses. Even though these far-field stresses may cause mode II or III shear loading, according to Kulander & Dean (1995) resultant crack-tip stresses in a brittle material are tensile.

We avoid repeating figures and definitions for the fractographic terms because these have been discussed and illustrated in the literature, especially in general views by Kulander *et al.* (1979, 1990), Pollard & Aydin (1988), Bahat (1991), Engelder *et al.* (1993), Kulander & Dean (1995) and Younes & Engelder (1999), and in early papers, e.g. Woodworth (1896), Hodgson (1961), Bankwitz (1965, 1966) and Bankwitz & Bankwitz (1984).

Key study areas of verified granite intrusion depth

Two key areas provide examples of a shallow and a deep intruded pluton to compare their first developed fracture sets. The shallow intruded Erzgebirge granites and the deep-seated emplaced SBP, of more than 150 km in length, have been well studied with regard to their geology, tectonics, petrology, mineralogy, gravity, magnetic, deep seismic sounding and, finally, also their fractography. Several thousand joints were interpreted by the authors. The Erzgebirge granites (granite (6) given earlier) intruded within the uppermost 3 km of the crust (Thomas & Klemm 1997), the exposed part of the SBP ((1) in the list given earlier) was solidified at 7.4 and 14.3 km (Bankwitz *et al.* 2004), in the northern and southern part, respectively.

Relative sequence of joints: Eibenstock Pluton of the Erzgebirge

A lot of plutons (e.g. Eibenstock, Aue and Geyer) intruded the Variscan metamorphic series of the Erzgebirge (SE Germany). Their joints were formed at relatively shallow crustal levels. The timing of the first fractures has yet to be determined and whether these first fractures formed at the intrusion depth or during uplift by unloading. The Eibenstock Pluton represents a large laccolithic intrusion (exposed area of 20 × 40 km) with complex fracturing. The orien-

Fig. 8. Eibenstock Pluton, Erzgebirge, Germany (Blauenthal quarry). Bars, 1 m in length. (**a**) Surface morphology of late E–W joints (10) terminating at previous N–S joints (4). (**b**) Fracture sets 4, 5 and 9 (Fig. 4).

tation of the joint sets and their successive formation is shown in Figure 4a. The synoptic diagram from various parts of the pluton is based on 28 single diagrams, each of them with 50–100 relationships between two, or sometimes three, interacting joints (containing 100–200 joint measurements). For each joint its termination at a pre-existing joint was determined. Only the spatial and sequential coinciding maxima of all 28 single diagrams were marked within the synoptic diagram (Fig. 4a). We managed to distinguish between eight dominating joint sets. The earliest formed joints are always flat-dipping 'lager' joints (Cloos 1921). Such subhorizontal joints are significantly different from post-uplift sheet fractures. Steeply dipping orthogonal joint systems consist mostly of joint sets formed at different times. The relative sequence of joint formation is similar in the three granites of Figure 4a. The Aue ((b) in Fig. 4a) and Geyer ((c) in Fig. 4a) granites occur as diapirs within NW-trending prominent fault zones of the Erzgebirge. The fault zones influenced the geometry of the granite bodies and probably the small difference of joint set orientations within the Aue (sets 2 and 3) and the Geyer (sets 2a and 2b) granites.

Each joint set has a definite spatial distribution and orientation. Figure 4b demonstrates schematically that only parts of the pluton fractured at one time (fracture phases 3–8). All joints are steeply dipping with the exception of set 1 and set 9. The jointing-free area within the diagrams of Figure 4b indicate two aspects. First, large regularly arranged and dominating joints were not recognized there. Second, joint relative age data are not equally distributed through

the whole pluton due to a lack of good exposures at some locations. At all outcrops within the pluton joints do occur, but not in all cases are these joints suitable for relative age determination. Even so, in certain areas that had regular jointing of defined relative age the age of the joints is seen to shift several times following the formation of a later joint set, mostly with changed orientation. Their temporal relationship could be proved, however; for example, the E–W joints developed first in the centre of the pluton (phase 3) and later in the eastern part (phase 5), determined by the occurrence of joint set 4. The N–S joints (set 4) are subsequent to set 3 in 'area 3', but pre-existing in 'area 5' at the time when joint set 5 was formed. After the earliest fracture formation during stage 1, the jointing shifted temporally in a 'wave-like' manner, first from the centre to the east and than back to the west (stage 2). Finally, sheet fractures (set 9) formed and latest vertical joints of smaller dimension (set 10, stage 3) abutted on set 9. The photographs (Fig. 8) show the closely spaced large N–S joints (set 4) as open fractures and the quarry face consisting of E–W joints (set 5) terminating at the former set.

The dense fracturing is completed by very narrow developed sheet fractures (set 9; 5–20 cm spacing). The earliest joints of set 1 and set 2 were formed at a greater distance of several metres, and in many places is filled with hydrothermal U-mica precipitate. In the Blauenthal quarry (dot in Fig. 4) the NE joints (set 4) are also covered by greenish micas that are not observed at the other younger joint sets.

Within the Eibenstock Pluton low surface morphology of joints is dominant. The joints are

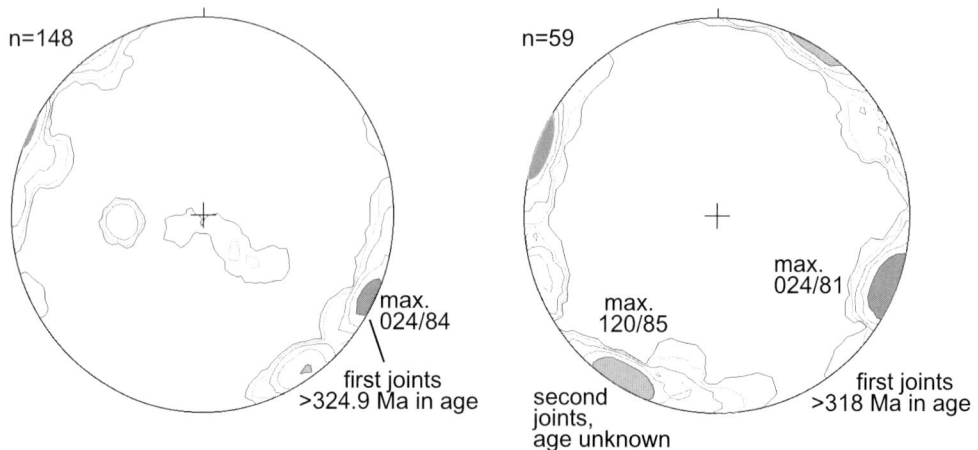

n=148

max.
024/84

first joints
>324.9 Ma in age

Earliest joint sets (Borsov quarry),
northern part of the SBP

n=59

max.
024/81

max.
120/85

second
joints,
age unknown

first joints
>318 Ma in age

Main joint sets, Austria (Friepeß quarry)
southern part of the SBP

(a)

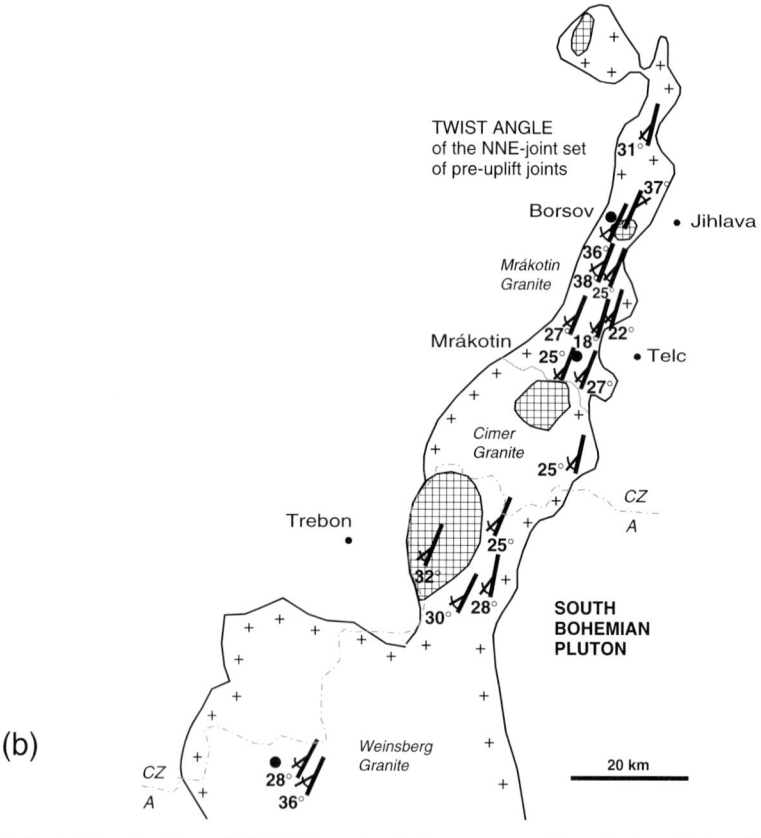

(b)

Fig. 9. South Bohemian Pluton (SBP): (**a**) lower-hemisphere stereographic projection of joint orientation within the Mrákotín Granite (Boršov quarry, CZ) and the Weinsberg Granite (Austria). (**b**) Geological sketch map of the SBP showing the regular orientation of first formed vertical fracture. Dots, locations of investigated fluid inclusions (Bankwitz *et al.* 2004) and diagrams in (**a**). Cross-hatched areas are granite bodies in the axial part of the SBP (Breiter 2001).

characterized by low twist angles, and undulations (rib marks) of the joint surface are lacking. At several locations tilt angles are between 30° and 45°, and more are developed.

Relative sequence of joints: SBP of the Bohemian Massif

Within the deep-seated SBP fracturing and distribution of joints are quite different from the shallow intruded Eibenstock Pluton. The dominant early subvertical fractures (025°) are regular and remarkably persistent through the whole pluton (Figs 1 and 9) suggesting that these fractures formed contemporaneously.

In the northern Mrákotín Granite of the SBP (e.g. in the Boršov or Mrákotín quarry) some appproximately E–W-trending joints could be recognized as being the oldest subvertical fractures that occur very sparsely, spaced at >50 m. These exceptional joints have not significantly influenced the predominant 025°-set. The NNE-trending joints are, in part, widely-spaced (distance 3–7 m, rarely 2 m), but often developed contemporaneously within the narrow fracture zone (c. 40 cm wide). There, several joints are closely spaced, each single joint between 5 and >60 m in size (Fig. 10b; e.g. traced at the free surface of the 120°-joint in Fig. 10a). These narrow initiated joints have interacted several times and during further propagation have moved closer to one another, finally forming a composite joint. The majority of the orthogonal ESE joint sets (c. 120°) were clearly formed later (Fig. 10a).

The vertical joint surfaces are seen to be decorated with plume structures, and, in some places, with frequent undulations (rib marks) and twist-hackle fringes. During a late uplift stage large subhorizontal sheet fractures developed close to the present erosional surface, spaced between 0.1 and >3 m apart, but this spacing decreases towards the surface.

Within the southern Weinsberg Granite (deepest exposed part of the SBP with a palaeodepth of 14.3 km), the NNE-trending joints (025°) represent the first formed set and consist of closely spaced large fractures, mostly arranged in zones. These joints were filled with fluids during, or immediately following, formation. The fluids deposited biotite, muscovite and chlorite within the joints. The 120°-joint set with a peculiar low surface morphology was clearly developed later and without any phyllosilicate deposits; it is always seen to terminate at the NNE joints.

In contrast to the shallow Eibenstock Pluton, a large variety of joint surface morphology exists within the SBP with respect to type and size of plume structures, frequency of undulations, and degree of tilt and fringe-crack twist. In general, the

Fig. 10. Mrákotín Granite of the northern SBP (Mrákotín quarry). Bars, 2 m in length. (**a**) E–W joint, later formed as the exposed traces of N–S joints at the face (height is c. 16 m). (**b**) Pre-existing N–S joint (60 m long in the photograph) with regard to (**a**); however, demonstrating three traces (1–3) of rare earlier formed E–W joints. Supposedly, both sets were interacting. Height of quarry faces is c. 22 m.

amount of fringe tilt ranges between 0° and 25°, which is much smaller than in the Erzgebirge. The fringe cracks show a wide range of twist angles.

Field data referring to regularities of tilt and twist angle combinations

Location and form of fringes

Fringes of granite joints can form around the entire boundary of the parent joint (Fig. 5), but we found that fringes frequently developed at the upper and lower edge of the joint plane. The early, subvertical fractures considered here formed large fringes, in part at great depth, mostly at the fracture edge perpendicular to the z-axis (z being depth). Usually, the previously laterally advancing fracture front leaves the parent joint propagating vertically (Fig. 10b), and is associated with breaking into gradial en echelon, but disconnected, twist hackles. However,

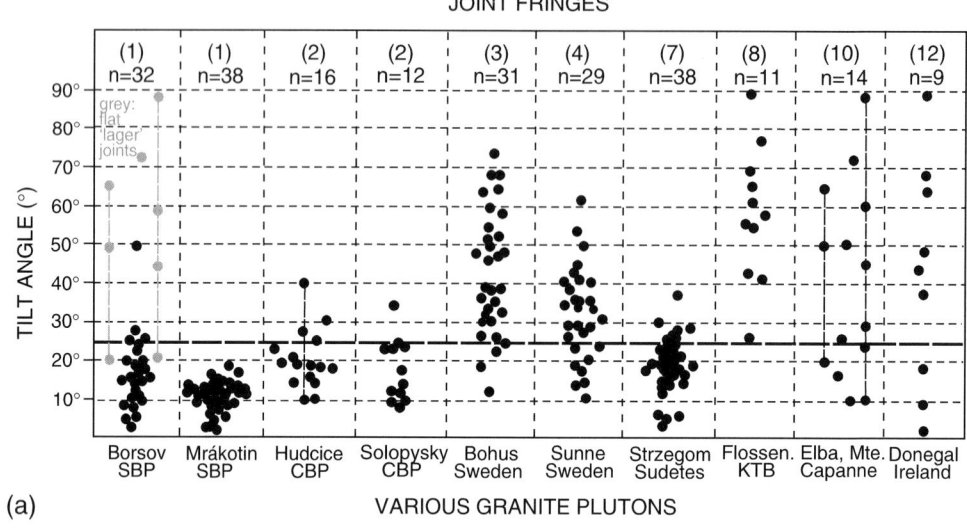

(a)

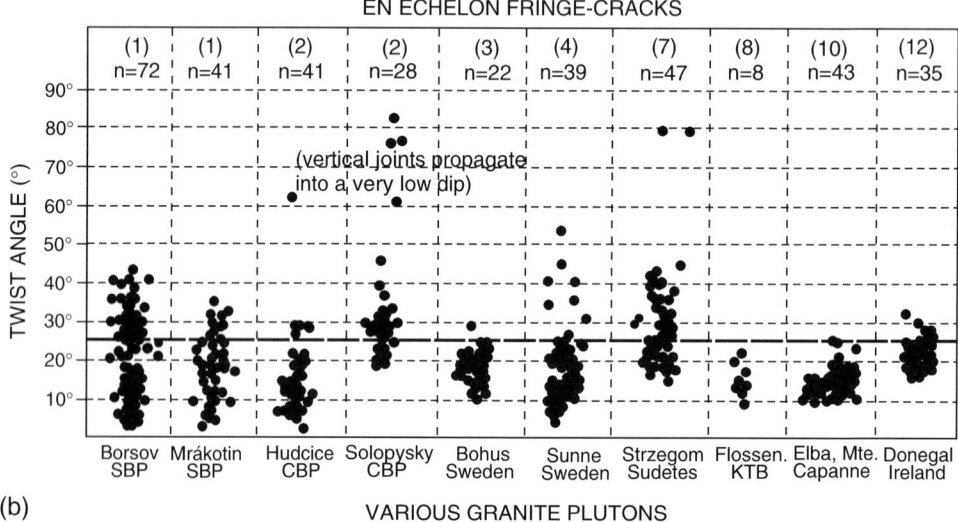

(b)

Fig. 11. Joint fringes from eight different plutons. Numbers in brackets are plutons according to the listed numbers in the paragraph section 'Plutons studied with regard to their joint features'. (**a**) Fringe tilt angles. Deviation of the fringe from the plane of the parent joint (grey: fringes of the earliest subhorizontal joints). Lines connect several fringes around the same initial joint. (**b**) Twist angles of en echelon fringe cracks. The angles define the rotation of the en echelon segments about the axis of joint propagation out of the fringe plane. SBP, South Bohemian Pluton; CBP, Central Bohemian Pluton; Flossen. KTB: Flossenbürg Granite (Bavaria), near the KTB (Continental Super-deep Borehole).

lateral fringes at the edges parallel to the *z*-axis were also developed (Fig. 6b), independent of whether the parent joint propagated laterally or vertically (Fig. 2a). The degree of tilt can slightly change at the various joint edges, but remains below the average tilt values of the defined fringe type.

Within the shallow intruded plutons we found a similar distribution of smaller fringes around the parent joints; however, a preference for vertical progress is absent and lateral fringes may be dominate. The trend of increased tilt angles in shallow intruded granites was recognized on joints of all ages.

Figure 11 summarizes the occurrence of fracture tilt and twist angles in various granites, demonstrating the remarkable differences between them. Also striking is the varying size of the fringes and their fringe–parent joint relationship. The trend of the different fringe/parent joint ratios which ranges in shallow intruded granites between 0.02 and *c.* 0.5, and in deep plutons between 0.2 and 1–>5.

Field data of fracture parameters

The size of granite joints hinder measurements of all twist-hackle orientations, and twist and tilt angles. Thus, Tables 1–4 offer some examples of individual first steeply dipping joints representing dominant types of fractures within the SBP and, in addition, within the CBP. Records of twist angles were sometimes completed using measurements of fringe tilt angles. Table 1 demonstrates the case of subparallel parent joints, initiated nearly in the same plane, that intermatch and form one composite quarry face (length, 60 m; Bahat *et al.* 2001) similar to the joint bundle (1–4) in Figure 2. The parent joints of various dimensions approach the same elliptical joint surface ratio after the formation of their fringes. Table 2 shows the complexity of surface structures of one individual joint that characterizes the predominant fracture set (025°) of the pluton. Tables 3 and 4 demonstrate the peculiarity of early subvertical joints in the CBP to tilt at the lower end into flat fringes. At localities where the deep-seated CBP (Solopysky and, in part, Hudčice) was poorly fractured, particular fringes in the vicinity of the circular or elliptical parent joints were recognized. From the lower edge of some subvertical and some oblique (*c.* 50°) dipping joints, in particular, the fringe was seen to tilt into a flat dip, thus forming elliptical niches several metres in diameter in the rock without undergoing interaction with other joints.

Remarks about plotted data

In Figure 11 tilt and twist data of joint fringes, that are characteristic of each granite, from various plutons in Europe are given. In spite of the numerous measurements, these data represent only a small insight into a number of existing explanations for the history of fracturing plutons. Here, the recorded tilt and twist angles of selected early steeply dipping joints are considered, together with other associated fracture patterns, in an attempt to determine whether defined combinations accumulate at special crustal positions of the plutons and at definable moments of the granite history.

The data in Figure 11 suggest significant variations of tilt and twist angles in different granites. Within the SBP and the CBP (columns 1–4; nos 1 and 2) predominant small tilt angles were observed, as in the Polish Strzegom Granite (no. 7). In addition

to the low tilt of the first steeply dipping joints, exceptionally large fringe tilt angles of early formed subhorizontal joints ('lager' joints) were plotted in grey in Figure 11a as anamalous values. All other granites show a remarkably wide range of tilt angles, including large tilt angles up to 90°, such as the Bohus and the Sunne granites (nos 3 and 4), the Flossenbürg Granite, NE Bavaria (no. 8), the Mte. Capanne Granite, Elba (no. 10), and the Main Donegal Pluton, Ireland (no. 12).

The twist-angle diagram (Fig. 11b) presents, however, the converse, that is larger twist angles in the Boršov, Solopysky and Strzegom granites, and in part in the Mrákotín Granite (nos 1, 2 and 7), and contrasting low angles in the Flossenbürg, Mte. Capanne and Donegal granites (nos 8, 10 and 11). These records suggest the combination of large tilt and low twist angles, and vice versa. However, two exceptions exist; joint measurements within the Hudčice and the Sunne granites do not well concide with these combinations.

Furthermore, the identification of this combination of angles involves only trends, indicated by the frequent occurrence of the two main combinations of angles, sometimes related to only one defined joint set. Other joint sets with formation at another time, however, may show another type of tilt–twist combination. In Figure 11 the first steeply dipping joints of a granite were considered. Relatively later formed sets of the whole joint sequence may follow another trend of tilt and twist angle combination, depending on a later moment of pluton history and, thereby, on different crustal-level conditions or, perhaps, on a change in the former stress anisotropy.

In this context it is of note that the joints within the Mrákotín Granite (South Bohemian Pluton, Fig. 9) vary significantly at two places that have a different geological position. Figure 9b shows the greater part of the SBP with the northern Mrákotín Granite and the southern Weinsberg Granite, which have nearly the same intrusion age (330–320 and 328–318 Ma, respectively; Breiter & Koller 1999). There are some other granitic bodies, but this chapter is not the place to report on them. Čiměř Granite is, in fact, a local name for the Mrákotín Granite and, accordingly, the Mrákotín quarry is located in the centre, and the Boršov quarry in the northern part, of this granite.

- The Mrákotín quarry is characterized by a predominantly orthogonal system of early subvertical joints, each set having a different surface morphology. Both sets have interacted in a few places, indicating, in part, a contemporenous development of both sets. But the NNE set was predominant in the first formation, characterized by relatively low tilt and low twist angles (Table 1). Only this first steeply dipping NNE set (025°) was considered in Figure 10.

Table 1. *Fracture parameters of one 025°-joint bundle, Mrákotín quarry (SBP)*

Joint	Main joint L × W (m)	Main joint ratio L/W (m)	Lower fringe L × W (m)	Upper fringe L × W (m)	Joint ratio L/W (m)	Mean twist angle of the fringe cracks
A	17.0 4.0	4.4	16.0 4.0	16.0 4.0	1.3	**26°**
C	13.0 1.6	8.1	13.0 8.0		1.4	**20°**
F	10.0 0.5	4.3	10.0 4.0	10.0 1.0	1.4	**25°**

Explanation: L × W designates (length × width), i.e. lengths of long and short axes of ellipses, respectively; L/W designates ratio of long/short axes. In Tables 1–4, bold numbers are values from Figures 13 and 14.

Table 2. *Fracture parameters of one individual 025°-joint, Boršov quarry (SBP)*

'Trefoil joint'	Fringe tilt angle		Twist angle of the en enchelon fringe cracks		Mean	Range (±2°)
Northern rim:	Fringe 1	+9°	Northern rim:	Fringe 1	–	
	Fringe 2	−8°		Fringe 2	–	
	Fringe 3	–		Fringe 3	**37°**	**26°–41°**
Southern rim:	Fringe 1	+3°	Southern rim:	Fringe 1	–	
	Fringe 2	+17°		Fringe 2	–	
	Fringe 3	+4°		Fringe 3	–	
	Fringe 4	+2°		Fringe 4	–	
	Fringe 5	–		Fringe 5	(a) **30°**	**25°–33°**
					(b) **41°**	**36°–44°**

Table 3. *Fracture parameters of selected Hudčice joints (CBP)*

Joint	Main joint orientation	Lower fringe orientation	Twist angle of echelon fringe cracks		Tilt angle of the fringe	
			Mean	Range	Mean	Range
A	261/55	F$_1$: 093/75, 080/75, 073/75			**18°**	**15°–25°**
		F$_2$: 097/87	–		**12°**	
B	282/40–45	*c.* 300/17–30	**6.5°**	**2°–18°**	**21.3°**	**10°–28°**
C	302/*c.* 90	342/75			*c.* **30°**	
D	299/*c.* 85	318/80			*c.* **20°**	

Table 4. *Fracture parameters of one of the Solopysky joints (CBP)*

Joint	Main joint orientation	Lower fringe orientation	Twist angle of the en echelon fringe cracks	Tilt angle of the fringe	
				Mean	Range

The en echelon segments (length: 8 m) initiated on the initial joint plane (length: >20 m) and tilted together with the fringe where they additionally twisted.

Joint	Main joint orientation	Lower fringe orientation	Twist angle of the en echelon fringe cracks	Mean	Range
1	110/80	F$_1$: 055/80	**18°–25°**	**25°**	**5°–28°**
		F$_2$: 050/56–29	**65°–70°**	**25°**	**20°–39°**
		F$_3$: 050/80	**24°–51°**	**30°**	**20°–35°**

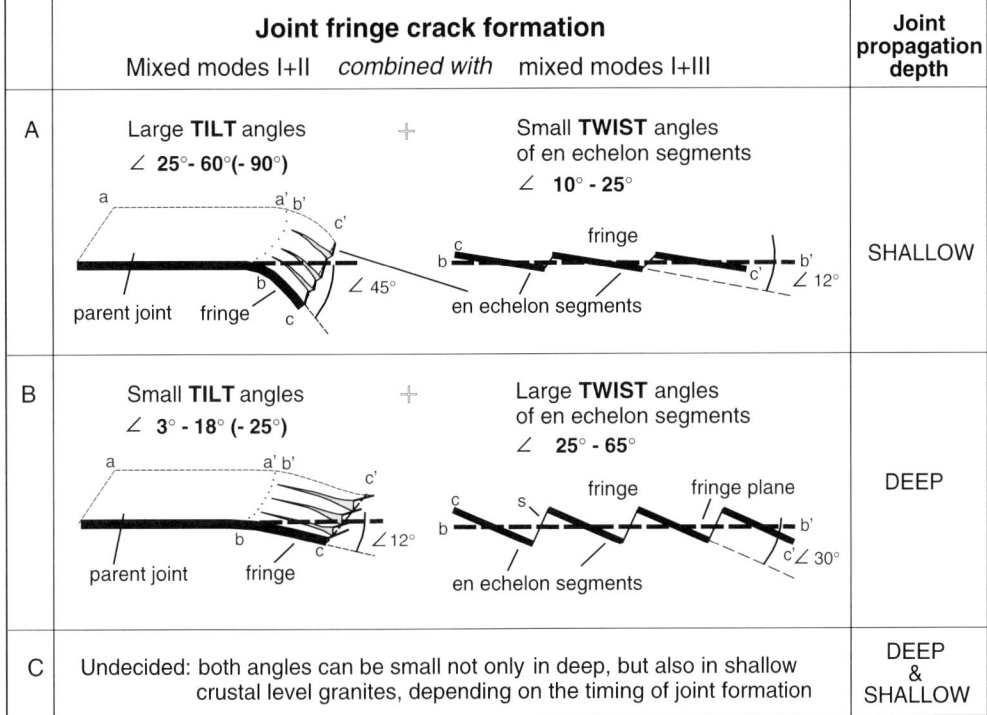

	Joint fringe crack formation	Joint propagation depth
	Mixed modes I+II *combined with* mixed modes I+III	
A	Large **TILT** angles ∠ 25°- 60°(- 90°) Small **TWIST** angles of en echelon segments ∠ 10° - 25°	SHALLOW
B	Small **TILT** angles ∠ 3° - 18° (- 25°) Large **TWIST** angles of en echelon segments ∠ 25° - 65°	DEEP
C	Undecided: both angles can be small not only in deep, but also in shallow crustal level granites, depending on the timing of joint formation	DEEP & SHALLOW

Fig. 12. Three groups of the tilt and twist angle relationship of granite joints can be distinguished. The parent joint plane a–a′ propagated out-of-plane into the fringe b–b′ to c–c′. Type A, high-angle tilted fringes (e.g. 45°) occur in most cases together with low-angle twisted en echelon segments (e.g. 12°). Type B, low-angle tilted fringes (e.g. 12°) are in most cases associated with high-angle twisted en echelon segments (e.g. 30°). The schemes illustrate sections of the joint. s, steps between the individual en echelon fringe cracks.

• The Boršov quarry to the north is governed only by a very regularly developed NNE set (025°/90°) that demonstrates a wide spectrum of fractographic features, including much higher twist angles, as found in the Mrákotín quarry. Obviously, tilt and twist angle are more uniform at the centre of the pluton. The map in Figure 9b shows larger twist angles, with average values of >35° in the southern and northern parts of the pluton.

In Boršov the joint features range between very low morphology associated with frequent undulations and joints with high-angle twisted fringe cracks (Bankwitz *et al.* 2000; Bahat *et al.* 2003). The Boršov quarry is located close to a former conduit of the ascending magma (Breiter in Bankwitz *et al.* 2001). But, in general, the trend of tilt and twist angle combination remains the same in both the Mrákotín and Boršov areas of the plutons.

Likewise within the Polish Sudetic granites, the early joint system occurs with different surface features depending on the orientation of the single sets. The dominating first N–S-trending set (0°–010°) is commonly characterized by small tilt angles and large fringe-crack twist angles (e.g. Strzegom, Fig. 11). By contrast, the relatively younger E–W joint planes are poorly covered with low fractographic indications, and in several places show only weak and smooth features. Such relationships between two early formed subvertical joint sets were recognized in various plutons.

Combinations of tilt and twist angles

One set of observations from this study involves the correlation of tilt and twist angles. Three main combinations were found (Fig. 12):

(A) strongly tilted joint fringes developed only en echelon cracks with very low twist angles within granites such as Eibenstock, Erzgebirge and Mte. Capanne, Elba (Fig.13);

(B) conversely, low tilted or almost untilted fringes are associated with fringe cracks that have high twist angles; the advancing fringe front remained within the initial fracture plane, e.g.

within the SBP (Czech Republic) and the Strzegom Granite (Poland);

(C) the third group of joints combines low tilt with low twist angles; these joints remain roughly within a plane.

The joints plotted in Figure 11 developed their characteristics (parent joint and gradual twist-hackle fringe) during the same fracture event and can be traced back to a single origin. We can consider the amount of the tilt and twist angles as the product of a single propagation event. However, Boršov joints and others with frequent undulations on the parent joint did not form in one pulse, instead they formed in increments within a period of slow propagation (e.g. Figs 3c and 5).

The occurrence of a special range of fringe angles within different granites suggests, in part, a correlation with the crustal depth of joint formation. The palaeodepth of pluton emplacement was determined for the SBP, and inferred from petrological, geological and geophysical investigations for the other granites. Taking this information into account, Figure 11 demonstrates depth-related tilt and twist relationships.

It is also important to consider when the fractures formed. In some cases, the first fractures may be initiated shortly after emplacement, during the early cooling of the granite. Evidence for that relationship comes from investigation of the SBP and the Bavarian Oberpfalz granites (Bankwitz et al. 2004). Mostly, the timing of joint formation is only estimated using associated features, such as fabric elements or the age determination of magmatic dykes, etc. Only the relative sequence of joint formation is commonly recognized. Nonetheless, one can ascertain that the deep-seated Bohemian plutons first fractured with small tilt and larger twist angles. However, the present-day shallowly intruded exposed parts of the Flossenbürg, Mte. Capanne and Donegal granites demonstrate first fractures with a large range of tilt angles and only small twist angles. Using this relationship, the trend derived from many investigated granites is noted in Figure 12 (right column).

Difficulties arise with the interpretation of the Bohus and the Sunne granites because neither their palaeodepth nor the timing of joint formation is well known. Both are intensely and complexly fractured, and may be related to the Precambrian age of the plutons and their long history. Concerning the trend derived from the plotted data (Fig. 11), it seems likely that the Bohus Pluton was fractured originally at more shallow crustal levels.

The graph in Figure 12 illustrates the geometry of two individual joints that are characteristic of the two main combinations of angles. We consider subvertical joints with fringes that formed both at the lateral edges and also at the lower and upper edges of the joint plane. The tilt and twist angle combinations associated with lateral or vertical propagating fringes are the same if the joints are part of the same set; although type A and type B are not related to a preferred location on the joint. It is of note that types A–C in Figure 12 have nothing to do with the fact that on one individual fringe, surrounding a quasi-circular joint, tilt and twist angles can change slightly producing sections with varying morphology. These variations range below the differences between fringe types A and B.

Significantly, changing occurrence of fringe types A and B within an identical joint set was not observed. More shallow initiated joints demonstrated a mixture of different fringe types because it is understood that such joints do not form only high tilt angles. Nevertheless, most of the large tilt angles associated with early subvertical joints are concentrated within shallow intruded granites, and fringe cracks with high twist angles are absent. Within deep granites fringe type B predominates, and fringe types A or C will only be connected to other joint sets of various relative ages, or those that also occur as first formed joints, within other plutons of various intrusion depth. Later formed joint sets could not develop in an homogeneous and isotropic material, which is the premise to recognize unconstrained mechanical behaviour of cracks. We are focused only on first formed joints, therefore we have to consider the depth-related variation of angles.

Figures 13 and 14 provide visuality for the data and the argument for depth-related variations by plotting tilt and twist and depth together. The two groups of plutons in Figure 13 show significantly different patterns of joint tilt v. twist. Only first formed steeply dipping joints were used. Both groups contain one pluton with an exactly determined palaeodepth of emplacement (Thomas & Klemm 1997; Bankwitz et al. 2004): the deep-seated SBP and the shallow intruded Eibenstock Granite (Erzgebirge, Germany), respectively. The detailed plotted joint pattern of both groups suggest that they are correlated with the depth of joint formation, noted in Figure 12, that low tilt and high twist form associates with deep intrusions, and, conversely, that a high tilt–low twist combination is present in shallow emplaced granites. Both groups also involve strikingly low values of tilt and twist within a wide range of data from each of the granites.

In general, this assumed relationship is contradictory to the stress (σ) conditions, because σ_{zz} (z being depth) should not influence the tilt angle of a vertical joint. The fringe is caused by rotation of the local maximum tensile stress in the $x–y$ plane (Fig. 4). The vertical stress component (σ_{zz}) is in the fracture plane and all shear components in this direction equal zero. Only shear stresses perpendicular to the

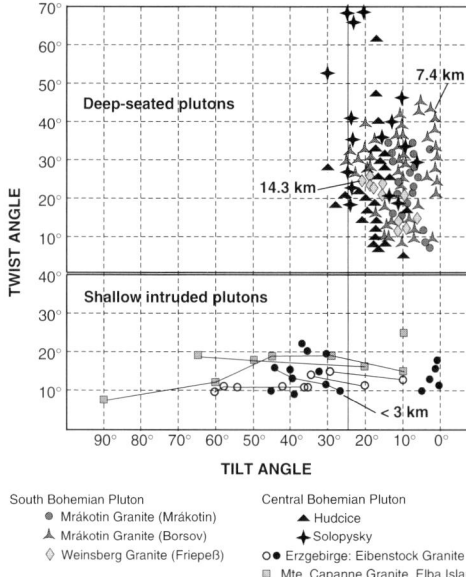

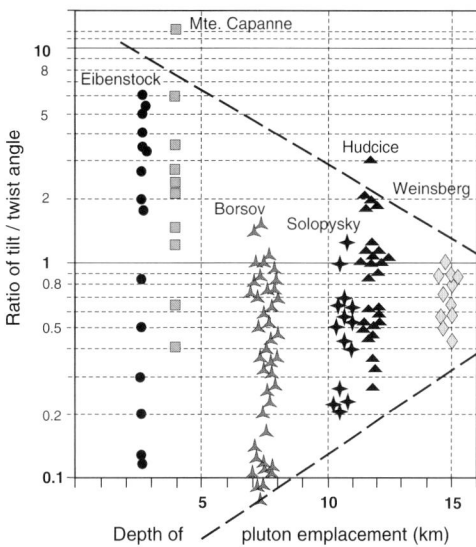

Fig. 13. Characteristic relationships between twist and tilt angle of early granite joints in different plutons. The trend occurs that in deep-seated plutons first formed joints developed small tilt and large twist angles. In contrast to this, in shallow intruded granites widely spaced early joints occur often with high-angle fringe tilt and low-angle twist of the fringe cracks.

Fig. 14. Logarithmic tilt/twist ratio v. palaeodepth of pluton emplacement for values are plotted in Figure 13. They show a noteworthy relation. The most shallowly intruded pluton (Eibenstock, Germany) represents the widest range of ratios, the deepest pluton (now exposed) is characterized by the smallest range of data between 0.5 and 1.0. Symbols as in Figure 13.

fracture plane could force it to tilt. Therefore, the degree of tilt should be independent of depth (Younes, pers. comm.). But, exposed steeply dipping joints are not always strictly vertical and often dip about 80°–70°, so that σ_{zz} can act at the joint plane. The dominance of type A or B associated with different intrusion depths of the granites is evident.

The favoured occurrence of fringe types A and B in defined plutons implies that the tilt/twist ratios should decrease from the shallow to the deep-seated granites. If further detail is required, Figure 14 reveals a specific trend of granite joints with respect to the depth of formation. The range of tilt/twist ratios differs widely and decreases, in fact, with increasing depth of the plutons and joint formation. The deepest formed joints (Weinsberg Granite, SBP) concentrates at a ratio of between 0.5 and 1. It is of note that the large ratios do not simply become smaller (Eibenstock to Weinsberg Granite), but the smallest ratios increase (Boršov to Weinsberg) and finally the plot field, limited by lower and upper broken lines, narrows at about a ratio of 1, or, to be exact, between a ratio of 0.5 and 1. Measurements of data leading to ratios below 0.1 for depths between 3 and 5 km are possible but need more attention. Obviously, a deviation of one order from ratio '1'

exists, not only when increasing from deep to shallow crustal levels but also when decreasing. This data distribution probably indicates the increasing influence of the confining pressure.

Considering only first formed joints, one can possibly try to use tilt/twist ratios in certain granites to infer the possible depth of first joint formation. However, the depth of joint formation is not the only factor that decides the amount of tilt and twist angles, but it should be taken into consideration.

Discussion

Dependence on intrusion depth

On the basis of measurements of joints, angles of fringes and fringe-crack series in the Eibenstock Pluton (Erzgebirge, Germany), and in the granite plutons of the Sudetes (Poland) and of Bohemia (Czech Republic) we recognized a characteristic difference in joint morphology. The tilt and twist of fringes differs in granites with a shallow emplacement of less than 3 km, such as the Erzgebirge plutons (Thomas & Klemm 1997), and in granites emplaced at deep crustal levels, such as the SBP at a depth of between 7.4 and 14.3 km (Bankwitz et al. 2004).

Based on the premise that the early joint sets developed at the depth of emplacement during an early stage of cooling, which proved correct in the SBP (Bankwitz et al. 2004), the fracture morphology was considered in an attempt to find a correlation with the depth of joint initiation.

We have to consider that the difference in surface morphology may depend on several factors:

- the *depth* of intrusion of the pluton, with regard to early formed first joints;
- the *timing* of the fracture initiation, with regard to the emplacement, cooling and uplift into the erosion level;
- the *driving mechanisms*, as defined by Engelder & Fischer (1996);
- the type of *stress sources*: only local stresses, or those associated with regional stresses, included within the pluton that cause a regionally consistent joint pattern;
- the *relative tensile stress* (average remote stress plus internal fluid pressure) according to Segall & Pollard (1983);
- the *magnitude* of differential stress;
- the *stress intensity factor* and fracture velocity estimated in comparison with the Wiederhorn curve (Bahat et al. 2003).

Depth

From field investigations evidence for joint initiation at different crustal levels has been found. The palaeo-intrusion depth of several plutons was determined by fluid and melt inclusion; however, the greater part of the intrusion depth was derived from gravity data (Meurers 1992; Breiter 2001), seismic reflection data (Bohus Granite, Sweden, Lusatian plutons and Bavarian granites), modelling (Vigneresse 1999) and petrological data (Zulauf 1993; Siebel 1998). The Boršov joints initiated very early at 7.4 km depth, the Weinsberg Granite joints possibly at about 14 km. We suppose that many granites intruded at a similar depth between approximately 7 and 15 km, supported by geophysical modelling (Vigneresse 1999). Other plutons of different age (e.g. Caledonian, Ireland; Variscan, Erzgebirge; Cenozoic age, Elba Island) are assumed to have been intruded into shallow crustal levels. The joints within these granites can be formed between 3 km and several hundreds of metres below the present surface, because these pre-existing joints were cross-cut by subhorizontal sheet fractures.

Timing

The moment of fracture initiation, with regard to the emplacement and cooling stage, and the magnitude of differential stress significantly influences the joint-surface pattern. The age of the vein-bearing twist hackles in Boršov is determined to be within the early period of granite cooling and thereby connected with its depth. Evidence for no loss of argon comes from the method used, where purified micas were ground carefully in pure alcohol to remove altered rims that might have suffered a loss of Ar or K. In this way, only the fresh cores of the muscovites were analysed. The excellent quality of the separates from the SBP is confirmed by the K_2O-content of more than 10% (Bankwitz et al. 2004).

Driving mechanism

The initiation of pre-uplift joints at deep crustal levels and, in addition within closed systems, were conducive to, in part, fluid-driven fracture formation in granite. Evidence for an internal fluid drive is the incremental propagation of joints (Lacazette & Engelder 1992) reflected among other features by frequent closely spaced rib marks in the form of tilted panels (undulations) of the initial planes and partly by multiple fringes (Bankwitz et al. 2000). Such features, known from sedimentary rocks that underwent compaction but were never deeply buried, are not common within granites and are best exposed locally in the Boršov (SBP) and Hudčice quarry (CBP), and the Bohus, Sunne, Bornholm and Mte. Capanne granites. Although igneous rocks have a very low permeability, fluids are present and not only as hydrothermal liquids. Fluids, including rest melts, are frequent in magmatic rocks (Thomas et al. 2000) and play a major role in the early period of cooling, e.g. after water separation from the crystallizing melt near the solidus, producing an overpressure (Bankwitz et al. 2004). This pressure increase in the cooling range between liquidus and solidus (Thomas 1994) can initiate fractures.

Secor (1969) predicted that fractures in impermeable rocks such as granites would be short and closely spaced, whereas the fractures in permeable rock would be long and widely spaced. However, the first formed joints in granite are often long and widely spaced, thus probably indicating that pressurized fluids were present at the moment of fracture formation.

Frequently, the fringe propagated into strongly twisted gradual fringe cracks (length of 2–4 to 8 m) with large overlap (metres), indicating the influence of regional stresses. This means that, after a first period of slow and incremental fracture growth, remote stresses increasingly governed the further propagation of the parent joint into such complex fringe cracks. Thus, the fluid-driven fracturing may operate together with other driving forces when progressing.

Stresses

Such joints indicate at least a relatively high magnitude of the remote stress and high differential stresses even at crustal levels of about 10 km in depth, as was also found at the KTB (Bavaria). (KTB stands for Kontinentales Tiefbohrprogramm der Bundesrepublik Deutschland, or German Continental Deep Drilling Programme.) Segall & Pollard (1983) calculated for joints in granitic rocks of the Sierra Nevada relative tensile stresses (average remote stress plus internal fluid pressure) of approximately 0.1–0.4 kbar. Within a solidifying magma the overpressure after water separation near the solidus amounted to 1.5 kbar (theoretically up to several 10 kbars for constant volume conditions: Burnham 1979). The increase in H_2O pressure of 1.5 kbar, measured in the Erzgebirge granites (Student & Bodnar 1996) and in the SBP (Bankwitz et al. 2004), can initiate fractures. The initial driving force within igneous rocks may be fluids, even, or particularly during the first period of fracturing. Younes & Engelder (1999) assumed the initial pressure during jointing in sedimentary rocks to be approximately 0.8 kbar, this is half of the pressure increase in solidifying magmas at the solidus. Thus, the fluid pressure near the solidus, within the early period of cooling, could initiate fractures.

However, jointing and the breakdown of the parent joint into regular twist hackles depends on the orientation and magnitude of the remote stress field, internal fluid pressure and the elastic properties of the rock. Supposedly, a large difference between σ_1 and σ_3 favours parallel joint sets and fracture-cleavage formation as recognized in Mrákotín. The more σ_1 is lowered, the more the joints will develop out of the main plane (Olson & Pollard 1989). The Mrákotín Granite (SBP, no. 1 on the list of granites given earlier in this chapter) provides examples for a high stress magnitude reflected by early parallel jointing of parent joints in bundles and of large en echelon fringe cracks.

Stress intensity factor and fracture velocity

Both stress intensity factor and fracture velocity correlate, as predicted by Wiederhorn et al. (1974) and, among other authors, by Rabinovitch & Bahat (1979). Compared to the parameters of the Wiederhorn curve, the position of the observed joint types can be estimated (Bahat et al. 2003). Obviously, the early formed pre-uplift joints of shallow intruded plutons are frequently low-energy fractures (significantly smooth, often covered only with faint plumes, both high- and low-angle fringes, and small fringe cracks), in spite of the general opposite assumption or expectation. Because shallow granites cool down faster and their fractures formed closer to the surface, one could expect that the joints should have higher energy for propagation, similar to unloading sheet fractures near the topographic surface. But numerous observations on first formed joints differ from this.

In contrast, within deep-seated plutons, high-energy fractures frequently occur with large fringes and dominant low tilt and high twist angles of the fringe cracks. Fracturing at great depth has not yet been completely studied. In part, first formed joints show characteristic high-energy style twist hackles, according to Bahat et al. (2001a, b) even hackle fringes were formed at some locations in the SBP.

However, locally restricted domains exist within the granites (e.g. Boršov) where initially cyclic growing low-energy fractures occur that after a period of slow propagation change the style, producing high-energy twist hackles.

Joint interactions

As far as possible we took care in this chapter to consider only first formed joints, and to avoid the research of joints that interacted with joints of other sets. Our purpose was to study the style of earliest joints in plutons. Therefore, the photographs mostly show joints that were not formed at their present location, even if their dominant upper fringes appear to correlate with the free surface. Within the Swedish plutons (Bohus and Sunne) Precambrian, Devonian and Permian dykes fill previously formed joints, thus confirming the ages and the defined palaeodepth of the hosting joints. These joint walls represent the same surface features as in Figures 5a, b and 6a, b at the present exposures.

Conclusions

Within granites, three frequent types of fringes can be distinguished by their tilt–twist relationship: Type A, high tilt angles correlate with low twist angles; type B, low tilt associates with high twist; type C, both angles are very small. Each of these fringe types, associated with first formed joints, occur preferentially in plutons with different intrusion depths. This points to depth being one factor in the development of various fringe styles. For three joints within the SBP the palaeodepth (7.4 and 14.3 km) and the timing of the joint formation (324.9 Ma) were determined, confirming the correlation of low tilt–high twist forms with deep joint propagation.

Deep-seated plutons frequently demonstrate high-energy fractures (fringe type A), but with typical differences of tilt and twist angles between the single sets in the joint sequence. The first formed joints have large, low tilted fringes with well-developed en echelon twist hackles and high twist angles. In contrast to this, the secondly formed joint surfaces are smooth, planar, and without or with low fringe angles.

In general, it seems to be a rule that in the case of high tilt angles the fringe plane breaks down only with a low-angle twist of its en echelon segments (low-energy fractures). Such a combination is characteristic of shallow plutons. If low-angle fringe tilt occurs, both variations of twist-hackle progress were observed, forming high or low twist angles. Which type of twist angle occurs depends on the driving mechanism and the fracturing condition (e.g. the magnitude of stresses, the time of jointing, pre- or syn-uplift related and the level of intrusion, deep-seated or shallow intruded pluton).

The tilt/twist angle ratio differs significantly in the various plutons. A wide range of ratios occur in shallow granites (0.1–8.0), which decreases towards a ratio of 1 with increasing depth of intrusion (to 0.5–1.0), thus indicating fringe tilt and twist formation under restrictive conditions due to the confining pressure at a depth of approximately 14 km.

We thank T. Engelder and J. Cosgrove for the Fracture Meeting in Weston-super-Mare, UK, in 2001. It was a good chance for fruitful discussions. This paper benefited from the constructive reviews of T. Engelder, B. R. Kulander and and A. J. Younes, which clarified and improved an earlier version.

References

BAHAT, D. 1991. *Tectonofractography*. Springer, Berlin, 1–354.

BAHAT, D. 1997. Mechanisms of dilatant en echelon crack formation in jointed layered chalks. *Journal of Structural Geology*, **19**, 1375–1392.

BAHAT, D., BANKWITZ, P. & BANKWITZ, E. 2001*a*. Joint formation in granite plutons: En echelon-hackle series on mirror fringes (Example: South Bohemian Pluton, Czech Republic). *Zeitschrift der deutschen geologischen Gesellschaft*, **152**, 593–609.

BAHAT, D., BANKWITZ, P. & BANKWITZ, E. 2001*b*. Changes in crack velocities at the transition from the parent joint through the en echelon fringe to a secondary mirror plane. *Journal of Structural Geology*, **23**, 1215–1221.

BAHAT, D., BANKWITZ, P. & BANKWITZ, E. 2003. Preuplift joints in granites: Evidence for subcritical and post-critical fracture growth. *Geological Society of America Bulletin*, **115**, 148–165.

BAHAT, D., BANKWITZ, P. & RABINOVITCH, A. 2001. Comparison of the new fracture areas created by the formation of en echelon and hackle fringes on joint surfaces. *Zeitschrift für Geologische Wissenschaften*, **30**, 1–12.

BANKWITZ, P., 1965. Über Klüfte. I. Beobachtungen im Thüringischen Schiefergebirge. *Geologie*, **14**, 241–253.

BANKWITZ, P. 1966. Über Klüfte II. Die Bildung der Kluftoberfläche und eine Systematik ihrer Strukturen. *Geologie*, **15**, 896–941.

BANKWITZ, P. & BANKWITZ, E. 1984. Die Symmetrie von Kluftoberflächen und Ihre Nutzung für eine Paläospannungsanalyse. *Zeitschrift für Geologische Wissenschaften*, **12**, 305–334.

BANKWITZ, P. & BANKWITZ, E. 1994. Event related jointing in rocks on Bornholm Island (Denmark). *Zeitschrift für Geologische Wissenschaften*, **22**, 97–114.

BANKWITZ, P. & BANKWITZ, E. 1995. Aspekte der Entwicklung von Klüften in postkinematischen Graniten des Erzgebirges (speziell Eibenstocker Massiv). *Zeitschrift für Geologische Wissenschaften*, **23**, 777–793.

BANKWITZ, P., BAHAT, D. & BANKWITZ, E. 2000. Granitklüftung – Kenntnisstand 80 Jahre nach Hans Cloos. *Zeitschrift für Geologische Wissenschaften*, **28**, 87–110.

BANKWITZ, P., BANKWITZ, E., THOMAS, R., WEMMER, K. & KÄMPF, H. (2004). Age and depth evidence for pre-exhumation joints in granite plutons: fracturing during the early cooling stage of felsic rock. *In*: COSGROVE, J. W. & ENGELDER, T. (eds) *The Initiation, Propagation, and Arrest of Joints and Other Fractures*. Geological Society, London, Special Publications, **231**, 25–47.

BREITER, K. 2001. South Bohemian pluton. Overview. *Exkursionsführer und Veröffentlichungen der GGW*, Berlin, **212**, 130–141.

BREITER, K. & KOLLER, F. 1999. Two-mica granites in the central part of the South Bohemian Pluton. *Abhandlungen der Geologischen Bundesanstalt*, Wien, **56**, 201–212.

BURNHAM, C. W. 1979. The importance of volatile constituents. *In*: YODER, H. S. (ed.) *The Evolution of the Igneous Rocks: Fiftieth Anniversary Perspective*. Princeton University Press, Princeton, NJ, 439–482.

BUSSY, F. 1991. Enclaves of the Late Miocene Monte Capanne granite, Elba Island, Italy. *In*: DIDIER, J. & BARBARIN, B. (eds) *Enclaves and Granite Petrology*. Elsevier, Amsterdam, 167–178.

CLOOS, H. 1921. *Der Mechanismus tiefvulkanischer Vorgänge*. Friedrich Vieweg, Braunschweig.

CRAIG, G. Y. 1991. *Geology of Scotland*, 3rd edn. Geological Society, London.

CRUIKSHANK, K. M. & AYDIN, A. 1995. Unweaving the joints in Entrada Sandstone, Arches National Park, Utah, U.S.A. *Journal of Structural Geology*, **17**, 409–421.

DYER, R. 1988. Using joint interactions to estimate paleostress ratios. *Journal of Structural Geology*, **10**, 685–699.

ELIASSON, T. M. 1992. Magma genesis and emplacement characteristics of the paraluminous Sveconorwegian Bohus granite, SW Sweden. *Geologiska Föreningens i Stockholm Förhandlingar*, **114**, 452–455.

ENGELDER, T. & FISCHER, M. P. 1996. Loading configurations and driving mechanisms for joints based on the Griffith energy-balance concept. *Tectonophysics*, **256**, 253–277.

ENGELDER, T. & GEISER, P. 1980. On the use of regional joint sets as trajectories of paleostress fields during the development of the Appalachian Plateau New York. *Journal of Geophysical Research*, **85**, 6319–6341.

ENGELDER, T., FISCHER, M. P. & GROSS, M. R. 1993. *Geological Aspects of Fracture Mechanics*. A Short Course Manual Note. Geological Society of America, Boston MA.

EIDAM, J. & KRAUSS, M. 2001. Outcrop Großschweidnitz. Contact between granodiorites and mafic rocks. *Exkursionsführer und Veröffentlichungen der GGW*, **212**, 123–126.

FERSHTATER, G.-B., MONTERO, P., BORODINA, N. S., PUSHKAREV, E. V., SMIRNOV, V. N. & BEA, F. 1997. Uralian magmatism: an overview. *Tectonophysics*, **276**, 87–102.

FÖRSTER, H.-J., TISCHENDORF, G., TRUMBULL, R. B. & GOTTESMANN, B. 1999. Late collisional granites in the Variscan Erzgebirge, Germany. *Journal of Petrology*, **40**, 1613–1645.

FREDEN, C. (ed.) 1994. Geology. *National Atlas of Sweden*, Geological Survey of Sweden. 1–208.

HODGSON, R. A. 1961. Classification of structures on joint surfaces. *American Journal of Science*, **259**, 493–502.

HOLUB, F. V., ROSSI, PH. & COCHERIE, A. 1997. Radiometric dating of granitic rocks from the Central Bohemian Plutonic complex (Czech Republic). *Comptes Rendus de l'Académie des Sciences Paris*, **325**, 19–26.

HUTTON, D. H. W. & ALSOPP, G. I. 1996. The Caledonian strike swing and associated lineaments in NW Ireland and adjacent areas: sedimentation, deformation and igneous intrusion patterns. *Journal of the Geological Society, London*, **153**, 345–360.

KULANDER, B. R., BARTON, C. C. & DEAN, S. C. 1979. *The Application of Fractography to Core and Outcrop Fracture Investigations*. Report to U.S. D.O.E., Morgantown Energy Technology Center, METC SP-79/3.

KULANDER, B. R. & DEAN, S. L. 1985. Hackle plume geometry and joint propagation dynamics. *In*: STEPHANSSON, O. (ed.) *Fundamentals of Rock Joints. Proceedings*. Cenetek Press, Lulea, Sweden, 85–94.

KULANDER, B. R. & DEAN, S. L. 1995. Observations on fractography with laboratory experiments for geologists. *In*: AMEEN, M. S. (ed.) *Fractography: Fracture Topography as a Tool in Fracture Mechanics and Stress Analysis*. Geological Society, London, Special Publications, **92**, 59–82.

KULANDER, B. R., DEAN, S. L. & WARD, B. J. 1990. *Fractured Core Analysis: Interpretation, Logging, and Use of Natural and Induced Fractures in Core*. AAPG Methods in Exploration Series, 8. American Association of Petroleum Geologists, Tulsa.

LACAZETTE, A. & ENGELDER, T. 1992. Fluid-driven propagation of a joint in the Ithaca siltstone, Appalachian basin, New York. *In*: EVANS, B & WONG, T.-F., *Fault Mechanics and Transport Properties of Rocks*. Academic Press, London, 297–370.

LARSEN, O. 1980. Geologisk alderbestemmelse ved isotopmålinger. *Dansk Natur-Dansk Skole. Årsskrift for 1980*, 89–106.

LAWN, B. R. & WILSHAW, T. R. 1975. *Fracture of Brittle Solids*. Cambridge University Press, London, 1–204.

LOBST, R. 2001. Der Intrusionsverband cadomischer Granodiorite am Klosterberg bei Demitz-Thumitz. *Exkursionsführer und Veröffentlichungen der GGW*, **212**, 67–69.

McCONAUGHY, D. T. & ENGELDER, T. 1999. Joint interaction with embedded concretions: joint loading configurations inferred from propagation paths. *Journal of Structural Geology*, **21**, 1637–1652.

MEURERS, B. 1992. *Korrigierte Bougueranomalie der Südlichen Böhmischen Masse*. Schwerpunktprogramm S47GEO, Präalpidische Kruste in Österreich, Salzburg.

MIERZEJEWSKI, M. P. 2001. Understanding the Karkonosze Mts Granite. *Exkursionsführer und Veröffentlichungen der GGW*, **212**, 70–74.

OLSON, J. E. & POLLARD, D. D. 1989. Inferring paleostress from natural fracture patterns: A new method. *Geology*, **17**, 345–348.

PITCHER, W. S. & BERGER, A. R. 1972. *The Geology of Donegal*. Wiley-Interscience, New York.

POLLARD, D. D. 2000. Strain and stress: Discussion. *Journal of Structural Geology*, **22**, 1359–1367.

POLLARD, D. D. & AYDIN, A. 1988. Progress in understanding jointing over the past century. *Geological Society of America Bulletin*, **100**, 1181–1204.

POLLARD, D. D. & SEGALL, P. 1987. Theoretical displacements and stresses near fractures in rock. *In*: ATKINSON, B. K. (ed.). *Fracture Mechanics of Rock*. Academic Press, London, 277–349.

POLLARD, D. D., SEGALL, P. & DELANEY, P. T. 1982. Formation and interpretation of dilatant echelon cracks. *Geological Society of America Bulletin*, **93**, 1291–1303.

RABINOVITCH, A & BAHAT, D. 1979. Catastrophe theory: A technique for crack propagation analysis. *Journal of Applied Physics*, **50**, 321–334.

SCHARBERT, S. 1998. Some geochronological data from the South Bohemian Pluton in Austria: a critical review. *Acta Universitatis Carolinae, Geologica*, **42**, 114–118.

SCHUST, F. 2000. Zum magmengeologischen Bau und zur Altersdatierung des Lausitzer prävariszischen Granitoidkomplexes. *Zeitschrift für geologische Wissenschaften*, **28**, 111–132.

SECOR, D. T., JR. 1969. *Mechanics of Natural Extension Fracturing at Depth in the Earth's Crust*. Geological Survey of Canada Special Papers, **68–52**, 3–47.

SEGALL, P. & POLLARD, D. D. 1983. Joint formation in granitic rock of the Sierra Nevada. *Geological Society of America Bulletin*, **94**, 563–575.

SIEBEL, W. 1998. Variszischer spät- bis postkollisionaler Plutonismus in Deutschland: Regionale Verbreitung, Stoffbestand und Altersstellung. *Zeitschrift für geologische Wissenschaften*, **26**, 329–358

STUDENT, J. J. & BODNAR, R. J. 1996. Melt inclusion microthermometry: Petrologic constraints from the H$_2$O-saturated haplogranite system. *Petrology*, **4**, 291–306.

STRÖMBERG, A. G. B., KARIS, L., ZACHRISSON, E., SJÖSTRAND, TH. & SKOGLUND, R. 1984. Bedrock geology of Jämtland County, Ser. Ca, **53**, 1:200 000. Sveriges geologiska undersökning.

THOMAS, R. 1994. Fluid evolution in relation to the emplacement of the Variscan granites in the Erzgebirge region: A review of the melt and fluid inclusion evidence. *In*: SELTMANN, R., KÄMPF, H. & MÖLLER, P. (eds) *Metallogeny of Collisional Orogens*. Czech Geological Survey, Prague, 70–81.

THOMAS, R. & KLEMM, W. 1997. Microthermometric study of silicate melt inclusions in Variscan granites from SE Germany. *Journal of Petrology*, **38**, 1753–1765.

THOMAS, R., WEBSTER, J. D. & HEINRICH, W. 2000. Melt inclusions in pegmatite quartz: complete miscibility between silicate melts and hydrous fluids at low pressure. *Contributions to Mineralogy and Petrology*, **139**, 394–401.

TICHOMIROVA, M. 2001. Altersdatierung von cadomischen Graniten und Anatexiten der Lausitz. *Exkursionsführer und Veröffentlichungen der GGW*, **212**, 97–99.

VIGNERESSE, J. L. 1999. Intrusion level of granite massifs along the Hercynian belt: balancing the eroded crust. *Tectonophysics*, **307**, 277–295.

WALTER, R. 1992. *Geologie von Mitteleuropa*, 5th edn. E. Schweizerbart'sche Verlagsbuchhandlung, Stuttgart.

WIEDERHORN, S. M., JOHNSON, H., DINESS, A. M. & HEUER, A. H. 1974. Fracture of glass in vacuum. *Journal of the American Ceramics Society*, **57**, 336–341.

WOODWORTH, J.B. 1895. Some features of joints. *Science*, **2**, 903–904.

WOODWORTH, J.B. 1896. On the fracture system of joints, with remarks on certain great fractures. *Boston Society Natural History Proceedings*, **27**, 63–184.

YOUNES, A. I. & ENGELDER, T. 1999. Fringe cracks: Key structures for the interpretation of the progressive Alleghanian deformation of the Appalachian plateau. *Geological Society of America Bulletin*, **111**, 219–239.

ZHENG, F. 1996. Tectonic development of the Bohus Granite (SW Sweden) and its adjoining areas. *Stockholm Contributions in Geology*, **44**, 1–208.

ZULAUF, G. 1993. Brittle deformation events at the western border of the Bohemian Massif. *Geologische Rundschau*, **82**, 489–504.

Differences between veins and joints using the example of the Jurassic limestones of Somerset

D. C. P. PEACOCK

Robertson Research International Ltd, Tyn-y-coed Site, Llanrhos, Llandudno LL30 1SA, UK (e-mail: dcp@robresint.co.uk)

Abstract: Although veins and joints are both extension fractures, they should be treated separately, especially in field analyses. For example, they commonly form at different times and under different conditions. Excellent exposures of the Liassic limestones at Kilve, Somerset, UK, are used as a case study. These rocks show calcite veins related to the Mesozoic extension and Alpine contraction of the Bristol Channel Basin, and joints that appear to have been caused by late- or post-Alpine stress release. The veins at Kilve show the following characteristics that distinguish them from joints. (1) They are mineralized. (2) Veins have measurable widths (up to tens of millimetres wide) and represent measurable strain. En echelon veins are common at Kilve, and indicate shear strains. (3) Vein widths may obey a power-law scaling relationship. (4) The veins were open for long enough and under the correct conditions for calcite to be deposited, indicating high fluid pressures. They are also typically en echelon. (5) Veins form sets with a narrow range of strikes, whereas the joints typically form complex networks. (6) The veins at Kilve are commonly clustered around faults, and are precursors to, or synchronous with, the faults.

Veins are mineral-filled fractures (e.g. Davis & Reynolds 1996), are commonly mode I fractures and usually form as en echelon segments in a vein array. *Joints* are mode I fractures across which there has been no measurable shear displacement (e.g. Pollard & Aydin 1988), and that are not mineral-filled. Veins and joints are mechanically similar (e.g. Thomas & Pollard, 1993), and so it may be appropriate to consider them together in mechanical analyses. There are, however, important geometric, scaling and genetic differences between veins and joints (e.g. Gillespie *et al.* 2001) and so it may not be helpful to group them together or to ignore their differences. For example, Petit *et al.* (1994) describe 'joints mineralised with calcite', while McGrath & Davison (1995) describe veins and joints together as 'tension fractures'. The aim of this chapter is to present the veins and joints in the Liassic limestones of the Somerset coast as a cautionary example of why veins and joints should be considered separately.

The approximately 18 km-long coastline around Kilve, between Hinkley Point and Blue Anchor Bay in Somerset, UK (Fig. 1), contains exceptional exposures of a wide range of structures. This area is a world-class natural laboratory for studying the geometry and evolution of veins, joints, normal and strike-slip faults, and reverse-reactivated normal faults. The structures are exposed in the Triassic marls and sandstones, and in the Liassic limestones and mudrocks. Whittaker & Green (1983) give a detailed description of the stratigraphy. The large tide range has produced a wide wave-cut platform, with fresh cliffs produced by rapid erosion of these relatively soft rocks. The structures were produced by the development of the Bristol Channel Basin (BCB) during the Mesozoic, and by contraction of the BCB during the Alpine orogeny. The area therefore provides an excellent example of the history of a sedimentary basin.

Deformation history of the Liassic rocks at Kilve

There is a relatively simple history of Mesozoic N–S extension and Tertiary (Miocene) N–S contraction between Kilve and Watchet (Table 1, Fig. 1). The Mesozoic extension is related to the opening of the BCB, and is represented by normal faults, with associated veins (Fig. 2b, c) and folds, all of which strike a few degrees clockwise of E–W. Tertiary contraction is related to the Alpine Orogeny and inversion of the BCB. It is represented by reverse-reactivated normal faults, E–W-striking thrusts, strike-slip faults conjugate about N–S, N–S-striking veins, and probably with tightening of the Mesozoic folds (Peacock & Sanderson 1992; Dart *et al.* 1995; Kelly *et al.* 1999). Joints post-date the other structures (Rawnsley *et al.* 1998; Peacock 2001). At Lilstock, about 3 km to the east of Kilve, there is a more complex deformation history (Peacock & Sanderson 1999), including: (1) 150° extension that produced 060°-striking veins and possibly faults; (2) approximately N–S extension with sinistral transtension that produced approximately 095°-striking normal faults and veins; (3) E–W contraction, with sinistral shear on some 095°-striking normal faults; (4) dextral

From: COSGROVE, J. W. & ENGELDER, T. (eds) 2004. *The Initiation, Propagation, and Arrest of Joints and Other Fractures.* Geological Society, London, Special Publications, **231**, 209–221. 0305-8719/04/$15 © The Geological Society of London 2004.

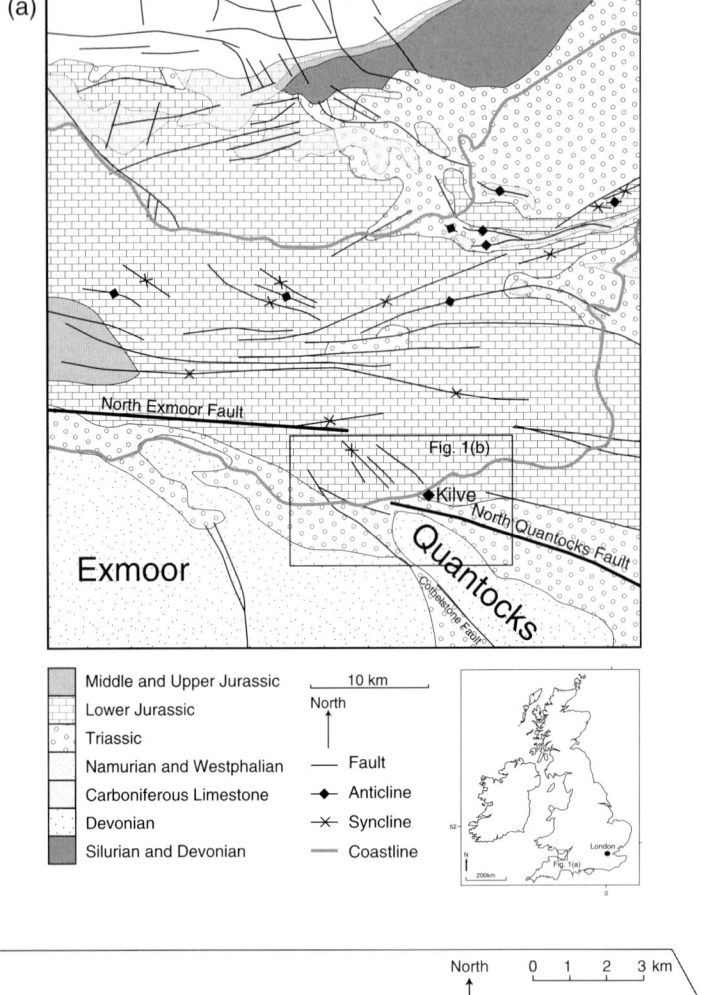

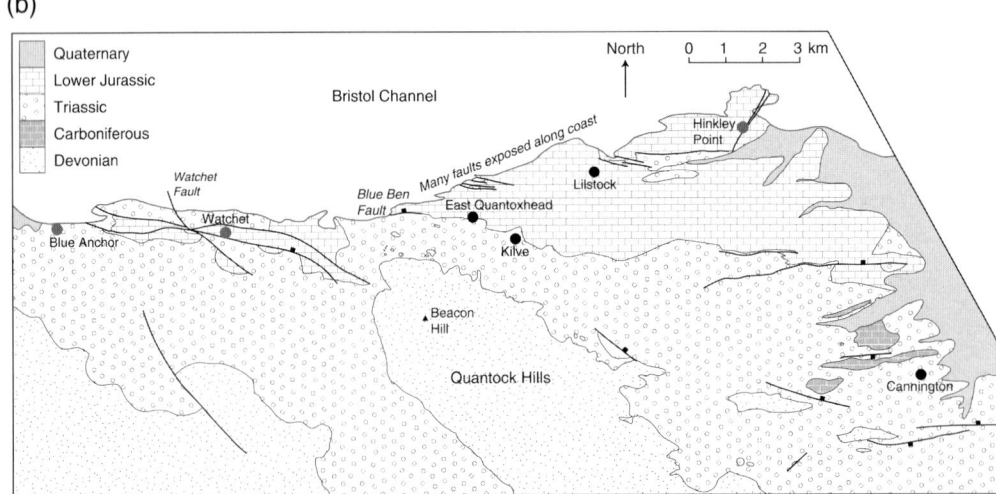

Fig. 1. (**a**) Map of the eastern Bristol Channel Basin, including the north coast of Somerset (based on the Bristol Geological Survey sheet 51°N-04°W, Bristol Channel) and the positions of the hypothesized North Quantocks and North Exmoor faults (Peacock & Sanderson 1999). (**b**) More detailed map of the Somerset coast between Hinkley Point and Blue Anchor Bay, based on the British Geological Survey 1:50 000 scale sheets.

Table 1. *Sequence of deformation events in the Liassic rocks at Kilve, on the Somerset coast. Note that a more complex sequence of events can be observed east of Lilstock (Peacock & Sanderson 1999)*

Tectonic event	Structure	Strike	Geometry	Associated structures	Figure	Reference
Late strike-slip	Single example of a fault at Lilstock	N–S	Unmineralized dextral fault with tens of millimetres of displacement	Appears to displace joints	Al-Mahruqi (2001, fig. 2.8)	Al-Mahruqi (2001)
Post-Alpine stress release	Joints	Full range	Network of joints, perhaps indicating anticlockwise stress reorientation	Cut across or abut normal and strike-slip faults	Fig. 7	Rawnsley *et al.* (1998), Engelder & Peacock (2001)
Alpine contraction	Calcite veins	N–S	Transtensional–transpressional arrays	Reverse-reactivated normal faults, strike-slip faults, thrusts, folds	Fig. 2(d)	Dart *et al.* (1995), Peacock & Sanderson (1999)
Mesozoic extension	Calcite veins	E–W	Transtensional arrays adjacent to normal faults	Precursors to normal faults	Fig. 2(c)	Peacock (1991)
Mesozoic extension	High-density calcite veins	E–W	Closely spaced, <1 mm-thick veins distributed through limestone beds	May be precursors to thicker E–W calcite veins	Fig. 2(b)	Rawnsley *et al.* (1998), Caputo & Hancock (1999)
Diagenesis?	Calcite veins	Full range	Commonly form triple junctions	Occur on millimetre-high mounds on tops of limestone beds	Fig. 2(a)	Peacock (1991, fig. 4)

reactivation of some of the 095° striking normal faults; (5) N–S contraction that produced reverse-reactivation of normal faults and the development of thrusts; and (6) joints that post-date the faults (see below). In addition, Al-Mahruqi (2001, fig. 2.8) showed a possible N–S-striking right-stepping strike-slip fault that appears to displace E–W-striking joints in a horizontal limestone bed at Lilstock. This is probably of a different age (probably later) than the strike-slip faults that are conjugate about N–S at Kilve, because it has a different strike and is not mineralized.

Vein sets in the Liassic rocks at Kilve

There are four sets of calcite veins in the Liassic rocks at Kilve, with cross-cutting relationships showing that these are an age sequence. First, relatively short and wide veins occur on millimetre-high mounds on the tops of limestone bedding planes (Fig. 2a), and they commonly form triple junctions. These are probably early structures related to diagenesis (Peacock 1991), with the softness of the limestone indicated by the high width/length ratio. These veins are cross-cut by all of the other structures.

Secondly, E–W-striking calcite veins less than

10 mm thick, and mostly less than 1 mm thick, are common throughout the Liassic limestones of the BCB (Rawnsley *et al.* 1998; Caputo & Hancock 1999). These thin, closely-spaced calcite veins are called *high-density veins* (HDVs) by Rawnsley *et al.* (1998). They predate, and are commonly exploited by, the E–W-striking calcite veins (related to the normal faults) and the joints. HDVs appear to be present everywhere in the limestone beds at Lilstock. West of Kilve, the HDVs have an average spacing of about 0.2 m, while around Lilstock they have average spacings of less than 10 mm (Rawnsley *et al.* 1998). They are typically only visible where erosion has enhanced the contrast between the calcite HDVs and the limestone country rock, especially around the edges of joints. In thin section, they can be seen to persist throughout the rock, and are commonly en echelon. Caputo & Hancock (1999) suggest that HDVs represent repeated opening and healing by cement, with the cement being stronger than the wall rocks, so new fractures develop. Caputo & Hancock (1999) compare HDVs with crack–seal veins (Ramsay 1980), which are also the result of repeated opening and healing of mode I veins.

The third set of veins is the E–W-striking veins related to the normal faults (Fig. 2c), while the

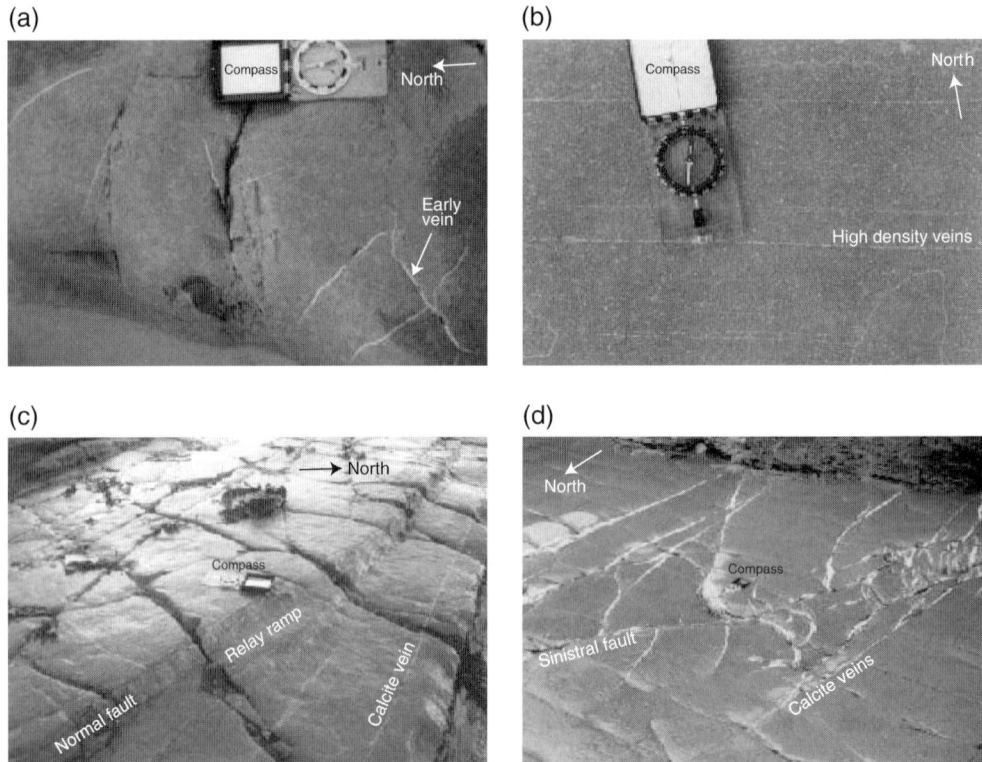

Fig. 2. Photographs of calcite veins exposed on gently dipping limestone bedding planes at Kilve. Joints cut all the veins. (**a**) Possible early, diagenetic veins, with a range of orientations. (**b**) High-density veins (see Rawnsley *et al.* 1998; Caputo & Hancock 1999). (**c**) E–W-striking veins adjacent to an E–W-striking normal fault zone, which itself is mineralized (see Peacock 1991). Note that the joints cut across the fault zone and the related veins. (**d**) N–S-striking veins adjacent to a strike-slip fault zone (see Willemse *et al.* 1997).

fourth set strike N–S and are closely related to the strike-slip faults (Fig. 2d), as described in the next section.

The relationship between veins and faults

Veins typically occur at and beyond the tips of normal (Peacock & Sanderson 1991) and strike-slip (Willemse *et al.* 1997) faults in the Liassic limestones of Somerset. For example, normal faults are commonly surrounded by several calcite veins, with veins extending beyond the fault tips. Each normal fault therefore appears to have developed along one or more veins when the veins reached a width of approximately 20 mm (Peacock & Sanderson 1992). Figure 3 shows an example of a very low displacement normal fault developing from a set of veins in a limestone bed. As the normal fault develops, the displacing vertical vein becomes a pull-apart in the limestone bed (Peacock & Sanderson 1992, figs 8a and 9; Peacock 2002, fig. 3). The pull-aparts overlap

as displacement increases enough for the limestone beds to be faulted past each other, resulting in a mineralized fault plane. The strike-slip faults at Kilve also appear to have developed by the linkage of en echelon veins, with calcite-filled pull-aparts developed as linked portions of each vein segment are sheared (Fig. 4b) (Peacock & Sanderson 1995; Willemse *et al.* 1997).

The relationship between joints and faults

Peacock (2001) interprets the joints at East Quantoxhead, about 1 km west of Kilve (Fig. 1b), as post-dating the normal and strike-slip faults. Evidence for this interpretation includes the following.

• Normal faults and associated calcite veins consistently strike E–W, while E–W-striking joints are only locally developed. The dominant (earliest and longest) joint set in the area strikes approximately NW–SE (e.g. Rawnsley *et al.*

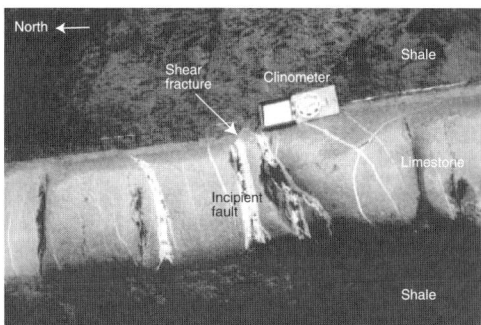

Fig. 3. Photograph of an incipient normal fault in a limestone bed at Kilve, Somerset. Calcite veins occur in the limestone bed, with shear fractures refracting out of the thickest vein at a lower angle into the surrounding mudrocks (see Peacock 2002, fig. 3).

1998). Later sets are shorter because they abut the earlier joints. This suggests that there was no consistent set of E–W-striking joints before the development of the E–W-striking normal faults and calcite veins, or before development of the NW–SE-striking joints.

- Joints consistently abut the E–W-striking normal faults that separate beds, and cut the smaller normal faults and the associated calcite veins (Fig. 4a). The normal faults and all associated fractures are mineralized, implying that all open fractures were affected by a phase of mineralization. The joints are not mineralized, suggesting that they post-date the mineralization and therefore the normal faults. Some 095°-striking joints run along and extend from the tips of some of the normal faults. C. Townsend (pers. comm.) argues that these faults followed pre-existing joints. While all normal faults die out along strike into calcite veins (e.g. Peacock & Sanderson 1991, 1992), not all normal faults pass into joints.

- Joints commonly cut across smaller strike-slip faults and are not displaced by them (Fig. 4b). These faults that are cut by joints do not separate beds and so did not represent a lithological barrier. Joints consistently abut larger strike-slip faults (Fig. 4c), which separates beds and therefore acted as barriers to joint propagation. Joints of all sets cross or abut the strike-slip faults, but the relationship is most obvious in joints that strike at a high angle to the faults. The strike-slip faults consistently offset the normal faults, so the joints must post-date the normal faults.

- Joints commonly curve into the strike-slip faults (Fig. 4d), indicating that stresses were perturbed around the strike-slip faults. The earliest joints abut the strike-slip faults at about 90°, indicating

that the strike-slip faults were traction-free surfaces at the time of jointing, so the principal axes of stress were parallel and perpendicular to the faults (Engelder & Gross 1993). Rawnsley *et al.* (1992, 1998) show joints curving into points along strike-slip faults at Lilstock and elsewhere in the BCB. It is possible that such a curving pattern of joints could form in a diffuse zone of shearing either as a precursor to, or during, the strike-slip faulting. Evidence against this, however, includes the fact that the fault-related calcite veins do not follow this curved trajectory, but consistently strike approximately N–S.

- Different patterns of joints occur in the same bed on either side of some strike-slip fault zones (Peacock 2001, fig. 10). This indicates that the strike-slip faults compartmentalized the stress field during joint formation, so the joints are either syn- or post-faulting.

- Joint frequency does not vary along scanlines that cross the strike-slip faults (Fig. 5b), implying that there is no genetic relationship of the type described by Mollema & Antonellini (1999).

This interpretation of post-fault joints is consistent with the suggestion of Rawnsley *et al.* (1998) that the joints in the Mesozoic sedimentary rocks of the BCB resulted from relaxation of stresses after the Alpine faulting and folding, during regional uplift and erosion. Peacock (2001, fig. 12) illustrates some of the criteria that can be used for distinguishing between joints that are pre-, syn- and post-faulting.

Evidence that joints are synchronous with folding

Field evidence therefore suggests that joints at Kilve post-date all of the other structures. Engelder & Peacock (2001) suggest, however, that some joints at Lilstock are synchronous with folding. This interpretation is based on the set of joints that strike approximately 105°–115°, which are rare in the horizontal and gently N-dipping beds, but more frequent in the steeper S-dipping beds. 105°–115°-striking joints are most frequent in the vicinity of the core of the anticline described by Engelder & Peacock (2001). The 105°–115°-striking joints are approximately parallel to the strikes of the fold hinge lines, and their poles tilt to the south when bedding is restored to horizontal. This southward tilt aims at the direction of σ_1 for Alpine inversion. Engelder & Peacock (2001) use finite-element analysis to explain the southward tilt of 105°–115°-striking joints. Tilted principal stresses are characteristic of limestone–shale sequences that are sheared during parallel (flexural-flow) folding. Shear tractions on the dipping beds may generate a

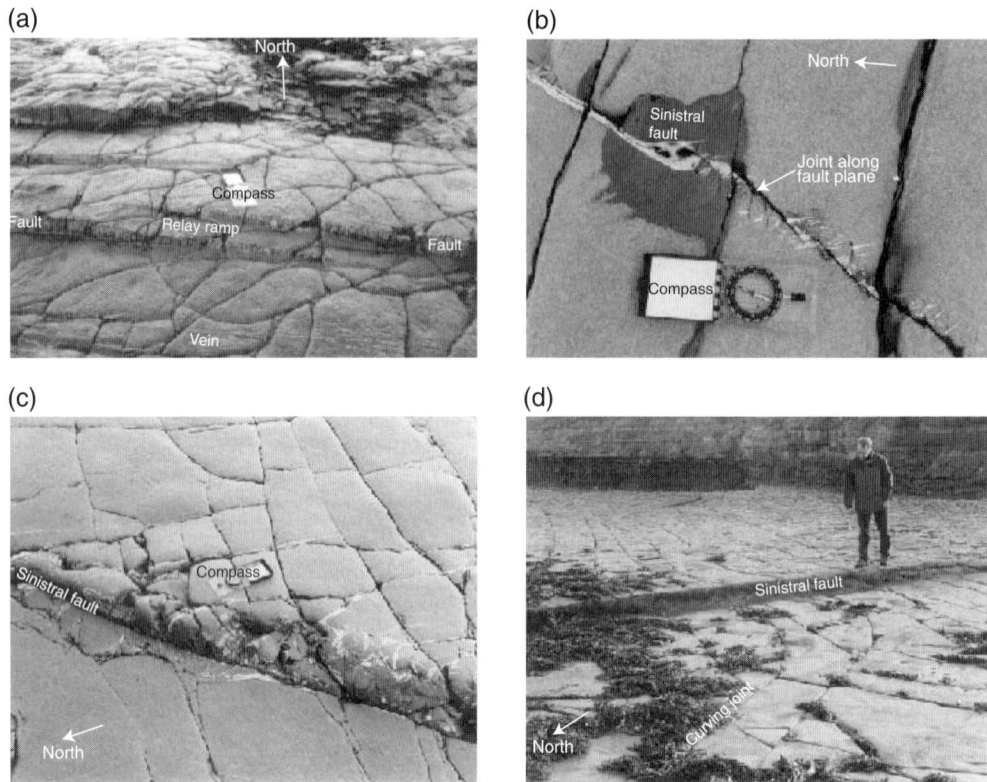

Fig. 4. Photographs illustrating the relationships between veins, faults and joints at Kilve and East Quantoxhead. (**a**) Joints cutting an E–W-striking normal fault and associated calcite veins. (**b**) Joints cutting a sinistral strike-slip fault (strike *c*. 030°) with millimetres of displacement. (**c**) joints abutting a larger (*c*. 1 m displacement) sinistral strike-slip fault, which separates beds and therefore represents a significant barrier to propagating joints. These joints cannot be matched across the fault zone and are not displaced by the fault. (**d**) Joints curving an approximately 030°-striking sinistral strike-slip fault that has tens of millimetres displacement at East Quantoxhead (see Peacock 2001, fig. 9). Beds have a separation across the fault because the slip vector is slightly oblique to the gently dipping bedding planes.

tensile stress in the stiffer limestone beds even when remote principal stresses are compressive. This favours the paradoxical opening of joints in the direction of the regional maximum horizontal stress. Engelder & Peacock (2001) conclude that 105°–115°-striking joints propagated during the Alpine contraction that caused tightening of the folds between Kilve and Lilstock. This interpretation based on numerical modelling is not, however, consistent with the field evidence described in previous sections.

Differences between veins and joints

The differences between veins and joints at Kilve are described in this section, with the discussion also including general differences between these structures.

Descriptive differences

The veins at Kilve are (by definition) mineral-filled, while the joints are not mineralized. Calcite is the filling mineral in these veins, this mostly being crystalline but not fibrous. Slickenside lineations and fibres are common along normal and strike-slip faults at this location (McGrath & Davison 1995), and fibres are more common in veins east of Lilstock (Peacock & Sanderson 1999, fig. 4). In addition, some of the reverse-reactivated normal faults show patches of radiating calcium carbonate minerals, these patches being up to hundreds of millimetres across. Some joints follow pre-existing veins and faults (Fig. 4b), but the joints themselves are not mineralized.

Joints can subsequently be filled by minerals (e.g. Petit *et al.* 1994; Rawnsley *et al.* 1998), but I suggest that a mineral-filled joint should be called a vein for field analysis.

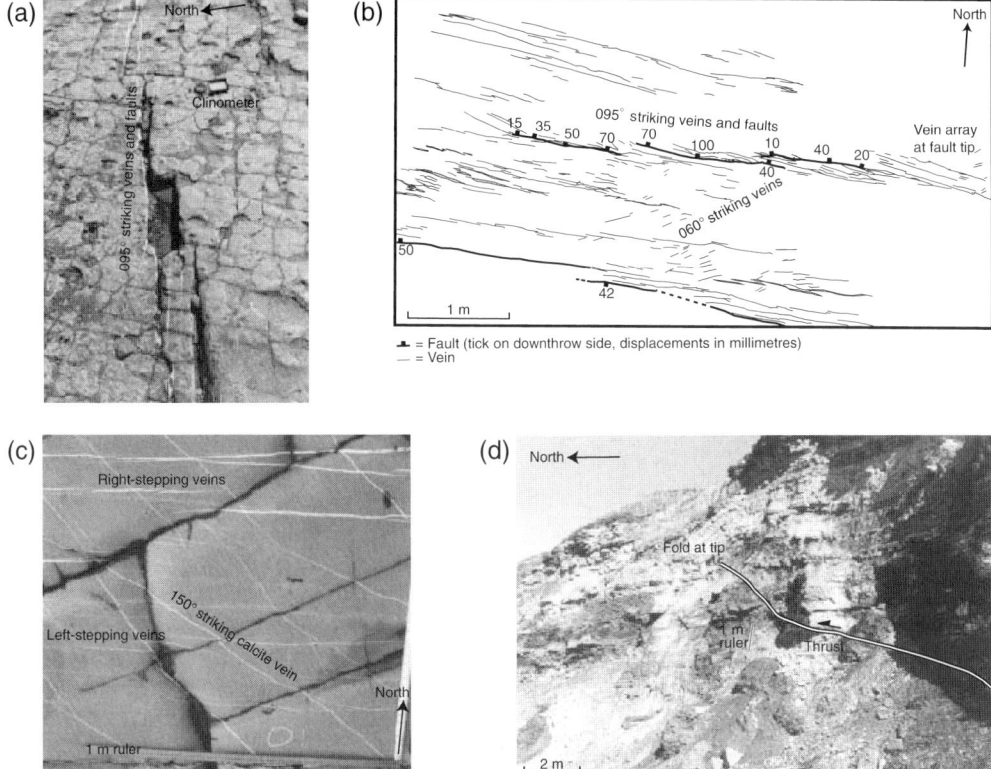

Fig. 5. Evidence for the relatively complex deformation history at Lilstock (see also Peacock & Sanderson 1999). (**a**) Photograph of an 095° striking normal fault zone and related veins, caused by approximately N–S extension with sinistral transtension, exposed on a bedding plane of Rhaetic limestone. (**b**) Map of the fault zone shown in Figure 3a, also illustrating the earlier 060° striking calcite veins. (**c**) Photograph of conjugate vein arrays exposed on a limestone bedding plane, indicating E–W contraction and N–S extension. Approximately 150° striking calcite veins occur, some of which trail through the 095° striking veins. (**d**) Photograph of a reverse fault caused by N–S Alpine contraction. This example appears to be developed at the tip of a reverse-reactivated normal faults.

Kinematic differences

The veins at Kilve have measurable width (up to tens of millimetres) and represent measurable strain. Bowyer & Kelly (1995) measured fault displacements along 15 scanlines at Kilve, and calculated an average of 7% extension. They also measured extension from veins with widths of >0.5 mm along two scanlines adjacent to normal faults (e.g. Fig. 5a), and calculated extensions of 7 and 12%, which is higher than the Bowyer & Kelly (1995) measurements of strain from faulting. Although attempts have been made to determine strain from joints (e.g. Swanson 1927; Chamberlin 1928), these strains would be very low, with the joints at Kilve usually <1 mm wide, and much of the joint width apparently related to weathering. Veins may therefore represent much larger strains than can joints.

The calcite veins at Kilve and Lilstock show a range of en echelon geometries, from extensional to transpressional (Fig. 6). Although joints show some stepping, this geometry is rare. Also, joint oversteps are narrow in proportion to joint-segment length, indicating that they are extensional to slightly transtensional.

Differences in scaling

Bowyer & Kelly (1995) suggest that the vein widths at Kilve obey a power-law scaling relationship. Vein widths or spacings have been shown to be fractal or multifractal in a wide range of settings (e.g. Velde *et al.*, 1991; Ledesert *et al.* 1993; Manning 1994; Sanderson *et al.* 1994; Magde *et al.* 1995; Johnston & McCaffrey 1996; Berkowitz & Hadad 1997; Hippertt & Massucatto 1998; Roberts *et al.* 1998; Bonnet *et al.* 2001; Gillespie *et al.* 2001; Monecke *et*

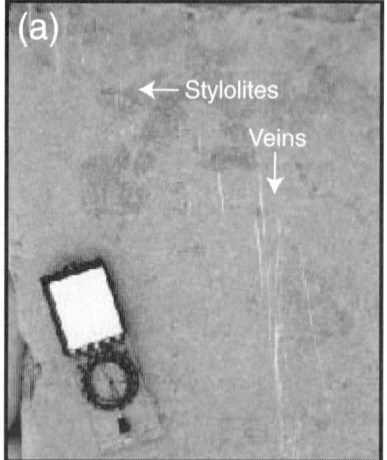

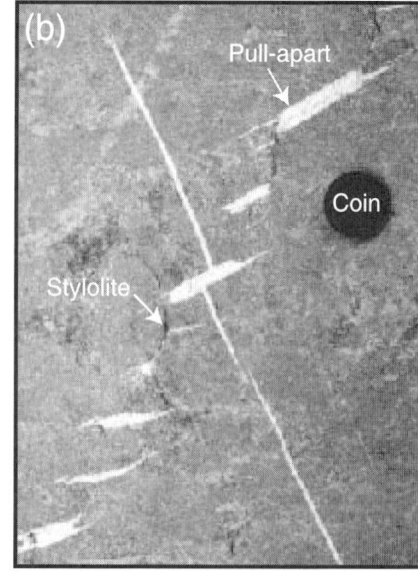

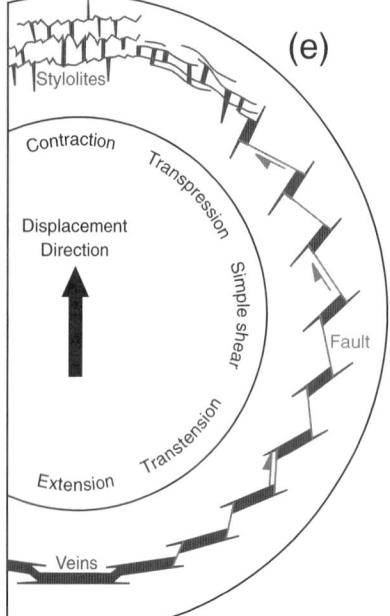

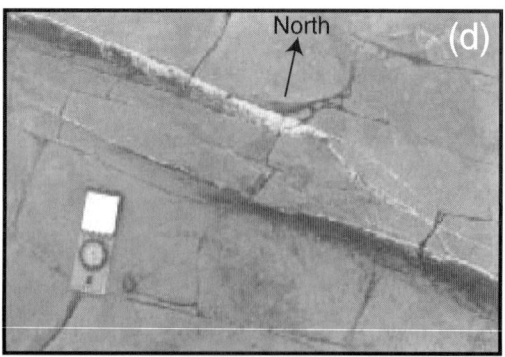

Fig. 6. (**a**)–(**d**) Vein and fault systems at Kilve and Lilstock, showing transpression, simple shear, transtension and extension, respectively. (**d**) A normal fault system with veins at the tips of the faults and a relay ramp between overstepping segments. Fibres in the veins and faults indicate opening perpendicular to the strike of the fault planes. Maximum displacement on the faults is <100 mm. (**e**) Annulus model for fault initiatiation and propagation to illustrate the variations in the geometries of fault tips with varying amounts of transtension or transpression (based on McCoss 1986; also see Ramsay & Huber 1987, fig. 26.42; Peacock & Sanderson 1995, fig. 7). There is a complete spectrum of geometries from veins with no pressure-solution seams (extension) to pressure-solution seams with few or no veins (contraction).

al. 2001). Some authors show, however, that vein spacings may be non-fractal (Nehlig *et al.* 1998; Simpson 2000; Brathwaite *et al.* 2001).

In contrast, it seems unlikely that joint widths are fractal, or they may be fractal but up to only very narrow widths. Some authors do, however, suggest that joint distributions can be fractal (King & Sammis 1992; Kulatilake *et al.* 1997*a*; Ehlen 2000), while joint surfaces may show fractal or self-affine geometries (e.g. Power & Tullis 1991; Sakellariou *et al.* 1991; Huang *et al.* 1992; Wang & Scholz 1993; Odling 1994; Brown 1995; Kulatilake *et al.* 1997*b*; Develi & Babadagli 1998; Borri-Brunetto *et al.* 1999; Du & Fan 1999; Kulatilake & Um 1999; Xie *et al.* 1999). Gillespie *et al.* (1993) and Rives *et al.* (1994) show that joint spacing in bedded sedimentary rocks commonly show negative exponential or log-normal frequency distributions, with single orientation joint sets being characterized by a regular (non-fractal) spacing.

Mechanical and environmental differences

The veins at Kilve are filled by calcite, so must have been open for long enough and under the correct environmental conditions for the mineral to be deposited. Fluid pressure would have exceeded the least compressive stress, and the fluids would have been rich in calcium carbonate and present in the open cracks for long enough for calcite to be precipitated. There are no published data on the depths, temperatures and pressures in the rocks when the veins formed, but stratigraphic evidence suggests depths of no more than about 2 km and there is no evidence of metamorphism.

In contrast, the joints at Kilve did not form under the correct mechanical or environmental conditions for minerals to be precipitated. Rawnsley *et al.* (1998) suggest that the joints formed as stress-release structures, during the reduction of Alpine stresses, and during uplift and erosion (also see Gillespie *et al.* 2001). Stress release and cooling may have reduced the compressive stress close to zero, so fluid pressures would not need to have been high for the joints to form. The joints at Kilve may therefore be similar to cooling joints in igneous rocks, which form as magma cools and contracts (e.g. DeGraff & Aydin 1993), or to mud cracks formed during desiccation (e.g. Weinberger 2001).

Geometric differences

The veins at Kilve form simple parallel sets, with E–W-striking veins related to Mesozoic extension and N–S-striking veins related to during Alpine contraction (Fig. 2). Although some complexity and

brecciation occur in fault zones (e.g. Figs 2d and 4b), the veins in each set have uniform strikes (Peacock & Sanderson 1992, figs 4 and 7). They would have formed perpendicular to the least compressive stress (e.g. Pollard & Aydin 1988), which would have been horizontal and N–S during Mesozoic extension, and horizontal and E–W during Alpine contraction. Elsewhere in the world, veins commonly form more *stockworks*, which are more complex networks of veins. For example, Cosgrove (2001) describes the network of gypsum veins in the Triassic marls at Watchet, Somerset (Fig. 1b). Windh (1995) shows that vein stockworks are indicative of fluctuating fluid pressures and low differential stresses.

The joints at Kilve form a complex network of fractures. Although it is possible to determine a sequence of joint events in each bed, adjacent limestone beds commonly have completely different joint patterns or spacings (e.g. Rawnsley *et al.* 1998; Engelder & Peacock 2001). For example, Figure 7 shows different joint sets and patterns in different thickness limestone beds. Rawnsley *et al.* (1998) and Engelder & Peacock (2001) describe several joint sets in the Liassic limestones at Lilstock, with such a range of strikes occurring that division of the joint population into sets becomes almost arbitrary. This complexity indicates a complex evolving stress system during joint formation, with Rawnsley *et al.* (1998) suggesting a general anticlockwise rotation of stresses during joint formation across the BCB. Gillespie *et al.* (2001) also describe a joint network that is related to uplift.

Chronological differences and relationships to faults

Evidence for the ages of the veins and joints, and their relationships to the faults at Kilve, are described above. Although some joints follow pre-existing veins and faults (Fig. 4b), the joints themselves are not mineralized. This indicates that the phase(s) of mineralization responsible for the veins occurred before the joints formed. It is possible that the joints had formed but were closed during veining, but this argument is not supported by the cross-cutting relationships illustrated in Figure 4. Gillespie *et al.* (2001) describe similar vein and joint geometries and chronologies in the Carboniferous Limestone of The Burren, Ireland. They show that veins are strongly clustered, obey a power-law scaling relationship and formed during orogenesis. Joints have regular spacings that scale with bed thickness, have log-normal length distributions and formed during uplift, under low differential stresses.

Veins at Kilve are commonly precursors to, or synchronous with, faulting, and so are clustered around faults (Fig. 5a). In contrast, joints post-date

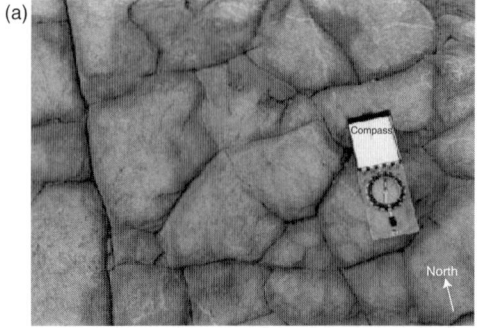

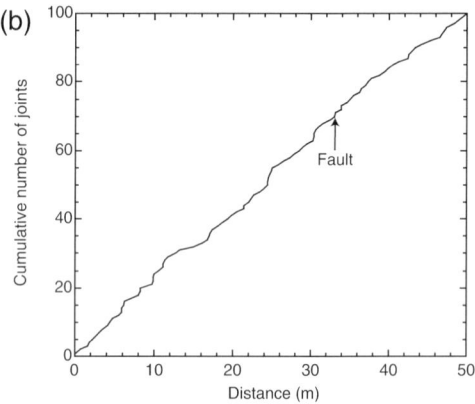

Fig. 7. Photographs of different patterns of joints on different limestone beds. (**a**) Irregular joints on an approximately 50 mm-thick limestone. (**b**) More systematic joints on an approximately 210 mm-thick limestone bed.

the faults, and do not tend to increase in frequency around faults (Fig. 5b). Clearly distinguishing between veins and joints therefore helps determine age and genetic relationships between the different structures at Kilve.

Conclusions

Mesozoic extension and development of the BCB is represented at Kilve by E–W-striking veins and normal faults. Contraction of the BCB during the Alpine orogeny (Miocene) is represented by N–S-striking calcite veins, reverse-reactivation of the normal faults, E–W-striking thrusts, strike-slip veins conjugate about N–S, and by locally-developed E–W-striking crenulation cleavage and pressure-solution cleavage. Joints formed during post-Alpine stress release (Rawnsley *et al.* 1998).

There are important differences between the veins and joints at Kilve (and elsewhere), including the following.

(1)   The veins at Kilve are (by definition) mineral-ized, while the joints are not mineralized.

Fig. 8. (**a**) Photograph of veins clustered around a normal fault zone at Kilve. (**b**) Graph of the cumulative number of joints against distance across a sinistral strike-slip fault zone with a displacement of less than 1 m at East Quantoxhead (scanline along a 288° trend). Joint frequency does not increase around the fault, indicating that the joints did not form at the same time as the fault (Peacock 2001).

(2)   The veins are up to tens of millimetres wide, and so have measurable width and represent measurable strain (e.g. Bowyer & Kelly 1995). The joints are usually less than 1 mm wide, even after weathering, and represent minute strains.

(3)   Vein widths can obey a power-law scaling rela-tionships, while joint widths do not.

(4)   The veins were open for long enough and under the correct conditions for minerals to be

deposited, indicating high fluid pressures. Also, the veins at Kilve are commonly en echelon, indicating significant shear. The joints did not form under the correct mechanical or environmental conditions for minerals to be precipitated. Joints do not necessarily form under conditions of high fluid pressures (e.g. cooling joints). Although some joints at Kilve show oversteps, they are dominantly purely extensional, with very little evidence of shear.

(5) The veins at Kilve form simple parallel sets, whereas the joints form complex fracture networks.

(6) Veins at Kilve are commonly precursors to, or synchronous with, faults (Willemse *et al.* 1997), while the joints post-date the faults (Peacock 2001). Veins are therefore commonly clustered around faults, while the joints do not increase in frequency near faults (Fig. 8).

These differences suggest that it is not helpful to consider veins and joints together simply as 'extension fractures', but to treat them as fundamentally different structures. This is especially significant in field analyses, where veins and joints may form at different times, in association with different structures, and under different conditions.

D. Sanderson is thanked for many discussions during the course of my work in Somerset, and E. Parfitt is thanked for her support. J. Cosgrove and T. Engelder are thanked for organizing the 'Mechanics of Jointing in the Crust' Meeting and the useful fieldtrips in Somerset.

References

AL-MAHRUQI, S. A. S. 2001. *Fracture patterns and fracture propagation as a function of lithology.* PhD thesis, Imperial College, University of London.

BERKOWITZ, B. & HADAD, A. 1997. Fractal and multifractal measures of natural and synthetic fracture networks. *Journal of Geophysical Research*, **102**, 12205–12218.

BONNET, E., BOUR, O., ODLING, N. E., DAVY, P., MAIN, I., COWIE, P. & BERKOWITZ, B. 2001. Scaling of fracture systems in geological media. *Reviews of Geophysics*, **39**, 347–383.

BORRI-BRUNETTO, M., CARPINTERI, A. & CHIAIA, B. 1999. Scaling phenomena due to fractal contact in concrete and rock fractures. *International Journal of Fracture*, **95**, 221–238.

BOWYER, M. O'N. & KELLY, P. G. 1995. Strain and scaling relationships of faults and veins at Kilve, Somerset. *Proceedings of the Ussher Society*, **8**, 411–415.

BRATHWAITE, R. L., CARGILL, H. J., CHRISTIE, A. B. & SWAIN, A. 2001. Lithological and spatial controls on the distribution of quartz veins in andesite- and rhyolite-hosted epithermal Au–Ag deposits of the Hauraki Goldfield, New Zealand. *Mineralium Deposita*, **36**, 1–12.

BROWN, S. R. 1995. Simple mathematical model of a rough fracture. *Journal of Geophysical Research*, **100**, 5941–5952.

CAPUTO, R. & HANCOCK, P. L. 1999. Crack-jump mechanism and its implications for stress cyclicity during extension fracturing. *Geodynamics*, **27**, 45–60.

CHAMBERLIN, R. T. 1928. The strain ellipsoid and Appalachian structures. *Journal of Geology*, **36**, 85–90.

COSGROVE, J. W. 2001. Hydraulic fracturing during the formation and development of a basin: a factor in the dewatering of low-permeability sediments. *AAPG Bulletin*, **87**, 737–748.

DART, C. J., MCCLAY, K. & HOLLINGS, P. N. 1995. 3D analysis of inverted extensional fault systems, southern Bristol Channel basin, U. K. *In*: BUCHANAN, J. G. & BUCHANAN, P. G. (eds) *Basin Inversion.* Geological Society, London, Special Publications, **88**, 393–413.

DAVIS, G. H. & REYNOLDS, S. J. 1996. *Structural Geology of Rocks and Regions.* Wiley, New York.

DEGRAFF, J. M. & AYDIN, A. 1993. Effect of thermal regime on growth increment and spacing of contraction joints in basaltic lava. *Journal of Geophysical Research*, **98**, 6411–6430.

DEVELI, K. & BABADAGLI, T. 1998. Quantification of natural fracture surfaces using fractal geometry. *Mathematical Geology*, **30**, 971–998.

DU, S. G. & FAN, L. B. 1999. The statistical estimation of rock joint roughness coefficient. *Chinese Journal of Geophysics*, **42**, 577–580.

EHLEN, J. 2000. Fractal analysis of joint patterns in granite. *International Journal of Rock Mechanics and Mining Sciences*, **37**, 909–922.

ENGELDER, T. & GROSS, M. R. 1993. Curving cross joints and the lithospheric stress field in eastern North America. *Geology*, **21**, 817–820.

ENGELDER, T. & PEACOCK, D. C. P. 2001. Joint development normal to regional compression during flexural-flow folding: the Lilstock buttress anticline, Somerset, England. *Journal of Structural Geology*, **23**, 259–277.

GILLESPIE, P. A., HOWARD, C. B., WALSH, J. J. & WATTERSON, J. 1993. Measurement and characterization of spatial distributions of fractures. *Tectonophysics*, **226**, 113–141.

GILLESPIE, P. A., WALSH, J. J., WATTERSON, J., BONSON, C. G. & MANZOCCHI, T. 2001. Scaling relationships of joint and vein arrays from The Burren, Co. Clare, Ireland. *Journal of Structural Geology*, **23**, 183–201.

HIPPERTT, J. F. & MASSUCATTO, A. J. 1998. Phyllonitization and development of kilometre-size extension gashes in a continental-scale strike-slip shear zone, north Goias, central Brazil. *Journal of Structural Geology*, **20**, 433–445.

HUANG, S. L., OELFKE, S. M. & SPECK, R. C. 1992. Applicability of fractal characterization and modeling to rock joint profiles. *International Journal of Rock Mechanics and Mining Sciences and Geomechanics Abstracts*, **29**, 89–98.

JOHNSTON, J. D. & MCCAFFREY, K. J. W. 1996. Fractal geometries of vein systems and the variation of scaling relationships with mechanism. *Journal of Structural Geology*, **18**, 349–358.

KELLY, P. G., MCGURK, A., PEACOCK, D. C. P. & SANDERSON, D. J. 1999. Reactivated normal faults in the Mesozoic of the Somerset coast, and the role of fault scale in reactivation. *Journal of Structural Geology*, **21**, 493–509.

KING, G. C. P. & SAMMIS, C. G. 1992. The mechanisms of finite brittle strain. *Pure and Applied Geophysics*, **138**, 611–640.

KULATILAKE, P. H. S. W. & UM, J. 1999. Requirements for accurate quantification of self-affine roughness using the roughness–length method. *International Journal of Rock Mechanics and Mining Sciences*, **36**, 5–18.

KULATILAKE, P. H. S. W., FIEDLER, R. & PANDA, B. B. 1997*a*. Box fractal dimension as a measure of statistical homogeneity of jointed rock masses. *Engineering Geology*, **48**, 217–229.

KULATILAKE, P. H. S. W., UM, J. & PAN, G. 1997*b*. Requirements for accurate estimation of fractal parameters for self-affine roughness profiles using the line scaling method. *Rock Mechanics and Rock Engineering*, **30**, 181–206.

LEDESERT, B., DUBOIS, J., GENTER, A. & MEUNIER, A. 1993. Fractal analysis of fractures applied to Soultz-Sous-Forets hot dry rock geothermal program. *Journal of Volcanology and Geothermal Research*, **57**, 1–17.

MAGDE, L. S., DICK, H. J. B. & HART, S. R. 1995. Tectonics, alteration and the fractal distribution of hydrothermal veins in the lower ocean crust. *Earth and Planetary Science Letters*, **129**, 103–119.

MANNING, C. E. 1994. Fractal clustering of metamorphic veins. *Geology*, **22**, 335–338.

MCCOSS, A. M. 1986. Simple constructions for deformation in transpression/transtension zones. *Journal of Structural Geology*, **8**, 715–718.

MCGRATH, A. G. & DAVISON, I. 1995. Damage zone geometry around fault tips. *Journal of Structural Geology*, **17**, 1011–1024.

MOLLEMA, P. N. & ANTONELLINI, M. 1999. Development of strike-slip faults in the dolomites of the Sella Group, Northern Italy. *Journal of Structural Geology*, **21**, 273–292.

MONECKE, T., GEMMELL, J. B. & MONECKE, J. 2001. Fractal distributions of veins in drill core from the Hellyer VHMS deposit, Australia: constraints on the origin and evolution of the mineralising system. *Mineralium Deposita*, **36**, 406–415.

NEHLIG, P., CASSARD, D. & MARCOUX, E. 1998. Geometry and genesis of feeder zones of massive sulphide deposits: constraints from the Rio Tinto ore deposit (Spain). *Mineralium Deposita*, **33**, 137–149.

ODLING, N. E. 1994. Natural fracture profiles, fractal dimension and joint roughness coefficients. *Rock Mechanics and Rock Engineering*, **27**, 135–153.

PEACOCK, D. C. P. 1991. A comparison between the displacement geometries of veins and normal faults at Kilve, Somerset. *Proceedings of the Ussher Society*, **7**, 363–367.

PEACOCK, D. C. P. 2001. The temporal relationship between joints and faults. *Journal of Structural Geology*, **23**, 329–341.

PEACOCK, D. C. P. 2002. Initiation, propagation and linkage in normal fault systems. *Earth-Science Reviews*, **58**, 121–142.

PEACOCK, D. C. P. & SANDERSON, D. J. 1991. Displacements, segment linkage and relay ramps in normal fault zones. *Journal of Structural Geology*, **13**, 721–733.

PEACOCK, D. C. P. & SANDERSON, D. J. 1992. Effects of layering and anisotropy on fault geometry. *Journal of the Geological Society, London*, **149**, 793–802.

PEACOCK, D. C. P. & SANDERSON, D. J. 1995. Pull-aparts, shear fractures and pressure solution. *Tectonophysics*, **241**, 1–13.

PEACOCK, D. C. P. & Sanderson, D. J. 1999. Deformation history and basin-controlling faults in the Mesozoic sedimentary rocks of the Somerset coast. *Proceedings of the Geologists Association*, **110**, 41–52.

PETIT, J.-P., MASSONNAT, G., PUEO, F. & RAWNSLEY, K. 1994. Mode-I fracture shape ratios in layered rocks: a case-study in the Lodeve Permian Basin (France). *Bulletin des Centres de Recherches Exploration–Production Elf Aquitaine*, **18**, 211–229.

POLLARD, D. D. & AYDIN, A. 1988. Progress in understanding jointing over the past century. *Geological Society of America Bulletin*, **100**, 1181–1204.

POWER, W. L. & TULLIS, T. E. 1991. Euclidean and fractal models for the description of rock surface roughness. *Journal of Geophysical Research*, **96**, 415–424.

RAMSAY, J. 1980. The crack–seal mechanism of rock deformation. *Nature*, **284**, 135–139.

RAMSAY, J. & HUBER, M. I. 1987. *The Techniques of Modern Structural Geology. Volume 2: Folds and Fractures*. Academic Press, London.

RAWNSLEY, K. D., PEACOCK, D. C. P., RIVES, T. & PETIT, J.-P. 1998. Jointing in the Mesozoic sediments around the Bristol Channel Basin. *Journal of Structural Geology*, **20**, 1641–1661.

RAWNSLEY, K. D., RIVES, T., PETIT, J.-P., HENCHER, S. R. & LUMSDEN, A. C. 1992. Joint development in perturbed stress fields near faults. *Journal of Structural Geology*, **14**, 939–951.

RIVES, T., RAWNSLEY, K. D. & PETIT, J.-P. 1994. Analogue simulation of natural orthogonal joint set formation in brittle varnish. *Journal of Structural Geology*, **16**, 419–429.

ROBERTS, S., SANDERSON, D. J. & GUMIEL, P. 1998. Fractal analysis of Sn–W mineralization from central Iberia: insights into the role of fracture connectivity in the formation of an ore deposit. *Economic Geology Society Bulletin*, **93**, 360–365.

SAKELLARIOU, M., NAKOS, B. & MITSAKAKI, C. 1991. On the fractal character of rock surfaces. *International Journal of Rock Mechanics and Mining Sciences and Geomechanics Abstracts*, **28**, 527–533.

SANDERSON, D. J., ROBERTS, S. & GUMIEL, P. 1994. A fractal relationship between vein thickness and gold grade in drill core from La Codosera, Spain. *Economic Geology and the Bulletin of the Society of Economic Geologists*, **89**, 168–173.

SIMPSON, G. D. H. 2000. Synmetamorphic vein spacing distributions: characterisation and origin of a distribution of veins from NW Sardinia, Italy. *Journal of Structural Geology*, **22**, 335–348.

SWANSON, C. O. 1927. Notes on stress, strain, and joints. *Journal of Geology*, **35**, 193–223.

THOMAS, A. L. & POLLARD, D. D. 1993. The geometry of echelon fractures in rock: implications from laboratory and numerical experiments. *Journal of Structural Geology*, **15**, 323–334.

VELDE, B., DUBOIS, J., MOORE, D. & TOUCHARD, G. 1991. Fractal patterns of fractures in granites. *Earth and Planetary Science Letters*, **104**, 25–35.

WANG, W. & SCHOLZ, C. H. 1993. Scaling of constitutive parameters of friction for fractal surfaces. *International Journal of Rock Mechanics and Mining Sciences and Geomechanics Abstracts*, **30**, 1359–1365.

WEINBERGER, R. 2001. Evolution of polygonal patterns in stratified mud during desiccation: The role of flaw distribution and layer boundaries. *Geological Society of America Bulletin*, **113**, 20–31.

WHITTAKER, A. & GREEN, G. W. 1983. *Geology of the Country Around Weston-super-Mare*. Memoir of the Geological Survey of Great Britain, Sheet 279 and parts of sheets 263 and 295.

WILLEMSE, E. J. M., PEACOCK, D. C. P. & AYDIN, A. 1997. Nucleation and growth of strike-slip faults in limestones from Somerset, U. K. *Journal of Structural Geology*, **19**, 1461–1477.

WINDH, J. 1995. Saddle reef and related gold mineralization, Hill End gold field, Australia: Evolution of an auriferous vein system during progressive deformation. *Economic Geology and the Bulletin of the Society of Economic Geologists*, **90**, 1764–1775.

XIE, H. P., WANG, J. A. & KWASNIEWSKI, M. A. 1999. Multifractal characterization of rock fracture surfaces. *International Journal of Rock Mechanics and Mining Sciences*, **36**, 19–27.

Fracture development within a stratovolcano: the Karaha–Telaga Bodas geothermal field, Java volcanic arc

M. NEMČOK[1], J. N. MOORE[1], R. ALLIS[2] & J. McCULLOCH[3]

[1]*Energy and Geoscience Institute, University of Utah, 423 Wakara Way, Suite 300, Salt Lake City, UT 84108, USA (e-mail: mnemcok@egi.utah.edu)*
[2]*Utah Geological Survey, 1594 W North Temple Street, Suite 3110, Salt Lake City, UT 84116, USA*
[3]*Coso Operating Company, LLC, 900 N. Heritage Drive, Ridgecrest, CA 93555, USA*

Abstract: Karaha–Telaga Bodas, a vapour-dominated geothermal system located in an active volcano in western Java, is penetrated by more than two dozen deep geothermal wells reaching depths of 3 km. Detailed paragenetic and fluid-inclusion studies from over 1000 natural fractures define the liquid-dominated, transitional and vapour-dominated stages in the evolution of this system.

The liquid-dominated stage was initiated by a shallow magma intrusion into the base of the volcanic cone. Lava and pyroclastic flows capped a geothermal system. The uppermost andesite flows were only weakly fractured due to the insulating effect of the intervening altered pyroclastics, which absorbed the deformation. Shear and tensile fractures that developed were filled with carbonates at shallow depths, and by quartz, epidote and actinolite at depths and temperatures over 1 km and 300°C. The system underwent numerous cycles of overpressuring, documented by subhorizontal tensile fractures, anastomosing tensile fracture patterns and implosion breccias.

The development of the liquid system was interrupted by a catastrophic drop in fluid pressures. As the fluids boiled in response to this pressure drop, chalcedony and quartz were selectively deposited in fractures that had the largest apertures and steep dips. The orientations of these fractures indicate that the escaping overpressured fluids used the shortest possible paths to the surface.

Vapour-dominated conditions were initiated at this time within a vertical chimney overlying the still hot intrusion. As pressures declined, these conditions spread outward to form the marginal vapour-dominated region encountered in the drill holes. Downward migration of the chimney, accompanied by growth of the marginal vapour-dominated regime, occurred as the intrusion cooled and the brittle–ductile transition migrated to greater depths. As the liquids boiled off, condensate that formed at the top of the vapour-dominated zone percolated downward and low-salinity meteoric water entered the marginal parts of the system. Calcite, anhydrite and fluorite precipitated in fractures on heating. Progressive sealing of the fractures resulted in the downward migration of the cap rock. In response to decreased pore pressure in the expanding vapour zone, walls of the fracture system within the vapour-dominated reservoir progressively collapsed. It left only residual permeability in the remaining fracture volume, with apertures supported only by asperities or propping breccia. In places where normal stresses acting on the fracture walls exceeded the compressive strength of the wall rock, the fractures have completely collapsed.

Fractures within the present-day cap rock include strike- and oblique-slip faults, normal faults and tensile fractures, all controlled by a strike-slip stress regime. The reservoir is characterized by normal faults and tensile fractures controlled by a normal-fault stress regime. The fractures show no evidence that the orientation of the stress field has changed since fracture propagation began.

Fluid migration in the lava and pyroclastic flows is controlled by fractures. Matrix permeability controls fluid flow in the sedimentary sections of the reservoir. Productive fractures are typically roughly perpendicular to the minimum compressive stress, σ_3, and are prone to slip and dilation within the modern stress regime.

This chapter introduces a study of the fracture development in an andesite-hosted geothermal system at Karaha–Telaga Bodas, Indonesia. It is based on the fracture logging from core, interpretation of Electrical Micro Imaging (EMI) and Formation Micro Scanner (FMS) images from deep geothermal production wells, and mineralogical and petrological work. Individual goals include:

- determination of whether far-field stresses changed among different fracture events;
- determination of fracture orientations and kinematics;
- determination of fracture seal/conduit properties;
- determination of the relative succession of fractures.

Existing studies of fracturing in thermal–elastic systems

During the past few decades, fracture studies have focused on various aspects of thermal–elastic

From: COSGROVE, J. W. & ENGELDER, T. (eds) 2004. *The Initiation, Propagation, and Arrest of Joints and Other Fractures.* Geological Society, London, Special Publications, **231**, 223–242. 0305-8719/04/$15 © The Geological Society of London 2004.

systems. Studies of potential nuclear waste repositories resulted in the formulation of governing equations of processes that link thermal gradients, hydrological flow and mechanical deformation in fractured rock (e.g. Tsang 1999, and references therein). Seismologists and petrophysicists have recognized that the bottoming of earthquakes related to the transition from brittle to plastic behaviour occurs at very shallow depths beneath active high-temperature geothermal fields such as The Geysers and the Imperial Valley, California (Gilpin & Lee 1978; Majer & McEvilly 1979; Sibson 1982). Geothermal deep wells drilled to depths where temperatures exceeded 370–400 °C have encountered a host rock with very little permeability and pore fluid pressures significantly higher than hydrostatic (e.g. Cappetti et al. 1985; Ferrara et al. 1985; Fournier 1991). Epithermal ore deposit studies have provided the evidence for the narrow transition from an overlying environment with hydrostatically pressured fluids to the underlying almost lithostatically pressured regime (e.g. Hedenquist et al. 1998; Fournier 1999). Conceptual models described this transition as the brittle–ductile transition (e.g. Fournier 1999). According to these models, magmatic fluids initially trapped in overpressured ductile rocks can escape into overlying hydrostatically pressured brittle rocks when they are released by brittle fracturing (Fournier 1999). Inside the brittle section with hydrostatic regime, these fluids interact with the dominating meteoric fluids. Owing to the progressive deepening of the brittle–ductile transition with cooling of the geothermal system, the release of the trapped magmatic fluids is understood as transient (e.g. Fournier 1999). Apart from cooling, fluid release from the ductile zone can be caused by increasing the pressure of the trapped fluid or extremely rapid stress release.

Mass-balance calculations of porphyry ore deposits, such as the Pine Grove molybdenum deposit in Colorado, demonstrate that large amounts of fluid from the underlying magmatic reservoir can be released through small intrusions (Keith & Shanks 1988). Further research on this subject resulted in an improved understanding of fluid transport in a convecting magma column (Shinohara et al. 1995). Recent experiments document that the mechanical properties of cooling intrusions related to stratovolcanoes depend on the melt fraction and transport properties (e.g. Shaw 1969; Murase & McBirney 1973; Renner et al. 2000). Other studies have shown the importance of fluid-assisted deformation in deep high-temperature environments (e.g. Blanpied et al. 1995), formulated fracture criteria and conditions for the overlying hydrostatically pressured environment (e.g. Secor 1965; Etheridge 1983; Sibson 1996), described the role of fluids in brittle faulting (e.g. Byerlee 1993; Malin 1994; Miller et al. 1996; Hardebeck & Hauksson 1999), and documented a

relationship between faults most likely to slip under modern stress and those likely to be conductive to fluids (e.g. Barton et al. 1995).

The literature addressing various subjects related to fracturing in a complex thermal–elastic setting is large and diverse; many papers deal with specific aspects of fracturing in this setting. They frequently lack data from well-explored present-day geothermal systems. In this chapter, we describe the fracture development of the Karaha–Telaga Bodas geothermal field of west Java, which is associated with an active stratovolcanic system. Karaha–Telaga Bodas is the most extensively cored volcanic-hosted active geothermal system in the world. The system is shallow and hot. Its wells reach the brittle–ductile transition, and contain pertinent temperature and pressure information. Maximum present-day temperatures are about 350 °C. A relatively simple thermal and structural history of the system makes it especially suitable as a natural laboratory for studying fracturing in a complex thermal–elastic setting.

Geological setting of the Karaha–Telaga Bodas geothermal field

Karaha–Telaga Bodas is one of several large geothermal systems occurring in west Java, Indonesia (Fig. 1a). This newly discovered geothermal field is situated within a N–S-trending andesitic ridge, which is perpendicular to the modern-day minimum principal stress (Nemčok et al. 2001). Galunggung Volcano, located at the southern end of the ridge, erupted five times between 1822 and 1984. Telaga Bodas, a shallow acid lake, is located approximately 5.2 km north of the crater at Kawah Galunggung. This ESE-facing horseshoe-shaped crater is believed to have formed 4200 years ago (Katili & Sudradjat 1984) in response to the formation of a large debris avalanche (Brantley & Glicken 1986). Deep drilling by the Karaha–Bodas Co. LLC in 1995–1997 defined a large, partially vapour-dominated system. This system consists of a locally thick cap rock characterized by steep temperature gradients and low permeabilities, an underlying vapour-dominated region that extends to depths below sea level, and a deep liquid-dominated region with measured temperatures up to 350 °C (Fig. 1b)(Allis et al. 2000). The vapour-dominated regime extends laterally for at least 10 km; and it is characterized by subhydrostatic pressures and low-temperature gradients.

The geothermal system is developed mainly in andesite–basaltic andesite lava flows, pyroclastics, epiclastic flows, tuffs and sediments, all deposited during the last 1.75 Ma (Katili & Sudradjat 1984; Ganda et al. 1985). Granodiorite has been encountered in several of the wells at depths of about 3 km, and locally mafic dykes are present. The granodiorite may provide the heat that drives the geothermal system.

Despite the large size of the system and the high measured temperatures, there are few surficial manifestations. These occur mainly at the southern and northern ends of the prospect. Features at the southern end include Telaga Bodas fumaroles (Kawah Saat), and springs that discharge acid sulphate–chloride (pHs to <2) and neutral pH bicarbonate waters. Surface manifestations at the northern end consist mainly of fumarolic activity and associated argillic alteration. ^{3}He/^{4}He ratios of 7.1–7.7 Ra indicate the presence of magmatic gas contributions in springs and fumaroles at both the Telaga Bodas (southern) and Karaha (northern) ends of the field.

The deep reservoir fluids appear to be mainly meteoric in origin with locally appreciable magmatic contributions (Powell *et al.* 2001). These deep liquids have relatively low salinities of 1–2 wt% total dissolved solids, and variable chloride/boron ratios of 2.4–20, which imply that the condensation of steam and gases may play an important role in controlling their chemistries. Shifts in the ^{18}O isotope compositions of the waters away from the meteoric water line suggest that a magmatic influence is strongest at the southern end of the field

The results of our investigations, which are described below, indicate that fluid flow within the geothermal system is predominantly fracture-controlled, although locally matrix permeabilities in the sedimentary units are high enough to contribute to fluid production. Only two faults with NW–SE strikes, one fault with a WNW–ESE strike and one with an ENE–WSW strike have been identified by surface geological mapping (Budhitrisna 1986) because of the extensive vegetation and young ash cover. However, numerous faults have been identified by Landsat image mapping (Nemčok *et al.* 2001). Of these faults 53% have strikes ranging between 102°–158° and 282°–338°, while the remaining 47% range between 44°–78° and 224°–258°.

Regional tectonics

The Indonesian island arc system is located at the junction of the Indian, Australian, Pacific and SE Asian plates. Java is characterized by active volcanism and deep earthquakes north of the island under the Java Sea (Hamilton 1974*a*, *b*, 1978). An outer arc, together with a forearc basin, is located south of Java (Hamilton 1974*b*; Baumann 1982). Sumatra is characterized by an outer island arc, a small amount of active volcanism, a lack of deep earthquakes east of the island and a dextral strike-slip component of deformation (Hamilton 1974*a*, 1978). This strike-slip system accommodates the SSW movement of the SE Asian Plate. The Sumatra and Java trenches

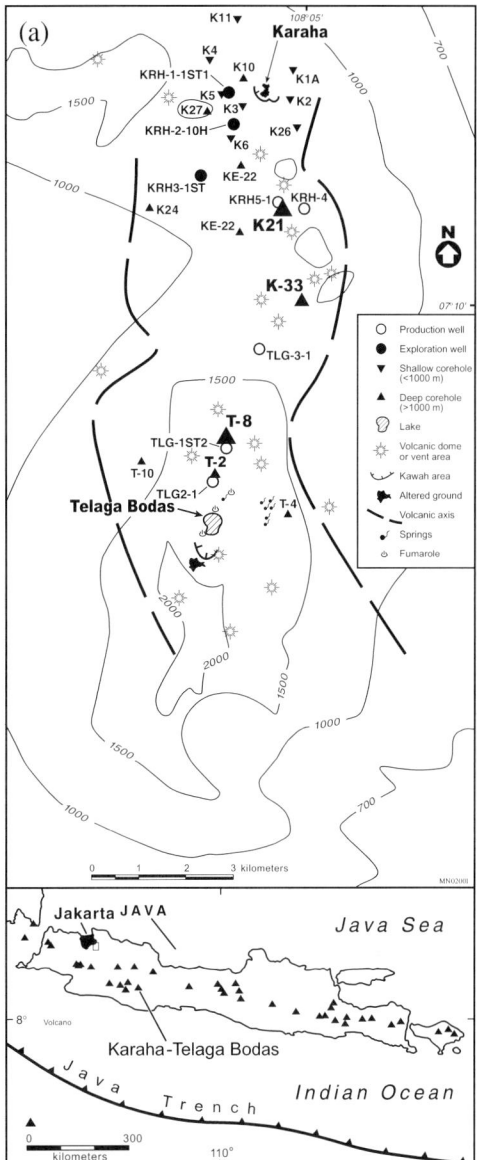

Fig. 1. (**a**) Map of the Karaha–Telaga Bodas geothermal system showing locations of wells and thermal features. Triangles in the inset indicate volcanoes active since 10 000 BP. (**b**) North–south cross-section through the Karaha–Telaga Bodas geothermal system with temperatures shown as isotherms in °C, fluid state and spot pore fluid pressure measurements in bars in the highlighted vapour zone (modified from Allis *et al.* 2000). (**c**) Pressure (*P*) and temperature (*T*) trends with depth from corehole K-33. Triangle and triangle/line patterns indicate lava flows and pyro-/epiclastic flows, respectively. Tuff and sediment horizons are in black.

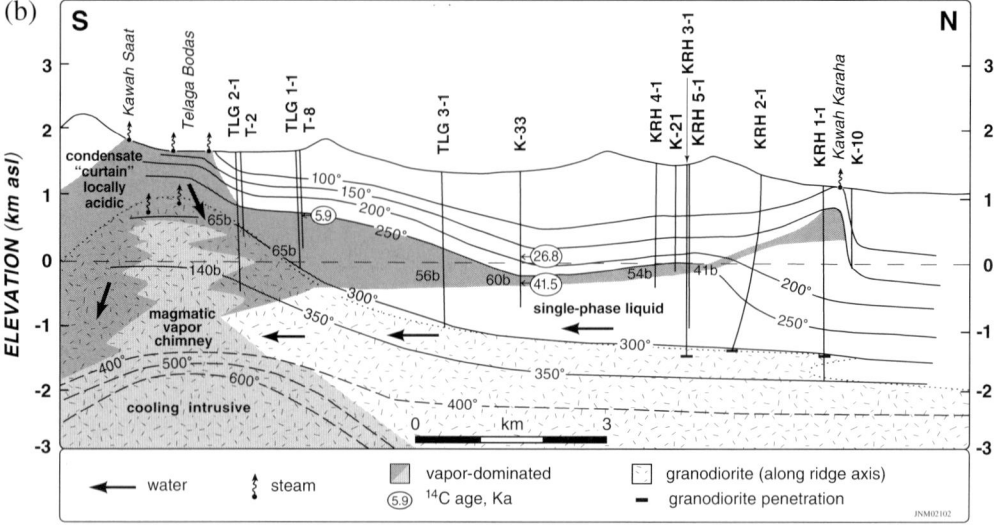

Fig. 1. continued

are zones of northward underthrusting of the Indian Plate.

Reconstruction of plate movements by Hall (1997) suggests that SE Asia was largely recognizable in its present form by 10 Ma, and that the modern-day stress regime was already established by Quaternary time. Since at least 10 Ma, convergence in Sumatra was partitioned into orthogonal subduction and dextral strike-slip motion. The fact that the plate movement vectors have been stable for the last 10 Ma implies that the orientation of the present-day regional stress field (Fig. 2) has been relatively stable throughout the evolution of Karaha–Telaga Bodas geothermal system, which is less than 6000 years old (Bronto 1989; Moore *et al.* 2002). This is an important conclusion because it suggests that the fault kinematics in the geothermal field can be predicted from present-day stress data and fracture geometries. This conclusion is also demonstrated by fracture analysis in corehole K-33 (Table 1), described below.

Local large-scale faults have the following orientations: dip directions of 90°–120° and 270°–300° for normal faults, 20°–40° and 200°–220° for dextral strike-slip faults, 40°–60° and 220°–240° for dextral slightly oblique–slip faults, 60°–90° and 240°–270° for dextral oblique–slip faults, 120°–130° and 300°–310° for sinistral oblique-slip faults, and 130°–150° and 310°–330° for sinistral strike-slip faults (Nemčok *et al.* 2001).

Methods

Fracture orientations and kinematics were studied in three wells, corehole K-33, and two deep production

tests, KRH 2-1OH and KRH 3–1ST. Data from the production tests are based on interpretation of electric image logs; data from K-33 are based on detailed logging of the core.

K-33 (Fig. 1a), drilled in the centre of the prospect, was studied in detail. Approximately 500 fractures, which allowed determination of their kinematics, were measured and described (Table 1). These data were compared with temperature and pore fluid pressure distributions in the well to determine the characteristics of producing and sealing fractures. Fracture wall structures have been studied in order to determine any changes in fracture driving stresses. Cross-cutting relationships of various fractures have been recorded in order to determine the relative succession of fractures.

Interpretation of EMI (Haliburton 1995) and FMS (Schlumberger 1992) electric images from wells KRH 2-1OH and KRH 3–1ST (Fig. 1a) was also conducted in order to determine the orientation of the modern principal stresses, fracture orientations and kinematics. Approximately 400 natural fractures of various kinds and orientations and 19 drilling-induced vertical tensile fractures were identified in the reservoir section of well KRH 3–1ST. One hundred and thirty-six natural fractures and 170 drilling-induced borehole breakout fractures were identified in the reservoir section of well KRH 2-1OH. Drilling-induced tensile and borehole breakout fractures were used to constrain the orientation of the principal stresses. These stresses allowed determination of fracture kinematics. Fracture orientation and kinematics data were compared with temperature and pore fluid pressure distributions in both wells to determine the characteristics of producing and sealing fractures.

(c)

DOWNHOLE SUMMARY PLOT, KARAHA COREHOLE K-33

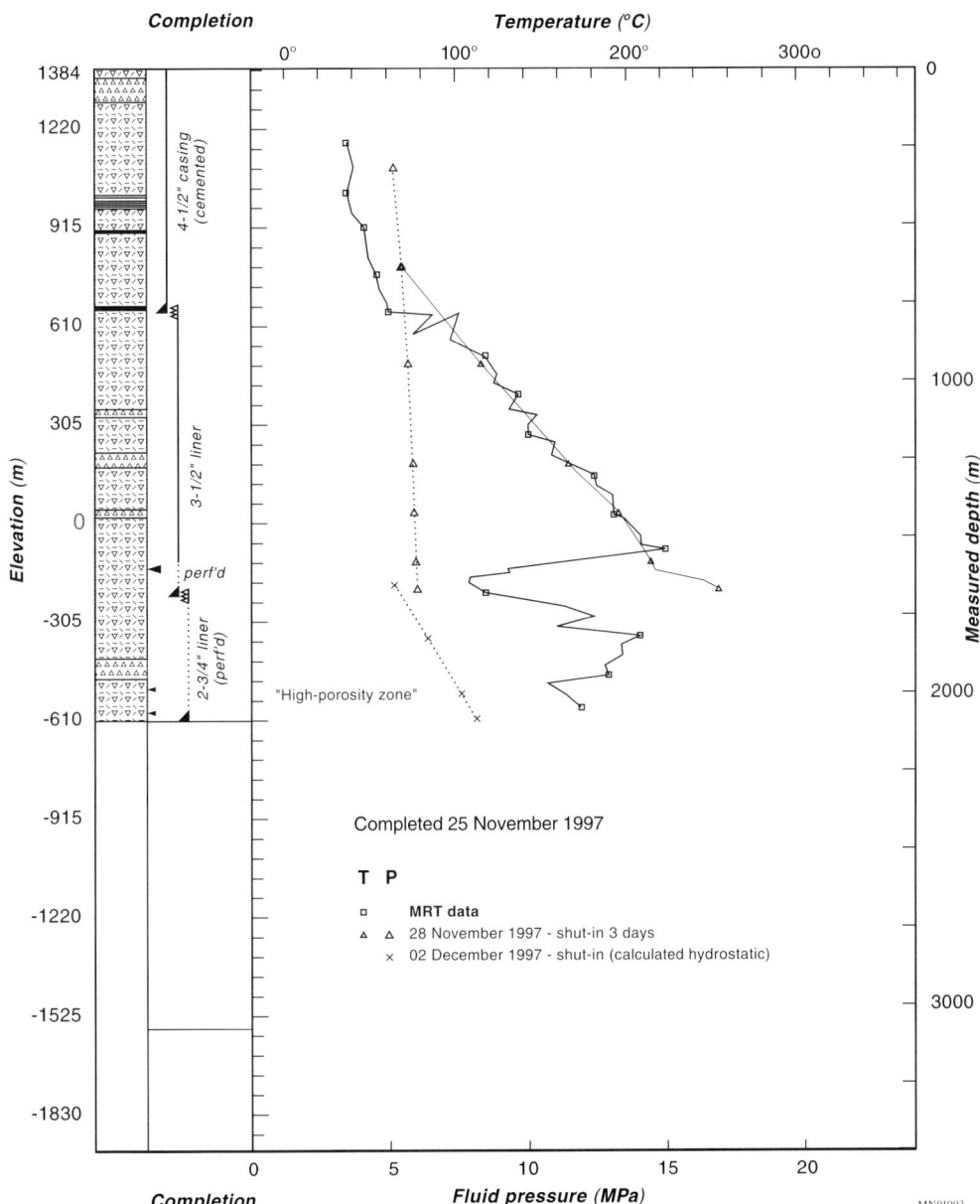

Fig. 1. continued

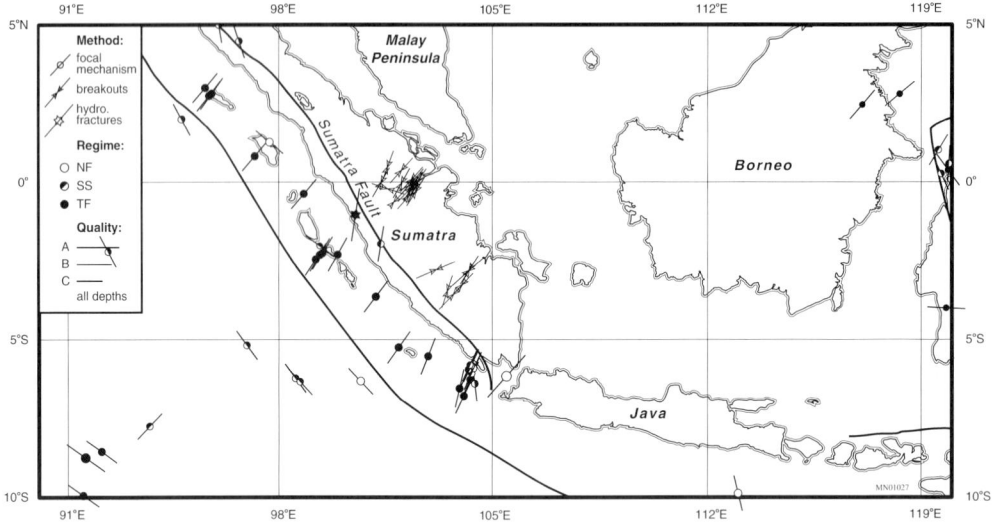

Fig. 2. *In situ* stress data from the Java region taken from the World Stress Map database (modified from Mueller *et al.* 1997). Data are based on focal-mechanism solutions, borehole breakouts and analysis of hydrofracturing results. In the case of focal-mechanism solutions, normal (NF), strike-slip (SS) and thrust (TF) faulting regimes are defined. Data quality, decreasing with the line length, is divided into three categories.

Detailed mineralogical, petrological and fluid-inclusion measurements have been made on coreholes K-33, T-8 and T-2 as part of an on-going study. These measurements provide insight into the sequence of mineralizing events (Allis & Moore 2000; Moore *et al.* 2000, 2001, 2002) and the relative timing of the associated fracturing.

Data

Fracture data from corehole K-33

K-33 was continuously cored from 754.3 to 2018.6 m with excellent recovery. The well encountered the top of the vapour-dominated reservoir at a depth of 1538 m. Temperatures within the reservoir section of the well range from 220 to 256 °C at total depth.

Fracture types occurring in K-33 are listed in Table 1. Normal faults, tensile fractures, sinistral and dextral strike-slip faults, and sinistral and dextral oblique-slip faults with a normal component of displacement are present within the cap rock. They have been controlled by the strike-slip stress regime with horizontal maximum and minimum compressional principal stresses, σ_1 and σ_3, and vertical intermediate stress, σ_2. Trends and plunges of principal stresses determined from well KRH 3–1ST, located just 3.4 km to the NW, are: $\sigma_1 = 188°/0°$, $\sigma_2 = 0°/90°$ and $\sigma_3 = 98°/0°$.

Strike-slip faults are totally missing in the reser-

voir. Instead, normal faults and tensile fractures are characteristic of the reservoir. They have been controlled by vertical σ_1, and horizontal σ_2 and σ_3. Stress trends and plunges are $\sigma_1 = 0°/90°$, $\sigma_2 = 188°/0°$ and $\sigma_3 = 98°/0°$.

The location of the deepest strike-slip fracture indicates that a boundary between the cap rock strike-slip and reservoir extensional regimes lies at a depth of about 1625 m. Magnitudes of the maximum principal compressional stress σ_1 and intermediate principal compressional stress σ_2 are equal at this depth. This is also the depth where the horizontal σ_1 characteristic of the cap rock changes to a vertical σ_1, which is characteristic for the reservoir.

Magnitudes of vertical σ_2 in the cap rock and vertical σ_1 in the reservoir are given by the relationship:

$$\sigma_V = \rho g h \tag{1}$$

where ρ is the density of overburden, g is the acceleration of gravity and h is the thickness of overburden. The change in horizontal σ_1 and σ_3 within the cap rock and horizontal σ_2 and σ_3 within the reservoir can be approximated by:

$$\Delta\sigma_H = \nu(\rho g \Delta h(1 = \nu)) \tag{2}$$

where ν is Poisson's ratio. This provides a maximum stress value because the fluid pressure can take a proportion of the overburden load (e.g. Rice & Cleary 1976; Mandl 1988; Engelder & Fischer 1994, and references therein).

Table 1. *Fracture types in corehole K-33*

Dip	s cr	s r	s Ttl	os cr	os r	os Ttl	d cr	d r	d Ttl	od cr	od r	od Ttl	n cr	n r	n Ttl	on cr	on r	on Ttl	t cr	t r	t Ttl	D
0	1		1																4	5	9	
5																					0	
10													1		1				2		2	
15																			2		2	
20	1		1	1		1							2		2				1	1	2	
25	2		2			2	2						2		2				5	3	8	
30			1	1	1		1	1				1	2		2	1		1	4	2	6	
35	1		1										3		3				4	3	7	
40	3		3			2	2	1				1	5	1	6	1	1	2	29	5	34	
45	7		7		1		1	1	1				6	1	7	1		1	14	5	19	1
50	7		7	3	1		4	2	2				9	4	13	1	2	3	10	9	19	
55	2		2	2	1		3	3	3				6	1	7	3		3	11	5	16	
60	2		2	2			2	1	1	1		1	8	1	9	3		3	11	10	21	2
65	5		5	1			1	3	3				10		10	3	1	4	18	10	28	
70	4		4	2		2	1	1				1	5	1	6	7	1	8	17	6	23	
75	6		6	1			1	3	3				2		2				12	10	22	
80	5		5	4		4	2		2	1		1	4	1	5				2	5	7	
85	2		2	4		4	4		4	2		2	5		5	3		3	38	27	65	
			48			24			25			7			80			28			290	

The number indicates the total number of fractures for each category.
Abbreviations: cr, cap rock; d, dextral strike-slip; D, dyke; n, normal fault; od, oblique dextral strike-slip; on, oblique normal fault; os, oblique sinistral strike-slip; r, reservoir; s, sinistral strike-slip; t, tensile fracture; Ttl, total amount.

These equations allow us to estimate the ratios of magnitudes of principal stresses upwards for the cap rock section, and downwards for the reservoir section from a depth of 1625 m in corehole K-33, where σ_1 equals σ_2 in magnitude. The stress estimate above this depth indicates a strike-slip stress regime with $\sigma_1 \geq \sigma_2 \gg \sigma_3$ in the lower portion of the cap rock as σ_1 magnitude approaches the σ_2 magnitude towards the cap rock–reservoir boundary. The boundary between the strike-slip stress regime and normal fault regime, roughly following the cap rock–reservoir boundary, is indicated by the presence of strike-slip fractures in the cap rock of the corehole K-33, and tensile fractures and normal faults in its reservoir section (Table 1).

Owing to the addition of the overburden load with depth, the strike-slip stress regime typical for the cap rock progressively changes with depth to the extensional stress regime characteristic of the reservoir. Within the reservoir, the σ_1 magnitude departs from the σ_2 magnitude with increasing depth towards $\sigma_1 \gg \sigma_2 \geq \sigma_3$ in the deepest parts of the reservoir.

None of the shear fractures observed in corehole K-33 (Table 1) displays any cross-cutting slip vectors. There is no evidence for the reactivation of tensile fractures as younger shear fractures, or for later tensile reactivation of earlier shear fractures. These observations indicate that the stress field did not change during their propagation; otherwise we would expect to see cross-cutting relationships on some of the more than 1000 fracture walls studied. This observation further supports the conclusion, derived from plate reconstructions of Hall (1997), that the local stress field in the Karaha–Telaga Bodas geothermal field was relatively stable during its development.

The temperature distribution in K-33 (Fig. 1c) indicates that faults and fractures in the cap rock section are sealed and poorly connected, because the dominant mode of heat transfer is by conduction. Figure 1c shows an efficient convective heat transfer in the reservoir section, suggesting that fluid flow occurs within a relatively well-connected fault–fracture system. The temperature–depth curve in the reservoir is perturbed at points where cold drilling fluid enters permeable zones.

Fracture data from well KRH 3–1ST

Well KRH 3–1ST has a measured depth of 3078 m. The top of the reservoir is encountered at a depth of about 1996 m and a temperature of 225 °C. The bottom hole temperature is 334 °C. An EMI log was obtained from 2117 to 3016 m.

The log shows that numerous normal faults/ tensile fractures and dextral strike-slip faults have strikes of 350°–30° and 120°–140°, respectively. The less numerous sinistral oblique-slip, sinistral strike-slip

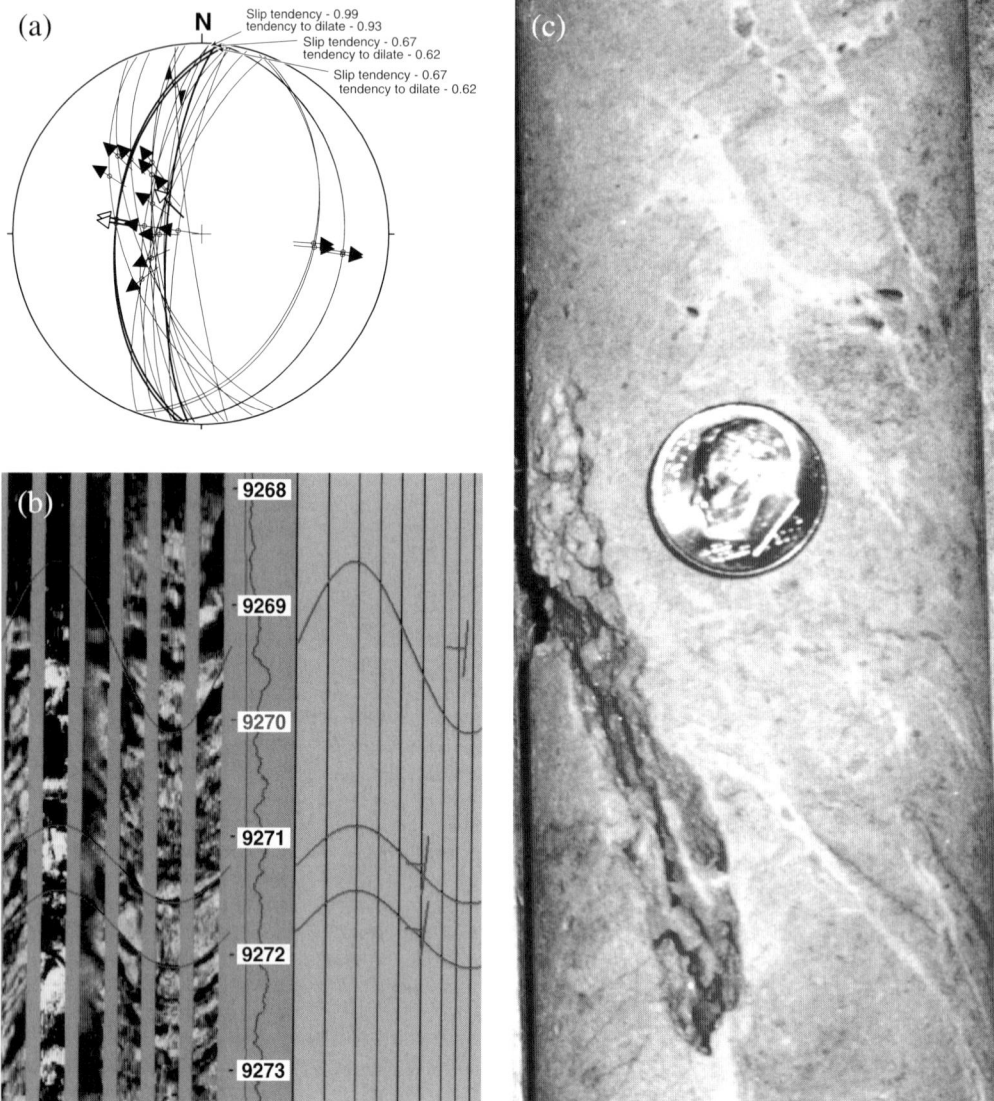

Fig. 3. (**a**) Great circle diagram of shear fracture types in the reservoir depth interval of 2816.1–2858.8 m (9233–9373 ft) in well KRH 3–1ST. Fractures at major fluid entry, documented also by FMS image (b), are in bold. Fracture density in fluid entry surroundings is five fractures per 10 m. Slip tendency is calculated from τ/σ_n, where τ is the shear stress and σ_n is the normal stress. Dilation tendency is calculated from $(\sigma_1 - \sigma_n)/(\sigma_1 - \sigma_3)$, where σ_1 and σ_3 are maximum and minimum principal compressional stresses. Fractures that are very close to being reactivated within the *in situ* stress regime have a slip tendency value near 1, whereas fractures that are unlikely to slip have a value near 0. The maximum and minimum tendency to dilate during slip, denoted by the 'tendency to dilate', is represented by values of 1 and 0, respectively. (**b**) Black and white areas in FMS image indicate conductivity and resistivity, respectively. Black spots located along productive fractures indicate presence of open cavities filled by fluid in dilatant regions. Example of such a fracture from a reservoir section of the corehole K-33 is shown in (**c**).

and dextral oblique-slip faults strike 30°–40°, 50°–70° and 140°–170°, respectively. There are few resistive fracture representations in well KRH 3-1ST. Electrically conductive fracture representations are dominant.

Resistive fracture representations are always associated with non-producing fractures. Their resistive response in the image log is caused by resistive fracture fill, which is interpreted as calcite or quartz, based on a comparison with fills observed in core from K-33 and their known electric representations (e.g. Adams & Dart 1998).

Conductive fracture representations are associated with both non-producing and producing fractures. EMI representations of some non-productive fractures have apertures at the threshold of detection, i.e. 1 cm; these representations contain large, highly conductive spots that locally coalesce into more extensive patches. Large patches of conductive spots also characterize fractures at minor fluid entries, while major fluid entries are characterized by an increase in the complexity and density of conductive spots within the fractured interval. For example, the densely spotted zone associated with the fluid entry at 2379.3–2380.5 m and the major fluid entry at 2343.3–2344.2 m are imaged as a plethora of fractures that include larger cavities. These cavities have a conductive representation when filled by fluid. Distinct fluid entries can sometimes be located along relatively small fractures, such as the entry at 2827–2828 m (Fig. 3a, b). Some productive fracture images contain sharply defined, highly conductive pull-aparts along the fracture planes. Fractures at all fluid entries, irrespective of fracture densities and the presence or absence of larger cavities, are all characterized by relatively small average apertures. EMI data from KRH 3–1ST also contain broad, highly conductive features that are not related to fluid entries. Their thicknesses range from 0.3 to 1.5 m. These fracture representations always contain numerous resistive spots related to host-rock clasts inside fault cores.

The best productive fractures are tensile fractures and normal faults with strikes typically roughly perpendicular to the minimum compressive stress σ_3, which trends 98°, but exceptions are possible. Such an exception is the fluid entry at 2032 m in KRH 2-1OH related to a fault with a 4.6 m-thick fault core. This dextral oblique-slip fault has a dip direction and dip of 325° and 89°, respectively, and therefore is not 'ideally' oriented. 'Ideally' oriented productive fractures, which have strikes around 8° and steep dips, have a tendency to achieve maximum possible apertures in the modern stress regime.

Calculation of the normal and shear stresses acting upon the fractures, using the equations of Wallace (1951) and Bott (1959), allowed determination of the tendency of the fracture to slip or dilate under the present-day stress regime (Fig. 3a). This calculation shows that among all fractures the majority of productive fractures are most prone to slip and/or dilation under the modern stress. It indicates that the fluid circulation in KRH 3–1ST is clearly affected by the present-day stress regime. All studied productive fractures are located in massive lava and pyroclastic flows, which are more than several metres thick.

The effects of the present-day stress field on reservoir behaviour are illustrated by the fluid entry at 2816.1–2858.8 m (Fig. 3a), which shows the favourable slip and dilation tendencies of productive fractures. The fracture density at this fluid entry is no higher than the average density of five fractures per 10 m determined for the logged portion of the reservoir. Similar fracture densities, no higher than the average, are associated with other production zones. This observation indicates that fracture density cannot be used for distinguishing productive zones from non-productive portions of the reservoir.

The EMI log also indicates that some fluid entries are not associated with fracture zones. These entries are located in moderately conductive, layered, rock sequences, representing epiclastic rocks, tuffs or sedimentary units. A lack of fractures at these entries suggests that matrix permeability controls the fluid flow in these cases.

Fracture timing

Mineralogical relationships. Mineralogical studies of the core and cuttings samples record three major stages in the evolution of the modern geothermal system at Karaha–Telaga Bodas (Moore *et al.* 2001, 2002, 2004). The mineralogical characteristics of these stages are summarized in Table 2.

The earliest hydrothermal activity is represented by pervasive argillic alteration and silicification of the rocks and minor veining. A variety of clay minerals and sheet silicates including smectite, illite–smectite, chlorite–smectite, illite and chlorite were deposited in the veins and wall rocks. These minerals are characteristic of low–moderate temperatures below 250 °C (Henley & Ellis 1983). Younger veins define two parageneses. At depths above 850 m, the veins are dominated by calcite, chlorite, pyrite, or haematite. At greater depths, the veins contain epidote, albite, amphibole and Fe–Cu sulphides. In the deepest explored parts of the system, biotite and clinopyroxene are also present. Amphibole is an important index mineral, occurring in active geothermal systems only at temperatures above 300 °C. In the southern part of the Karaha–Telaga Bodas field, this mineral is locally abundant below 950 m. The widespread occurrence and mineral parageneses of these veins demonstrate that the early geothermal system was liquid dominated and that temperatures were higher in the past than they are today.

Silica deposited as chalcedony and saline fluid inclusions in quartz record the transition from a liquid-dominated stage to a vapour-dominated stage. In the shallow veins, chalcedony replaces the carbonates and fills open spaces. In the deeper veins, chalcedony post-dates the formation of epidote and amphibole (Table 2), occurring mainly as cores of euhedral quartz crystals. Evidence of chalcedony

Table 2. *Relative timing of mineralization events with approximate location*

	Time				
	Penetrative	Early liquid-dominated veins	High-temperature liquid-dominated veins	Transitional veins	Vapour-dominated veins
Near surface	argillic alt. silicification	sericite/chlorite			advanced argillic
Shallow	argillic alt. silicification	sericite/chlorite	chlorite + pyrite calcite/haematite	chalcedony/quartz	calcite + anhydrite
Deep	argillic alt. silicification	quartz sericite/chlorite	epidote + albite + pyrite ± actinolite ± biotite ± clinopyroxene	chalcedony/quartz	anhydrite + pyrite ± wairakite ± calcite ± fluorite

deposition has been found throughout much of the modern geothermal reservoir.

Quartz crystals associated with chalcedony display unusual growth forms. Twinning, curved '*c*'-axes and epitaxial growth are common. No workable fluid inclusions were found in chalcedony. However, fluid inclusions in quartz from depths where chalcedony is present yielded average temperatures of 234 °C, based on 95 inclusions from a depth of 793 m in corehole T-2, and 315 °C, based on 72 inclusions from a depth of 1203 m in corehole T-8. These data and the absence of intervening low-temperature mineral assemblages suggest that chalcedony formed at temperatures above 180 °C. Fournier (1985) has shown that the formation of chalcedony at these temperatures requires extreme supersaturation of silica and rapid decompression of the hydrothermal fluids.

Vapour-rich fluid inclusions, indicative of boiling, are dominant in many of the quartz crystals. Liquid-rich inclusions have yielded homogenization temperatures up to about 350 °C and salinities ranging mainly from 3 wt% NaCl eq. to 24 wt% $NaCl/CaCl_2$ eq. in coreholes T-2 and T-8. Moore *et al.* (2002) have suggested that the high salinities are the result of extreme boiling and concentration of the hydrothermal fluids. In contrast, lower salinity fluids with less than 1–2 wt% NaCl eq. are common in corehole K-33. These fluids are interpreted as steam condensate.

The age of depressurization is constrained by ^{14}C dating of lake beds from a depth of 978 m in corehole T-8 (Moore *et al.* 2002). These sedimentary deposits predate the formation of amphiboles, epidote and chalcedony. Organic carbon in the lake beds has yielded an age of 5910 ± 76 years. Thus, the data suggest that the depressurization is related to an event that is younger than 5910 years old.

The deposition of chalcedony and quartz was followed by anhydrite, calcite, fluorite, pyrite and wairakite (Table 2). Anhydrite, calcite and fluorite have

retrograde solubilities and deposit as fluids are heated. The simple parageneses of these veins suggest that mineral deposition occurred in response to the heating of downward-percolating condensate. Fluid inclusions trapped within these minerals record the evolution of the fluids. At shallow depths in T-2, the downward heating is represented by increasing homogenization temperatures from 160 to 225 °C and decreasing salinities due to mineral deposition. At greater depths in T-8, where fluid inclusion homogenization temperatures exceed 225 °C, the apparent salinities increase with temperature, reflecting the boiling off of the fluids. Fluids with salinities as high as 31 wt% NaCl eq. were trapped in anhydrite. Scales of halite, sylvite and Fe-chlorides on the hydrothermal vein minerals demonstrate that the condensate eventually boils off completely within the modern vapour-dominated portion of the system. Thus, the fluid inclusion and mineral data document the drying out of the early liquid-dominated geothermal system.

The low salinities of 1–2 wt% of the present-day reservoir fluids indicate that meteoric waters are presently recharging the geothermal system. However, the occurrence of high $^3He/^4He$ ratios from 7.1 to 7.7 *R*a throughout the field and large oxygen isotope shifts of the deep liquids suggest the presence of a magmatic contribution to the fluids (Powell *et al.* 2001).

Fractures active during the liquid-dominated stage of geothermal activity. Fractures active during the liquid-dominated stage were coeval with penetrative argillic alteration and silicification. The liquid-dominated system was capped by the upper parts of the stratovolcanic complex, which were affected by argillic alteration. Detailed logging of K-33 indicates that andesite lava flows in the shallow portion of the system have not been extensively fractured, perhaps due to the insulating effect of the intervening altered

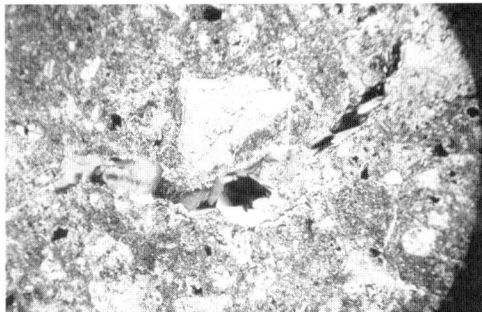

Fig. 4. Thin section documenting a penetrative argillic alteration followed by quartz veining.

pyroclastic and epiclastic rocks, which absorbed the deformation. The end result of the argillic alteration in the more porous parts of the section was a significant reduction in the porosity of the cap rock (Fig. 4). Although lava flows in the cap rock were locally fractured, low overall permeabilities are demonstrated by high-temperature gradients indicative of conductive heat transfer (Fig. 1c).

Penetrative fluid events have changed the initial rheological properties of the geothermal system. For example, core from K-33 documents cases of initially low porosity lava and pyroclastic flows of the reservoir that underwent a distinct porosity increase through dissolution of feldspars and only partial fill by later epidote (Fig. 5a). Younger shear fractures in these altered rocks resemble deformation of coarse-grained sandstone (e.g. Antonellini & Aydin 1994). Rocks are not deformed by discrete fractures, which are typical of unaltered lava and pyroclastic flows. They are deformed by densely spaced shear bands.

Later fractures were synchronous with fluid flow. These fluids mainly affected the walls of the fractures with only minor penetration of the wall rocks. The majority of the observed fractures indicates that fluid-assisted failure was dominant. Local exceptions indicate that failure caused by cooling-related stress release was possible (Fig. 6a).

Subhorizontal tensile fractures, opened against the overburden weight, anastomosing tensile fracture patterns, and implosion breccias in the quartz and calcite vein fills are frequent (Fig. 6b). Twenty-three subhorizontal tensile fractures, eight anastomosing tensile fracture patterns and 18 implosion breccias were observed in K-33. They document numerous episodes when local fluid pressures exceeded hydrostatic pressure.

The geothermal system is cut by several faults. Fault cores are usually thinner than 2 m, although one with a thickness of 4.6 m was observed in KRH 2-1OH. Their damage zones are a few times wider than the cores. Fault cores are characterized by

Fig. 5. Example of the penetrative dissolution of feldspars from the pyroclastic flow at a depth of 1750.7 m in the corehole K-33. Note that matrix around the less permeable clasts is preferentially affected. Holes after feldspars are just barely lined by epidote, which did not destroy an enhanced porosity.

cemented or uncemented fault breccia and gouge, which are developed by competing cataclastic flow and fluid-assisted deformation mechanisms. The cataclasis is indicated by frictional wear of clasts (Fig. 6c). Fluid-assisted deformation is indicated by zoned fractures within the damage zones (Fig. 6d). Adjacent parts of the host rock are frequently altered or cemented. There is evidence for hydraulic fracturing that reopened fractures previously cemented by quartz, and for younger cementation by quartz (Fig.

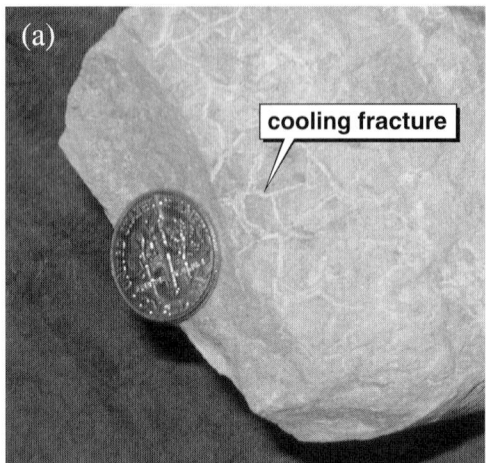

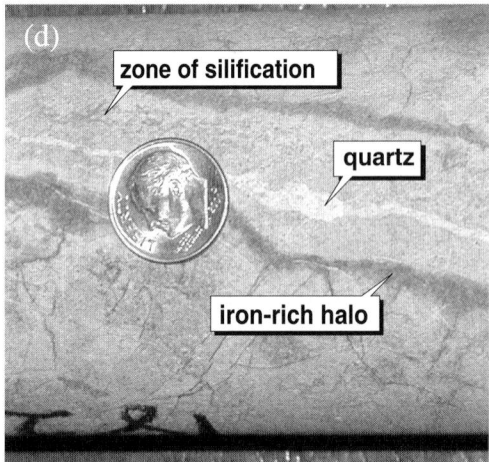

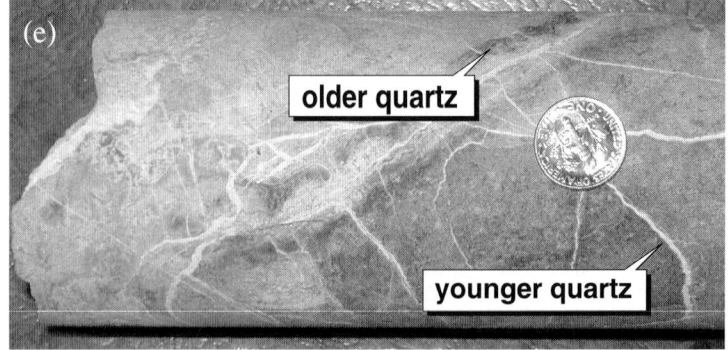

6e). Figure 6e shows how the increased pressure of the fluid trapped in the fracture sometimes resulted in small hydraulic fractures propagating away from this fracture into the adjacent host rock in a relatively random fashion, generating temporary storage for fluids, before the main fracture ruptured.

Fractures active during the transition from a liquid-to vapour-dominated stage. Fractures active during this stage are coeval with chalcedony and quartz deposition. They are present at all depths. The chalcedony and associated quartz are found only in tensile fractures and shear fractures with normal displacement. These fractures are characterized by large apertures and steep dips. They were highly prone to slip and dilation. As discussed later, the deposition of chalcedony in these fractures is significant because the chalcedony occupies a unique paragenetic position within the evolution of the hydrothermal system. This position can be correlated across the field. Chalcedony has been found within all of the core-holes over a strike length of over 6 km, and to depths in excess of 2 km. In the shallow portion of the geothermal system, the chalcedony post-dates the calcite associated with the earlier liquid-dominated stage. At greater depths the chalcedony post-dates epidote, amphibole and pyrite deposited by higher temperature liquids (Fig. 7).

Fractures associated with the modern geothermal system. Fractures that were propagated or active after the deposition of chalcedony and quartz are filled by anhydrite, calcite, pyrite and fluorite (Fig. 8a–d).

Fig. 8a shows anhydrite post-dating earlier mineralization in the cap rock. The anhydrite completely fills the only remaining space in a dilatant region of a strike-slip fracture. Figure 8b, c shows anhydrite and calcite that post-date earlier mineralization in the cap rock and reservoir. The calcite completely seals the residual aperture of the pre-existing fracture in Figure 8c. In contrast, the fracture in Figure 8b is only

Fig. 7. Thin section with a chalcedonic vein in the older actinolite, epidote and pyrite paragenesis within the reservoir at a depth of 1139.5 m in the corehole T-8 (Moore *et al.*, 2000).

partially sealed by the anhydrite. Figure 8d demonstrates that this stage of calcite deposition post-dates chalcedony in the uppermost part of the reservoir.

The final stage of mineral deposition in these fractures is represented by precipitates of Na-, K- and Fe-chlorides on the mineral surfaces. The high solubilities of these precipitates indicate that no mobile water is present and that essentially complete dry-out of the fractures within the vapour-dominated portion of the system has occurred (Fig. 9). These fractures contain frequent voids. Their apertures are supported by cement bridges, host-rock asperities and propping breccias.

Fig. 6. (**a**) Fracturing related to the release of the residual stress accumulated by cooling of the andesite above the cooling granodiorite intrusion at a depth of 367.6 m in the corehole T-2. Fractures are developed in the altered andesite and filled by quartz. (**b**) Implosion breccia with altered andesite inside a fine calcite matrix at a depth of 1653 m in the corehole K-33. The implosion character can be documented by the fact that boundaries of the neighbour clasts match together. Facts such that there is no deposition of denser clasts on the bottom of the vein and that clasts are not even a little gravitationally sorted indicate a high viscosity of the initial fluid. Such an implosion breccia originates when a rupturation connects areas with higher and lower pore fluid pressure. A pore pressure gradient between such areas is quickly balanced by a rapid fluid flow, and a viscous fluid 'freezes' below the certain threshold flow rate and preserves a pocket of the implosion breccia. (**c**) Fault core from a depth of 1545 m in the corehole K-33 with fault breccia and gouge. The gouge is made by size reduction of andesitic clasts and reaction with fluids rich in silica. Clasts of different sizes are progressively more rounded with smaller clast size, indicating frictional wear. (**d**) Example of the single fracture from a depth of 1827.5 m in the corehole T-2. The fracture is filled by quartz and surrounded by the zone of silicification and iron-rich halo, documenting that silica-rich fluids flowing along the fracture also altered the adjacent portion of the host rock. (**e**) Earlier fracture reopened by hydraulic fracturing from a depth of 389 m in the corehole T-2. Earlier fracture was filled by grey quartz, reopened by hydraulic fracturing and filled by white quartz. Note the small hydraulic fractures filled by white quartz fed by the main hydraulic fracture. They are propagated in a seemingly random manner into the adjacent host rock, generating storage for overpressured fluid.

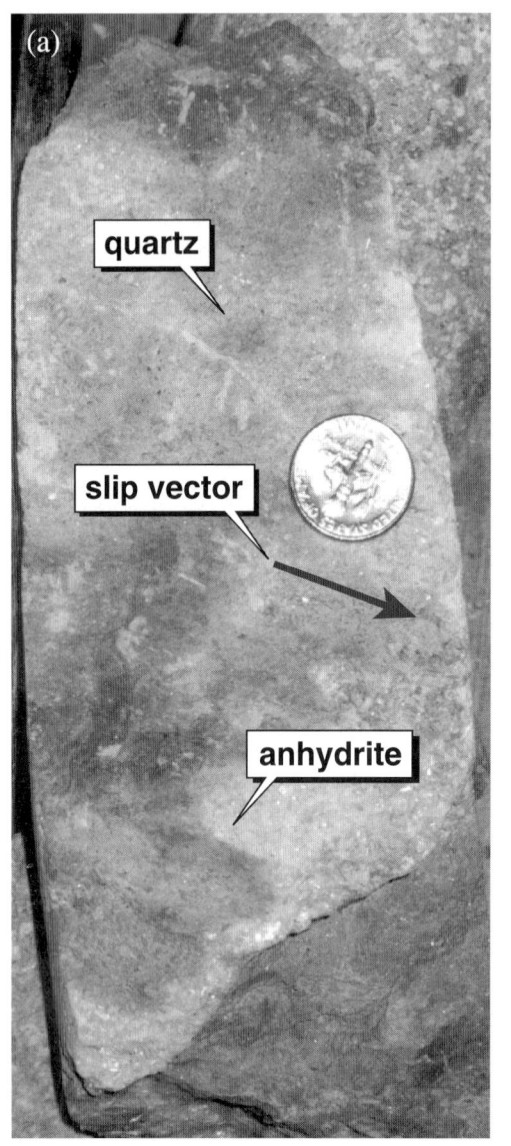

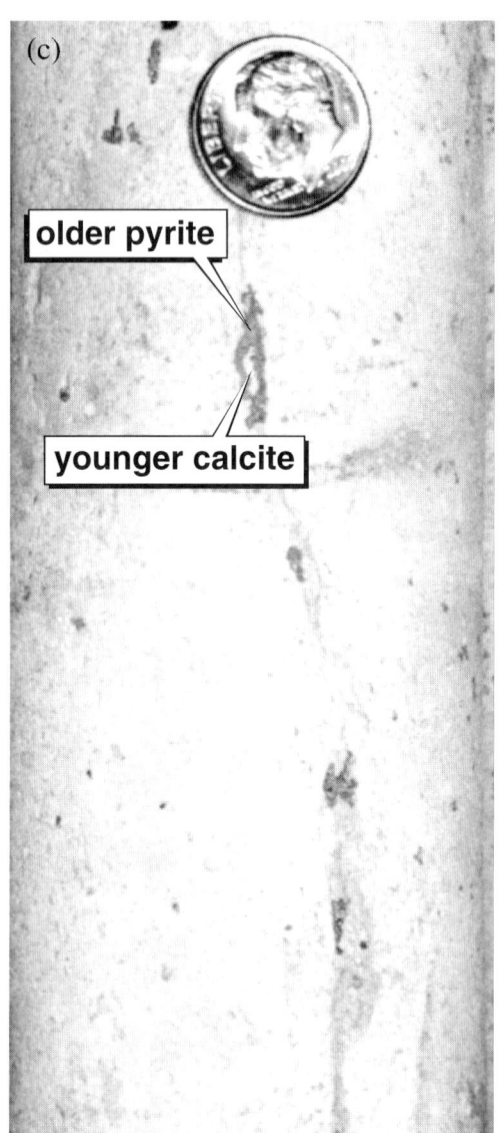

Fig. 8. (**a**) Core from a depth of 1308.8 m in the corehole K-33 showing the sinistral strike-slip fracture. The slip vector is defined by quartz. A silicified clay zone surrounds the fracture. Remnant dilatant regions of the fracture are filled by younger idiomorphic anhydrite precipitated from downward percolating condensate. (**b**) Core from a depth of 305.5 m in the corehole T-8 showing the tensile fracture. Fracture is lined by idiomorphic crystals of the older high-temperature epidote–chlorite–pyrite suite precipitated into the open space. A fractal of the remnant space was used for the precipitation of just one nest of the idiomorphic anhydrite. The remaining space remained open because there was no more condensate to feed the continuing filling of the space left. (**c**) Core from a depth of 2017.4 m in the corehole K-33 with a normal fracture with several dilatant regions. Dilatant regions are lined with the earlier pyrite and sealed by the younger calcite that precipitated from downward percolating condensate. (**d**) Scanning electron microscope (SEM) backscattered electron image with a vein filled by the earlier chalcedonic quartz and younger calcite from a depth of 794 m in the corehole T-2. Calcite fills the remaining space in the vein.

Fig. 9. SEM backscattered electron image of the vein from a depth of 1044.9 m in the corehole T-8 (Moore *et al.* 2000). Image shows the anhydrite crystals post-dating actinolite crystals. Subhorizontally oriented anhydrite crystal is coated with peeling-off titanium-rich scale, which indicates extremely dry conditions.

Fracture development within the geothermal system – interpretation and discussion

Liquid-dominated stage

Fractures of this stage are either newly formed or reactivated pre-existing fractures developed in the country rock. Our data (Table 1, Fig. 6b, 6e) suggest that these fractures were controlled by the stress regime, which was a result of plate movements, gravity forces and pore fluid pressures. The stress regime, with horizontal σ_1 and σ_3 stresses trending roughly N–S and E–W, respectively, controlled shear and tensile fractures in the shallow portion of the system. In the deeper portion of the system, σ_1 was vertical, whereas σ_3 was horizontal and roughly

E–W trending. Cycles of fluid overpressure are indicated by implosion breccias (Fig. 6b), subhorizontal tensile fractures and anastomosing patterns of hydraulic fractures (Fig. 6e). Subhorizontal tensile fractures indicate that the fluid pressure was approaching the value of the effective overburden load. Implosion breccias indicate rapid bursts of fluid flow immediately post-failure, balancing fluid pressure gradients among various compartments of the fracture system. Random orientations of anastomosing tensile fracture patterns document fluid-assisted failure.

High pore fluid pressures, characteristic of this stage, were driven by the field-scale and local mechanisms. The field-scale mechanism involves a combination of a magmatic fluid source below the reservoir section and an effective cap rock above. Although the granodiorite underlying the geothermal system has not been dated directly, several lines of evidence indicate that the intrusion played a key role in the development of the liquid-dominated system. These include: (1) the presence of amphibole and epidote, which indicate temperatures in excess of 300 °C in rocks younger than 6000 years BP; (2) the simple vein parageneses suggesting a single major heating event; and (3) the parallelism of the isotherms with the intrusion geometry. Taken together, these relationships suggest that the initiation of the liquid-dominated system and the emplacement of the granodiorite occurred during the last 6000 years.

Magmatic fluids and gases can accumulate inside the magma column by a combination of the diffusion-driven fractionation (McBirney 1995 and references therein), compaction-driven fractionation (Shirley 1986, 1987; Sparks *et al.* 1985; Ortoleva *et al.* 1987)

and convection-driven fractionation (Tait *et al.* 1984; Morse 1986; Tait & Kerr 1987; Shinohara *et al.* 1995). Fluids released from magma can have a volume much larger than the volume of the intrusion itself. These volume relationships are demonstrated, for example, by mass-balance calculations made for the Pine Grove ore deposit, Colorado (Keith & Shanks 1988).

Our observations indicate that the cap rock initially consisted of permeable tuffs, pyroclastic and epiclastic horizons intercalated with impermeable lava flows.

The tuffs could have had initial porosities as high as 22%, as indicated by rock data compiled by Lama & Vutukuri (1978). This would have made them good reservoir rocks prior to their alteration and compaction. We do not have porosity data for the pyroclastic and epiclastic rocks, but their densities of approximately 2360 kg m^{-3} compared with lava flow densities of about 2430 kg m^{-3} indicate higher porosities than those of andesitic lava flows. As with the tuffs, the pyroclastic and epiclastic rocks were prone to penetrative shallow argillic alteration, and later compaction, which progressively changed them into effective seals during the liquid-dominated stage.

Unfractured brittle lava flows are sandwiched among the less competent tuffs, pyroclastic flows and epiclastic deposits. The lava flows initially formed seals inside the porous horizons. These flows show little alteration, indicating that the matrix porosity was too low to host the early penetrative fluids. Rock mechanics analogues (e.g. Lama & Vutukuri 1978) indicate an initial porosity of 8% or less for andesitic lava flows. The surrounding porous horizons absorbed the deformation, allowing lava flows to escape fracturing and retain their sealing properties.

Local-scale mechanisms capable of generating high fluid pressures include (e.g. Fournier 1967, 1999; Cunningham 1978; Byerlee 1993):

• fault self-sealing due to cementation combined with compaction, resulting in overpressured compartments;
• development of fractured zones sealed by incompetent horizons with porosity destroyed by alteration;
• boiling of upward migrating fluids, resulting in rapid volume expansion.

Observed textural relationships indicate that all three mechanisms operated. Evidence for the first mechanism is documented in Figure 6e. Here, fracture sealing by quartz cementation was followed by re-opening of the fracture and re-cementation by quartz. Self-sealing combined with compaction could have occurred in faults with horizontal component of displacement. Their cores are formed by

tectonic gouge produced by cataclasis (Figure 6c). Conductive electric images of some of the cores (Christensen *et al.* 2002) indicate fluid saturation. Their cores are locally silicified and have rather irregular geometries. These complex geometries can result in the occurrence of local restraining bends, which lead to compaction, and to local releasing bends, which lead to dilation (e.g. Nur & Booker 1972; Segall & Pollard 1980; Sibson 1985, 1989). The presence of different deformation domains within the observed fault zones could have produced a complex system of pore fluid pressure gradients before faulting and their rapid equilibration during faulting (e.g. Muir Wood & King 1993; Muir Wood 1994).

There are frequent instances for the second mechanism in corehole K-33. The presence of incompetent horizons with porosity destroyed by alteration is shown in Figure 4. These horizons are significantly less affected by fracturing than the intercalated layers of brittle lava flows. They act as effective seals among the more fractured horizons, separating the rock section into numerous compartments with different pore fluid pressures. A similar behaviour occurs in multilayer sedimentary sequences (e.g. Nemčok *et al.* 1995, 2002).

There is abundant evidence for boiling. The core material from K-33 reveals that numerous fractures contain blade-shaped calcite crystals, which indicate boiling. The separation of the vapour phase during boiling results in rapid volume expansion.

Transition to vapour-dominated stage

The transition to vapour-dominated conditions was marked in the fractures by the deposition of chalcedony (Table 2). The presence of chalcedony throughout the reservoir at temperatures exceeding 235 °C implies extreme supersaturation of silica with respect to quartz. As discussed by Fournier (1985), catastrophic depressurization and boiling is the most likely cause of silica supersaturation at these temperatures. In T-8, rocks containing chalcedony overlie organic-bearing lake beds dated at 5910 ± 76 BP (Moore *et al.* 2002). However, no massive volcanic eruptions that could have depressurized the system are known to have occurred during the last 6000 years and the absence of older episodes of chalcedony implies that earlier volcanic eruptions were also of insufficient magnitude to cause extensive silica supersaturation and widespread boiling. We therefore conclude that depressurization must have been caused by the flank collapse that led to the formation of the volcano's crater, Kawah Galunggung at 4200 years BP (Bronto 1989). We suggest that this collapse was triggered by a combination of increased pore fluid pressures and penetrative alteration, initiated during the liquid-dominated stage. Elevated N_2/Ar ratios of fluid inclusions trapped in quartz overprinting chalcedony in T-2, T-8 and K-33, combined with the high $^3He/^4He$ ratios of the modern gases and the oxygen isotope shifted compositions of the deep liquids, imply that the granodiorite continued to provide heat and fluids during the transition and later stages of the geothermal system's evolution.

Flank collapse must have been a major venting event for accumulated pore fluid pressures. The data described earlier demonstrate that chalcedony precipitation took part preferentially along steeply dipping fractures with the largest possible apertures. The orientations of these fractures indicate that the escaping overpressured fluids used the shortest possible paths to the surface. The choice of the shortest paths documents a rapid post-failure fluid redistribution.

The modern geothermal system

The fluid inclusion and mineral record demonstrates that vapour-dominated conditions were initiated following the transitional stage. As the vapour-dominated region expanded, the condensate that formed at the top of the steam zone percolated downward, modifying the cap rock–reservoir boundary. Anhydrite, calcite and pyrite were deposited as the fluids were heated. Figure 8a–d documents the downward shift of the cap rock–reservoir boundary caused by mineral deposition in the fractures. Although voids are present in the cap rock, thermal profiles (Fig. 1c) are indicative of conductive gradients and low overall permeabilities. The top of the present-day reservoir is indicated by the first occurrence of fractures with interconnected voids (Fig. 8b).

Some of the reservoir fractures have large cavities preserved in their dilatant areas (Figs 3c and 8b). Closed parts of reservoir fractures are formed by rock–rock or cemented rock–rock contacts, which support the open space among them against collapse. Locally, the open fractures are propped by breccias, which contain large voids among the clasts. Propping breccias support residual fracture apertures, but the average residual fracture porosity remains significant.

There is no evidence for pore fluid pressures equal or higher than hydrostatic in fractures within the vapour-dominated zone. Fractures related to the modern system do not contain implosion breccias or anastomosing offshoots. The fracturing in the geothermal system is now driven only by tectonic stresses and gravity. Pore fluid pressure is no longer capable of shifting the differential stress towards the failure envelope. Stress build-ups are driven only by

plate motions interplaying with overburden changes due to new volcanic eruptions or erosion.

The liquid reservoir underlying the vapour-dominated zone is underpressured by several tens of bars relative to the surrounding regional groundwater system (Allis *et al.* 2000). During the time between reservoir fracture events, the residual fracture apertures were supported only by rock asperities or propping breccias. There is no evidence for increased pore fluid pressure to support the fracture apertures. Therefore, we conclude that the base of the fractured reservoir is controlled by the value of the normal stress acting on fracture walls. If it exceeds the compressive strength of asperities or propping breccias, collapse of the fractures will occur.

Rock mechanics data (e.g. Lama & Vutukuri 1978) indicate that porosity and density of andesitic lava flows are about 8% and 2430 kg m^{-3}, respectively. The porosity of andesitic tuffs is higher at about 22% and their density slightly lower at about 2160 kg m^{-3}. There are no data on porosities of andesitic epiclastic and pyroclastic rocks, but their average density of about 2360 kg m^{-3} indicates a higher porosity than that of andesitic lava flows. The average density of a sandy tuff is 1910 kg m^{-3}, suggesting even higher porosities. These data are consistent with the observation that fluid entries in lava flows and pyroclastics are fracture controlled. In contrast, there is a lack of fractures associated with fluid entries in the more porous epiclastic and tuff deposits. Fluid entries in these zones are controlled by high matrix permeabilities. Fracture permeability thus competes with matrix permeability in the reservoir. It dominates in massive and low permeability volcanic sequences, whereas it is subordinate in the more porous layered volcanoclastic and sedimentary sections.

Conclusions

- The stress field during the development of the Karaha–Telaga Bodas geothermal system has remained relatively stable. Fractures in the cap rock have been controlled by a strike-slip stress regime. The orientations of the principal stresses are $\sigma_1 = 188°/0°$, $\sigma_2 = 0°/90°$ and $\sigma_3 = 98°/0°$. Fractures in the reservoir have been controlled by the extensional stress regime. The orientations of the principal stresses are $\sigma_1 = 0°/90°$, $\sigma_2 = 188°/0°$ and $\sigma_3 = 98°/0°$.
- The fracture system within the geothermal field consists mainly of normal faults and tensile fractures with strikes of 350°–30° and dextral strike-slip faults with strikes of 120°–140°. Other less numerous sets include sinistral oblique-slip faults with strikes of 30°–40°, sinistral strike-slip faults with strikes of 50°–70° and dextral

oblique-slip faults with strikes of 140°–170°. Productive fractures are steep and have strikes around 8°, although large productive faults with different orientations are present. Their size apparently makes up for their less favourable orientation.

- Conductive temperature profiles characterize the cap rock, whereas the reservoir is characterized by convective profiles. The cap rock consists of penetratively deformed horizons of incompetent rocks and competent rocks with poorly connected fracture systems. Sealed fractures are cemented by calcite, quartz and anhydrite. Productive fractures within the reservoir are characterized by large interconnected voids located in dilatant regions. These voids are supported by rock asperities, cement bridges or propping breccias.
- The geothermal system developed rapidly within the last 6000 years. Three stages in its development can be recognized.

The earliest stage was liquid-dominated. It is characterized by penetrative alteration that enhanced the cap rock. Fracturing was controlled by tectonic stress interacting with the overburden load and modified by pore fluid pressures. Cooling-related residual-stress development contributed to this control only locally. Pore fluid pressures above hydrostatic locally developed and were driven by interacting mechanisms such as fault–fracture compaction, fault–fracture cementation, and input of magmatic fluids and magmatic heating.

Stage 2 represents the transition from the liquid- to a vapour-dominated stage. It is characterized by a catastrophic drop in pore fluid pressure, triggered by a flank collapse of Galunggung volcano 4200 years ago. Chalcedony and quartz were deposited in fractures with the shortest possible path to the surface. Fracturing was strongly affected by pore fluid pressures.

Stage 3 represents the modern vapour- and deep liquid-dominated regimes. The liquid represents the inflow of low-salinity meteoric waters. Fracturing during this stage was driven by the tectonic stress interacting with the overburden load. There was no evidence for high pore fluid pressures. Precipitation from condensed vapour progressively sealed the fracture system downwards, resulting in expansion of the cap rock. Pore fluid pressures are below hydrostatic in the modern reservoir. Fractures have collapsed to their residual apertures, supported only by rock asperities, cement bridges and propping breccias. The normal stress acting upon the fracture walls controls the base of the productive reservoir. The fracture apertures collapse completely where normal stresses exceed the compressive strength of the host rock.

M. Nemčok and J. N. Moore were supported by DOE grants to EGI (01-0059-5000-55800164 and 01-0059-5000-55900160). The management and staff of the Karaha–Bodas Company are thanked for providing data in support of this research. Authors wish to thank D. Jensen for help with drafting, and K. M. Kovac and S. Moore for help with editing. The final version of the paper was considerably improved by constructive criticism from A. Gudmundsson.

References

ADAMS, J. T. & DART, C. 1998. The appearance of potential sealing faults on borehole images. *In*: JONES, G., FISHER, Q. J. & KNIPE, R. J. (eds) *Faulting, Fault Sealing and Fluid Flow in Hydrocarbon Reservoirs*. Geological Society, London, Special Publications, **147**, 71–86.

ALLIS, R. & MOORE, J. N. 2000. Evolution of volcano-hosted vapor-dominated geothermal systems. *Geothermal Resources Council Transactions*, **24**, 24–27.

ALLIS, R., MOORE, J.N., MCCULLOCH, J., PETTY, S. & DEROCHER, T., 2000. Karaha–Telaga Bodas, Indonesia: a partially vapor-dominated geothermal system. *Geothermal Resources Council Transactions*, **24**, 217–222.

ANTONELLINI, M. & AYDIN, A. 1994. Effect of faulting on fluid flow in porous sandstones: Petrophysical properties. *AAPG Bulletin*, **78**, 355–377.

BARTON, C. A., ZOBACK, M. D. & MOOS, D. 1995. Fluid flow along potentially active faults in crystalline rock. *Geology*, **23**, 683–686.

BAUMANN, P. 1982. Depositional cycles on magmatic and back arcs; an example from western Indonesia. *Revue de l'Institut Francais du Petrole*, **37**, 3–17.

BLANPIED, M. L., LOCKNER, D. A. & BYERLEE, J. D. 1995. Frictional slip of granite at hydrothermal conditions. *Journal of Geophysical Research*, **100**, 13045–13064.

BOTT, M. H. P. 1959. The mechanics of oblique slip faulting. *Geological Magazine*, **96**, 109–117.

BRANTLEY, S. & GLICKEN, H. 1986. Volcanic debris avalanches. *In*: *Earthquakes and Volcanoes*. US Geological Survey Report, EV, **5**, 195–206.

BRONTO, S. 1989. *Volcanic geology of Galunggung, West Java, Indonesia*. PhD thesis, University of Canterbury.

BUDHITRISNA, T. 1986. *Peta Geologi Lembar Tasikmalaya, Jawa Barat*, scale 1:100000. Pusat Penelitian Dan Pengembangan Geologi.

BYERLEE, J. 1993. Model for episodic flow of high-pressure water in fault zones before earthquakes. *Geology*, **21**, 303–306.

CAPPETTI, G., CELATI, R., CIGNI, U., SQUARCI, P., STEFANI, G. & TAFFI, L. 1985. Development of deep exploration in the geothermal areas of Tuscany, Italy. *In*: STONE, C. (ed.) *International Symposium on Geothermal Energy, International Volume*. Geothermal Resources Council, Davis, CA, 303–309.

CHRISTENSEN, C., NEMČOK, M., MCCULLOCH, J. & MOORE, J. 2002. The characteristics of productive zones in the Karaha–Telaga Bodas geothermal system.

Geothermal Resources Council Transactions, **26**, 623–626.

CUNNINGHAM, C. G. 1978. Pressure gradients and boiling mechanisms for localizing ore in porphyry systems. *US Geological Survey Journal of Research*, **6**, 745–754.

ENGELDER, T. & FISCHER, M. P. 1994. Influence of poroelastic behavior on the magnitude of minimum horizontal stress, S_h, in overpressured parts of sedimentary basins. *Geology*, **22**, 949–952.

ETHERIDGE, M. A. 1983. Differential stress magnitudes during regional deformation and metamorphism: Upper bounds imposed by tensile fracturing. *Geology*, **11**, 231–234.

FERRARA, G. C., PALMERINI, G. C. & SCAPPINI, U. 1985. Update report on geothermal development in Italy. *In*: STONE, C. (ed.) *International Symposium on Geothermal Energy, International Volume.* Geothermal Resources Council, Davis, CA, 95–105.

FOURNIER, R. O. 1985. The behavior of silica in hydrothermal solutions. *In*: BERGER, B. R. & BETHKE, P. M. (eds) *Geology and Geochemistry of Epithermal Systems. Reviews in Economic Geology*, **2**, 45–61.

FOURNIER, R. O. 1991. The transition from hydrostatic to greater than hydrostatic fluid pressure in presently active continental hydrothermal systems in crystalline rock. *Geophysical Research Letters*, **18**, 955–958.

FOURNIER, R. O. 1999. Hydrothermal processes related to movement of fluid from plastic into brittle rock in the magmatic–epithermal environment. *Economic Geology*, **94**, 1193–1212.

GANDA, S., BOEDIHARDI, M., RACHMAN, A. & HANTONO, D. 1985. *Geologi daerah Kawah dan Sekitarnyia Kabupaten Tasikmalaya, Garut, Majalengka dan Sumedang, Proponsi Jawa Barat.* Divisi Geotermal, Pertamina Pusat.

GILPIN, B. & LEE, T. C. 1978. A microearthquake study of the Salton Sea geothermal area, California. *Seismological Society of America Bulletin*, **68**, 441–450.

HALL, R. 1997. Cenozoic tectonics of SE Asia and Australasia. *In*: HOWES, J.V.C. & NOBLE, R.A. (eds) *Proceedings of an International Conference on Petroleum Systems of SE Asia and Australasia, Jakarta, Indonesia, 21–23 May 1997.* Indonesian Petroleum Association, Jakarta, 47–62.

HALLIBURTON, 1995. *Electrical Micro Imaging Service, Brochure, EL1076.*

HAMILTON, W. 1974a. *Earthquake Map of the Indonesian Region.* Department of the Interior, US Geological Survey, Reston.

HAMILTON, W. 1974b. *Map of Sedimentary Basins of the Indonesian Region.* Department of the Interior, US Geological Survey, Reston.

HAMILTON, W. 1978. *Tectonic map of the Indonesian Region.* Department of the Interior, US Geological Survey, Reston.

HARDEBECK, J. L. & HAUKSSON, E. 1999. Role of fluids in faulting inferred from stress field signatures. *Science*, **285**, 236–239.

HEDENQUIST, J. W., ARRIBAS, A., JR & REYNOLDS, T. J. 1998. Evolution of an intrusion–centered hydrothermal system: Far Southeast-Lepanto porphyry and epithermal Cu–Au deposits, Philippines. *Economic Geology*, **93**, 373–404.

HENLEY, R. W. & ELLIS, A. J. 1983. Geothermal systems ancient and modern; a geochemical review. *Earth-Science Reviews*, **19**, 1–50.

KATILI, J. A. & SUDRADJAT, A. 1984. *Galunggung: the 1982–83 Eruption.* Volcanological Survey of Indonesia, Bandung, 102.

KEITH, J. D. & SHANKS, W. C. 1988. Chemical evolution and volatile fugacities of the Pine Grove porphyry molybdenum and ash-flow tuff system, southwestern Utah. *In*: TAYLOR, R. P. & STRONG, D. F. (eds) *Recent Advances in the Geology of Granite-related Mineral Deposits.* Canadian Institute of Mining and Metallurgy, Special Volume, **39**, 402–423.

LAMA, R. D. & VUTUKURI, V. S. 1978. *Handbook on Mechanical Properties of Rocks.* Transtech Publications, Bay Village.

MAJER, E. L. & McEVILLY, 1979. Seismological investigation at The Geysers geothermal field. *Geophysics*, **44**, 246–269.

MALIN, P. 1994. The seismology of extensional hydrothermal system. *Geothermal Resources Council Transactions*, **18**, 17–22.

MANDL, G. 1988, *Mechanics of Tectonic Faulting. Models and Basic Concepts.* Elsevier, Amsterdam.

McBIRNEY, A. R. 1995. Mechanisms of differentiation in the Skaergaard intrusion. *Journal of Geological Society, London*, **152**, 421–435.

MILLER, S. A., NUR, A. & OLGAARD, D. L. 1996. Earthquakes as a coupled shear stress – high pore pressure dynamical system. *Geophysical Research Letters*, **23**, 197–200.

MOORE, J. N., ALLIS, R. G. & McCULLOCH, J. E. 2001. The origin and development of vapor-dominated geothermal systems. *In*: *Eleventh Annual V.M. Goldschmidt Conference, 20–24 May 2001, Hot Springs, Virginia, USA.* Lunar and Planetary Institute, Houston, TX.

MOORE, J. N., CHRISTENSEN, B., BROWNE, P. R. L. & LUTZ, S. J. 2004. The mineralogic consequences and behavior of descending acid-sulfate waters: an example from the Karaha–Telaga Bodas geothermal system, Indonesia. *Canadian Mineralogist* (in press).

MOORE, J. N., LUTZ, S. J., RENNER, J. L., McCULLOCH, J. & PETTY, S. 2000. Evolution of a volcanic-hosted vapor-dominated system: petrologic and geochemical data from corehole T-8, Karaha–Telaga Bodas, Indonesia. *Geothermal Resources Council Transactions*, **24**, 24–27.

MOORE, J. N., ALLIS, R. G., RENNER, J., MILDENHALL, D. & McCULLOCH, J. 2002. Petrologic evidence for boiling to dryness in the Karaha–Telaga Bodas geothermal system, Indonesia. *In*: *Proceedings of the 27th Stanford Workshop on Geothermal Reservoir Engineering, 28–30 January 2002.* Stanford University, Stanford, CA, 223–232.

MORSE, S. A. 1986. Convection in aid of adcumulus growth. *Journal of Petrology*, **27**, 1183–1214.

MUELLER, B., WEHRLE, V. & FUCHS, K. 1997. The 1997 release of the World Stress Map. http://www-wsm. physik.uni-karlsruhe.de/pub/Rel97/wsm97.html.

MUIR WOOD, R. 1994. Earthquakes, strain-cycling and the mobilization of fluids. *In*: PARNELL J. (ed.) *Geofluids; Origin, Migration and Evolution of Fluids in Sedimentary Basins.* Geological Society, London, Special Publications, **78**, 85–98.

MUIR WOOD, R. & KING, G. C. P. 1993. Hydrological signatures of earthquake strain. *Journal of Geophysical Research*, **98**, 22035–22068.

MURASE, T. & MCBIRNEY, A. R. 1973. Properties of some common igneous rocks and their melts at high temperatures. *Geological Society of America Bulletin*, **84**, 3563–3592.

NEMČOK, M., GAYER, R. A. & MILIORIZOS, M. 1995. Structural analysis of the inverted Bristol Channel Basin: Implications for the geometry and timing of the fracture porosity. *In*: BUCHANAN, J. G. & BUCHANAN, P. G. (eds) *Basin Inversion*. Geological Society, London, Special Publications, **82**, 355–392.

NEMČOK, M., HENK, A., GAYER, R. A., VANDYCKE, S. & HATHAWAY, T. M. 2002. Strike-slip fault bridge fluid pumping mechanism: insights from field-based palaeostress analysis and numerical modelling. *Journal of Structural Geology*, **24**, 1885–1901.

NEMČOK, M., MCCULLOCH, J., NASH, G & MOORE, J. N. 2001. Fault kinematics in the Karaha-Telaga Bodas, Indonesia, geothermal field: an interpretation tool for remote sensing data. *Geothermal Resources Council Transactions*, **25**, 26–29.

NUR, A. & BOOKER, J. R. 1972. Aftershocks caused by pore fluid flow? *Science*, **175**, 885–887.

ORTOLEVA, P., MERINO, E., MOORE, C. & CHADAM, J. 1987. Geochemical self-organization, I. Reaction-transport feedbacks and modeling approach. *American Journal of Science*, **287**, 979–1007.

POWELL, T., MOORE, J., DEROCHER, T. & MCCULLOCH, J. 2001. Reservoir geochemistry of the Karaha–Telaga Bodas prospect, Indonesia. *Geothermal Resources Council Transactions*, **24**, 363–367.

RENNER, J., EVANS, B. & HIRTH, G. 2000. On the rheologically critical melt fraction. *Earth and Planetary Science Letters*, **181**, 585–594.

RICE, J. & CLEARY, M. 1976. Some basic stress diffusion solutions for fluid-saturated elastic porous media with compressible constituents. *Reviews of Geophysics and Space Physics*, **14**, 227–241.

SCHLUMBERGER, 1992. *FMI Fullbore Micro Imager, Brochure SMP 9210*.

SECOR, D. T. 1965. Role of fluid pressure in jointing. *American Journal of Science*, **263**, 633–646.

SEGALL, P. & POLLARD, D. D. 1980. Mechanics of discontinuous faults. *Journal of Geophysical Research*, **85**, 4337–4350.

SHAW, H. R. 1969. Rheology of basalt in the melting range. *Journal of Petrology*, **10**, 510–535.

SHINOHARA, H., KAZAHAYA, K. & LOWENSTERN, J. B. 1995. Volatile transport in a convecting magma column: Implications for porphyry Mo mineralization. *Geology*, **23**, 1091–1094.

SHIRLEY, D. N. 1986. Compaction of igneous cumulates. *Journal of Geology*, **94**, 795–809.

SHIRLEY, D. N. 1987. Differentiation and compaction in the Palisades Sill. *Journal of Petrology*, **28**, 835–865.

SIBSON, R. H. 1982. Fault zone models, heat flow, and the depth distribution of seismicity in the continental crust of the United States. *Seismological Society of America Bulletin*, **72**, 151–163.

SIBSON, R. H. 1985. Stopping of earthquake ruptures at dilational fault jogs. *Nature*, **316**, 248–251.

SIBSON, R. H. 1989. Earthquake faulting as a structural process. *Journal of Structural Geology*, **11**, 1–14.

SIBSON, R. H. 1996. Structural permeability of fluid-driven fault-fracture meshes. *Journal of Structural Geology*, **18**, 1031–1042.

SPARKS, R. S. J., HUPPERT, H. E., KERR, R. C., MCKENZIE, D. P. & TAIT, S. R. 1985. Postcumulus processes in layered intrusions. *Geological Magazine*, **122**, 555–568.

TAIT, S. R. & KERR, R. C. 1987. Experimental modelling of interstitial melt convection in cumulus piles. *In*: PARSON, I. (ed.), *Origins of Igneous Layering*. NATO ASI Series C, **196**, 569–587.

TAIT, S. R., HUPPERT, H. E. & SPARKS, R. S. J. 1984. The role of compositional convection in the formation of adcumulate rocks. *Lithos*, **17**, 139–146.

TSANG, C. F. 1999. Linking thermal, hydrological, and mechanical processes in fractured rocks. *Annual Reviews of Earth and Planetary Sciences*, **27**, 359–384.

WALLACE, R. E. 1951. Geometry of shearing stress and relation to faulting. *Journal of Geology*, **59**, 118–130.

Palaeostress orientation inferred from surface morphology of joints on the southern margin of the Bristol Channel Basin, UK

MANDEFRO BELAYNEH

Department of Earth Science and Engineering, Royal School of Mines, Imperial College of Science, Technology and Medicine, Prince Consort Road, London SW7 2BP, UK (e-mail: m.belayneh@imperial.ac.uk)

Abstract: A field study of the orientation of surface features on joint planes has been carried out on the north Somerset coast along the southern margin of the Bristol Channel Basin (BCB), with the aim of inferring the palaeostress orientations. Surface features are very delicate structures on joint surfaces and when preserved give important clues regarding the direction of propagation of a joint and the interaction between the remote and local stresses.

In this chapter surface features (i.e. plumose structures and twist-hackle fringes) observed on joints in Liassic limestone beds interbedded with shales from the BCB are described and their implications regarding the evolution of the stress field operating during their formation considered. The results are then compared with the generally accepted model for the region of an anticlockwise rotation of the stress field. The incompatibility between N–S-directed remote compression and continually changing local stress caused by the local structures (folds and faults), inhomogeneities and the anisotropy created by rapidly alternating limestone and shale beds resulted in the master joints breaking down into en echelon cracks (twist-hackle fringes). Joints in thicker limestone beds show well-developed plumose structures and twist-hackle fringes, whereas in thinner beds the fringe zones extend from the top to the bottom of the bed. It is found that regardless of the orientation of the parent joint in which the twist-hackle fringes are contained, they generally strike E–W and show both right- and left-stepping arrangements.

The study area (Fig. 1) is located on the southern margin of the Bristol Channel Basin (BCB). This rift-related basin is an exhumed, E–W-trending basin that formed as a result of the N–S extension across southern England that began in the Permian (Kamerling 1979; Whittaker & Green 1983; Van Hoorn 1987; Dart *et al.* 1995). The oldest beds in the basin are represented by the Permo-Triassic succession, which is made up of continental redbeds. These are overlain by Lower Triassic strata (the arenaceous Sherwood Sandstone Group) that pass upward into the Mercia mudstone, which consists of siltstone, mudstone and evaporites. The overlying Blue Anchor Formation consists of grey–greenish marls (the Tea Green Marls) and this is overlain by the Westbury Formation, which comprises limestones, sandstones and dark grey shales. Black shales of the Westbury Formation mark the stratigraphic contact between the Triassic and the Lias (Dart *et al.* 1995). The major facies change that occurred at the end of Triassic was the result of arid continental conditions giving way to marine conditions as isolated rift basins became linked to each other and to the sea as Pangaea became progressively more fractured. The Westbury Formation is overlain by interbedded grey mudstones, siltstones and limestones of the Lilstock Formation, and by Lower Lias rocks consisting of alternating beds of grey mudstone and light grey limestone. The surface features of joints in the Lower Lias limestones and their palaeostress implications are the subject of this paper.

Structures in the Bristol Channel Basin

The contact between the Uppermost Triassic and Lower Lias is exposed along the southern margin of the Bristol Channel coast (Fig. 1). The Triassic marls contain only poorly defined bedding planes, and are relatively homogeneous and isotropic. The faults that form within them occur as diffuse fault zones, most of which dip north at approximately 60° (i.e. they are synthetic to the major normal faults bounding the southern margin of the BCB). The rapidly alternating micritic limestones and marly shales of the Lower Lias show a distinctly different rheology. The Lower Lias is markedly anisotropic, and the faults within it are more localized and form discrete fault planes.

The study area has been the focus of many stratigraphic, palaeontological, sedimentological and structural investigations. Kamerling (1979) and Whittaker & Green (1983) give a detailed description of the stratigraphy of the study area, and Van Hoorn (1987) discusses basin formation and inversion of the South Celtic Sea and the BCB. The area has been used to study various structures such as normal faults (Davison 1995; Nemčok & Gayer 1996), the

From: COSGROVE, J. W. & ENGELDER, T. (eds) 2004. *The Initiation, Propagation, and Arrest of Joints and Other Fractures.* Geological Society, London, Special Publications, **231**, 243–255. 0305-8719/04/$15 © The Geological Society of London 2004.

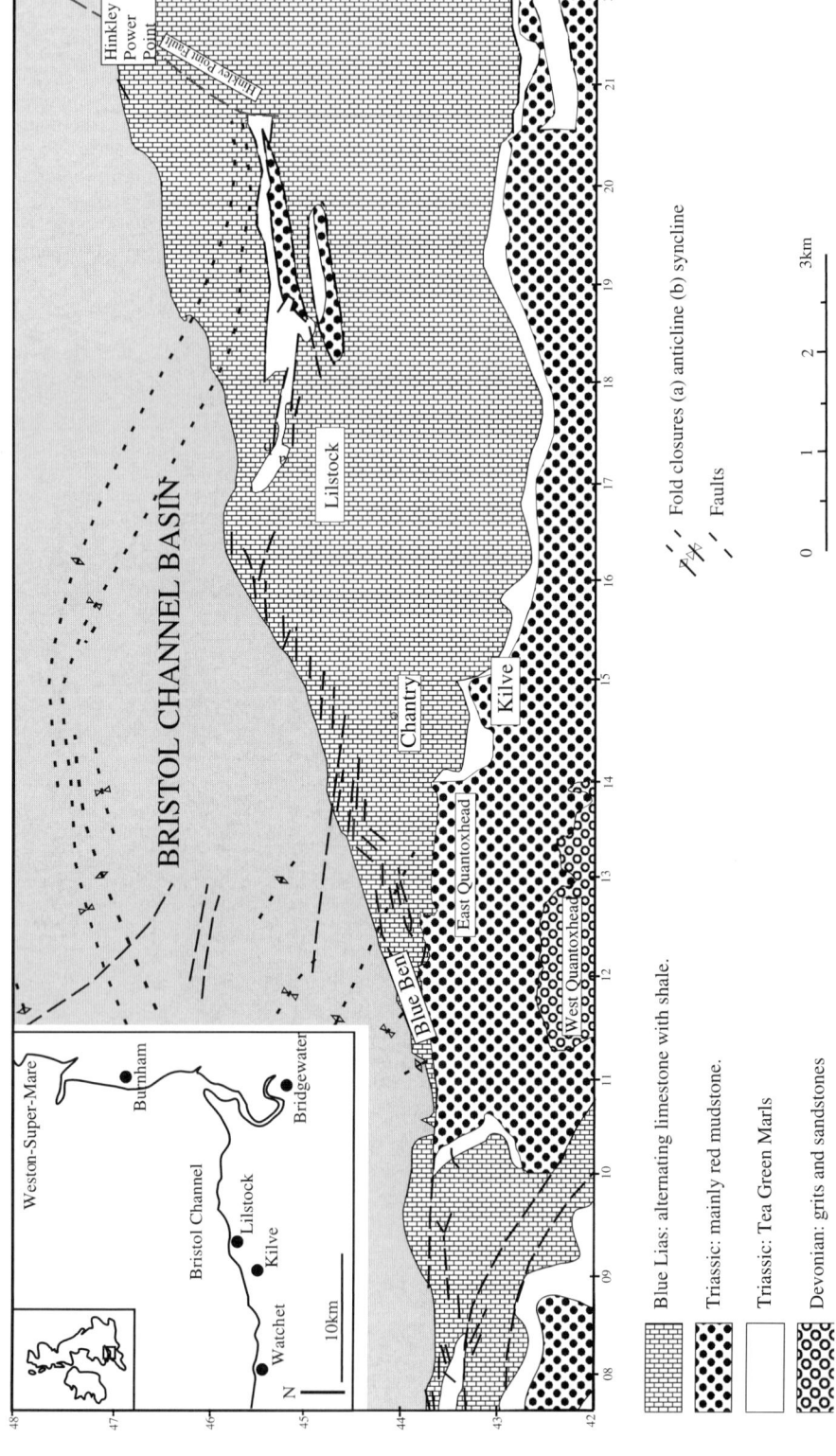

Fig. 1. Geology and major structures in Lilstock–Kilve on the southern margin of the BCB, north Somerset, UK (from 1:50000 scale geological maps of England and Wales, sheets 279, and parts of sheets 263 and 295). The inset map shows the location of the BCB.

development of relay ramps in association with normal faults (Peacock & Sanderson 1991 & 1994), inverted normal faults (Chadwick 1993; Dart *et al.* 1995; Nemčok *et al.* 1995; Kelly *et al.* 1999) and strike-slip faults (Peacock & Sanderson, 1995; Willemse *et al.* 1997; Kelly *et al.* 1998). Veins from the area have been studied by Caputo & Hancock (1999) and Cosgrove (2001), and joints by Loosveld & Franssen (1992), Rawnsley *et al.* (1998), Al-Mahruqi (2001), Engelder & Peacock (2001) and Belayneh (2003).

The excellent cliff and foreshore exposures allow a complete three-dimensional (3D) visualization of the various structures (i.e. folds, faults, joints sets and joint networks). The most abundant structures are joints and they have been divided by the author into three groups based on the time of their formation. These are burial joints that were formed during the basin opening, Alpine joints that were formed during basin inversion, and post-Alpine joints associated with uplift and exhumation of the basin (Belayneh 2003). Despite the abundance of joints in the area, those decorated with plumose structures and twist-hackle fringes, the subject of this chapter, are rare. The reason for this may relate to the fine-grained nature of the rocks and/or to their susceptibility to solution. Engelder (1987) argued that the formation of plumose structures is highly dependant on lithology and pointed out that very fine-grained rocks do not generally develop these patterns. This suggestion has been supported by subsequent field observations (McConaughy & Engelder 2001), which show that joints in shale rarely carry a marked plume morphology, whereas slightly coarser-grained siltstones have a very well-developed plume morphology.

Despite the carbonates of the study area being fine grained, a sufficient number of plumose structures are present to enable data concerning joint initiation, the direction of joint propagation, and the interaction between the remote and local stresses to be obtained. In this chapter, surface features on joints from the study area are described and analysed, and the evolution of the stress regime deduced from this work is compared with the model of an anticlockwise rotation of the stress field documented by Rawnsley *et al.* (1998).

'Discontinuity' is the general term used to describe a break of continuity in a mass of rock. These breaks occur on all scales from the microscopic to the regional. They begin to form in a rock mass as soon as it is capable of brittle failure, whether near surface or deeply buried. These processes take place in all types of rocks and, as a result of early lithification, limestone beds are particularly susceptible to fracturing. Joints initiate from flaws where geochemical, thermal and mechanical processes all tend to be active during the geological evolution of a rock.

There are two principal approaches to joint description. The first is fractography that deals with the surface morphology of individual joints (Bahat 1991). The second deals with generalized attributes, such as dominant orientation and styles, and takes into account large populations of joints throughout a region (Hancock 1985, 1994). In the latter approach, joints are described quantitatively by defining their spacing, length, orientation, distribution, continuity (extent), thickness of the fractured layer (i.e. height of the joint), nature of any infill, aperture, roughness and the age relationship between different joint sets making up the network. The history of displacement across a joint may be difficult to determine because of the small amount of movement involved, the lack of kinematic indicators such as slickensides on the fracture surface and the infilling of joints by dykes or veins.

There is no simple and universally accepted definition of a joint. Some define joints as those 'fractures with field evidence for dominantly opening displacements' (Pollard & Aydin 1988). This definition implies that joints are the result of extensional failure. Others define joints in terms of 'a surface along which there has been no appreciable displacement' regardless of the mode of origin of the fracture (Price 1966; Blés & Feuga 1986; Hancock 1994). This latter definition admits the possibility of hybrid shear fractures, a phenomenon for which there is no experimental evidence (see Engelder 1999). Most workers accept that joints are fractures across or along which very little displacement has occurred, and Pollard & Aydin (1988) point out that, even if a joint closes after formation, the existence of the two surfaces demonstrates that some displacement remains. Faults are defined as shear failure planes in which the two surfaces are displaced relative to each other along a direction parallel to that plane. Pollard & Segall (1987) argued that joints and faults are mechanically different because they have unique stress, strain and displacement fields.

From the point of view of fracture mechanics, crack tips have been related to three modes of displacement (Lawn 1975) (Fig. 2a–c), namely extensional or mode I displacement, and shear fractures of modes II and III. A crack tip may propagate under either shear or normal surface displacement:

Mode I (opening mode). In this mode the crack tip is subjected to a normal stress (σ_n) and the crack faces separate at the crack front so that the displacement of the crack surfaces is perpendicular to the crack plane (Fig. 2a). Pure mode I loading produces the maximum propagation energy for an increment of crack growth oriented in the plane of the joint leading to 'in-plane propagation'. A joint oriented normal to σ_3 satisfies this condition. If the principal stress direction remains the same throughout the extent of the

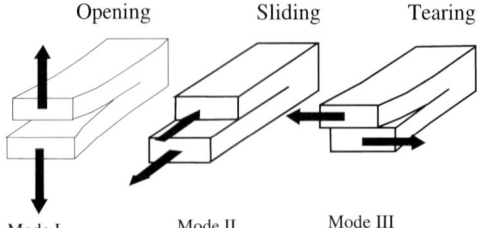

Opening   Sliding   Tearing

Mode I   Mode II   Mode III

Fig. 2. The three basic modes of crack displacement (after Lawn 1975). Note that the arrow indicates the direction of displacement of the blocks.

joint plane, in-plane propagation will produce a planar crack.

Mode II (sliding). This mode of crack-tip displacement occurs when the crack tip is subjected to shear and the crack faces slide past one another so that the displacement of the crack faces lies in the crack plane and is perpendicular to the crack front (Fig. 2b).

Mode III (tearing). In this mode of crack-tip displacement, the crack tip is subjected to shear and the crack faces move relative to one another so that the displacement and the crack front are parallel (Fig. 2c.)

Mixed-mode failure. This involves a combination of any of the above three modes of crack displacement (i.e. mode I and II, mode I and III, mode II and III, or mode I, II and III). Joints formed by mixed-mode propagation lead to displacements in both tensile mode and by shear failure (Fig. 3). If the principal stress direction changes as the joint propagates, shear stresses are generated at the joint tip. In both cases the preferred joint-propagation path is out-of-plane and tends to be oriented perpendicular to the local maximum tensile stress (σ_3).

Fractography

Introduction

Fractography is defined as the science that deals with the description, analysis and interpretation of fracture-surface morphologies (fracture topographies), and links them to the causative stresses, mechanisms and subsequent evolution of the fractures (Ameen 1995). The interpretation of the surface morphology of cracks requires knowledge of fracture mechanics, and has a wide application in material science, engineering and the geosciences. Fractography has been extensively used to study

brittle failure in materials, and significant contributions have been made by Bahat (1991) who discusses fracture processes in glass, ceramics and rocks. Bahat describes fracture-surface morphology as 'the fingerprint of fractography'. A detailed account of fractography as a tool in fracture mechanics and stress analysis has been given by Ameen (1995).

The rate at which cracks propagate during earthquakes is supersonic, a fact that is also evident from seismicity accompanying rock failure incurred during blasting. It is extremely difficult to determine the age of a single crack in a rock mass. Caputo & Hancock (1999) argue that the lifetime of each joint propagation event is not very long. They pointed out that extension fractures do not form instantaneously, but incrementally and repeatedly over a time-span of 10^5 10^6 years, and argue that this is because joint propagation depends on many factors including the amount of deformation, the strain rate and the mechanical properties of the jointed rocks. In addition, younger joints with the same orientation could be superimposed on pre-existing joints, making the interpretation extremely difficult.

Joints inherit remarkable surface markings or ornamentations from which vital information regarding joint initiation, direction of propagation and positions of arrest can be gathered. This rendered fractography a powerful tool that can be used to better understand the relationship between the mode and direction of crack propagation, and the remote and local stresses.

Woodworth (1896) was one of the first to recognize and report fracture-surface morphologies such as plumose structures on joint faces (Fig. 4). Woodworth suggested that joints are opening-mode fractures (i.e. mode I); features that occurred on joint faces were carefully documented and described, and interpreted as inherent characteristics of joints that fail by extension. Woodworth described joint initiation at a point and its outward propagation to form plumose structures. Several other investigators later applied fractographic principles to the analysis of joint-surface features in order to deduce the points from which joints initiated, the direction of propagation and stages of arrest (e.g. Corte & Higashi 1960; Hodgson, 1961; Bahat & Engelder 1984; Pollard & Aydin 1988; Bahat 1991; Helgeson & Aydin 1991; Bankwitz & Bankwitz 1995; Younes & Engelder 1999; Müller & Dahm 2000; Bahat *et al.* 2001; Belayneh 2003). Kulander & Dean (1995) suggest that fractographic features can be used in any brittle material to determine the origin of the failure plane, changes in propagation velocities, stress directions and stress magnitudes at the crack tip during failure.

As a result of solubility of limestones well-preserved fracture-surface morphologies on joints in the limestones of the Bristol Channel are compara-

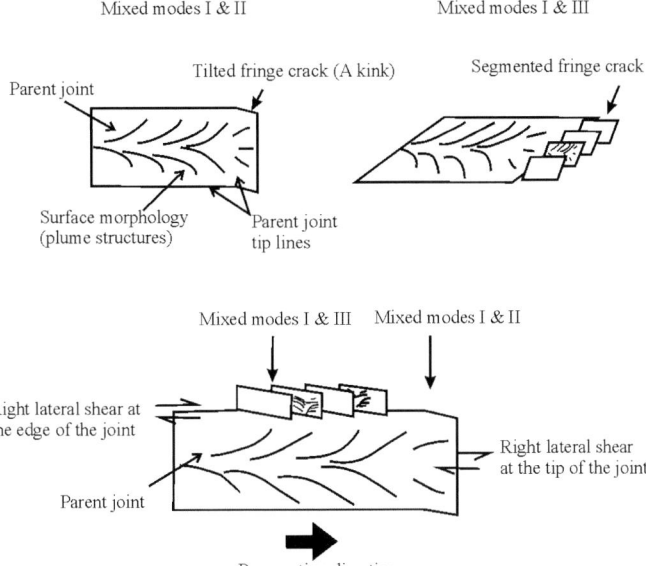

Mixed modes I & II

Mixed modes I & III

Parent joint

Tilted fringe crack (A kink)

Segmented fringe crack

Surface morphology
(plume structures)

Parent joint
tip lines

Mixed modes I & III   Mixed modes I & II

Right lateral shear at
the edge of the joint

Right lateral shear
at the tip of the joint

Parent joint

Propagation direction

Fig. 3. Joint propagation in mixed-mode loading (from Younes & Engelder, 1999).

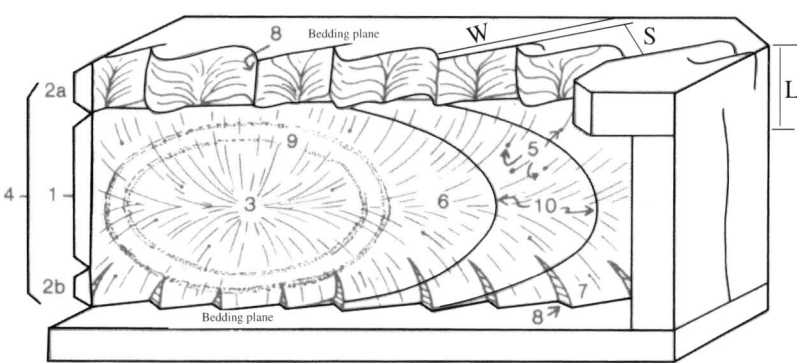

1. main joint face 2. twist hackle fringe 3. origin 4. hackle plume 5. inclusion hackle 6. plume
axis 7. twist hackle face 8. twist hackle step 9. arrest lines and 10. constructed fracture front lines

Fig. 4. Surface features on joint surface (modified from Hodgson 1961). S, L and W denote spacing, length and width
of twist-hackle fringes that are arranged in and en echelon manner.

tively rare and difficult to find. Fortunately, some of
the joints do have preserved surface markings.

Types of fracture-surface morphology

The suite of surface features displayed by joint surfaces is summarized in Figure 4. In the area of
the BCB studied by the present author, the most
common fracture surface markings are plumose
structures and twist-hackles fringes. 'Plumose structure' is the name given to the feather-like markings

that appear to radiate from some central point (3 in
Fig. 4), which is the point of fracture initiation. They
are the result of mode I failure and the direction of
propagation is parallel to the plumose lineation. The
position of the fracture front at different times in the
fracture history can be obtained by studying fracture
front lines normal to the plumes (10 in Fig. 4). These
lines are often present on the face as arrest lines, i.e.
lines that mark the position of the fracture front
when the fracture temporarily stopped propagating
or the fracture front decelerated (9 in Fig. 4). The
fracture shown in Figure 4 is bounded by a 'twist-

hackle fringe' with a systematic morphology, the topographic amplitude of which is usually much higher than that of the central zone (or plume) of the joint.

Younes & Engelder (1999) divided fringe cracks into three types based on their study of surface features in siltstone and shale intercalations in the Appalachian Plateau. These are gradual twist hackles, abrupt twist hackles and kinks. Gradual twist hackles (twist-hackle fringes in this work) are curviplanar en echelon fringe cracks that initiate in a bed that contains a parent joint. When it forms, a twist hackle twists away from the parent joint (either gradually to form a twist hackle or abruptly to form an abrupt twist hackle). Experiments on analogue materials such as starch–water mixtures (Müller 2001) indicate that in a twist-hackle fringe zone, joints propagate more slowly than in the central zone (i.e. in a region defined by plumose structure). Müller argued that a similar process will operate in a rock layer in which the twist-hackle fringe forms along the layer interface (Fig. 4). Individual en echelon cracks that are planar and that emanate abruptly from the edge of a parent joint are known as abrupt twist hackles. Abrupt twist hackles propagate as planar features within the beds above and below the bed containing the parent joint. In map view they appear to grow across the edge of the parent joint. Kinks occur as mode I cracks and the tilt in the crack path is driven by shear traction superimposed on the parent joint after arrest following a finite amount of mode I propagation. Kinks occur at the termination of a straight joint, adopting a strike that is abruptly different from that of the initial joint. A kink that turns clockwise with respect to the direction of the main joint is the result of right-lateral shear, whereas a kink that turns an anticlockwise is the result of left-lateral shear (Cruikshank *et al.* 1991).

A specific feature of fracture surface markings, namely the border plane, or twist-hackle fringe, has been used to determine the orientation of stresses under which the twist-hackle structures develop. Plumose structures indicate mode I failure (opening) and the twist-hackle fringe (border planes) indicate the out-of-plane propagation from the main joint. The recognition of twist-hackle planes or dilatant echelon cracks described by Pollard *et al.* (1982) is important because they form when a propagating fracture abruptly enters a region of different stress orientation. The fracture breaks (twists) into a series of individual en echelon lance-like twist-hackle faces forming in steps, each perpendicular to the resultant tension, that are generally more closely spaced than the adjacent main joint planes (Kulander & Dean 1995; Roberts 1995). The breakdown of the crack into en echelon segments is indicative of different mode III/mode I (e.g. Cooke & Pollard 1996; Bahat 1997) or mode II/mode I ratios depending on

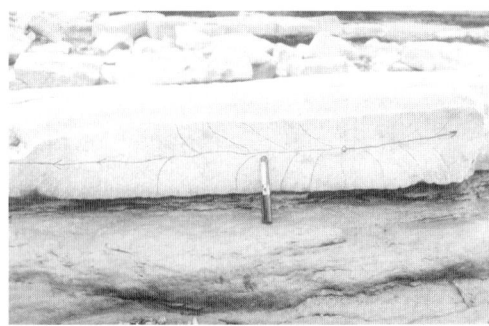

Fig. 5. Plumose structures indicating the initiation of a joint from a point source in the middle of 17 cm-thick limestone bed at Kilve anticline. Note that the plume axis is approximately symmetrical with respect to the edge of the bed. The upper fringe zone has been weathered back from the joint face shown because of the enhanced weathering (chemical and mechanical) resulting from the higher intensity of fracturing.

the direction of crack propagation and sense of displacement. Younes & Engelder (1999) argue that fringe cracks originate when an advancing joint encounters a shear couple. They propose that if the propagating crack enters a volume of rock subject to a stress field misaligned in relation to that guiding the parent joint, transverse adjustments will occur as a result of mixed-mode crack propagation, thus causing the joint to break down into fringe cracks.

A crack subjected only to the opening-mode deformation will dilate and propagate in its own plane (Pollard *et al.* 1982). The breakdown of a parent joint into an echelon cracks is initiated by spatial and temporal rotation of the remote principal stresses about an axis parallel to crack propagation direction. The tendency for breakdown increases with increase in stress intensity ratio K_{III}/K_I (Cooke & Pollard 1996; Bahat 1997).

Müller & Dahm (2000) studied fracture-surface morphology, mainly plumose structures on tensile cracks, and the rupture velocity of cracks formed by desiccation of starch–water mixtures. Based on rupture velocity they distinguished joints, which they argued were spontaneously nucleating dynamic cracks, from quasi-static cracks. They deduced that plumose structures are indicative of rupture direction and drew an analogy between plumose structures and rupture direction, on the one hand, and seismic rays and seismic wave velocity, on the other.

Bankwitz & Bankwitz (2001) have demonstrated the applicability of the science of fractography to geology. They used fractographic features on early fracture sets in granite plutons as evidence of the cooling stages and, in some cases, the conditions operating in the crust at the time of joint formation. They argue that the fractographic features of joints

Fig. 6. Photograph showing multiple initiation points and complex propagation directions on a joint face at Lilstock. Black marker on the lower side is 15.5 cm for scale.

formed at depth are not a typical reflection of simple thermoelastic contraction in the granite. They present evidence that these joints are fluid driven, and estimate the depth of formation of these early fractures to lie between 7 and 15 km.

Present work

A study of surface features on joints has been carried out by the author on the southern coast of the BCB at Lilstock, Kilve, East Quantoxhead and Blue Ben (Fig. 1). Although there are numerous sets of joints in the area, joints decorated with plumose structures and twist-hackle fringes are rare. However, excellent exposures can be seen on some of the wave-cut platforms and in the cliffs. The majority of observations of plumose structures in the jointed Liassic limestones at these locations indicate that these joints were initiated from points in the middle of the beds (Fig. 5). Many thick limestone beds (thickness generally >20 cm) show plume axes approximately parallel to the bedding plane, indicating that joint propagation was bed parallel. However, some of the beds have multiple initiation points and show complex plumose structures (Fig. 6). As noted earlier, the formation of twist-hackle fringe occurs when the joint propagates towards the bedding-plane interface with the shale above and below. Here it encounters a local stress field that is neither normal nor parallel to the joint plane. The incompatibility between the remote and local stresses distorts the crack-tip stress field causing the joint to breakdown into twist-hackle fringes. Propagation of the twist hackle is controlled by the local stress field, which is

not necessarily in the same orientation as the remote stress field.

An interesting field observation in the study area relating to the hackle fringe is that the individual twist hackles strike approximately E–W regardless of the strike of the parent joint. Thus, in plan, depending on the strike of the parent joint with which they are associated, they show a clockwise (CW) or an anticlockwise (ACW) sense of rotation. Joints whose strike is between 0°–090° show a left-stepping (clockwise) sense of rotation; joints striking 090°–180° show right-stepping (anticlockwise) sense of rotation (Fig. 7). The reason for the variety of orientations of the Alpine joint sets is not clear and is discussed in Belayneh & Cosgrove (2004). The position of the plume axis is often offset from the central position in the bed by heterogeneities in the local stress field (Engelder & Savalli 2001).

The microveins in the study area, like the twist hackles, strike approximately E–W. They were formed during basin subsidence, and were filled with fine-grained crystalline calcite (Fig. 8). The twist-hackle fringes were formed during the Alpine inversion of the BCB and are not filled with any mineralization.

In the study area it was found that the width of the twist-hackle fringes (w in Fig. 4) depends on the angle of deviation of the parent joint from the general E–W direction. As the strike deviation of the parent joint increases, the width of the twist hackle fringe decreases from 3 cm to a few mm. The range of twist angle (β) measured was between 0° and 80° (Table 1).

As discussed earlier, the Alpine joints were formed during the Alpine compression that caused

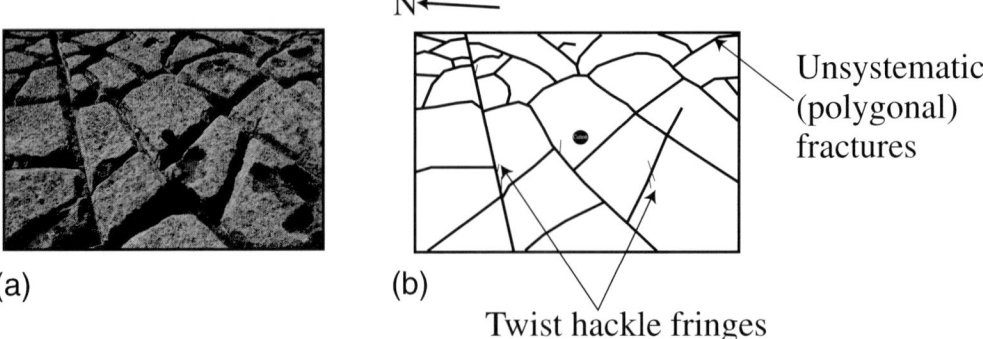

Fig. 7. (**a**) Photograph and (**b**) line drawing showing a plan view of twist-hackle fringes observed at Lilstock. Note that the orientation of the fringes remains approximately E–W regardless of the strike of the parent joint, confirming constant N–S orientation of the remote stress field during Alpine inversion of the BCB. Note that the twist-hackle fringes are left-stepping and right-stepping for the parent joints striking between N–S and 090°, and 090° and 180°, respectively. The camera cap in the middle is 5 cm in diameter for scale.

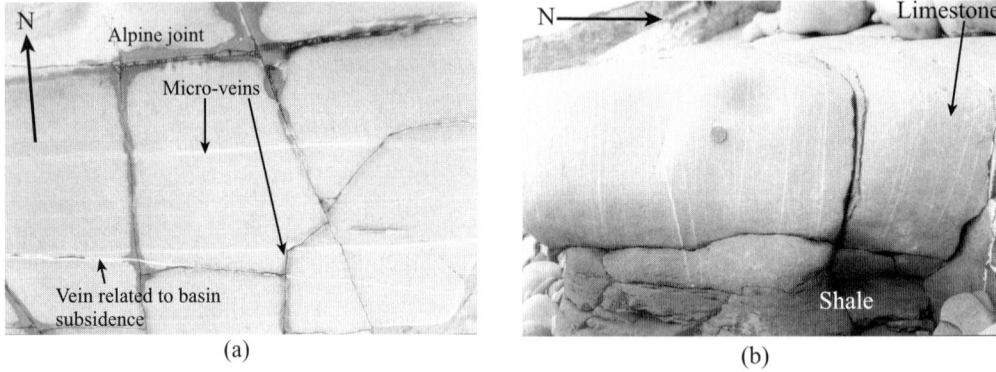

Fig. 8. (**a**) Plan and (**b**) section view of microveins that were formed during basin subsidence on the southern margin of the BCB. Although these have the same strike as the twist hackles (Figs 5–7), the latter are barren and were formed during different stages of basin evolution.

the amplification of pre-existing rollover folds such as the Lilstock and Kilve anticlines. At both localities, in the hinge region of these folds, the joints are subparallel to the fold axes and are therefore considered to be fold-related. Twist-hackle fringes strike approximately E–W regardless of the orientation of parent joint (Fig. 7), probably indicating a consistent N–S orientation of the local least principal compressive stress along the hinge zone of the fold at the time of joint formation. It was noted earlier that the local stress field is controlled by the position of the joint relative to pre-existing structure or heterogeneity. In this example, the local stress is compatible with outer arc extension in the hinge region of the fold.

Field observations made by the author in the Bristol Channel area indicate that the fringe zones (2a and 2b in Fig. 4) are mostly asymmetric, i.e. smaller in length (L, Fig. 4) along one margin of the

Table 1. *Twist-hackle fringes measured on joints from Lilstock*

Strike of parent joint (°)	Twist angle (β)(°)	Sense of rotation	Steps
360	81	CW	Left stepping
12	65	CW	Left stepping
14	80	CW	Left stepping
85	Subparallel	Difficult to determine	?
102	Subparallel	Difficult to determine	?
105	11	ACW	Right stepping
115	35	ACW	Right stepping
116	20	ACW	?

CW, clockwise; ACW, anticlockwise.

Fig. 9. Asymmetrical twist-hackle fringe observed at Kilve. Note the difference in length of the fringe zone at the top and bottom of the bed. The top part is made up of more compact shaly limestone than the remainder of the bed, and has different mechanical properties. The twist angle is also higher along the lower portion of the bed.

Fig. 10. Photograph showing a plumose structure with an approximately symmetrical twist-hackle fringe at Blue Ben. Note that the plumose structure in the middle is approximately twice the length of the twist-hackle fringe. The propagation direction of the joint is from right to left, as deduced from the feather on the joint surface. The hammer on the left is 40 cm for scale.

parent joint than on the other (Fig. 9). The contact between the fractured limestone and the adjacent shale bed is highly variable and ranges from sharp to gradational (i.e. grading from mudstone to marls or marly limestone). It is observed that if the contact between the fractured limestone and the adjacent shale bed is sharp, the length of the fringe zone is long. Alternatively, if the contrast in mechanical properties between the limestone and the adjacent bed is either low or gradational, the length of the fringe zone is short. These empirical observations indicate that the length of the fringe zone depends on the properties of the interface between the limestone

and shale beds, and indicates that the difference in material properties on either side of the interface is an important factor in determining the length of the fringe zones. In general, the width of the plumose structure (1 in Fig. 4) is found to be approximately twice the length of the twist-hackle fringe (Fig. 10 and Table 2).

Discussion and conclusions

Based on the work of Dart *et al.* (1995) and the work described in the present chapter, the tectonic evolution

Table 2. *Thickness of the plumose structure and twist-hackle fringe (i.e. the combined lower and upper twist-hackle fringes) in limestone beds from the BCB*

Total thickness of fractured limestone bed (cm)	Length of twist-hackle fringe (cm)	Thickness of plumose structure (cm)	Locality
45	27	18	Lilstock
42	20	22	Lilstock
47	30	17	Lilstock
40	20	20	Lilstock
40	17	23	Lilstock
45	19	26	Lilstock
44	25	19	Lilstock
48	25	23	Lilstock
50	26	24	Kilve-Lilstock
42	19	23	Kilve-Lilstock
39	14	25	Kilve-Lilstock
45	19	26	Kilve-Lilstock
26	6	20	Kilve-Lilstock
40	20	20	Kilve-Lilstock
38	16	22	Kilve-Lilstock
40	19	21	Kilve
37	17	20	Kilve
37	21	16	West of Kilve
44	26	18	East Quantoxhead
39	18	21	East Quantoxhead
40	22	18	East Quantoxhead
41	18	23	East Quantoxhead
41	20.5	20.5	East Quantoxhead
41	25	16	East Quantoxhead
44	20	24	Kilve
42	17	25	East of Blue Ben
44	24	20	West of Blue Ben
42	26	16	West of Blue Ben

of the north Somerset coast has been divided into three stages. (1) The development of an E–W-striking extensional fault system and associated rollover folds and joints in response to a N–S-directed extension linked to the opening of the BCB. (2a) A N–S compression that lead to the formation of thrusts, the partial reactivation of normal faults as reverse faults, the amplification of rollover folds and the formation of Alpine joints. During this event, the partially reactivated normal faults acted as WNW–ESE-trending buttresses that imposed this orientation on the amplifying rollover folds (e.g. Nemčok et al. 1995; Engelder & Peacock 2001). This deformation event resulted in fracturing of the limestone beds and the formation of Alpine joints. The maximum and intermediate principal stresses were horizontal during folding and thrusting, the least compressive stress being vertical. The vertical thickening of the sequence resulting from the tectonic compression increased the overburden until either the vertical stress became the

intermediate principal stress and conditions for the formation of new strike-slip faults were created or conditions for re-shear on existing strike-slip faults in the Variscan basement were satisfied. During stage (2b) NW–SE-striking dextral and NE–SW-striking sinistral strike-slip faults (i.e. conjugate shears linked to the N–S compression) developed.

During the Alpine inversion of the BCB a mismatch existed between the remote compression and the local stress generated by the presence of local heterogeneities, pre-existing structures (such as faults and folds), and the mechanical anisotropy of the alternating limestones and shales sequence. One of the effects of the local stress fields is to cause a breakdown of the main joints into a series of en echelon fractures as it propagates towards the layer boundary and enters the local stress field linked to the change in mechanical properties at this interface. As noted earlier, it is found that within the study area the strike of the twist hackles is generally E–W, regardless of the orientation of the parent joints with which they are associated. It follows therefore that the configuration of the en echelon hackles, whether right-stepping or left-stepping, depends on the strike of the parent joints. Thus, for joints striking between 0° and 090° the sense of rotation is clockwise, and for joints striking 090° and 180° the sense of rotation is anticlockwise.

Based on abutting relationships, it has been suggested by Rawnsley et al. (1998) that the various joint sets in the study area were formed by opening mode and provides evidence of an anticlockwise stress rotation during the Alpine inversion of the basin. However, work by the present author indicates that deformation linked to the inversion occurred with a constant N–S remote compression. As noted in the previous paragraph, the twist hackles strike consistently E–W regardless of the orientation of the parent joint. Consequently, whether they relate to the parent joint in a clockwise or anticlockwise manner depends only on the strike of the parent joint. Thus, it is inappropriate to interpret the right-stepping twist-hackle fringes only as an indication of anticlockwise stress rotation. In addition, a number of observations made in the study area support the suggestion that basin inversion occurred with a consistently N–S compression. These include the following. (i) Most Alpine joints, with very few exceptions, strike approximately E–W (Al-Mahruqi 2001; Engelder & Peacock 2001; Belayneh 2003). (ii) Left-stepping twist-hackle fringes (clockwise sense of stress rotation) can be observed on parent joints striking between 0° and 090° (see Fig. 11). This is not compatible with an anticlockwise sense of rotation of the stress field. In addition, Al-Mahruqi (2001) and Belayneh (2003) have described a local clockwise rotation of the stress field deduced from the chronology of joints at Lilstock, and East of Lilstock res-

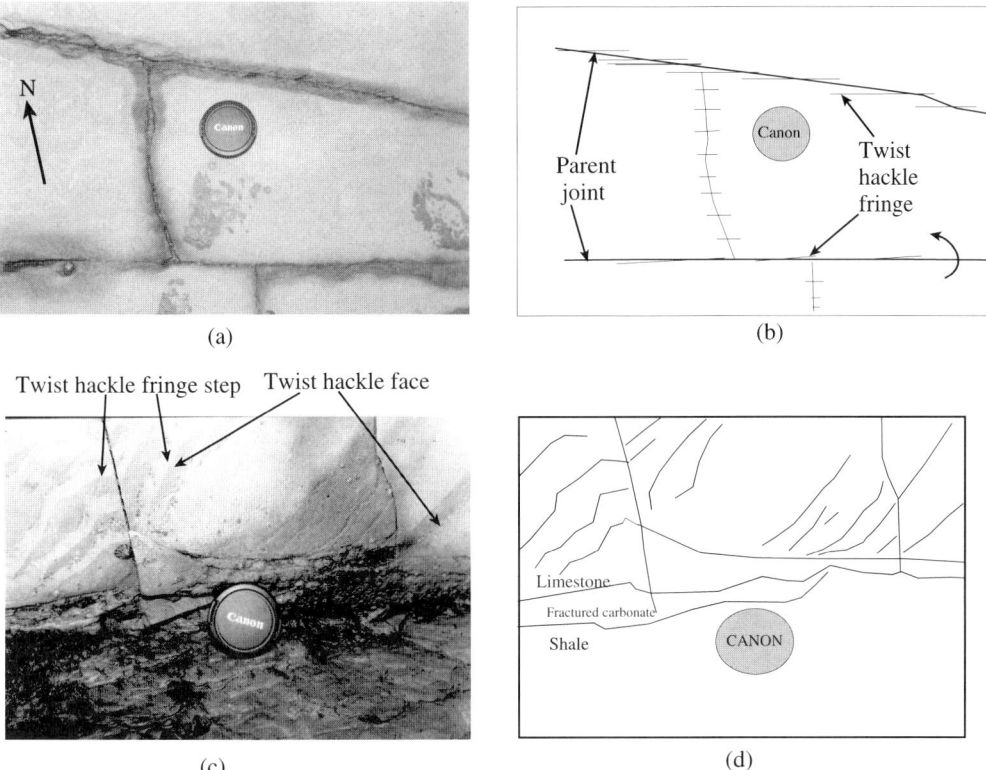

Fig. 11. (**a**) and (**c**) are photographs in plan and section views of a parent joint, (**b**) and (**d**) are line drawings of twist-hackle fringes on a 16 cm-thick limestone at Lilstock. Note that the entire joint is made up of a series of twist-hackle faces and fringe steps which gives an undulating surface on the faces of the parent joint. The twist-hackle fringes are right stepping for the parent joint striking between 090° and 180°, and give an anticlockwise sense of displacement during the formation of the joint. The camera cap is 5 cm for scale.

pectively. (iii) An E–W-striking pressure-solution cleavage post-dates the joint development at Kilve, Blue Ben and East Quantoxhead (Fig. 12) suggesting that the remote maximum principal compressive stress in the post-Alpine jointing phase of inversion of the BCB was approximately N–S. (iv) The present author has observed N–S-striking extensional veins linked to the strike-slip faults at Kilve and East Quantoxhead. In addition, strike-slip faults displacing the joints at Lilstock have been described by Al-Mahruqi (2001) and have also been observed by the present author at Lilstock and Kilve. The faults strike NW–SE and NE–SW, and form a conjugate sets compatible with an approximately N–S maximum principal compression. (v) Chadwick (1993), in his study of the Wessex Basin, confirms that the mid-Tertiary compressive stress was approximately N–S. (vi) The joint set formed around the hinge zone of folds in the area (e.g. Lilstock and Kilve anticlines) is consistent with outer arc extension associated with folding during N–S compression. The

Fig. 12. Pressure-solution cleavage striking approximately E–W and post-dating joint development in the area is evidence that the remote stress remained N–S during basin inversion. The pencil is 15.5 cm long for scale, and is aligned N–S.

Alpine joints and post-Alpine joints in the hinge region of the folds display a characteristic ladder pattern. (vii) Palaeostress determinations based on observations of fracture-surface markings in the interbedded limestones and mudstones of Liassic age between Lavernock Point and Saint Mary's Well Bay at Penarth, South Wales (Roberts 1995) suggest that σ_1 was oriented approximately N–S at the time of joint propagation.

The author would like to thank Mrs C. Thomas for financial support during his study at Imperial College. The work has benefited from the discussions and field trips to the north Somerset coast during the Paul L. Hancock memorial meeting at Weston-Super-Mare on 'The Mechanisms of Jointing in the Crust'. J. Cosgrove and T. Engelder are thanked for organising the conference and field trips to the area. A. Younes and K. Rawnsley, both from Shell, are thanked for their helpful comments in the early version of the manuscript. I would like thank T. Engelder for reviewing the paper, and for his detailed and helpful suggestions.

References

AL-MAHRUQI, S. A. S. 2001. *Fracture patterns and fracture propagation as a function of lithology.* PhD thesis, Imperial College, University of London.

AMEEN, M. S. (ed.). 1995. *Fractography: Fracture Topography as a Tool in Fracture Mechanics and Stress Analysis.* Geological Society, London, Special Publications, **92**.

BAHAT, D. & ENGELDER, T. 1984. Surface morphology on cross-fold joints of the Appalachian Plateau, New York and Pennsylvania. *Tectonophysics*, **104**, 299–313.

BAHAT, D. 1991. *Tectonofractography.* Springer, Heidelberg.

BAHAT, D. 1997. Mechanisms of dilatant en echelon crack crack formation in jointed layered chalks. *Journal of Structural Geology*, **19**, 1375–1392.

BAHAT, D., BANKWITZ, P. & BANKWITZ, E. 2001. Changes of crack velocity at the transition from the parent joint through en echelon to secondary mirror plane. *Journal of Structural Geology*, **23**, 1215–1221.

BANKWITZ, P. & BANKWITZ, E. 1995. Fractographic features on joints of KTB drill cores, Bavaria, Germany. *In*: AMEEN, M. S. (ed.) *Fractography: Fracture Topography as a Tool in Fracture Mechanics and Stress Analysis.* Geological Society, London, Special Publications, **92**, 39–58.

BANKWITZ, P. & BANKWITZ, E. 2001. *Differences of mirror and fringe pattern of primary joints in granitic plutons.* Mechanisms of Jointing in the Crust. Paul L. Hancock Memorial Meetings 1–4 August 2001. Program and abstracts, 27–30.

BELAYNEH, M. 2003. *Natural fracture networks in massive and well-bedded carbonates and the impact of these networks in fluid flow in dual porosity modelling.* PhD thesis, Imperial College, University of London.

BELAYNEH, M. & COSGROVE, J. W. (2004). Fracture-pattern variations around a major fold and their implications regarding fracture prediction using limited data: an example from the Bristol Channel Basin. *In*: COSGROVE, J. W. & ENGELDER, T. (eds) *The Initiation, Propagation, and Arrest of Joints and Other Fractures.* Geological Society, London, Special Publications, **231**, 89–102.

BLÉS, J. L. & FEUGA, B. 1986. *The Fracture of Rocks.* North Oxford Academic, Oxford.

CAPUTO, R. & HANCOCK, P. L. 1999. Crack-jump mechanism and its implications for stress cyclicity during extension fracturing. *Geodynamics*, **27**, 45–60.

CHADWICK, R. A. 1993. Aspects of basin inversion in southern Britain. *Journal of the Geological Society*, London, **150**, 311–322.

COOKE, M. L. & POLLARD, D. D. 1996. Fracture propagation paths under mixed mode loading with rectangular blocks of polymethyl methacrylate. *Journal of Geophysical Research*, **101**, 3387–3400.

CORTE, A. & HIGASHI, A. 1960. *Experimental Research on Desiccation Cracks in Soils.* SIPRE (Snow, Ice and Permafrost Research Establishment) Report **SIPRE-RR-66**, 1–71.

COSGROVE, J. W. 2001. Hydraulic fracturing during the formation and deformation of a basin: a factor in the dewatering of low-permeability sediments. *AAPG Bulletin*, **85**, 737–748.

CRUIKSHANK, K. M., ZHAO, G. & JOHNSON, A. D. 1991. Analysis of minor fractures associated with joints and faulted joints. *Journal of Structural Geology*, **13**, 865–886

DART, C. J., MCCLAY, K. & HOLLINGS, P. N. 1995. 3D analysis of inverted extensional fault systems, southern Bristol Channel Basin, UK. *In*: BUCHANAN, J. G. & BUCHANAN, P. G. (eds) *Basin Inversion.* Geological Society, London, Special Publications, **88**, 393–413.

DAVISON, I. 1995. Fault slip evolution determined from crack-seal veins in pull-aparts and their implications for the general slip models. *Journal of Structural Geology*, **17**, 1025–1034.

ENGELDER, T. 1987. Joints and shear fractures in rock. *In*: ATKINSON, B. K. (ed.) *Fracture Mechanics in Rocks.* Academic Press, London, 27–65.

ENGELDER, T. 1999. Transitional-tensile fracture propagation: a status report. *Journal of Structural Geology*, **21**, 1049–1055.

ENGELDER, T. & PEACOCK, D. C. P. 2001. Joint development normal to regional compression during flexural-flow folding: the Lilstock buttress anticline, Somerset, England. *Journal of Structural Geology*, **23**, 259–277.

ENGELDER, T. & SAVALLI, L. 2001. *Some of the peculiar characteristics of joint surface morphology: questions that they raise about the mechanical conditions for in-plane propagation joints.* Mechanisms of Jointing in the Crust. Paul L. Hancock Memorial Meeting, 1–4 August 2001. Program and abstracts, 31–35.

HANCOCK, P. L. 1985. Brittle microtectonics: principles and practice. *Journal of Structural Geology*, **7**, 437–457.

HANCOCK, P. L. 1994. From joints to paleostresses. *In*: ROURE, F. (ed.) *From Joints to Paleostress.* Technip Editions, Paris, 141–158.

HELGESON, D. E. & AYDIN, A. 1991. Characteristics of joint propagation across layer interfaces in sedimentary rocks. *Journal of Structural Geology*, **13**, 897–911.

HODGSON, R. A. 1961. Regional study of jointing in Comb Ridge–Navajo area, Arizona and Utah. *AAPG Bulletin*, **45**, 1–38.

KAMERLING, P. 1979. The geology and hydrocarbon habitat of the Bristol Channel Basin. *Journal of Petroleum Geology*, **2**, 75–93.

KELLY, P. G., PEACOCK, D. C. P., SANDERSON, D. J. & McGURK, A. C. 1999. Selective reverse-reactivation of normal faults, and deformation around reverse-reactivated faults in the Mesozoic of the Somerset coast. *Journal of Structural Geology*, **21**, 493–509.

KELLY, P. G., SANDERSON, D. J. & PEACOCK, D. C. P. 1998. Linkage and evolution of conjugate strike-slip fault zones in limestones of Somerset and Northumbria. *Journal of Structural Geology*, **20**, 1477–1493.

KULANDER, B. R. & DEAN, S. L. 1995. Observation on fractography with laboratory experiments for geologists. *In*: AMEEN, M. S. (ed.), *Fractography: Fracture Topography as a Tool in Fracture Mechanics and Stress Analysis*. Geological Society, London, Special Publications, **92**, 59–82.

LAWN, B. 1975. *Fracture of Brittle Solids*. Cambridge University Press, Cambridge.

LOOSVELD, R. J. H. & FRANSSEN, R. C. M. W. 1992. Extensional vs. shear fractures: implications for reservoir characterisation. Society of Petroleum Engineers, **SPE25017**, 23–30.

McCONAUGHY, D. T. & ENGELDER, T. 2001. Joint interaction with embedded concretions: joint loading configurations inferred from propagation paths. *Journal of Structural Geology*, **21**, 1637–1652.

MÜLLER, G. 2001. Experimental simulation of joint morphology. *Journal of Structural Geology*, **23**, 45–49.

MÜLLER, G. & DAHM, T. 2000. Fracture morphology of tensile cracks and rupture velocity. *Journal of Geophysical Research*, **105**, 723–738.

NEMČOK, M. & GAYER, R. 1996. Modelling palaeostress magnitude and age in extensional basins: a case study from the Mesozoic Bristol Channel Basin, UK. *Journal of Structural Geology*, **18**, 1301–1314.

NEMČOK, M., GAYER, R. & MILIORIZOS, M. 1995. Structural analysis of the inverted Bristol Channel Basin: implications for the geometry and timing of fracture porosity. *In*: BUCHANAN, J. G. and BUCHANAN, P. G. (eds), *Basin Inversion*. Geological Society, London, Special Publications, **88**, 355–392.

PEACOCK, D. C. P. & SANDERSON, D. J. 1991. Displacement, segment linkage and relay ramps in normal fault zones. *Journal of Structural Geology*, **13**, 721–733.

PEACOCK, D. C. P. & SANDERSON, D. J. 1994. Geometry and development of relay ramps in normal fault systems. *AAPG Bulletin*, **78**, 147–165.

PEACOCK, D. C. P. & SANDERSON, D. J. 1995. Strike-slip relay ramps. *Journal of Structural Geology*, **17**, 1351–1360.

POLLARD, D. D. & AYDIN, A. 1988. Progress in understanding jointing in the last century. *Geological Society of American Bulletin*, **100**, 1181–1204.

POLLARD, D. D. & SEGALL, P. 1987. Theoretical displacements and stresses near fractures in rocks, with application to faults, joints, veins, dikes and solution surfaces. *In*: ATKINSON, B. K. (ed.) *Fracture Mechanics in Rocks*. Academic Press, New York, 277–347.

POLLARD, D. D., SEGALL, P. & DELANEY, P. T. 1982. Formation and interpretation of dilatant *echelon* cracks. *Geological Society of American Bulletin*, **93**, 1291–1303.

PRICE, N. J. 1966. *Fault and Joint Development in Brittle and Semi-brittle Solids*. Pergamon Press, Oxford.

RAWNSLEY, K. D., PEACOCK, D. C. P., RIVES, T. & PETIT, J.-P. 1998. Joints in the Mesozoic sediments around the Bristol Channel Basin. *Journal of Structural Geology*, **20**, 1641–1661.

ROBERTS, J. C. 1995. Fracture surface markings in Liassic limestone at Lavernock Point, South Wales. *In*: AMEEN, M. S. (ed.) *Fractography: Fracture Topography as a Tool in Fracture Mechanics and Stress Analysis*. Geological Society, London, Special Publications, **92**, 175–186.

VAN HOORN, B. 1987. The South Celtic Sea/Bristol Channel Basin: origin, deformation and inversion history. *In*: ZEIGLER, P. A. (ed.) Compressional Intra-Plate Deformations in the Alpine Foreland. *Tectonophysics*, **137**, 309–334.

WHITTAKER, A. & GREEN, G. W. 1983. *Geology of the Country around Weston-Super-Mare*. Memoir of the Geological Survey of Great Britain, Sheet 279, and parts of sheets 263 & 295.

WILLEMSE, E. J. M., PEACOCK, D. C. P. & AYDIN, A. 1997. Nucleation and growth of strike-slip faults in limestones from Somerset. *Journal of Structural Geology*, **19**, 1461–1477.

WOODWORTH, J. 1896. On the fracture system of joints, with remarks on certain great fractures. *Boston Society of Natural History Proceedings*, **27**, 163–184.

YOUNES, A. I. & ENGELDER, T. 1999. Fringe cracks: key structures for the interpretation of the progressive Alleghanian deformation of the Appalachian Plateau. *Geological Society of American Bulletin*, **111**, 219–239.

Eight distinct fault–joint geometric/genetic relationships in the Beer Sheva syncline, Israel

DOV BAHAT

Department of Geological and Environmental Sciences, Ben Gurion University of the Negev, Beer Sheva, Israel

Abstract: Extended research over two decades reveals more than 20 distinct fracture episodes in the Beer Sheva syncline. This paper focuses on eight fault–joint systems that differ from each other in their genetic affiliation and/or their geometric relationship and fracture properties. Two systems are linked to burial, whereas six others relate to various syntectonic–uplift associations. Joint sets within these systems are categorized into three, pre-, syn- and post-fault groups. Correspondingly, synfault early uplift and post-fault early uplift events can be distinguished from prefault and syn-fault late uplift events. Water drainage may be considerably improved along certain fault–joint systems. Accordingly, in formulating fracture-network models the particular distribution of fault–joint systems and their properties need to be taken into consideration.

Joints are the most ubiquitous structures in sedimentary rocks. Therefore, timing of fracture development relative to the history of synforms (and basins) is based on the temporal relationship of joints with the fold. Engelder (1985) and Bahat (1991) proposed genetic classifications of joints. According to the latter classification, joints may be divided into four groups: burial, syntectonic, uplift and post-uplift. These can be distinguished in the field by both geometric and fractographic criteria. The four groups reflect four characteristic stages in basin evolution. In retrospect, the geometric and genetic classifications of joints are difficult, mostly due to the great complexity of joint formation. Joints may be formed under small differential stresses, and, therefore, most environments are conducive to their development, both systematically and non-systematically (Hodgson 1961).

The last 50 years of fault investigation enjoyed the benefit of the pioneering classification of faults into three classes by Anderson (1951). The great power of Anderson's grouping stemmed from its simplicity. It helped to demonstrate the dependence of the three main fault types on systematic changes in the stress field that also reflected on the most likely tectonic conditions. However, fault classification cannot be applied to the timing of basin histories. On the other hand, results of extended research on the temporal fault–joint relationship (FJR) (e.g. Sterns 1968; Pohn 1981; Segall & Pollard 1983; Cruikshank et al. 1991; Martel 1997; Kattenhorn et al. 2000; Eyal et al. 2001; Wilkins et al. 2001) can be used in refining the fracture histories of basins. The work by Peacock (2001) is a recent summary of some key observations on FJR.

Unlike many other folds that exhibit relatively uniform structures, the Beer Sheva Eocene syncline is divided into four distinct rock units that display

different fracture characteristics and store a record of many (more than 20) different fracture episodes since the Lower Eocene (Bahat et al. 2004). One manifestation of this multifracture complex is the occurrence of at least eight distinct FJR. Various aspects of the FJR of this syncline were introduced in previous publications, including, pre- and post-fault burial joints, synfault syntectonic joints and post-fault uplift joints that partly transform into faults (e. g. Bahat 1991, pp. 241, 275 and 282, respectively). A synthesis of these observations is a major part of the present study.

Eight types of temporal fault–joint relationships in the Beer Sheva syncline

Eight FJR types from the Beer Sheva syncline (Fig. 1) are characterized below. Table 1 summarizes FJRs that take into account different: (1) temporal relationships; (2) geometric contacts; (3) relative ages; (4) stresses; and (5) interaction styles. Six types relate to normal faults, one type concerns the terminal zone of a strike-slip fault and the eighth type is about vertical faults (Fig. 2a–f).

Type 1 occurs in the Mor Formation from the Lower Eocene and is associated with the burial stage of the syncline. The fault strikes approximately 292° and dips c. 45°N, cutting alternating chalk layers (about 90 cm thick) with chert beds (about 10 cm thick) (Fig. 2a). The chalk is dissected by two orthogonal, vertical, single-layer joint sets oriented 328° and 059°. Hence, the joint orientations are unrelated to the fault attitude. During the burial stage of the syncline differential stresses gradually increased, such that the formation of the cross-fold joints and some strike-parallel joints preceded the creation of the fault. The difference in orientation

From: Cosgrove, J. W. & Engelder, T. (eds) 2004. *The Initiation, Propagation, and Arrest of Joints and Other Fractures.* Geological Society, London, Special Publications, **231**, 257–267. 0305-8719/04/$15 © The Geological Society of London 2004.

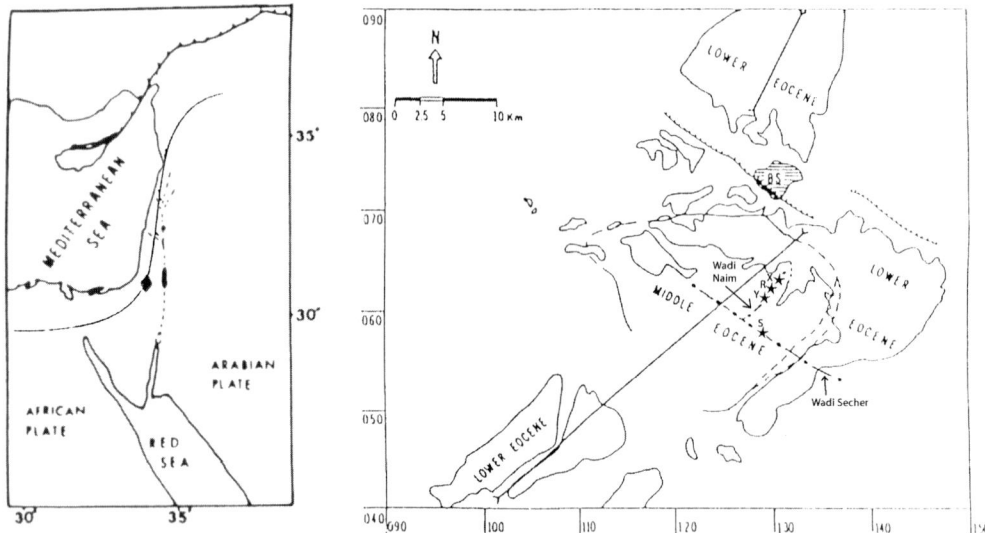

Fig. 1. The map of part of the northern Negev in Israel to the right. Wadi Naim and Wadi Secher occur in the northern part of the Beer Sheva syncline, which is situated south of Beer Sheva city (BS). The Shephela syncline is at the north of the city. The map is marked by a rhombus at the centre of the regional map on the left.

between the joint and fault strikes requires that the remote stresses had to change from the conditions under which the joints were formed to new conditions that induced the fault. As a rule, joint orientation and spacing is almost the same in the four sectors of the fault (Fig. 3a, d).

Type 2 relates to the left fault, shown with its conjugate associate in Figure 2a, but it concerns a local spacing deviation from the above rule, i.e. there is a reduction in joint spacing in sector I along one chalk layer (bottom of Fig. 2a and Fig. 3). Gross *et al.* (1997) found that due to the dragging of one of the chert beds, an unrelieved strain remained stored in the hanging wall from the early stages of the fault slip. This unrelieved strain was subsequently released in the form of additional 059°-oriented joints. Hence, the fracture sequence during the burial stage was two early orthogonal joint sets, normal faults and some subsequent additional fold strike-parallel joints.

Two more joint-spacing peculiarities were observed in the same outcrop (Bahat 1988). First, joint spacing is wider in thinner chalk beds, i.e. the layer thickness–joint spacing plot shows a negative slope for both cross-fold joints and strike joints, which is in contrast to most observations by other investigators that show positive plots. An exceptional finding of a negative slope was recently made by Eyal *et al.* (2001) in the chalky limestone beds from Nahal Neqarot in southern Israel. Second, for a given layer thickness, spacing is a little wider in the strike joints (oriented 059°) than in the cross-fold joints (oriented 328°). A resolution of the latter two peculiarities

requires further studies. Currently, it may be speculated that spacing is wider in the strike joints because they were formed in response to reduced reoccurring stress relaxation, compared to the cross-fold joints that reflected greater reoccurring regional stresses (Bahat *et al.* 2004).

Types 3 and 4 occur along Wadi Secher (Fig. 1), and relate to the Horsha Formation from the Middle Eocene chalks of the syncline. Type 3 concerns a set of inclined joints that are parallel to the dip of the Secher Fault (Fig. 2b). Some 12 joints of this type are spatially confined to sector I of the Secher Fault (one or two joints also occur in sector IV). Generally, the damage volume around faults contains secondary structures (mostly fractures) and it has the shape of an ellipsoid (Fig. 3b). The tipline forms an elliptical line that marks the loci on the fault plane (the 'parallel plane') on which displacement has decreased to zero (Fig. 3c). The 'normal plane' is orthogonal to the parallel plane and is divided into four sectors by the trace of the fault (Fig. 3d). For normal faults the extensional sectors, I and III are widest (where rock is weak) and narrowest in the contractional sectors II and IV (where rock is strong) (e.g. Knott *et al.* 1996). Unexpectedly, the width of sector I was found to be about 60% of the length of the Secher Fault (study in progress), implying a strong deviation of the normal plane from an elliptical shape (Fig. 3e), resulting in a 'quasi-ellipsoid damage volume'.

The joints of type 3 resemble the joints that are parallel to the dip of the normal fault described by

Table 1. *Eight types of fault–joint relationships (FJR) in the Beer Sheva syncline, Israel*[*]

FJR type[†]	Temporal relationship of joints	Joint attitude in relation to the fault one	Tectonic setting	Joint formation by stresses remote/local	Fracture system
1 (n)	Prefault	Different, independent	Burial early	Remote	Closed, non-interacting joints, regular spacing
2 (n)	Post-fault	Different, independent	Burial late	Remote/local	Closed, non-interacting joints, decreased spacing
3 (n)	Synfault	Correlateable, dependent	Syntectonic early uplift	Local	Local fracture multiplication (in sector I)
4 (n)	Synfault	Corelateable, dependent	Syntectonic early uplift	Local/remote	Local twisted secondary faults (in sector III)
5 (n)	Post-fault	Different, independent	Uplift late uplift	Remote	Regular joint spacing
6 (n)	Synfault	Correlateable, dependent	Uplift late uplift	Remote	Decreased joint spacing
7 (s)	Post-fault	Secondary stress field	Syntectonic early uplift	Local	Non-interacting joints, variable spacing along primary and secondary faults.
8 (v)	Transitional	parallel	Uplift	Remote/local	Deep penetrating network producing vertical joints and faults

[*]Elaborations on the various FJR types are Bahat (1991, 1999; Bahat *et al.* 2004) and Gross *et al.* (1997).
[†](n), normal fault; (s), strike-slip fault; (v), vertical fault.

Stearns (1968) (Fig. 4). They also resemble the FJR of large strike-slip faults in which joints cut parallel to the dip (and strike) of the primary fault close to the fault termination zone (Chinnery 1966). In order to obtain the contours of these joints the local directions of maximum shear stress need to be translated to the directions of principal strresses in a simple 45° rotation exercise (Means 1979, problem 12.4). Type 3 relates to joints that formed in new, local stress fields that were created when displacements occurred along the existing primary faults (Chinnery 1966). The joints of type 3 formed while the local minimum principal stress σ_3 was normal to the fault plant (Fig. 3a). This type is associated with an early uplift stage in the Beer Sheva syncline (see below).

Type 4 concerns a set of three fractures (f_1–f_3) that stemmed from the Secher Fault in sector III (faults 1–3 in Fig. 2b). They exhibit changes in orientation relative to that of the fault. The changes reflect combined strike twist and a gradual dip tilting from an attitude relatively close to that of the fault to an almost vertical fracture that is totally divorced from the trend of the fault. The fault strikes 040° and dips 51° NW, f_1 strikes 035° and dips 68° NW, f_2 strikes 38° and dips 69° NW, and f_3 strikes 088° and dips 85° NW. Apparently, the mixed-mode loading was intensified with distance of the fractures from the fault (up to about 3.5 m), producing a series of acute angles between frac-

tures f_1–f_3 and the fault. This reflects a gradual increase of the influence of the remote stresses at the expense of the local stresses that were effective in initiating fractures f_1–f_3. There is some displacement along these fractures, which distinguishes them from the joints of the other types. Type 4 resembles the fracture set that occurs at a large angle ($> 45°$ but $< 90°$) relative to the fault in the Stearns model (Fig. 4). However, in the latter the angles created between the fault and joints are larger than in type 4. Type 4, like type 3, is thought to be genetically related to displacements along the fault that took place in association with an early uplift stage in the Beer Sheva syncline (see below).

Type 5 resembles type 1 in showing vertical joints with strike orientation unrelated to the strike of Naim fault A (Fig. 2c). There are no visible differences in the behaviour of joints from both sides of the fault. The strike of the joint set (recognized by characteristic plumes) forms an angle larger than 45° with the strike of the fault. However, unlike the system described by Kattenhorn *et al.* (2000), there are no genetic relations between the fault and joints. The young fault cuts single-layer uplift joints that were associated with an earlier stage of the uplift in the syncline (Bahat 1999). Therefore, the fault of type 5 was associated with a late uplift stage of the syncline.

Type 6 consists of vertical joints that subparallel

(a)

Fig. 2. Structural relationships of eight (**a**–**f**) fault–joint types, in section, from the Beer Sheva syncline. (**a**) Types 1 and 2. Top: conjugate faults; bottom: focus on the left fault. Vertical single-layer two joint sets with their unrelated strikes to those of two conjugate normal faults in the Lower Eocene. Joints and faults were formed in different remote stress fields. Type 2 occurs in the hanging wall of the fault, to the left (it is difficult to distinguish type 1 from type 2 in this exposure). (**b**) Type 3 is represented by inclined 'joints' that parallel the dip of a normal fault in Wadi Secher at the upper hanging wall (in sector I). Type 4 is shown as inclined 'fractures' at various acute angles to the fault strike at the lower footwall (marked 1–3, respectively, in sector III). (**c**) Type 5. Vertical joints with unrelated orientation relative to the strike of a normal fault in Wadi Naim. Joints and faults were formed at different times in distinct remote stress fields. (**d**) Type 6. Vertical joints that subparallel the strike of a normal fault (in sectors II and III). All fractures were developed in the same stress field. (**e**) Type 7 is represented by a fault-termination zone of a strike-slip fault consisting of three major structures: (1) a primary fault, Pr; (2) secondary faults, S1–S3; and (3) a syntectonic joint set, 344°, associated with the faults (after Bahat 1999). (**f**) Three sections (from stations (A) 20, (B) 63 and (C) 58, after Bahat 1991, Fig. 5.29) display type 8 in the Lower Eocene chalks of the Beer Sheva area. Inclined solid lines are normal faults, vertical solid lines are multilayer vertical faults (marked by Arabic numbers) and dashed lines (marked by Roman numerals) are multilayer joints.

(b)

(c)

Fig. 2. continued

(d)

(e)

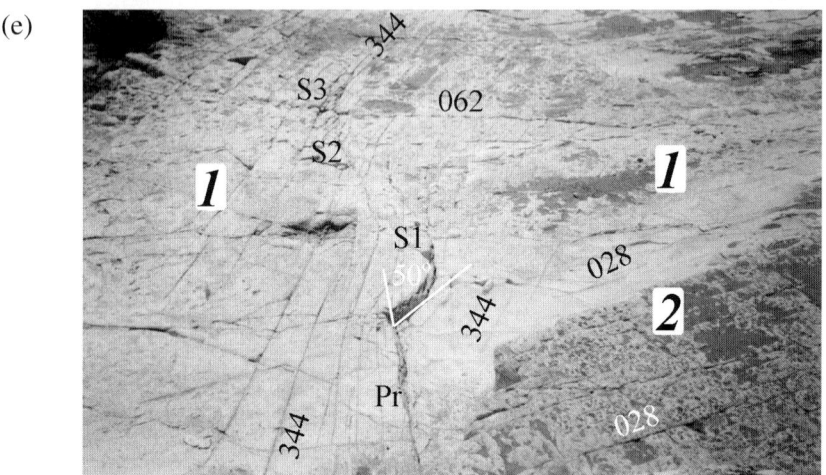

(f)

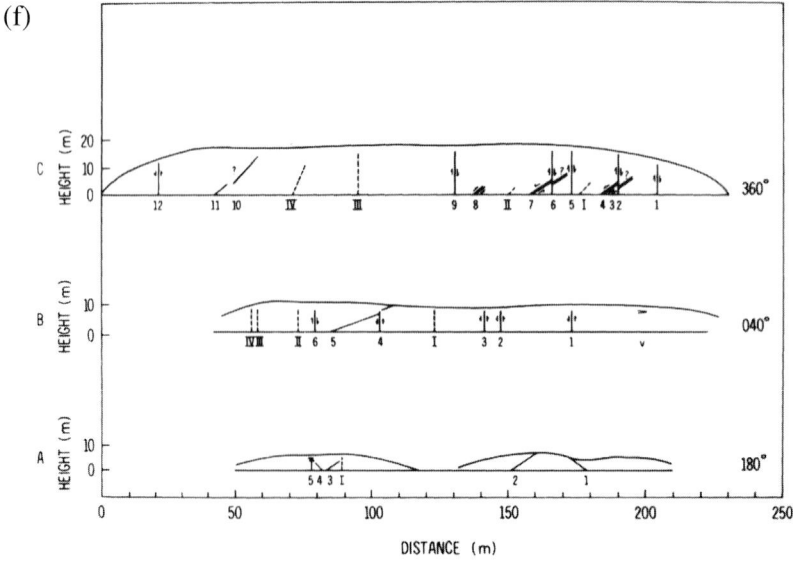

Fig. 2. continued

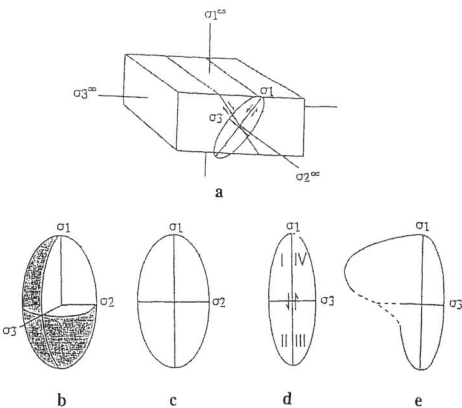

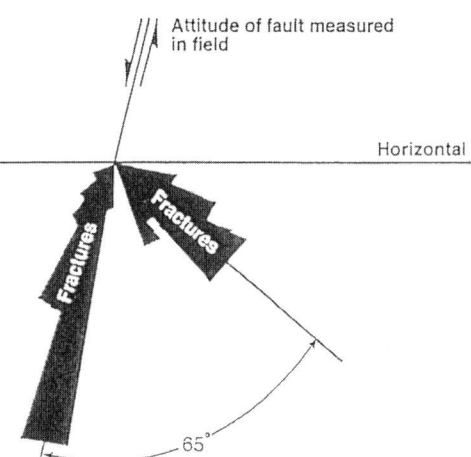

Fig. 3. Diagram of primary normal fault and secondary fractures in three dimensions. (**a**) The normal plane (ellipse) that describes the arrangement of local principal stresses is superposed on one of two conjugate faults in a rectangular block that describes the remote principal stresses σ_1^∞–σ_3^∞. Orientations of remote and local principal stresses differ significantly. (**b**) Ellipsoid in perspective, showing the fault plane σ_1–σ_2, which is termed the 'parallel plane' with its elliptical tipline boundary, and the 'normal plane' σ_1–σ_3. (**c**) The fault plane with its elliptical tipline boundary, and (**d**) the 'normal plane' divided, into four sectors I–IV. (**e**) The approximate shape of sector I in the normal plane of the Secher Fault. These figures contain geometric elements from Means (1979) and Park (1983).

Fig. 4. Rose diagram showing the distribution of dips of two sets of fractures associated with a normal fault (after Sterans 1968; and Twiss & Moore 1992).

the fault strike. The joints show a significant decrease in spacing in sector III (and to a lesser extent in sector II) close to the lower fault tip of Naim fault B (Fig. 2d). Compared to types 1 and 5, type 6 represents an assemblage of joints and a fault that were formed in the same local stress field. Thus, even if the faults from types 5 and 6 cut the same chalk layers, and are probably genetically linked to the same faulting episode, these two types have different FJR manifestations (Fig. 2c and d, respectively). Whereas the fault of type 5 is not associated with syntectonic joints that formed due to fault slip, the fault of type 6 created a local stress field that formed syntectonic joints. The reason for this difference is not clear. Possibly, the syntectonic joints of type 6 are the youngest fractures among the various fracture types characterized in this study.

Type 7 is represented by a fault-termination zone consisting of three major structures: (a) a primary fault; (b) secondary faults; and (c) a joint set associated with the faults (Bahat 1999) (Fig. 2e). (a) The primary fault is a vertical right-lateral strike-slip fault, striking 318°; (b) a set of three partly curved secondary faults initiate at the tip of the primary fault; and (c) a joint set striking 344° that curves in sympathy with the primary and secondary faults is confined to

close proximities with the faults. Therefore, the joints of set 344° are considered to be genetically associated with the faults, and they are termed syntectonic joints (Bahat 1991, fig. 5.13). Joint set 028° is ubiquitous in Wadi Naim, cutting layers 1 and 2 in Fig. 2e. However, where these joints approch the termination zone they arrest or interact (hooking style) with set 344°, but they never penetrate into the termination zone. Thus, joint set 028° post-dated the above three structures in the same chalk layer and belongs to a series of rotating sets that formed during an advanced stage of uplifting (Bahat *et al.* 2004). One of these sets is displaced by the fault of type 5 (Fig. 2c).

Type 8 is found in the Lower Eocene chalks of the Beer Sheva area, single-layer burial joints (about 12 cm regular spacing), which are occasionally displaced by normal faults. Vertical multilayer joints and vertical multilayer faults that are associated with them (in irregular 5–15 m spacing) often traverse the entire outcrops and may run alongside or displace normal faults (Fig. 2f). These vertical fractures are therefore considered to have been formed after the burial stage, during single- or multiuplift episodes of the area (Bahat 1999). There is, however, no clear age relationship between uplift joints cutting Lower Eocene and those cutting Middle Eocene chalks.

Discussion

Genetic relationships

The five fault–joint types, 3–7, occur in the Middle Eocene. Types 3 and 4 crop out in older chalk layers

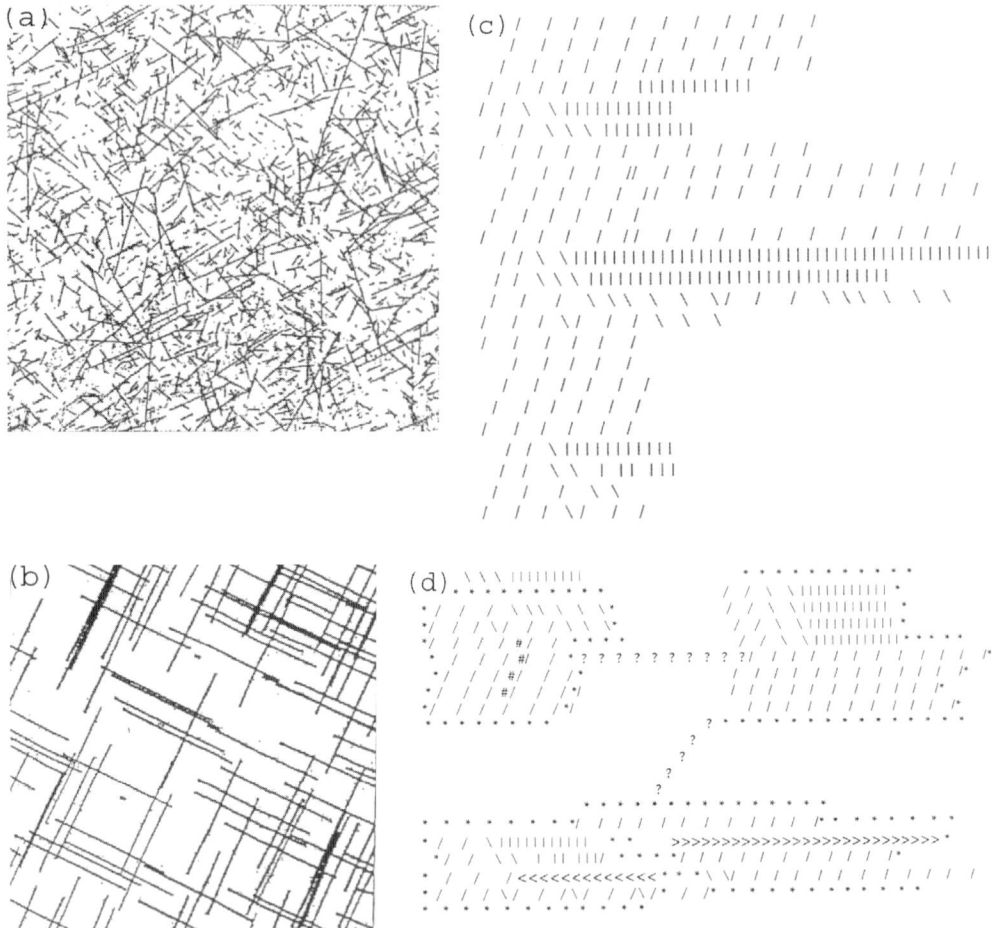

Fig. 5. Hypothetical maps of fracture networks. (**a**) Random distribution of joints (slightly modified from a simulation by Odling 1992). (**b**) Orthogonal two fracture joints sets above percolation threshold (slightly modified from a simulation by Odling *et al.* 1999). (**c**) A continuous hierarchical network of three joint sets, while the set oriented NNE–SSW dominates. (**d**) A discontinuous network is represented by three domains that are separated by unknown transitional zones, (e.g. due to talus?), where faults are marked by # and >, and domain boundaries are shown by small stars.

at the flank of the fold (Wadi Secher, Fig. 1) and exhibit synfault–early uplift fracture. The fault was created by remote stresses, but the 'joints' and 'fractures' were formed in new, local stress fields that were created in association with slip (or slips) along the fault (Fig. 3a). Types 5–7 are exposed in younger layer, close to the synclinal centre (Wadi Naim, Fig. 1) where fracture started in the formation of a strike-slip fault that was followed by jointing at the termination zone of the fault (post-fault–early uplift fracture exhibited by type 7) (Bahat 1999). A series of rotating joint sets developed after a certain time interval, and were then displaced by a normal fault (post-fault–late uplift

fracture exhibited by type 5) (Bahat 1999). Finally, a syntectonic joint set was formed in association with a normal fault (synfault–late uplift fracture exhibited by type 6). There is no clear field evidence that would indicate what the age relationships are between types 3 and 4 and types 5–7. However, the fractures associated with types 3 and 4 were formed in pristine chalks – in which no earlier fracture had taken place. On the other hand, types 5 and 6 occurred in chalk layers that had already been affected by previous fracture episodes (like type 7). Therefore, it is tentatively speculated that types 3 and 4 preceded types 5–7.

The importance of temporal fault–joint relationships

The temporal fault–joint relationships are important for a variety of reasons that are specified below:

- sequences between faults and joints may help to decipher the tectonic history of the region (including the occurrence of earthquakes) and provide information on the stress history of the region (critical for petroleum exploration);
- faults may affect later joint development and joints may affect later fault growth, and their interactions are particularly mechanically intriguing;
- distinct joint–fault genetics may be linked with different fracture parameters (such as aperture, length, spacing and interaction) (Bahat 1998);
- fluid flow (water, hydrocarbon liquids and gas) depends on fault–joint relationships and their fracture parameters. For instance (Bahat 1991):
  - burial pre-joints may be closed due to their development in a process of increasing lithostatic pressure;
  - some syntectonic, synfault joints may be sealed into veins, while others strongly enhance water flow, like the strike-parallel joints (062°) in the Beer syncline;
  - unlift joints that form after burial-normal faults are opened fractures;
- fault–joint systems may form important parts of fracture networks that are the subject matter of the next section.

Network-fracture models

There have been extended studies on water flow in the Ramat Hovav area (e.g. Dahan *et al.* 1999, 2000; Weisbrod *et al.* 1999), which is situated in the northern part of the Beer Sheva syncline (Fig. 1) (Bahat 1991, fig. 5.29). Such studies commonly involve the construction of network-fracture models (e.g. Berkowitz *et al.* 1988). The present study points out the ubiquity of fault–joint systems in the syncline. Joints and faults cutting chalks (as well as other rocks) may provide poor drainage systems to water flow due to secondary material, termed fault gouge or fault-zone breccia, that often forms along them (Avigur & Bahat 1990; Dahan *et al.* 1999, 2000). However, it is generally well accepted among hydrologists that quite frequently faults enhance rock permeability and water flow (e.g. Birkeland & Larson 1978). According to Shan *et al.* (1995), in groundwater hydrology and petroleum engineering faults are classified according to their hydraulic response during a field pumping test. Many faults are 'leaky', for them the hydraulic conductivity, K,

of the fault (Witherspoon *et al.* 1967) is given $0 < K < \infty$, and they provide important flow paths for liquids to move underground. Furthermore, liquid drainage may be considerably improved along fracture intersections (e.g. Hsieh *et al.* 1985). Hence, the increased liquid flow along faults combined with the improved conductivity due to fault–joint intersections is expected to create important drainage in fault–joint systems that need to be taken into consideration in hydraulic analyses. Such systems may occasionally be rejuvenated by earthquake reoccurrence.

The maps in Figure 5 compare several hypothetical models of fracture networks. These include a random distribution of joints (Fig. 5a), two orthogonal joint sets (Fig. 5b) and an hierarchical network of three joint sets, one of which is the dominant set (Fig. 5c). The fourth model displays a continuous network consisting of separated, adjacent domains of fault–joints (Fig. 5d).

The random network (Fig. 5a) is unrealistic because it ignores the tendency of most fractures (joints) to form sets. There are many reports in the literature on orthogonal sets (Fig. 5b). Quite intriguingly, orthogonal joint sets are common in the Mor Formation at the tips of the two Eocene synclines north and south of Beer Sheva (Fig. 1), but they are rare in outcrops of Lower Eocene and Middle Eocene that occur closer to the fold centres (Bahat *et al.* 2004). The hierarchical network (Fig. 5c) shows schematically three joint sets that trend NNW–SSE, N–S and ENE–WSW with a particular emphasis on the ENE–WSW set. This diagrammatically represents previous actual interpretations of water flow in the area. Apparently, most water flow occurred along a particular joint set (the 062° set), rather than 'averaging' the flow among the various sets cutting the chalk layers (Rophe *et al.* 1992; Bahat & Adar, 1993).

The faults invariably exhibit various geometrical relationships with their neighbouring joints and form distinct domains, according to their genetic affiliations (Table 1, Fig. 5d). These domains are separated from each other by narrow boundaries (small stars in the figure) or transitional zones. The domains are thought to frequently have a greater water flow than their surroundings. Accordingly, future estimates of water flow in the area will have to distinguish between the contributions of fault–joint domains and their associated boundaries/transitional zones, and other areas of joint networks.

In conclusion, the problematics of groundwater flow in Ramat Hovav has attracted an international attention, resulting in intensive research by 'field hydrologists' and 'model hydrologists', all seeking to meet at the interface between their different approaches. Future research will show how fault–joint systems will be taken into consideration.

Summary

Sequences between faults and joints may help to decipher the tectonic history of basins and provide information on the stress history of the region. Joint–fault genetics may be linked with different fracture parameters and are important in the construction of fracture networks.

The single-layer joint v. fault age relation in the uplifted chalk layers of the Horsha Formation from the Middle Eocene differs from the fracture succession in the burial stage of the Mor Formation from the Lower Eocene.

- Joint set 328°, and partially joint set 059° (types 1 and 2, respectively), preceded normal faults during the burial stage in Lower Eocene in the synclinal tip.
- Vertical multilayer joints and associated multilayer faults formed after the burial stage in the Lower Eocene chalks, during single- or multi-uplift episodes of the area.

Types 3 and 4 exhibit synfault–early uplift fracture. Types 5–7 initiated in the formation of a strike-slip fault that was followed by jointing at the termination zone of the fault (post-fault–early uplift fracture exhibited by type 7). Then, a series of rotating joint sets developed after a certain time interval, and were then displaced by a normal fault (post-fault–late uplift fracture exhibited by type 5). Finally, a syntectonic joint set was formed in association with a normal fault (synfault–late uplift fracture exhibited by type 6).

The random fracture network model is unrealistic because it ignores the tendency of most joints to form sets. There are many reports in the literature on orthogonal sets. The hierarchical network represents actual constraints on water flow in the area. The domain network is of importance because it frequently has a greater water flow than its surroundings.

The description of types 3 and 4 benefitted from information obtained by R. Shavit and Z. Levy for their 'Student projects'. Personal communication from S. P. Neuman is greatly appreciated.

References

ANDERSON, E. M. 1951. *The Dynamics of Faulting*. Oliver and Boyd, Edinburgh.

AVIGUR, A. & BAHAT, D. 1990. Chemical weathering of fractured Eocene chalks in the Negev, Israel. *Chemical Geology*, **89**, 149–156.

BAHAT, D. 1985. Low angle normal faults in Lower Eocene chalks near Beer Sheva, Israel. *Journal of Structural Geology*, **7**, 613–620.

BAHAT, D. 1988. Early single-layer and late multi-layer joints in the lower Eocene chalks near Beer-Sheva, Israel. *Annales Tectonicae*, **2**, 3–11.

BAHAT, D. 1991. *Tectonofractography*. Springer, Berlin.

BAHAT, D. 1998. Four joint genetic-groups and their distinct characteristics. *In*: ROSSMANITH, H. P. (ed.) *Mechanics of Jointed and Faulted Rock*. Balkema, Rotterdam, 211–216.

BAHAT, D. 1999. Single-layer burial joints versus single-layer uplift joints in Eocene chalk from the Beer Sheva syncline in Israel. *Journal of Structural Geology*, **21**, 293–303.

BAHAT, D. & ADAR, E. M. 1993. Comment on 'Analysis of subsurface flow and formation anisotropy in a fractured aquitard using transient water level data' by B. ROPHE, B. BERKOVITZ, M. MARGARITE and D. RONEN. *Water Resources Research*, **29**, 4171–4173.

BAHAT, D., RABINOVITCH, A. & FRID, V. 2004. *Tensile Fracturing in Rocks*. Springer, Berlin.

BERKOWITZ, B., BEAR, J. & BRAESTER, C. 1988. Continuum models for contaminant transport in fractured porous formations. *Water Resources Research*, **24**, 1225–1236.

BIRKELAND, P. W. & LARSON, E. E. 1978. *Putnam's Geology*, 3rd edn. Oxford University Press, New York.

CHINNERY, M. A. 1966. Secondary faulting I. theoretical aspects. *Canadian Journal of Earth Science*, **3**, 163–174.

CRUIKSHANK, K. M., ZHAO, G. & JOHNSON, A. M. 1991. Analysis of minor fractures associated with joints and faulted joints. *Journal of Structural Geology*, **13**, 865–886.

DAHAN, O., NATIV, R., ADAR, E. M., BERKOWITZ, B. & RONEN, Z. 1999. Field observation of flow in a fracture intersecting unsaturated chalk. *Water Resources Research*, **35**, 3315–3326.

DAHAN, O., NATIV, R., ADAR, E. M., BERKOWITZ, B. & WEISBROD, N. 2000. On fracture structure and preferential flow in unsaturated chalk. *Ground Water*, **38**, 444–451.

EYAL, Y., GROSS, M. R., WACKER, M., ENGELDER, T. & BECKER, A. 2001. Joint development during fluctuation of the regional stress field in southern Israel. *Journal of Structural Geology*, **23**, 279–296.

ENGELDER, T. 1985. Loading paths to joint propagation during cycle: an example of the Appalachian Plateau, U.S.A. *Journal of Structural Geology*, **7**, 459–476.

GROSS, M. R., BAHAT, D. & BECKER, A. 1997. Relations between jointing and faulting based on fracture-spacing ratios and fault-slip profiles: A new method to estimate strain in layered rocks, *Geology*, **25**, 887–890.

HODGSON, R. A. 1961. Regional study of jointing in Comb Ridge–Navajo Mountain area, Arizona and Utah. *AAPG Bulletin*, **45**, 1–38.

HSIEH, P. A., NEUMAN, S. P., STILES, G. K. & SIMPSON, E. S. 1985. Field determination of the three dimensional hydraulic conductivity tensor of anisotropic media 2. Methodology and application to fractured rocks. *Water Resources Research*, **21**, 1667–1676.

KATTENHORN, S. A., AYDIN, A. & POLLARD, D. D. 2000. Joints at high angles to normal fault strike: an explanation using 3-D numerical models of fault-perturbed stress fields. *Journal of Structural Geology*, **22**, 1–23.

KNOTT, S. D., BEACH, A., BROCKBANK, P. J., BROWN, J. L., MCCALLUM, J. E. & WELBON, A. I. 1996. Spatial and mechanical controls on normal fault populations. *Journal of Structural Geology*, **18**, 356–372.

MEANS, W. D. 1979. *Stress and Strain*. Springer, Heidelberg.

ODLING, N. E. 1992. Network properties of a two-dimensional natural fracture pattern. *Pure and Applied Geophysics*, **138**, 95–114.

ODLING, N. E., GILLESPIE, P., BOURGINE, B., CASTAING, C., CHILES, J.-P., CHRISTENSEN, N.P., FILLION, E., GENTER, A., OLSEN, C., THRANE, L., TRICE, R., AARSETH, E., WALSH, J. J. & WATTERSON, J. 1999. Variations in fracture system geometry and their implications for fluid flow in fractured hydrocarbon reservoirs. *Petroleum Geosciences*, **5**, 373–384

MARTEL, S. 1997. Effects of cohesive zones on small faults and implications for secondary fracturing and fault trace geometry. *Journal of Structural Geology*, **19**, 835–847.

PARK, R. G. 1983. *Foundations of Structural Geology*. Chapman & Hall, New York.

PEACOCK, D. C. P. 2001. The temporal relationship between joints and faults. *Journal of Structural Geology*, **23**, 329–341.

POHN, H. A. 1981. Joint spacing as a method of locating faults. *Geology*, **9**, 258–261.

ROPHE, B., BERKOWITZ, B., MAGRITZ, M. & RONEN, D.

1992. Analysis of subsurface flow and anisotrophy in a fractured aquitard using transient water level data. *Water Resources Research*, **28**, 199–207.

SEGALL, P. & POLLARD, D. D. 1983. Joint formation in granitic rock of the Sierra Nevada. *Geological Society of America Bulletin*, **94**, 563–575.

SHAN, C., JAVANDEL, I. & WITHERSPOON, P. A. 1995. Characterization of leaky faults: Study of water flow in aquifer–fault–aquifer systems. *Water Resources Research*, **31**, 2897–2904.

STEARNS, D. W. 1968. Certain aspects of fracture in naturally deformed rocks. *In*: RIECKER, R. E. (ed.) *Rock Mechanics Seminar, Vol.1, Special Report*, Terrestrial Sciences Lab. Air Force, Cambridge Research Labs, Bedford, MA, 97–116.

TWISS, R. J. & MOORES, E. M. 1992. *Structural Geology*. Freeman, New York.

WEISBROD, N., NATIV, R., ADAR, E. M. & RONEN, D. 1999. Impact of intermittent rain water and wastewater flow on coated fractures in chalk. *Water Resources Research*, **35**, 3211–3222.

WILKINS, S. J., GROSS, M. R., WACKER, M., EYAL, Y. & ENGELDER, T. 2001. Faulted joints: kinematics, displacement–length scaling relations and criteria for their identification. *Journal of Structural Geology*, **23**, 315–327.

WITHERSPOON, P. A., JAVANDEL, I., NEUMAN, S. P. & FREEZE, R. A. 1967. *Interpretation of Aquifer Gas Storage Conditions From Water Pumping Tests*. American Gas Association, New York.

Probabilistic–mechanistic simulation of bed-normal joint patterns

WENDY HOFFMANN[1], WILLIAM M. DUNNE[1] & MATTHEW MAULDON[2]

[1] Department of Geological Sciences, University of Tennessee, Knoxville, TN 37996–1410, USA
(e-mail: wdunne@utk.edu)
[2] Department of Civil & Environmental Engineering, Virginia Tech, Blacksburg, VA 24061, USA

Abstract: Mechanistic and probabilistic methods are individually used to characterize and predict joint networks. Combining these two approaches yields a method where the mechanical controls are honoured and implemented probabilistically in order to efficiently model joint development at the scale of the entire network with a useful ease of implementation. For this approach, bed-normal joints are characterized not with fracture trace geometries, but rather with intersection geometries to bedding. T-intersections represent joint termination at bedding, X-intersections represent joints crossing bedding and E-intersections are those intersections at the sample window edge. Using the intersection counts as input, a new computer program was developed that uses mechanically constrained probabilities to simulate and predict the spatial distribution of bed-normal joints in profiles across bedding. Initially, simulations are compared to ideal joint geometries for one or two lithologies with one or two bed thickness values, and found to match well. Simulations are then compared to joint geometries in four natural profiles from Llantwit Major, Wales, UK and Huntingdon, PA, USA. Simulations visually resemble the natural profiles and reasonably match the natural values of the joint network for density and mean joint height. We also extend the methodology to predicting joint networks beyond sample windows by investigating the minimum count of intersections needed to produce a representative result. Based on the five natural profiles with typical joint geometries, a sample size of about 50–100 intersection counts is sufficient to produce a reasonable prediction of the expected count and, hence, the joint geometry in a rock volume.

This chapter presents a new combined probabilistic–mechanical approach to simulating representative fracture pattern characteristics in a rock volume from a limited sample. The method simulates the spatial distribution of joints, and uses density and mean joint height as control variables to test simulation predictions.

Previous approaches to simulating fracture patterns include probabilistic decision models (Stone 1984; LaPointe 1993; Pascal *et al.* 1997) and mechanical analyses (e.g. Segall & Pollard 1983; Gross *et al.* 1995). Stochastic descriptions of joint patterns replicate geometric characteristics of fracture networks via probability distributions such as Poissonian or fractal (e.g. LaPointe 1993; Pascal *et al.* 1997). In contrast, mechanical analyses incorporate boundary conditions, material behaviour and state parameters to predict joint patterns (e.g. Pollard & Segall 1987; Cooke & Underwood 2001). The basic limitation of a purely statistical approach is that it applies a quantitative characterization to attributes without reference to their origin. In contrast, in most geological settings, insufficient information about forces, material behaviour, and initial state precludes mechanical analyses from uniquely predicting fracture patterns. Also, when investigating joint development in inhomogeneous media with pre-existing discontinuities, such as a profile across bedding, mechanical analyses tend to consider a very limited volume due to computational difficulties rather than a larger region on the scale of the fracture network (e.g. Bai & Pollard 2000a; Cooke & Underwood 2001; Engelder & Peacock 2001).

Owing to these limitations, we blend the two approaches to minimize their weaknesses. Combining them is appropriate because: (1) initial and boundary conditions for a mechanical analysis are not predicted uniquely from a final geometry, creating uncertainty; and (2) probabilistic methods are ideally suited for dealing with uncertainties. Also, the combination of the two approaches efficiently applies key understandings from computationally intensive mechanical modelling of small volumes to large volumes enclosing entire fracture networks.

This new approach is applied to the common case of bed-normal joints in sedimentary rocks, which has been the focus of numerous geometric and mechanical studies (e.g. Parker 1942; Hobbs 1967; Narr & Lerche 1984; Hancock 1985; LaPointe & Hudson 1985; Olson & Pollard 1989; Gross & Engelder 1991; Helgeson & Aydin 1991; Gross *et al.* 1995; Becker & Gross 1996; Narr 1996; Pascal *et al.* 1997; Ruf *et al.* 1998; Bai & Pollard 2000a). It should be noted that the joint patterns predicted from this new approach would be consistent with the expected results from a detailed mechanical analysis because the same deterministic rules are applied.

From: COSGROVE, J. W. & ENGELDER, T. (eds) 2004. *The Initiation, Propagation, and Arrest of Joints and Other Fractures.* Geological Society, London, Special Publications, **231**, 269–284. 0305-8719/04/$15 © The Geological Society of London 2004.

a)

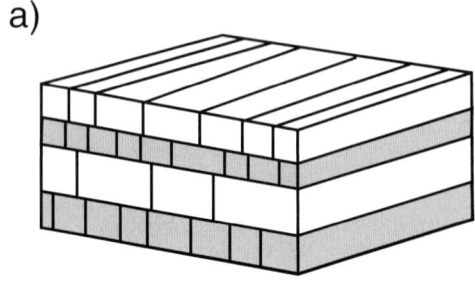

b)

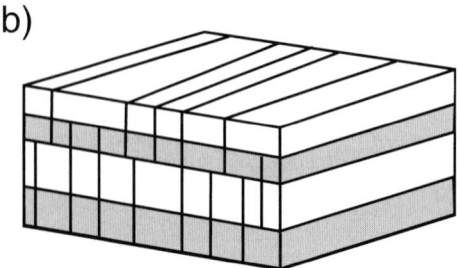

Fig. 1. Geometry of bed-normal joints that are (**a**) bed-contained and (**b**) not contained within single beds.

Fracture-pattern characteristics

Network characteristics

Before explaining the simulation approach, the observational framework and salient geometric characteristics need to be established. A single set of bed-normal joints in sedimentary rocks forms a simple fracture pattern (Helgeson & Aydin 1991; Gross *et al.* 1995; Becker & Gross 1996; Pascal *et al.* 1997; Gillespie *et al.* 2001). Data about these patterns are typically gathered from rock exposures such as natural pavements, roadcuts, cliffs, tunnels or boreholes. These exposures can be sampled by constructing trace maps, or using straight or circular scanlines (e.g. Terzaghi 1965; LaPointe & Hudson 1985; Priest 1993; Wu & Pollard 1995; Becker & Gross 1996; Narr 1996; Gillespie *et al.* 2001; Mauldon *et al.* 2001).

A typical characteristic for a single set of bed-normal joints is that joints are much longer parallel to bedding than perpendicular to bedding (Figs 1 and 2), with a typical aspect ratio of one to three orders of magnitude. As such joints are laterally extensive parallel to bedding, we can analyse their development in two-dimensional cross-sections normal to both joints and bedding (i.e. the exposure of Fig. 2a). In these cross-sections, bedding planes and fractures

a)

b)

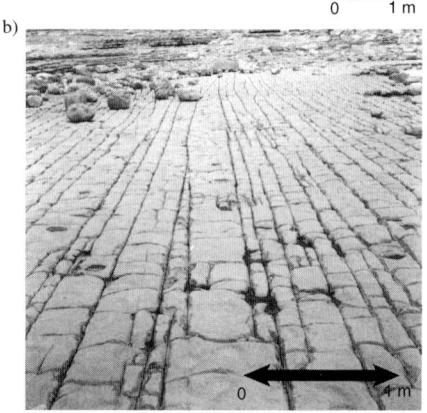

Fig. 2. Bed-normal joints in interbedded limestone and mudstone of Llantwit Major, Wales. (**a**) Cliff face. (**b**) Bedding pavement beneath the cliff face, arrow length of 1 m refers to the foreground.

both appear as line segments or *traces*. The geometry of both synthetic and natural bed-normal joint traces are described in this chapter in terms of density and mean joint height (Dershowitz & Herda 1992; Mauldon & Dershowitz 2000; Mauldon *et al.* 2001; Rohrbaugh *et al.* 2002), and are used to test simulation results.

Two-dimensional fracture density is the number of trace centres per unit area, where a trace centre is the true centre point of the entire trace and not just the geometric centre of the visible portion of a trace in the sample window (Mauldon 1998). Not all trace centres are identifiable in a sample window because full fracture tracelengths are not always visible within the cross-section or window, an effect known as censoring (Baecher & Einstein 1977). As a result, a common practice is to determine the apparent density, which is the total number of partial and whole traces visible in the outcrop window. Apparent density, however, overestimates density because it counts all traces, including those trace centres that lie outside the sample window and should not be counted (Mauldon *et al.* 2001). Here, we use an unbiased estimate of density, given by half

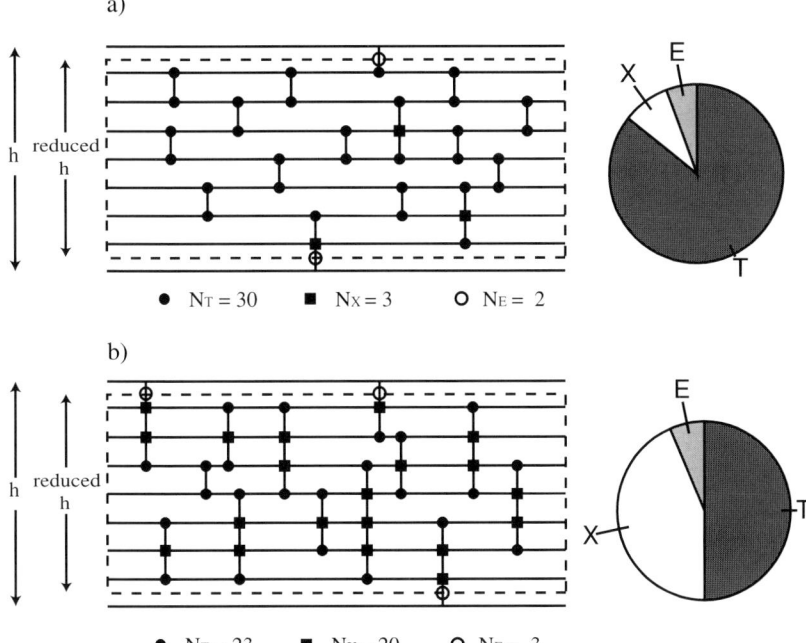

Fig. 3. Bedding profiles with bed-normal joints ornamented for intersection types. Pie charts show relative proportions of intersection types. (**a**) Mostly bed-contained joints are dominated by T-intersections. (**b**) Joints that cross bedding surfaces have a higher proportion of X-intersections. Dashed lines bound sample windows and solid horizontal lines are traces of bedding surfaces.

the number of fracture endpoints per unit area, which serves as an accurate proxy for trace centres (Mauldon 1998). Where bedding surfaces coincide with window boundaries, the window height is reduced by half the bed thickness, in order to avoid ambiguities in endpoint counts (e.g. Fig. 3)

Mean tracelength, μ, or joint height, in our case is the average uncensored length of all fracture traces that intersect a specified sample plane or window. Given the competing sampling biases of censoring that curtails traces and length bias which favours the sampling of longer traces, determination of μ can be problematic (Baecher & Einstein 1977). In practice, apparent mean tracelength (μ'), the average length of all visible traces and portions thereof, is often measured, but tends to underestimate true mean tracelength (Baecher & Einstein 1977; Mauldon *et al.* 2001). We use a mean tracelength estimator (Mauldon 1998) for parallel traces that corrects for censoring and length bias:

$$\frac{1}{\tilde{\mu}} = \frac{1}{\mu'} - \frac{1}{h} \qquad (1)$$

where $\tilde{\mu}$ is the estimated mean tracelength, μ' is the apparent mean tracelength and h is the window height in the direction of the traces.

Joint–bedding plane intersections

As both a convenient way of tracking joint height and termination relationships, and for guiding simulation logic, we define three types of intersection point (Fig. 3):

- T-intersection: a joint terminating at a bedding surface;
- X-intersection: a joint crossing a bedding surface;
- E-intersection: a joint trace intersecting the edge of the sampling window.

Counting the three types of intersections is simple, and is advantageous because the data are zero-dimensional points. Counting intersections is easier and more accurate than measuring tracelengths when dealing with limited exposures (e.g. natural pavements or man-made boreholes) because the points are uncensored and not subject to length bias, unlike the traces.

Abundant T-intersections (Fig. 3a) are indicative of bed-contained joints, whereas abundant X-intersections (Fig. 3b) are indicative of joints that cut multiple beds. The degree of bed-containment may reflect the variability of lithological material parameters or the degree of bedding contact strength during joint formation (Underwood *et al.* 2003).

Modelling approach

Mechanistic methods

Mechanics-based simulations of joints apply linear elastic material properties, boundary conditions and initial state conditions to predict fracture formation and geometry. For example, *in situ* stress, confining pressure, fluid pressure, strain rate, Young's modulus, fracture toughness and bedding contact strength influence the formation and geometry of bed-normal joints (Segall 1984; Pollard & Segall 1987; Helgeson & Aydin 1991; Gross *et al.* 1995; Engelder & Fischer 1996; Olson 1997; Underwood *et al.* 2003). If bed-normal joints form in response to remote extension, the applied strain produces stress in rock layers as a function of elastic properties that drives joint formation (Hobbs 1967; Gross *et al.* 1995; Bai & Pollard 2000*b*).

More than one mechanical explanation can exist for the same fracture geometry because different combinations of material properties and loading conditions can yield similar patterns. At the same time, the fractures may not contain features that differentiate between possible mechanical origins. For example, Olson & Pollard (1989) inferred differential stress from the degree of tip curvature of en echelon joints as a function of the relative magnitudes of remote and crack-tip stress. Yet, curvature may also be influenced by subcritical fracture growth and fracture roughness (Renshaw & Pollard 1994). Thus, straight joints may develop as a result of large differential stress, by subcritical growth due to decreasing the necessary driving stress or from limitations of shear displacement due to fracture surface roughness. Natural joints may lack features to distinguish between these causes, creating uncertainty about the mechanical origin of a joint characteristic.

Uncertainty in mechanical modelling also arises from limited knowledge of geometric conditions prior to natural joint propagation. For example, many mechanical models assume initial size and spatial distributions for the pre-existing flaws that serve as loci for joint initiation (e.g. Segall 1984; Rives *et al.* 1992; Gross *et al.* 1995). The sizes of pre-existing flaws influence the stress necessary for joints to initiate, as larger flaws require less stress to propagate (Segall 1984; McConaughy & Engelder 2001; Weinberger 2001). Yet, despite recognition of conditions where flaw distribution favours joint formation, little is actually known about flaw size and spatial distributions in nature, or the nature of stress

concentrations around flaws (e.g. McConaughy & Engelder 2001).

Probabilistic–mechanistic approach

Given the uncertainties about mechanical origin and initial geometric conditions, we believe that it is useful to apply mechanical principles within a probabilistic framework. Rock fracture patterns have been studied extensively using probabilistic methods since the 1970s (Baecher & Einstein 1977; Cruden 1977; Warburton 1980; Pahl 1981; Hudson & Priest 1983; LaPointe & Hudson 1985; Dershowitz & Einstein 1988; Kulatilake 1988; Dershowitz & Herda 1992; LaPointe 1993; Priest 1993; Mauldon 1994, 1998; Odling 1997; Zhang & Einstein 1998). The probabilistic view begins by recognizing that fracture location within a layer, as observed in thin section, borehole, outcrop or geological map, can be viewed as a snapshot of a random process, akin to a photograph of cars on a motorway. In this context, process does not refer to the mechanical and geological processes that led to fracture formation, but to the overall spatial pattern of which the observed fractures (or cars) are a sample. Random in this sense does not mean arbitrary or unconstrained, but instead means that a fracture characteristic such as location can be usefully viewed as being governed, at least in part, by probability laws. Such probability laws may take on a variety of forms, and can themselves depend on geological or other factors at any desired level of detail.

In this study, simple mechanistic rules are linked to a simple probabilistic decision model for joint initiation and propagation. Deterministic rules enforce mechanical principles. For example, requiring the driving stress to exceed a critical value precludes fracture initiation inside stress shadows, which would not be the case for an unrestricted Poissonian fracture distribution. The probabilistic decision model deals explicitly with uncertainties. For example, knowing the ratio of X- and T-intersections can guide the evolution of the simulated fracture geometry even when the exact state of initial conditions, such as flaw distributions or locations of weak bedding surfaces, is uncertain. Thus, combining probabilistic modelling with mechanical constraints minimizes the weaknesses of each approach, while exploiting their strengths. Our intention is to predict representative final bed-normal joint geometry from a limited sample based on mechanical principles.

Computer program logic

The analysis uses a Visual Basic™ computer program for which the main input parameters are

Table 1. *Simulation input parameters*

	LM	Hunt1	Hunt2	Hunt3
Number of beds	13	23	21	12
Window height (m)	1.77	3.62	4.00	3.99
Window length (m)	9.00	12.20	8.20	7.20
Window area (m²)	15.89	44.16	32.80	28.73
Reduced window height (m)	1.65	3.425	3.935	3.765
Reduced window area (m²)	14.85	41.79	32.27	27.11
Total joint height	50.50	43.29	22.25	18.54
Number of traces	250	202	108	84
T-intersections*	422 (74)	392 (12)	216 (0)	157 (11)
X-intersections*	94 (4)	65 (0)	1 (0)	3 (0)
E-intersections	78	12	0	11
Limestone (K) (MPa-m$^{-1/2}$)	1.25	na	na	na
(E) (MPa)	45	na	na	na
(υ)	0.2	na	na	na
Sandstone (K) (MPa-m$^{-1/2}$)	na	1.2	1.2	1.2
(E) (MPa)	na	50	50	50
(υ)	na	0.25	0.25	0.25
Siltstone (K) (MPa-m$^{-1/2}$)	na	1	1	1
(E) (MPa)	na	20	20	20
(υ)	na	0.2	0.2	0.2
Siltstone 2 (K) (MPa-m$^{-1/2}$)	na	na	na	1
(E) (MPa)	na	na	na	17
(υ)	na	na	na	0.2
Mudstone (K) (MPa-m$^{-1/2}$)	0.9	0.9	0.9	0.9
(E) (MPa)	16	16	16	16
(υ)	0.14	0.14	0.14	0.14

*Numbers in parentheses represent E-intersections that are known to be T- or X-intersections.
LM, locality at Llantwit Major, Wales. Hunt1, Hunt2 and Hunt3, localities at Huntingdon, PA, USA.
(K), fracture toughness; (E), Young's modulus; (υ), Poisson's ratio.
Lithology parameters compiled from Atkinson & Meredith (1987) and Gross *et al.* (1995).

lithologies, mechanical properties as a function of lithology, bed thicknesses and distribution of weak bedding surfaces, and the target (T-, X-, E-) intersection counts normalized by outcrop area (Table 1). Once parameters are defined, fractures are initiated and propagated in a series of iterative steps (Fig. 4, Table 2).

During decision loops, the program determines if stopping parameters are met and, if not, proceeds to the next initiation or propagation. The program stops when both the target N_X and N_T counts are reached (Table 2). The decision to initiate a new joint or propagate an existing joint is controlled by a probability based on the joint–bedding intersection counts. As a new T-intersection requires a new fracture, the probability of initiation is set proportional to the difference between the current N_T count in the simulation and the target N_T. At every X-intersection, a fracture has propagated across a bedding surface, so the probability to propagate is set proportional to the difference between the current N_X count and the target N_X. These two probabilities govern the decision to initiate or propagate during the current step.

Initiation

The probability of initiating a new fracture in a particular simulation is a function of the existing distribution of fractures and their stress shadows, plus whether the remote strain yields a sufficient driving stress for flaw growth in a particular bed. Flaws have a default size, c_i, of 0.005 m, and are assumed to be ubiquitous (Segall & Pollard 1983; Gross *et al.* 1995; McConaughy & Engelder 2001). Remote strain is set to the minimum value that triggers fracture initiation for all lithologies that contain joints in the natural profile. For simplicity, the driving stress is the result of a uniaxial extension with no vertical strain, and no gravitational load. Inclusion of gravity would have changed the magnitude of the remote strain, but would not have changed key results.

For initiation, stress shadows are treated as finite zones where flaws cannot grow to form new fractures, rather than as regions of reduced stress, because any stress drop precludes initiation of the flaws as joints in almost all cases (Hobbs 1967;

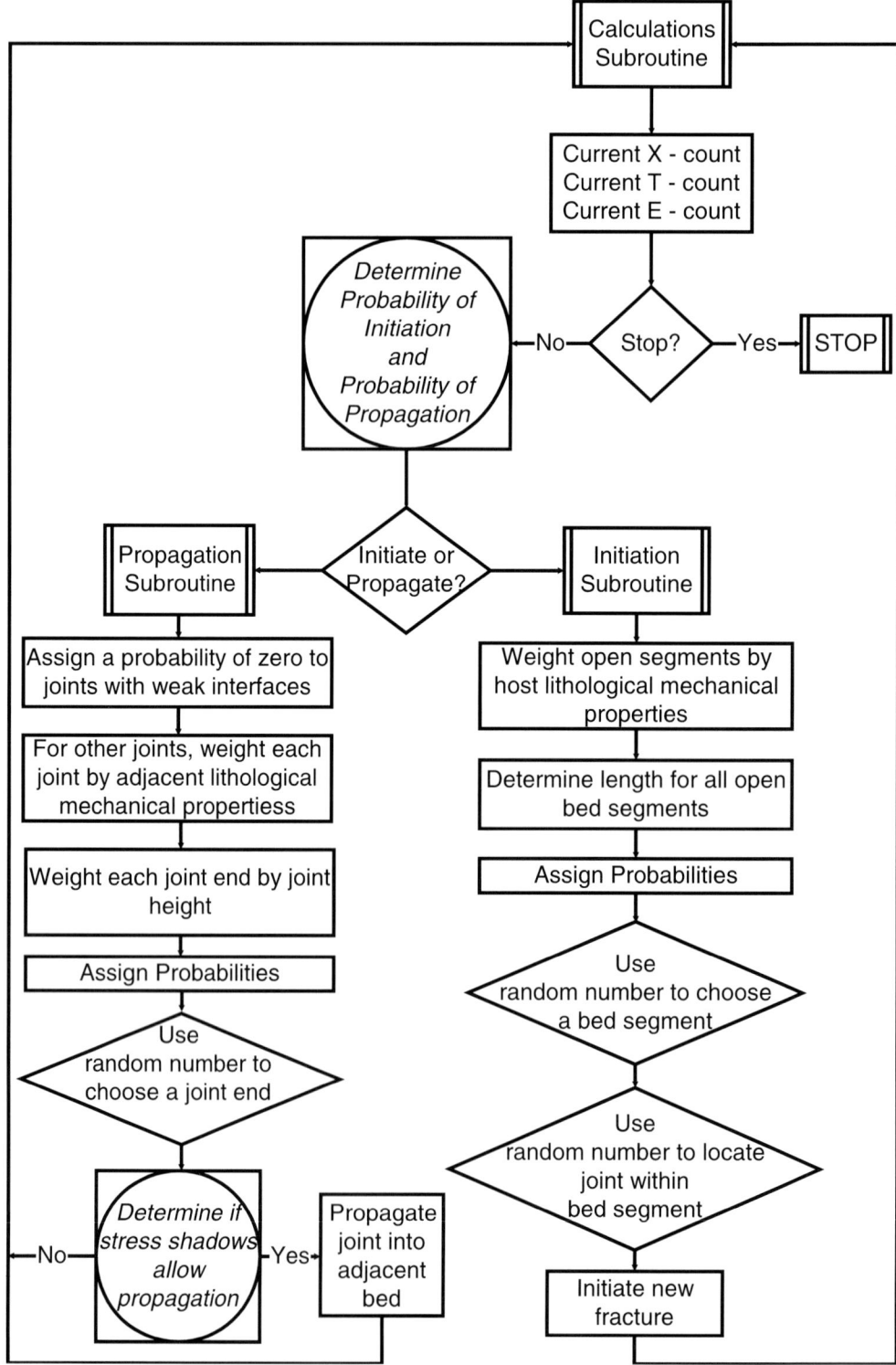

Fig. 4. Flow chart of program logic.

Table 2. *Parameters for calculations in program*

Parameters for determining whether to propagate or initiate	
Current N_T	current T count /area
Target N_T	natural T count /area
Current N_X	current X count /area
Target N_X	natural X count /area
Probability initiate	$\alpha \dfrac{\text{Current } N_T}{\text{Target } N_T}$
Probability propagate	$\alpha \dfrac{\text{Current } N_T}{\text{Target } N_T}$
Critical stress	$\sigma_f = K_{Ic} \sqrt{\pi c}$
Driving stress	$\sigma_h = E(1+\upsilon)[(1+\upsilon)(1-2\upsilon)]\varepsilon_h$
Initiation parameters	
Lithology weight	$(1/K_{Ic_i}) \times E_i$
Flaw height weight	$\sqrt{\pi c_i}$
Bed length weight	S_i
Probability of segment	$P(\text{init})_i = \dfrac{1}{K_{Ic_i}} \times E_i \times \sqrt{\pi c_i} \times S_i$
Probability of segment i given initiation	$P(i|\text{Initiation}) = \dfrac{P(\text{init})_i}{\sum\limits_{i=1}^{m} P(\text{init})_i}$
Propagation parameters	
Lithology weight	$(1/K_{Ic_j}) \times E_j$
Joint height weight	$\sqrt{\pi c_j}$
Probability of end	$P(\text{prop})_j = \dfrac{1}{K_{Ic_j}} \times E_j \times \sqrt{\pi c_j}$
Probability of end j given propagation	$P(j|\text{Propagation}) = \dfrac{P(\text{prop})_j}{\sum\limits_{j=1}^{n} P(\text{prop})_j}$

K_{Ic_i} is the fracture toughness for initiation; K_{Ic_j} is the fracture toughness for propagation; E_i is the Young's or stiffness modulus of the host bed; E_j is the modulus of the adjacent bed; c_i is the initial flaw size half-length; c_j is the existing joint half-length; m is the number of bed segments; n is the number of joint ends; σ_f is the critical stress; σ_h is the horizontal driving stress; ε_h is the horizontal extensional strain; and υ is Poisson's ratio.

Gross *et al.* 1995). The shadows have width to each side of a fracture equal to joint height (Bai & Pollard 2000*a*). This width can be varied or even set to zero for simulating different geological situations.

We use the term 'open segment' to refer to portions of beds free of stress shadows (Fig. 5). The next step is to identify open segments in fractured lithologies where the available driving stress equals or exceeds the critical stress for triggering a flaw to initiate (Table 2). The probability of a joint initiating in an open segment where failure can occur is a function of flaw size, c_i, open segment length, S_i, and material properties, E_i, υ and K_{Ic_i} (Fig. 4, Table 2). Once an open segment is chosen, joint location

within the segment is assigned using a uniform distribution.

Propagation

The likelihood of a fracture end propagating is a function of bed contact strength, joint size, c_j, material properties, E_j, υ and K_{Ic_j}, and the presence of stress shadows (Fig. 4, Table 2). Weak bedding interfaces prevent propagation and strong ones allow propagation (Helgeson & Aydin 1991; Cooke & Underwood 2001; Rijken & Cooke 2001). Moderate contact strengths or the presence of mudstone beds less than a

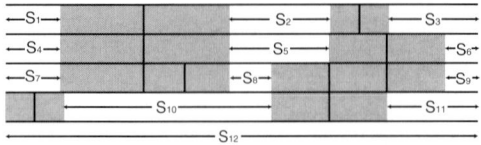

Fig. 5. Length S_i of open segments between stress shadows (shaded) for locating a new initiation site.

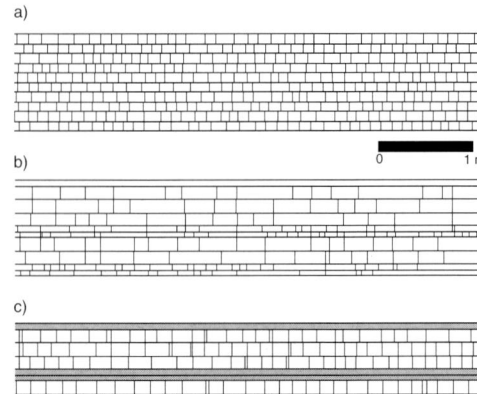

Fig. 6. Ideal-case profiles. (**a**) All beds are the same lithology, thickness and have weak bedding contacts. (**b**) All beds are the same lithology and bedding surfaces are strong, but two bed thicknesses are used. (**c**) Beds have one of two thickness values with the thinner beds being a different lithology (grey) that prevents joint initiation and propagation.

few centimetres thick favour stepovers to form composite joints (Helgeson & Aydin 1991; Cooke & Underwood 2001; McConaughy & Engelder 2001). Stepovers were not simulated at present, so only weak and strong bedding surfaces are considered.

For strong bedding surfaces, the probability that a particular fracture end propagates is weighted by fracture size and the elastic properties of the bed into which propagation will occur (Fig. 4, Table 2). Once an end is selected, it is tested to see if a stress shadow is present in the adjacent bed and if the fracture has sufficient driving stress to propagate. Stress shadows are treated for propagation simply as rectangular regions with widths to either side equal to fracture tracelength and with a linear increase in stress reduction from 0% at the perimeter to 100% at the joint.

Application to simple cases

The first step in testing the combined probabilistic–mechanistic joint simulation logic is to apply it to simple end-member cases. Three example profiles were simulated to demonstrate the effects on joint geometry of varying bed thickness, lithology and bed contact strength. Simulation A (Fig. 6a) represents the case of a single lithology with constant bed thickness and weak bedding contacts, where joints are confined to single beds. Simulation B (Fig. 6b) has a single lithology with two different bed thicknesses and strong contacts so that joints can propagate across bedding boundaries. In simulation C (Fig. 6c) the thinner beds are assigned a smaller stiffness that prohibits joint propagation or initiation.

For simulation A, as joints were unable to propagate across bed boundaries due to weak bedding contacts, the mean tracelength is equal to the average bed thickness. New joints are prohibited from forming within the stress shadows of existing joints, resulting in fairly constant fracture spacing. For simulation B, joints were able to propagate across layer boundaries, resulting in several fractures that cross multiple beds. The joints, particularly in the thinner beds, still exhibit a fairly regular spacing. Once joints have begun to propagate, however, cracklength may become sufficiently large to allow propagation into or through existing stress shadows,

resulting in some fractures being closer than was possible for case A. In simulation C, joints were confined to the thicker beds. Again, sufficiently long fractures allow some joints to propagate close to existing joints. Although these simulations do not represent actual joint networks, they are similar in appearance to natural fracture patterns (Gross *et al.* 1995, 1997; Becker & Gross 1996; Pascal *et al.* 1997; Ruf *et al.* 1998), and indicate that the program is capable of yielding realistic fracture patterns.

Field data sets

Four natural profiles were analysed (Figs 7a–10a). One profile is a sequence of interbedded limestone and mudstone at Llantwit Major, Wales (Fig. 7). Three profiles are from a sequence of interbedded sandstone, siltstone and mudstone at Huntingdon, Pennsylvania, USA (Figs 8–10). The profiles represent fairly typical geometries for bed-normal joints in sedimentary rocks with different interbedded lithologies and a range of termination relationships.

Llantwit Major

The Llantwit Major profile (LM, Fig. 7a) is from a cliff containing a set of bed-normal master joints (strike 170°), primarily bed-contained, cutting horizontally layered rocks of the Jurassic Porthkerry Formation at Llantwit Major, Wales. The Formation consists of micritic limestone beds that are decime-

a)

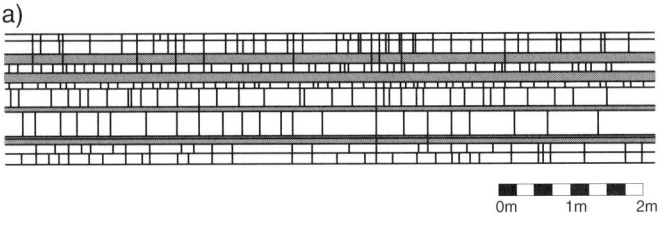

0m   1m   2m

b)

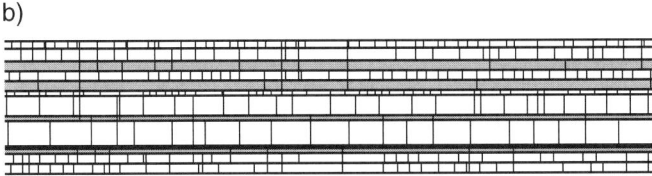

Fig. 7. LM natural profile and simulation results. (**a**) Natural profile. (**b**) Average simulation for this profile. Grey represents mudstone and white is limestone.

tres thick and mudstone interbeds that are centimetres thick. Another location at Llantwit Major is the site of a previously published two-dimensional probabilistic model for bed-normal joints in interbedded sedimentary rocks (Pascal *et al.* 1997). Their analysis used as input the total number of joints visible in the outcrop, the number of limestone beds, the number of joints in each limestone bed, the number of joints common to a limestone bed and the closest limestone bed below it, the thickness of each bed and the outcrop length. Although their analysis did produce a reasonable match between the synthetic pattern and the natural pattern, their approach has three limitations. First, because the analysis matches the number of fractures on a bed-by-bed basis, the number of fractures in each bed must already be known. Thus, the methodology cannot easily be extrapolated beyond the sample area. Second, the analysis underestimates the number of joints confined to single beds, reducing fracture density while increasing mean tracelength for the population, as a result of not correlating propagation probabilities for successive beds (Pascal *et al.* 1997). Third, their analysis does not allow for fractures extending into or out of the sample window, creating a simulation that is artificially isolated from surrounding beds. The present analysis does not have these limitations.

Huntingdon

An outcrop of the Devonian Brallier Formation near Huntingdon, Pennsylvania (Ruf *et al.* 1998), provided three additional profiles: Hunt1, Hunt2 and Hunt3 (Figs 8–10). The joints in these profiles are approximately 0°–10° from perpendicular to beds,

which may result from tilting of principal stresses by shear tractions such as during flexural-slip/flow folding (Engelder & Peacock 2001). In other respects, the joint networks exhibit typical bed-normal joint characteristics. A significant number of fractures cross one or more bedding surfaces in Hunt1 (Fig. 8a), while joints are primarily confined to single beds in Hunt2 and Hunt3 (Figs 9a and 10a).

The Brallier Formation is a distal turbidite sequence of interbedded fine sandstone, siltstone and mudstone in beds up to about 50 cm thick. The oldest joint set (strike *c.* 050°), as determined by termination relationships, was the focus of this study because fracture propagation in this set was not influenced by pre-existing fractures.

Fracture characteristics

For each profile the estimate of joint density is half the number of T-intersections (N_T), divided by the reduced sample area (Mauldon 1998). As the top and bottom window boundaries coincide in all cases with bedding planes, the reduced sample area eliminates half the thickness of the top and bottom bed multiplied by sample width. The mean tracelength estimate is given by equation (1) (Mauldon 1998).

Results

Simulation of natural profiles

Data from the natural profiles (Figs 7a–10a) were used to generate synthetic profiles. Input data for

a)

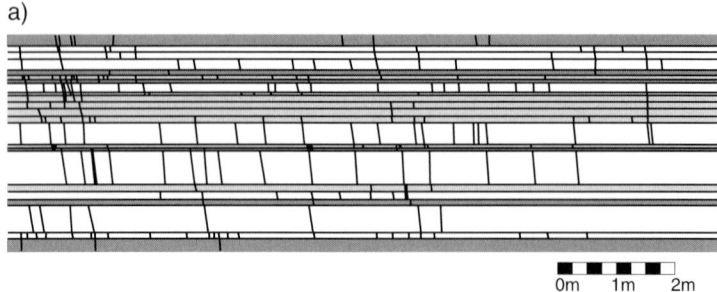

b)

Fig. 8. Hunt1 natural profile and simulation results. (**a**) Natural profile. (**b**) Average simulation for this profile. Dark grey represents mudstone, light grey is siltstone and white is sandstone.

a)

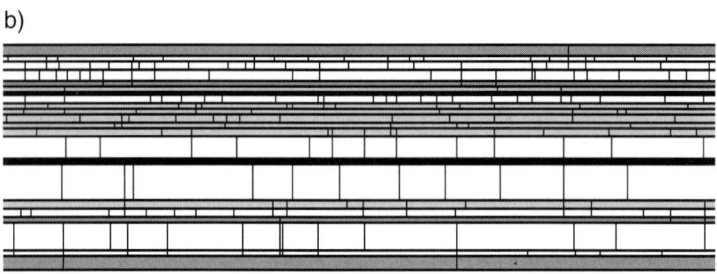

b)

Fig. 9. Hunt2 natural profile and simulation results. (**a**) Natural profile. (**b**) Average simulation for this profile. Dark grey represents mudstone, light grey is siltstone and white is sandstone.

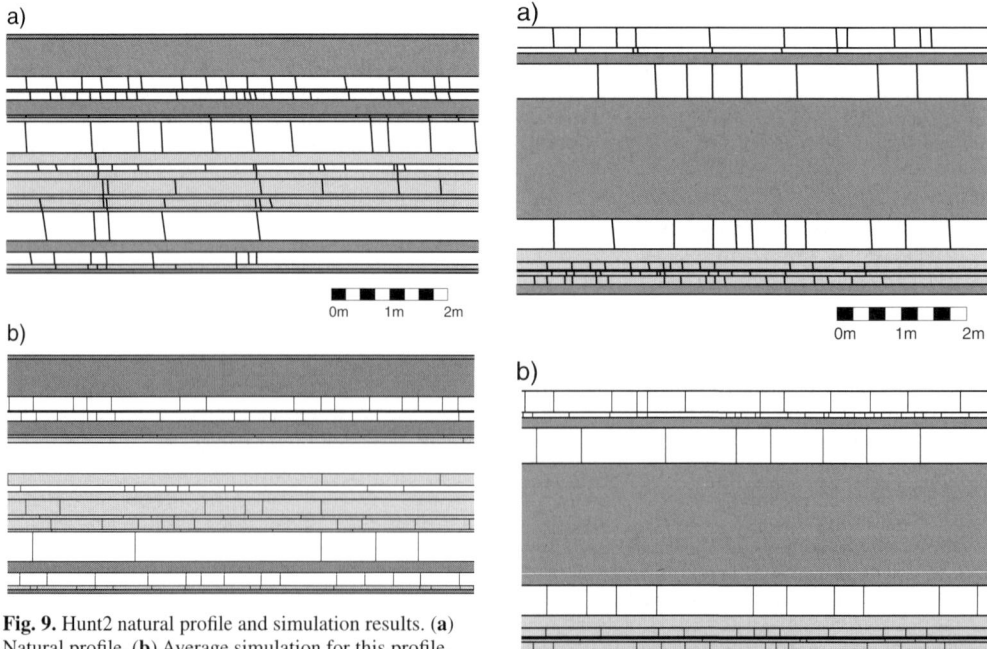

Fig. 10. Hunt3 natural profile and simulation results. (**a**) Natural profile. (**b**) Average simulation for this profile. Dark grey represents mudstone, light grey is siltstone and white is sandstone.

Table 3. *Comparison of joint characteristics between natural case and simulation results*

	LM	Hunt1	Hunt2	Hunt3
Natural profile				
Estimated density	14.21	4.69	3.35	2.90
Estimated joint height	0.23	0.23	0.22	0.23
Simulation values				
Average estimated density	14.22 ± 0.02	4.7 ± 0.01	3.35 ± 0.00	2.9 ± 0.01
Coefficient of variation	0	0	0	0
Average estimated joint height	0.21 ± 0.00	0.23 ± 0.01	0.17 ± 0.02	$0.23 + 0.01$
Coefficient of variation	2	3	13	4
Comparison				
% Difference estimated density	0	0	0	0
% Difference estimated joint height	9	2	23	2

simulations include (T-, X-, E-) intersection counts and outcrop dimensions, bed thicknesses, lithologies and mechanical parameters (Table 1). Also, a bedding surface was defined as weak if no joints crossed it. The remote strain was chosen to yield sufficient driving stress to fracture those lithologies that were observed in the field to contain fractures (Figs 7a–10a). Fifty simulations were run for each profile. The estimated density and mean joint height of each simulation were compared to the corresponding values for the natural profiles. As the number of endpoints (N_T) is a control parameter and the estimated density is calculated using N_T, the simulated density will match the natural density.

Match quality for mean joint height is influenced by two factors: (1) which beds, thin v. thick, contain joints compared to the natural profiles; and (2) the number of fractures crossing multiple beds (Figs 7–10, Table 3). A given bed may have more or fewer joints than the natural pattern because the program does not match the exact number of fractures on a bed-by-bed basis. Requiring such a bed-by-bed fit would improve the match, but would destroy the predictive capability of the approach for rock volumes where fracture geometries are not known on a bed-by-bed basis. Thus, the potential for a lesser match quality is necessitated by the need to develop a predictive tool.

The quality of simulations can be judged by whether they visually resemble the natural profiles, and by comparison to key network characteristics. The simulations that most closely match the average values for density and mean joint height from each group of 50 simulations were visually compared to the natural profile (Figs 7b–10b, Table 3). The LM profile (Fig. 7b) is a good visual match to the natural profile and has mean joint height within 8% of the natural value. The Hunt1 profile (Fig. 8b) has more joints in the upper thin beds in the simulation with

less clustering than the natural profile, but overall the visual match is good. Mean joint height is <2% from the natural value for Hunt1. However, Hunt2 (Fig. 9b) has an overabundance of joints in the thinner beds of siltstone in the synthetic compared to the natural profile, which causes mean joint height to differ about 20% from the natural profile. The Hunt3 profile (Fig. 10b) has more fractures in the thin sandstone bed near the top and fewer fractures in the lower siltstone beds in the simulation compared to nature, yet mean joint height is within 2% of the natural value.

The simulations for LM, Hunt1 and Hunt3 produce close matches to the natural mean joint heights, although LM simulations slightly underpredict the mean joint height (Fig. 11). In contrast, Hunt2 simulations almost all underestimate mean joint height with the largest variation (Fig. 11, Table 3).

Simulation quality as a function of heterogeneity in bed thickness and material properties

When considering fracture initiation in beds of only one lithology for the same height of sample window, thinner beds can be more abundant than thicker beds (e.g. five beds of 0.2 m thickness v. one bed of 1 m thickness in a window with 1 m of height). Therefore, thin beds may have more open segments for simulated joint initiation and, hence, on a random basis may develop more joints than found in an equivalent natural example, particularly when natural fractures are less abundant (e.g. Fig. 8a v. 8b, or Fig. 9a v. b). The potential for more joints in these beds is enhanced when bed thickness correlates to lithology, such as in the Huntingdon profiles (Figs 8–10) where sandstones occur in thicker beds and

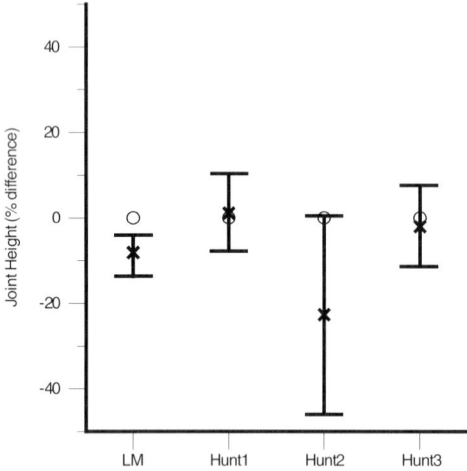

Fig. 11. Simulation performance for mean joint height as a per cent difference from natural value. Open circle, natural value; cross, mean value for 50 simulations with bars showing the range of simulation values.

dance in a given natural profile. For example, Hunt2 has 73% of joints in sandstone but only 38% of the beds are sandstone, whereas Hunt3 has 44% of joints in sandstone but 57% of the beds are sandstone. Thus, unless material properties are defined for each individual bed, rather than overall for a given lithology, the simulations will not exactly duplicate the relative lithological distributions of joints. However, even with the present-day measurement of elastic constants, investigators are unlikely to establish with confidence the elastic constants for every bed during fracturing.

Overall, the synthetic profiles appear similar to natural profiles and simulation characteristics are comparable to natural characteristics. Differences do occur, but are also expected in mechanical models. Therefore, this approach successfully simulates natural examples and is applicable to predicting bed-normal joint characteristics beyond a sample window.

Simulation quality as a function of sample size for intersections count

The quality of predictions is partially a function of sample size. To investigate the effects of sample size we considered the four natural profiles discussed previously, plus a fifth profile from Pascal *et al.* (1997). This fifth profile (LMP) is also from near Llantwit Major, Wales, and has a large number of X-intersection counts, unlike the other four localities (Fig. 12). Hunt2 and Hunt3 have negligible X-intersection counts while LM and Hunt1 have less than 100 X-intersections each (Table 1). Total X and T counts for the natural profiles range from 160 to 596 (Table 1, Fig. 12).

Subsamples of the natural profiles were used to investigate the sample size necessary to obtain reasonable estimates of X- and T-intersections previously measured for whole profiles. Thus, for this analysis, the subsamples are the known regions and the natural profiles are the prediction regions.

A subsample must be representative of the whole network to accurately predict the joint pattern. Wu & Pollard (1995) found that scanline samples provided a good match with area samples for well-developed

siltstones in thinner beds. For Hunt2, simulations favour joints in thinner beds, particularly siltstone beds (Fig. 9). A similar effect occurs for Hunt1 (Fig. 8), but is counteracted by longer simulated joint heights where joints propagated to or from a thicker sandstone bed. This counteracting tendency indicates the relative ease of propagation between simulated sandstone and siltstone compared to the natural case, which suggests that simulated material properties are less different for the two lithologies than in the natural case. LM shows a similar effect (Fig. 7) with simulated profiles that have more joints crossing limestone–mudstone bedding surfaces than in the natural profile, particularly for the thinner mudstones, leading to modest underestimates of mean joint height.

One approach to improving the match between simulations and nature would be to refine the stiffness coefficients and fracture toughnesses of the different lithologies. Yet, Huntingdon yields a cautionary note with respect to this approach because the relative abundance of joints in sandstones is not solely a function of sandstone abun-

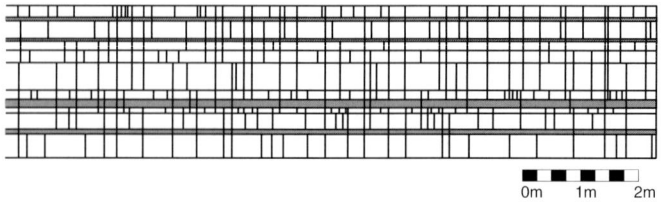

Fig. 12. LMP natural profile. $N_E = 82$, $N_X = 326$, $N_T = 270$. Grey represents mudstone and white is limestone.

Table 4. *Coefficient of variation (COV) for subsample intersection counts*

	LMP	LM	Hunt1	Hunt2	Hunt3
X-intersection count – COV	0.44	0.99	1.61	na	na
T-intersection count – COV	0.38	0.3	0.73	0.71	0.72

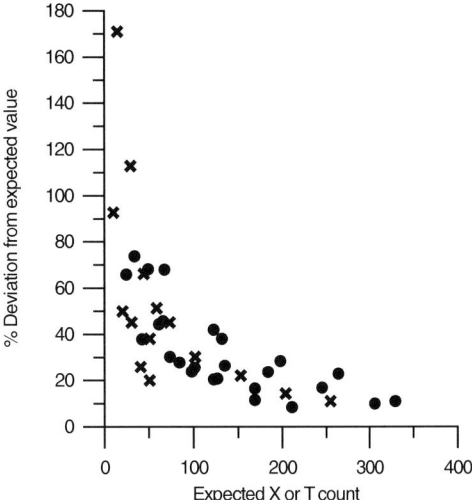

Fig. 13. Percentage deviation from the expected count for the T-intersections for all five natural profiles and X-intersections for LMP, LM, and Hunt1 v. the normalized expected X and T counts. Xs represent the X counts and circles represent the T counts. Increased sample count is typically the result of increased sample area.

(homogeneous) fracture patterns, but that the quality of the match declined for poorly developed networks with greater heterogeneity. Therefore, joint network heterogeneity can greatly affect a subsample and, hence, prediction quality. The heterogeneity of the natural patterns (Figs 8a–11a and 12) was examined by dividing each profile into 32 equal width columns that each encompass the full outcrop height and 1/32 of the outcrop width. The X- and T-intersections were counted for each 1/32 of a profile and their variability from the mean was expressed as coefficients of variation (Table 4). LMP is the most homogeneous overall and LM is homogeneous for T-intersection counts. Hunt1 shows the greatest heterogeneity, with Hunt2 and Hunt3 also showing a great deal of heterogeneity for T-intersection counts. X-intersection count variations were not investigated for Hunt2 or Hunt3 because the profiles have essentially zero X-intersections, making these nearly homogeneous by default.

The effects of heterogeneity and sample size were explored by using independent randomly selected groups of five, 10, 15, 20 and 25 of the 1/32 columns for a profile, with no column repetition within a group. The selection process was carried out 50 times for each size group, yielding a range of counts. Again, X-intersection counts for Hunt2 and Hunt3 were not considered as they are essentially zero. Counts were normalized to their respective areas and plotted against the corresponding expected X or T count (Fig. 13). Expected sample counts are determined by normalizing the whole outcrop count to the size of the subsample. For example, the ideal T-intersection count for a 5/32 sample of LM is $422 \times 5/32$ or 66, as the whole outcrop count is 422.

Prediction quality shows a non-linear improvement with increasing sample size (Fig. 13). Rate of improvement in prediction quality drops noticeably in the range of 50–100 intersections, where predicted counts are within 20–40% of natural values. Therefore, counts of 50–100 for each for X- and T-intersections should yield a reasonable prediction of bed-normal joint characteristics in regions outside a sample area. Although the natural profiles are not exhaustive of all possible bed-normal joint net-

works, they have a large range of heterogeneity and illustrate joints contained in one or more beds.

Discussion

This combined mechanistic–probabilistic simulation approach is not meant to replace the use of numerical simulations of mechanical conditions to deduce new details of fracturing behaviour, but it has several advantages. First, it is quick and easy to implement. Second, it is easy to modify the Visual Basic™ code compared to the code for a numerical model to include or modify mechanical behaviour. Third, this approach provides investigators with a means of quickly testing for primary mechanical controls. Fourth, the program provides an easier pathway for student understanding and experimentation with the primary controls on formation of bed-normal joints, compared to implementing a numerical model.

Conclusions

- We present a new approach to predicting the geometry of bed-normal joints, which combines a probabilistic decision model with mechanistic principles. The approach allows results from small-scale mechanical simulations to be applied to the scale of entire joint networks, while yielding results that should be expected from a more computationally intensive mechanical analysis, because the same mechanical rules govern. Also, this approach allows both easy testing of key mechanical assumptions and an opportunity for students to investigate primary controls on joint development.

- We use counts of different types of joint–bed intersections as input for simulating natural fracture patterns. These data are readily obtained from field data or boreholes.
- Simulations using this combined probabilistic–mechanistic approach produced a reasonable match for mean joint height of natural profiles, as well as a reasonable visual similarity. Therefore, this approach can simulate and predict natural joint patterns.
- Joint-pattern heterogeneity and small sample size can reduce the quality of predictions, although intersection counts of about 50–100 appears to limit this effect, aiding the prediction of bed-normal joint geometries beyond a sample region.

This research would not have been possible without partial funding from the donors of the Petroleum Research Fund, administered by the ACS; the Geological Society of America; the American Association of Petroleum Geologists; and the Pennsylvania State Survey. We wish to thank T. Engelder (Pennsylvania State University, Geological Sciences) for his recommendation of the Huntingdon, Pennsylvania field site. Also, we would like to thank B. Rohrbaugh for his assistance with fieldwork and Visual Basic™. Reviews by M. Cooke, A. Whitaker and T. Engelder improved the manuscript.

References

ATKINSON, B. K. & MEREDITH, P. G. 1987. Experimental fracture mechanics data for rocks and minerals. *In*: ATKINSON, B. K. (ed.) *Fracture Mechanics of Rock*. Academic Press, New York, 477–525.

BAI, T. & POLLARD, D. D. 2000a. Fracture spacing in layered rocks a new explanation based on the stress transition. *Journal of Structural Geology*, **22**, 43–57.

BAI, T. & POLLARD, D. D. 2000b. Closely spaced fractures in layered rocks: initiation mechanism and propagation kinematics. *Journal of Structural Geology*, **22**, 1409–1425.

BAECHER, G. B. & EINSTEIN, H. H. 1977. Statistical description of rock properties and sampling. *In*: WANG, C. (ed.) *Proceedings for the 18th US Symposium on Rock Mechanics*. Colorado School of Mines Press, Golden, CO, 5C1.1–5C1.8.

BARTON, C. C. 1995. Fractal analysis of scaling and spatial clustering of fractures. *In*: BARTON, C. C. & LaPOINTE, P. R. (eds) *Fractals in the Earth Sciences*. Plenum Press, New York, 141–178.

BECKER, A. & GROSS, M. R. 1996. Mechanism for joint saturation in mechanically layered rocks: an example from southern Israel. *Tectonophysics*, **257**, 223–237.

COOKE, M. L. & UNDERWOOD, C. A. 2001. Fracture termination and step-over at bedding interfaces due to frictional slip and interface opening. *Journal of Structural Geology*, **23**, 223–238.

CRUDEN, D. M. 1977. Describing the size of discontinuities. *International Journal of Rock Mechanics and Mining Sciences and Geomechanical Abstracts*, **14**, 133–137.

DERSHOWITZ, W. S. & EINSTEIN, H. H. 1988. Characterizing rock joint geometry with joint system models. *Rock Mechanics and Rock Engineering*, **21**, 21–51.

DERSHOWITZ, W. S. & HERDA, H. H. 1992. Interpretation of fracture spacing and intensity. *In*: TILLERSON, W. (ed.) *Proceedings of the 33rd US Symposium on Rock Mechanics*, Balkema, Rotterdam, 19–30.

ENGELDER, T. & FISCHER, M.P. 1996. Loading configurations and driving mechanisms for joints based on the Griffith energy-balance concept. *Tectonophysics*, **256**, 253–277.

ENGELDER, T. & PEACOCK, D. C. 2001. Joint development normal to regional compression during flexural-flow folding: the Lilstock buttress anticline, Somerset, England. *Journal of Structural Geology*, **23**, 259–277.

GILLESPIE, P. A., WALSH, J. J., WATTERSON, J., BONSON, C. G. & MANZOCCHI, T. 2001. Scaling relationships of joint and vein arrays from The Burren, Co. Clare, Ireland. *Journal of Structural Geology*, **23**, 183–201.

GROSS, M. R. & ENGELDER, T. 1991. A case for neotectonic joints along the Niagra escarpment. *Tectonics*, **10**, 631–642.

GROSS, M. R., BAHAT, D. & BECKER, A. 1997. Relations between joint and faulting based on fracture-spacing ratios and fault-slip profiles: A new method to estimate strain in layered rocks. *Geology*, **10**, 887–890.

GROSS, M. R., FISCHER, M. P., ENGELDER, T. & GREENFIELD, R. J. 1995. Factors controlling joint spacing in interbedded sedimentary rocks: integrating numerical models with field observations from the Monterey Formation, USA. *In*: AMEEN, M. S. (ed.) *Fractography: Fracture Topography as a Tool in Fracture Mechanics and Strain Analysis*. Geological Society, London, Special Publications, **92**, 215–233.

HANCOCK, P. L. 1985. Brittle microtectonics: principles and practice. *Journal of Structural Geology*, **7**, 437–457.

HELGESON, D. E. & AYDIN, A. 1991. Characteristics of joint propagation across layer interfaces in sedimentary rocks. *Journal of Structural Geology*, **13**, 897–911.

HOBBS, D. 1967. The formation of tension joints in sedimentary rocks: an explanation. *Geological Magazine*, **104**, 550–556.

HUDSON, J. A. & PRIEST, S. D. 1983. Discontinuity frequency in rock masses. *International Journal of Rock Mechanics and Mining Sciences and Geomechanical Abstracts*, **20**, 73–89.

KULATILAKE, P. H. S. W. 1988. Corrections for sampling biases in joint surveys; state-of-the-art. *In*: FRAGASZY, R. J. (ed.) *Proceedings of the 24th Symposium on Engineering Geology and Soils Engineering*. University of Idaho, Boise, ID, 359–374.

LaPOINTE, P. R. 1993. Pattern analysis and simulation of joints for rocking engineering. *In*: HUDSON, J.A. (ed.) *Comprehensive Rock Engineering, Volume 3 – Rock*

Testing and Site Characterization. Pergamon Press, New York, 215–239.

LaPointe, P. R. & Hudson, J. A. 1985. *Characterization and Interpretation of Rock Mass Joint Patterns*. Geological Society of America, Special Papers, **199**.

Mauldon, M. 1994. Intersection probabilities of impersistent joints. *International Journal of Rock Mechanics and Mining Sciences and Geomechanical Abstracts*, **31**, 107–115.

Mauldon, M. 1998. Estimating mean fracture trace length and density from observations in convex windows. *Rock Mechanics and Rock Engineering*, **31**, 201–216.

Mauldon, M. & Dershowitz, W. S. 2000. A multi-dimensional system of fracture abundance. *Geological Society of America Annual Meeting, Abstracts with Programs*, **32**, A474.

Mauldon, M., Dunne, W. M. & Rohrbaugh, M. B., Jr 2001. Circular scanlines and circular windows: new tools for characterizing the geometry of fracture traces. *Journal of Structural Geology*, **23**, 247–258.

McConaughy, D. T. & Engelder, T. 2001. Joint initiation in bedded clastic rocks. *Journal of Structural Geology*, **23**, 203–221.

Narr, W. 1996. Estimating average fracture spacing in subsurface rock. *AAPG Bulletin*, **80**, 1565–1586.

Narr, W. & Lerche, I. 1984. A method for estimating subsurface fracture density in core. *AAPG Bulletin*, **68**, 637–648.

Odling, N. E., Gillespie, P., Bourgine, B., Castaing, C., Chilés, J-P., Christenson, N. P., Fillion, E., Genter, A., Olsen, C., Thrane, L., Trice, R., Aarseth, E., Walsh, J. J. & Watterson, J. 1999. Variations in fracture system geometry and their implications for fluid flow in fractured hydrocarbon reservoirs. *Petroleum Geoscience*, **5**, 373–384.

Olson, J. E. 1997. Natural fracture pattern characterization using a mechanically-based model constrained by geologic data – moving closer to a predictive tool. *International Journal of Rock Mechanics and Mining Sciences and Geomechanical Abstracts* **34**(3–4), 391.

Olson, J. E. & Pollard, D. D. 1989. Inferring paleostresses from natural fracture patterns: A new method. *Geology*, **17**, 345–348.

Pahl, P .J. 1981. Estimating the mean length of discontinuity traces. *International Journal of Rock Mechanics and Mining Sciences and Geomechanical Abstracts*, **18**, 221–228.

Parker, L. M. 1942. Regional systematic jointing in slightly deformed sedimentary rocks. *Geological Society of America Bulletin*, **53**, 381–408.

Pascal, C., Angelier, J., Cacas, M. & Hancock, P. L. 1997. Distribution of joints: probabilistic modeling and case study near Cardiff (Wales, U.K.). *Journal of Structural Geology*, **19**, 1273–1284.

Pollard, D. D. & Segall, P. 1987. Theoretical displacements and stresses near fractures in rock: with applications to faults, joints, veins, dikes, and solution surfaces. *In*: Atkinson, B. K. (ed.) *Fracture Mechanics of Rock* Academic Press, New York, 277–349.

Price, N. J. 1966. *Fault and Joint Development in Brittle and Semi-brittle Rock*. Pergamon Press, New York.

Priest, S. D. 1993. *Discontinuity Analysis for Rock Engineering*. Chapman & Hall, New York.

Renshaw, C. E. & Pollard, D. D. 1994. Are large differential stresses required for straight fracture propagation paths? *Journal of Structural Geology*, **16**, 817–822.

Rijken, P. & Cooke, M. L. 2001. Role of shale thickness on vertical connectivity of fractures: application of crack-bridging theory of the Austin Chalk, Texas. *Tectonophysics*, **337**, 117–133.

Rives, T., Razack, M., Petit, J.-P., Hencher, S. R. & Lumsden, A. C. 1992. Joint spacing: analogue and numerical simulation. *Journal of Structural Geology*, **14**, 925–937.

Rohrbaugh, M. B., Jr, Mauldon, M. & Dunne, W. M. 2002, Estimating joint trace intensity, density and mean length using circular scanlines and circular windows. *AAPG Bulletin*, **86**, 3089–2104.

Ruf, J. C., Kelly, K. R. & Engelder, T. 1998. Investigating the effect of mechanical discontinuities on joint spacing. *Tectonophysics*, **295**, 245–257.

Segall, P. 1984. Formation and growth of extensional fracture sets. *Geological Society of America Bulletin*, **95**, 454–462.

Segall, P. & Pollard, D. D. 1983. Joint formation in granitic rock of the Sierra Nevada. *Geological Society of America Bulletin*, **94**, 563–575.

Stone, D. 1984. Sub-surface fracture maps predicted from borehole data: an example for the Eye–Dashwa pluton, Atikokan, Canada. *International Journal of Rock Mechanics and Mining Sciences and Geomechanical Abstracts*, **21**, 183–194.

Terzaghi, R. D. 1965. Sources of error in joint surveys. *Geotechnique*, **15**, 287–304.

Underwood, C. A., Cooke, M. L., Simo, J. A. & Muldoon, M. A. 2003. Stratigraphic controls on vertical fracture patterns in Silurian dolomite, northeastern Wisconsin. *AAPG Bulletin*, **87**, 121–142.

Warburton, P. M. 1980. A stereological interpretation of joint trace data. *International Journal of Rock Mechanics and Mining Sciences and Geomechanical Abstracts*, **17**, 181–190.

Weinberger, R. 2001. Joint nucleation in layered rocks with non-uniform distribution of cavities. *Journal of Structural Geology*, **23**, 1241–1254.

Wu, H. & Pollard, D. D. 1995. An experimental study of the relationship between joint spacing and layer thickness. *Journal of Structural Geology*, **17**, 887–905.

Zhang, L. & Einstein, H. H. 1998. Estimating the mean trace length of rock discontinuities. *Rock Mechanics and Rock Engineering*, **31**, 217–234.

The orientation distribution of single joint sets

TERRY ENGELDER[1] & JEAN DELTEIL[2]

[1] *Department of Geosciences, Pennsylvania State University, University Park, PA 16802, USA*
(e-mail: engelder@goesc.psu.edu)
[2] *Geosciences Azur, UMR 6526, Université de Nice – Sophia Antipolisu, 250 Rue Albert Einstein, Sophia Antipolis, F-06560 Valbonne, France*

Abstract: Poles from line samples of systematic joint sets scatter about a mean pole because joints are neither perfectly planar nor parallel, and because measurement instruments are imprecise. Definition of a single joint set can be based solely on its orientation distribution and this distribution is assessed using two statistical parameters: square root of the circular variance (approximately equal to the standard deviation σ for two-dimensional (2D) data) and cone of confidence (α_{95} for 3D data). The distribution for joints generated in the absence of tectonic deformation is well clustered with $\sigma = 1.7°$ and $\alpha_{95} = 0.48°$ based on a bootstrap sample of 50. Jointing associated with various fold styles show less clustering: the kink of a fault-bend fold ($\sigma = 6.1°$ and $\alpha_{95} = 1.7°$), basement-cored anticline ($\sigma = 3.5°$ and $\alpha_{95} = 1.5°$), regional joint set transected by a basement-cored anticline ($\sigma = 5.2°$ and $\alpha_{95} = 1.8°$) and a buttress anticline ($\sigma = 4.3°$ and $\alpha_{95} = 1.7°$). Jointing associated with local faulting tends to show even less clustering: a Cretaceous marl ($\sigma = 8.3°$ and $\alpha_{95} = 2.4°$) and a glauconitic sandstone ($\sigma = 8.6°$ and $\alpha_{95} = 2.2°$). The latter sample was drawn from two overlapping joint sets, indicating that distribution data greater than $\alpha_{95} = 2.2°$ may signal overlapping joint sets.

How can geologists identify joint sets from line samples such as tunnels, drill cores, borehole logs and outcrop faces along road and stream cuts? In line samples (especially in the case of borehole logs), orientation and position of joints may be the only data available. On outcrop faces and along drill core, joint-surface morphology may be of further help in distinguishing joint sets (e.g. Savalli, 2003). Even if orientation and position are the only data, identification of different joint sets can be important for detailed modelling of fractures for fluid-flow and stability analyses.

To answer the question posed above, geologists require knowledge of the orientation distribution of joints populating a single set. If joints propagate within an isotropic, homogeneous rock under the influence of an homogeneous remote driving stress (an homogeneous stress field has straight stress trajectories at the scale of the sample; Means 1976), the joint set is a collection of systematic joints, planar and parallel, with poles projecting to one point on a stereonet (Hodgson 1961). In the Earth, however, stress fields are commonly spatially and temporally inhomogeneous (i.e. their stress trajectories curve as demonstrated by elastic analyses, such as those of Ode 1957, and at the microscopic and smaller mesoscopic scales rock is quite inhomogeneous, e.g. Kranz 1979). The presence of joints and joint tips further complicates the local stress field (e.g. Olson & Pollard 1989). Although many joints have fairly smooth surfaces, possibly a direct consequence of subcritical crack growth, others are irregular because they formed by dynamic crack growth.

Joints of a set are neither completely planar nor perfectly parallel (e.g. Dyer 1988; Rawnsley *et al.* 1992). For this reason, the definition of a collection of joints populating a single joint set must be based on some statistical inferences. In geological media, there will never be absolute agreement on how to identify members of a single joint set nor will there be absolute certainty that all members of a collection of joints belong to the same set. In these instances, statistical inferences are helpful if it is necessary to define joint sets based on orientation data alone. Even then, a line sample crossing curving joints will yield orientation data that cannot be easily grouped into one single set.

The objective of this chapter is to present some statistical criteria that characterize the orientation distribution of joints in a set. This exercise then allows us to examine the degree of clustering of the poles to joints propagating in different tectonic settings, particularly from line samples in which the orientation and position of joints are the only available data. First, we consider the precision of the measurement instrument, usually a geological compass of some sort if samples are collected from scanlines along outcrop faces. Next, we make use of two statistics, kurtosis and the cone of confidence, to compare the degree of joint clustering from various tectonic settings. For a control sample we use data from relatively isotropic, homogeneous rock where joints are driven under the influence of a reasonably homogeneous stress field. Then, we calculate these two statistics for joint sets from tectonically complex settings where the driving stress might have been inhomogeneous in space or

From: Cosgrove, J. W. & Engelder, T. (eds) 2004. *The Initiation, Propagation, and Arrest of Joints and Other Fractures.* Geological Society, London, Special Publications, **231**, 285–297. 0305-8719/04/$15 © The Geological Society of London 2004.

changing orientation in time. Finally, we use our statistical analysis to test whether or not a joint set correlates with the orientation of other structures.

Analysis of directional data

There are many ways to present joint orientation
data. However, there are some instances when three-
dimensional (3D) sampling is not possible (e.g. an
outcrop that is a pavement surface). Rose diagrams
for 2D data and stereonets for 3D data are two of the
most commonly employed plots (e.g. Davis 1986). A
qualitative measure of the clustering of orientation
data is readily apparent on both plots, although
neither allows for testing of statistical hypotheses
and neither gives a sense of spatial variation in joint
orientation because all data are plotted as if taken
from one point. To illustrate spatial variation, ball
and stick diagrams are presented on well logs (Narr
1991). Azimuth v. traverse distance plots serve the
same purpose for outcrop data (Wise & McCrory
1982). The advantage of the latter, in particular, is
that it is very helpful for identification of a single
joint set that changes orientation by over several tens
to hundreds of metres in terrain where outcrop is
limited. Such dramatic swings in orientation are,
indeed, a reality for joint sets that have propagated
near fault contacts (e.g. Rawnsley et al. 1992).

For statistical testing of directional data, the probability model that is most useful is the von Mises distribution, an equivalent to the normal distribution
(Davis 1986). Like the normal distribution, the von
Mises distribution is characterized by two parameters, the mean direction and a concentration parameter equivalent to the standard deviation. To get a
sense of the degree of clustering in 2D directional
data, each datum (i.e. the strike of a joint) is assumed
to be unit vector at an angle, θ, relative to the x-axis:

$$x_i = \cos\theta_i \text{ and } y_i = \sin\theta_i. \quad (1)$$

The resultant vector from a data set is

$$R = \sqrt{(\Sigma_{i=1}^n \cos\theta_i)^2 + (\Sigma_{i=1}^n \sin\theta_i)^2} \quad (2)$$

where n is the number of strike measurements. The
mean resultant vector, $\bar{R}$, is calculated according to

$$\bar{R} = \sqrt{\bar{C}^2 + \bar{S}^2} \text{ where } \bar{C} = \frac{1}{n}\sum_{i=1}^n \cos\theta_i$$

$$\text{and } \bar{S} = \frac{1}{n}\sum_{i=1}^n \sin\theta_i. \quad (3)$$

Dispersion of a joint set measured on a pavement
surface is measured by the complement of $\bar{R}$, the circular variance,

$$s_o^2 = 1 - \bar{R} \quad (4)$$

This leads directly to a confidence angle around the
mean strike of the sample (Davis 1986). For data that
are normally distributed about a mean, with sample
ranges having a small angular variation, the square
root of the circular variance is similar to standard
deviation in a normal distribution (Mardia 1972).

The clustering of 3D data is characterized by a
cone of confidence, a statistic that was developed to
describe dispersion on a sphere (Fisher 1953). The
cone of confidence measures the probability, P, that
the actual mean pole of a joint set sits outside the
cone measured from the calculated mean pole, based
on the sample size of N joints. α_{95}, the cone outside
of which there is only a 5% probability that the true
mean pole sits, is calculated from

$$\cos\alpha = 1 - \frac{(N-R)}{R}\left\{\left(\frac{1}{P}\right)^{1/1-N} - 1\right\} \quad (5)$$

where $N = n$. R, as in the 2D case, is the sum of the
vectorially added individual unit vectors, such that
$R \leq N$.

While the clustering of joint-orientation data is
best assessed using the cone of confidence (equation
5), we also reduce scanline data to 2D sets as if we
measured the strike of joints on a pavement surface where the third dimension was inaccessible.
Assuming a von Mises distribution with an angular
variation that is equivalent to standard deviation, σ,
we calculate σ and kurtosis, κ, of these 2D data using
the standard statistical package on Microsoft Excel.
Kurtosis is a measure of the deviation of a data set
from an ideal normal distribution. A positive κ means
that the data distribution is more peaked than a normal
distribution, whereas a negative κ means the data are
flatter than a normal distribution (Walpole & Myers
1993). This measure of the distribution of joint orientation data is worth exploring despite being somewhat
ad hoc because systematic joint sets, planar and parallel, will have a large positive κ. A negative κ will be
the signal that the joints are not systematic or that
more than one joint set is being measured.

Calibration of the Freiberger compass

Because sampling techniques are not perfectly reproducible, sampling will affect the distribution of joint
orientation data even if the joint population is perfectly systematic. Orientation data collected for this
chapter were taken along scanlines using a Freiberger
compass. The flat plate on the back of the Freiberger
compass is perfectly suited for measuring planar surfaces with both dip and strike data collected during a
single contact between compass and outcrop. Under

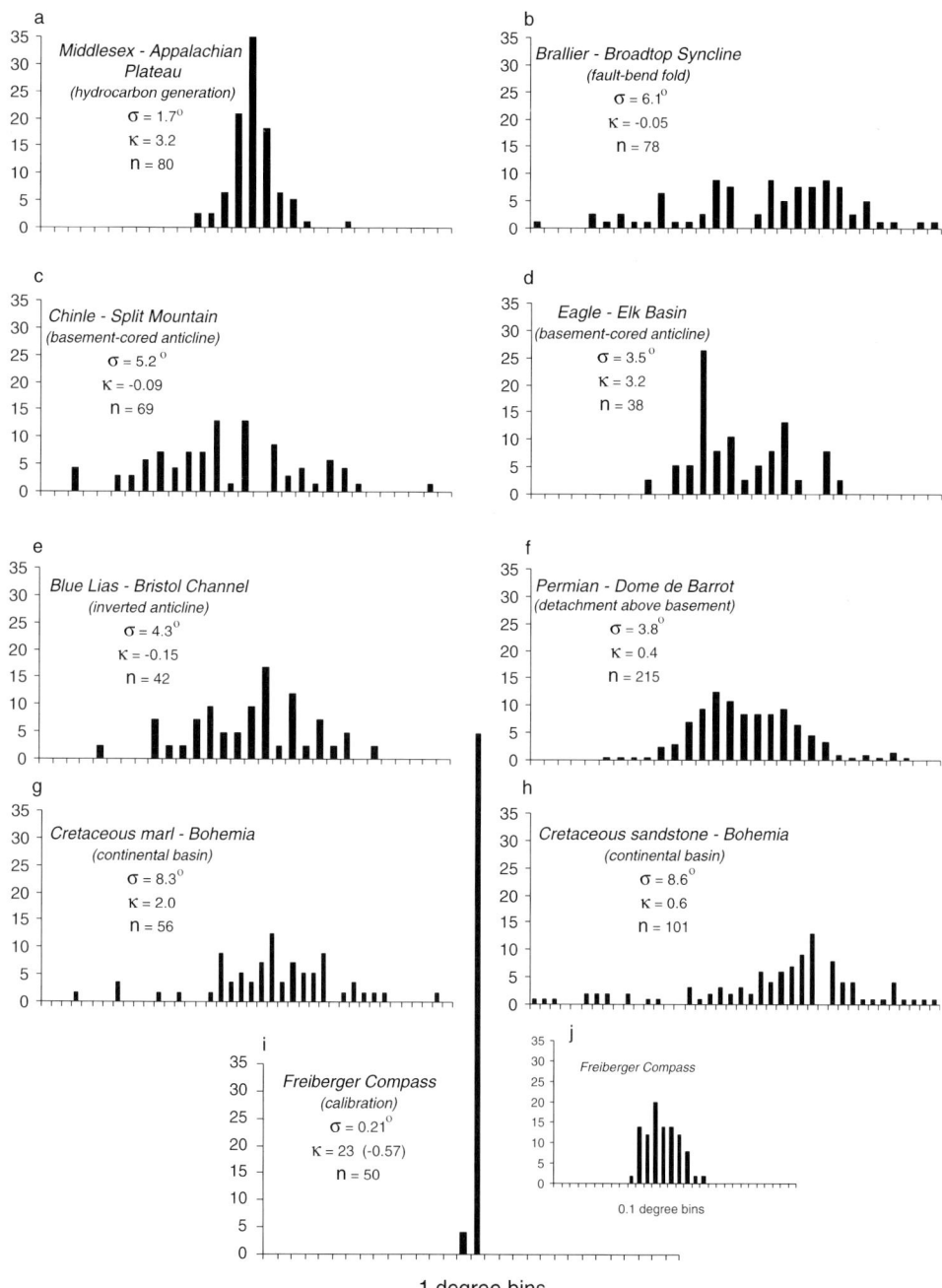

Fig. 1. Histograms of the strike of joints from one bed in eight exposures representing different tectonic environments throughout the world. The data are normalized to 100% so that the vertical scale on each histogram is divided into 5% intervals. Each bin is 1° except for the insert for data from the Freiberger compass which is divided into 0.1° bins. σ, standard deviation; κ, kurtosis; n, number of data. Source of the data: (**a**) Devonian Middlesex Formation, New York (Hagan 1997); (**b**) Devonian Brailler Formation, Pennsylvania (Ruf *et al.* 1998); (**c**) Triassic Chinle Formation, Utah (Silliphant *et al.* 2002); (**d**) Cretaceous Eagle Formation, Wyoming (Engelder *et al.* 1997); (**e**) Jurassic Blue Lias, UK (Engelder & Peacock 2001); (**f**) Permian of the Dome de Barrot (this paper); (**g**) and (**h**) Cretaceous Jizera Formation of Bohemia (this paper). (**i**) and (**j**) The Freiberger compass (used for all data sampling except that of Hagan 1997, who used a Silva).

field conditions, strike (or dip direction) is commonly read to the nearest degree but it can be read, arguably, to the nearest tenth of a degree with a steady hand under laboratory conditions (i.e. sitting at a desk with no flies and no wind). With a Freiberger compass under field conditions, the dip of a surface is commonly read to the nearest degree but under laboratory conditions dip can be read to the nearest half degree.

To develop a sense for sampling error, the orientation of the inside plaster wall of a 19th century house in Bizanos, France, was measured 50 times. The hinged plate of the Freiberger compass (7×7 cm) was positioned at the same point (± 1 cm) on a slightly tilted wall. Strike was read to $0.1°$ and dip read to $0.5°$. A 2D version of these data (i.e. strike data only) is presented as a histogram in both field format with $1°$ bins and in laboratory format with $0.1°$ bins (Fig. 1i, j). A $\sigma = 0.2°$ is the statistic that we use to describe the clustering of repeated measurements using a Freiberger compass and a measure of the precision of the Freiberger compass. Because this is such a tight cluster relative to the smallest sample interval (i.e. bin size) used for field measurement, we conclude that the instrument error has negligible impact on field sampling.

Kurtosis, κ, is another instructive statistic (Walpole & Myers 1993). When the compass calibration data is binned in $1°$ intervals, it possesses a strong positive κ (>23). This, then, would be a characteristic of a systematic joint set propagating in an isotropic, homogeneous medium driven under the influence of an homogeneous stress field. The κ ($= -0.57$) arising from reading the instrument to $0.1°$ is negative. A flat orientation distribution is consistent with the inference that strike data cannot be consistently read to the nearest $0.1°$ on the Freiberger compass, otherwise these data would also possess a positive κ.

There is the question about how much data are enough to define clustering within the population of a single joint set. Again, there is no right or wrong answer to this question. We can start to develop a sense of the degree of clustering using the cone of confidence (McElhinny 1964; Silliphant et al. 2002). A cone of confidence indicates the angular distance from the mean pole within which the true pole is found to a certain level of confidence. For example, the statistically averaged dip of the interior wall in Bizanos is $86.01°$ after 50 measurements, but the true average vector mean dip is somewhere in the range $86.01° \pm 0.14°$ as measured by the 95% cone of confidence (α_{95}). The statistic, $\alpha_{95} = 0.14°$, is a measure of the precision of our sampling instrument based on 50 samples (Fig. 2i). We have no independent measure of the orientation of the wall in Bizanos, so the accuracy of the measurement instrument cannot be assessed.

We can see how a smaller sample impacts on our assessment of instrument precision and then the clustering of a population of poles to joints. For the compass calibration, we selected a small number of data from our randomly sorted collection of 50 measurements. We repeated this random sort 21 times, sampled the same number of data (i.e., four or seven or 10 or 15 or 25 or 35 or 50) after each sort, and calculated a 95% cone of confidence (α_{95}) for each draw. The α_{95} data for each set of 21 similar samples are then sorted for plotting as a box and whisker diagram showing the highest α_{95}, the 75th percentile α_{95}, the 25th percentile α_{95} and the lowest α_{95} (Fig. 2i). With just four random samples 21 times, the α_{95} for our compass calibration varies anywhere from $1.1°$ to $0.14°$ with a median of $0.49°$. This range of α_{95} from $1.1°$ to $0.14°$ is one measure of the degree of clustering of our calibration data. Thus, if a perfectly planar, systematic joint set was encountered in the field, the probability of calculating an $\alpha_{95} \leq 0.49°$ with four measurements using a Freiberger compass is 50%. The probability of calculating $\alpha_{95} \leq 0.49°$ with 10 measurements is $\geq 95\%$. This method of estimating errors is similar to those techniques known in statistics as 'bootstrap techniques' (Efron & Gong 1983; Fisher et al. 1987).

The cone of confidence can be compared with 2D strike data for which $\sigma = 0.2°$. When just four samples are taken, most of the clusters of four data will have an α_{95} outside one standard deviation (1 SD) from the mean for the strike data. With 25 samples of our randomly sorted data, more than half of the draws yield an α_{95} within 1 SD of the mean from the 2D data. With 35 samples, each draw yields an α_{95} within the 1 SD. In effect, after 35 samples we have defined the orientation of the wall of the house in Bizanos at the 95% confidence level within the limits of the precision of the instrument.

Clustering of a single joint set independent of tectonic deformation

We wish to define the degree of clustering of poles to a joint set associated with propagation in a rock that is as nearly isotropic and homogeneous as possible. The remote stress field must also be as close to homogeneous as possible and the crack-tip stress fields of pre-existing members of a joint set cannot interfere with the propagation of infilling joints (cf. Olson & Pollard 1989). Natural hydraulic fractures infill without interference from the crack-tip stress field of neighbours (Fischer et al. 1995). For this purpose we look to a joint set that cuts black shale of the Devonian Middlesex Formation in the Appalachian Basin (Sheldon 1912). Within this black shale an ENE joint set is the most prominent and it propagated independently of Alleghanian structures (Engelder et al. 2001). These are hydraulic fractures with pressures generated during initial maturation of hydrocarbons within the black

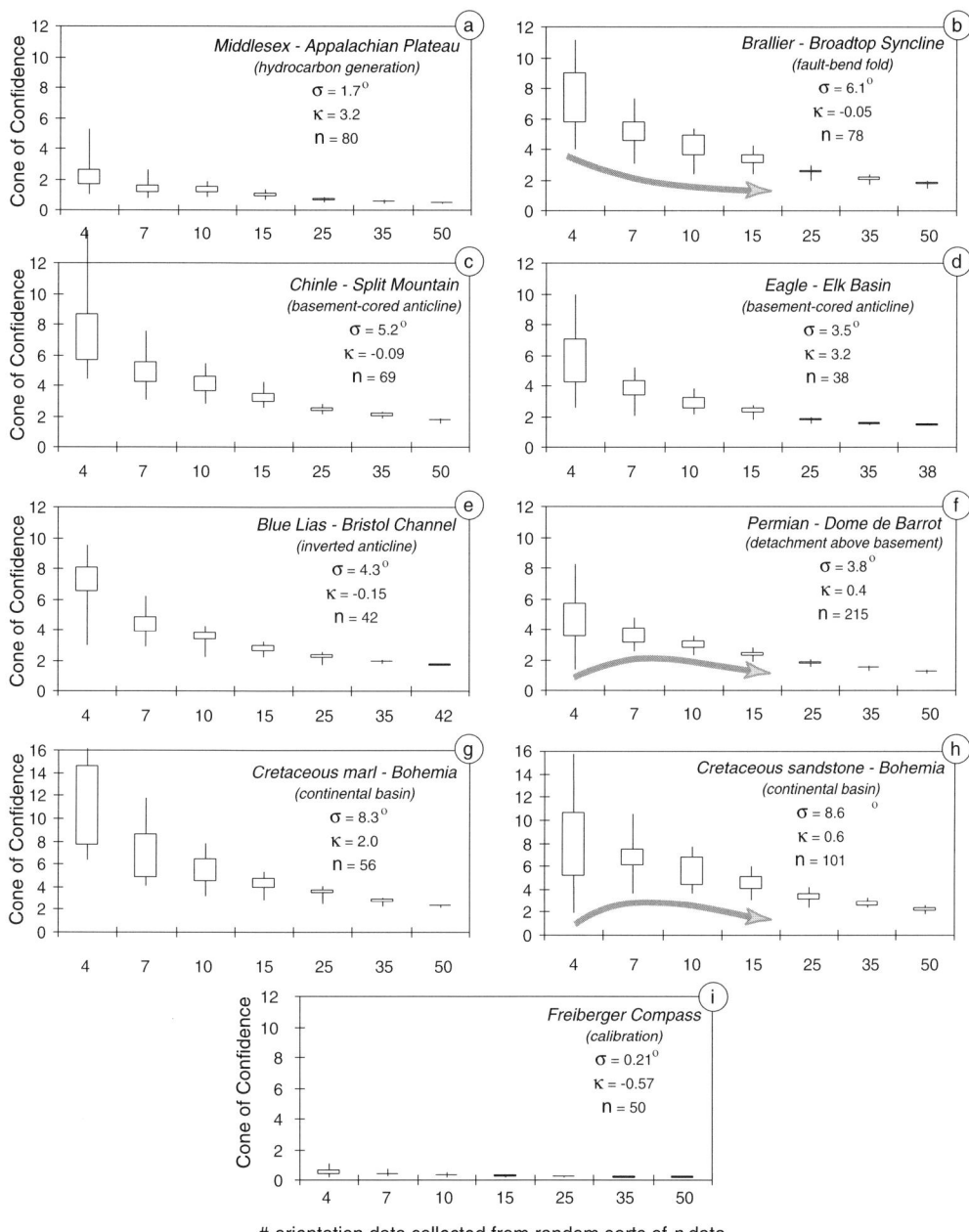

Fig. 2. Box and whisker diagram for the distribution of the α_{95} cone of confidences for 21 draws of random samples (4 $\leq n \leq$ 50). Each box defines the α_{95} for the 25th and 75th percentile. The tips of the whiskers define the maximum and minimum α_{95}. For the sources of the data see caption to Figure 1.

shale prior to Alleghenian deformation (Lash *et al.* 2004). Other than a bedding fissility and concretions (e.g. McConaughy & Engelder 1999), this black shale was devoid of internal structures at the time of joint propagation. We examine orientation data col-

lected by Hagen (1997) from this ENE joint set within the Finger Lakes District of New York State.

A scanline, placed on the pavement surface of a streambed, crossed 80 joints of the ENE set in a distance of 33 m (outcrop STE-01–AY). The 2D statistics include a $\sigma = 1.7°$ and $\kappa = 3.2$ (Fig. 1a). Anytime a sample of joint orientations displays a positive κ, there is a stronger probability that the sample comes from a single joint set. The natural variability of joint orientation due to subtle rock inhomogeneities and a stress field that is not perfectly homogeneous gives a $\sigma > 1°$ and tends to blunt an otherwise sharply peaked κ (i.e. >10).

These measurements were converted to a 3D set by arbitrarily assigning a dip of 90° to all joints. This is not unreasonable based on dip measurements from outcrop faces. Twenty-one draws of four data from the randomly sorted set of 80 data gives an α_{95} of between 1.0° and 5.2°, with a median of 2.0° (Fig. 2a). More than 75% of the time a sample of seven from randomly sorted data gives an $\alpha_{95} < \sigma$. Thus, for a single joint set like the ENE joints in the Middlesex Formation, very little information on clustering of orientation is gained by taking more than seven measurements along a scanline. There may, however, be other reasons for collecting more data. Each random draw of 25 data gives an α_{95} less than the bin size used in field sampling.

A sampling program in a rock as homogeneous as the Middlesex black shale might start with a scanline of nearly 100 measurements. Once σ is defined with this larger sample, then sampling at other outcrops may be limited to the sample size required to define an $\alpha_{95} \leq \sigma$. In the case of the relatively homogeneous Middlesex Formation, seven data are enough. For a perfectly systematic set of planar joints (i.e. the wall in Bizanos) very little information is gained by collecting more than four orientation data provided that a Freiberger compass is used.

Characteristics of joint sets generated during folding

The tectonic conditions during joint propagation within the Middlesex Formation on the Appalachian Plateau favours a population displaying a peaked κ. This joint set propagated by hydraulic fracturing in a stress field consistent with Acadian or early Alleghanian deformation to the east (Engelder et al. 2001; Lash et al. 2004). The black shale was undisturbed by Acadian deformation. There are other tectonic settings where the stress field is not expected to be as homogeneous nor is the host rock as isotropic and homogeneous as the black shale of the Middlesex Formation. Local inhomogeneities in the stress field are expected near faults and folds.

Limb of a large detachment fold

In the Appalachian Valley and Ridge of Pennsylvania, the Brallier Formation is the distal turbidite

sequence sitting at about the same stratigraphic position as the Middlesex Formation on the Appalachian Plateau. Near Huntington, Pennsylvania, the Brallier has been folded into the foreland limb of the Broadtop syncline (Ruf et al. 1998). Folds of the Valley and Ridge have sharp hinges (e.g. Faill 1973) typical of fault-bend folds (e.g. Suppe 1983). The fine-grained sandstone beds of the Brallier carry a well-developed joint set that parallels the strike of the larger-scale structures. The population of strike joints have a vector mean pole that makes an angle of roughly 85° to pole to bedding in the NW limb of the Broadtop syncline (Ruf et al. 1998). This non-orthogonality is interpreted to reflect a stress field that developed below a neutral fibre as the beds were passing through the sharp synclinal hinge going into the hanging-wall ramp at the trailing edge of a horse in lower Palaeozoic carbonate rocks of the Valley and Ridge (Ruf et al. 1998).

The stress field in the hinge of chevron folds like those of the Appalachian Valley and Ridge is controlled by bending stresses and, thus, is local rather than being regional, as was the case during propagation of joints in the Middlesex black shale. Depending on exactly when and where the joints propagated with respect to the hinge of the chevron fold, a population of strike joints could have a much larger σ than found in the black shale of the Appalachian Plateau. Strike joints in one bed of the Brallier display $\sigma = 6.1°$ and $\kappa = -0.05$ (Fig. 1b). A negative κ and a large σ are both consistent with jointing in an inhomogeneous stress field. There is no sign that this collection of joints comes from more than one set.

Twenty-one draws of four data collected at random from the Brallier measurements give an α_{95} between 4.0° and 11.2°, with a median of 7.2° (Fig. 2b). For 21 draws of 10 data collected at random, $\alpha_{95} < \sigma$. Although 10 data are enough to define the mean pole for strike joints in the Brallier based on $\sigma = 6.1°$, there are instances in which larger samples are appropriate. For example, we may wish to know if the strike joint set is coaxial with the fold axis. The answer to this question may require an orientation data set large enough to reduce the α_{95} to within $\pm 1°$. Fifty data reduces α_{95} to $\pm 1.4°$, which may be good enough because we see that the mean pole is within 1.4° of the local fold axis as defined by the strike of bedding at Huntington.

Limbs of two basement-cored folds

Basement-cored folds of the Laramide style found in the Rocky Mountains develop in a manner that is consistent with the trishear model for folding (Erslev 1991). The difference between a trishear fold and fault-bend folds of the Appalachian Valley and Ridge is that fold hinges are not as sharp and the fold develops incrementally through time (Fischer & Wilkerson 2000). Here the stress field may be thought of as varying temporally rather than spatially, as is the case of a fixed hinge fold. We examine two cases for Laramide folds. In one example a pre-existing joint set is transected by later folding and in the other example jointing is coaxial with folding.

Split Mountain anticline. The Split Mountain anticline is a W-plunging forced fold that formed during the Eocene uplift of the eastern Uinta Mountains (Hansen 1986). Transected joints (i.e. systematic joints that strike at an angle to the present fold axis trend) occur on the flanks of Split Mountain (Silliphant *et al.* 2002). The common orientation on both flanks for these WNW-striking joints is inconsistent with joints driven by a synfolding stretch normal to the direction of highest curvature. Silliphant *et al.* (2002) concluded that these joints propagated as a systematic set prior to Laramide folding because a smaller dispersion of the poles to these transected joints occurs when they are rotated with bedding to their 'prefold' orientation. Post-folding joints are wing cracks growing from the tips of prefolding joints (Wilkins *et al.* 2001).

Prefolding joints are particularly well developed in the Triassic Chinle Formation, a fluvial clastic rock with interbedded overbank mud deposits and channel sandstones of irregular thickness. Relative to both the black shale of the Middlesex Formation and the regular bedding of the distal turbidites in the Brallier Formation, the sandstone beds in Chinle Formation are discontinuous and quite inhomogeneous. Even if regional jointing in the Chinle were driven by a homogeneous stress field, a relatively larger σ is expected for these joints because of the irregular bedding. Indeed, the transected joints in the Chinle Formation at Split Mountains display $\sigma = 5.2°$ and $\kappa = -0.09$ (Fig. 1c). Again, this bed was selected because more than 50 joints were sampled in a single scanline.

Twenty-one draws of four orientation data collected at random from the Chinle data set give an α_{95} of between 4.4° and 14.1°, with a median of 6.8°. As with the Middlesex and Brallier, when 10 samples were collected at random in the Chinle, $\alpha_{95} < \sigma$ for a collection of strike data (Fig. 2c).

Elk Basin anticline. Elk Basin is a breached anticline, typical of Laramide (i.e. Late Cretaceous–Eocene) basement-involved folds in the northern portion of the Big Horn Basin, Montana–Wyoming (McCabe 1948). Structural relief on the Elk Basin anticline is about 1500 m. Deep seismic sections show a basement thrust to the ENE, causing an asymmetric drape of cover rocks with a forelimb dip in excess of 30° to the ENE, and the backlimb dipping 23° to the WSW (Bally 1983; Stone 1993).

The structural evolution of the Elk Basin anticline extends from late Cretaceous sedimentation through Palaeocene to Eocene basement thrusting. Of 72 outcrops sampled within sandstone beds, 67 are cut by one or more systematic joint sets, with systematic strike joints being far more common than dip joints (Engelder *et al.* 1997). In some outcrops two joint sets develop with dihedral angles of 10°–25° and abutting relationships that indicate a clockwise reorientation of the bedding-parallel stress axes during fold development. Joints coaxial to the strike of the fold at Elk Basin are considered to have a synfolding origin as a consequence of fold-induced stretching over the crest of the fold.

The strike joints in one bed of the Eagle Formation of Elk Basin display $\sigma = 3.5°$ and $\kappa = 3.2$ (Fig. 1d). When comparing these statistics to those found in the Brallier Formation or the Appalachian Valley and Ridge and the Chinle Formation of Split Mountain, we might conclude that sample population of joints within the Eagle are drawn from one joint set and that this set propagated more or less at one time during axis-normal stretching accompanying fold growth.

Twenty-one draws of four orientation data collected at random from the Eagle data set give an α_{95} between 2.6° and 9.9°, with a median of 5.4° (Fig. 2d). The axis of the Elk Basin is curved, reflecting the spoon-shaped basement fault over which the Cretaceous clastic rocks are draped. Because of the curved axis, it is not appropriate to use the α_{95} statistic to test whether or not the joint set is coaxial to the fold.

Comparing our two Laramide folds, the range of α_{95} for draws of four data from the Chinle at Split Mountain is the larger. This was not expected because prefolding joints sets like those found in the Middlesex have a relatively small α_{95} (Fig. 2a) One source for the large α_{95} at Split Mountain is infilling by joint growth during folding, as documented by Wilkins *et al.* (2001). Renewed growth of prefolding joints as wing cracks adds to the dispersion of the orientation data. If so, then it might be argued that our sample of Chinle is an example of an incipient second joint set (i.e. wing cracks) overprinting a single, early set. This would also be the reason for the slightly negative κ at Split Mountain.

Limb of a buttress anticline

Inversion in the Bristol Channel Basin during Pyrenean tectonics includes reverse-reactivated normal faults with hanging-wall buttress anticlines (Nemčok *et al.* 1995). At Lilstock Beach, one of several joint sets (i.e. the J_3 set of Engelder & Peacock 2001) in Lower Jurassic Blue Lias limestone beds clusters about the trend of the hinge of the

Lilstock buttress anticline. In horizontal and gently N-dipping beds, J_3 joints (c. 295°–285° strike) are rare, while other joint sets indicate an anticlockwise sequence of development (Engelder & Peacock 2001). In the steeper S-dipping beds, J_3 joints are the most frequent in the vicinity of the reverse-reactivated normal fault responsible for the buttress anticline. The J_3 joints strike parallel to the fold hinge, and their poles tilt to the south when bedding is restored to horizontal. This southward tilt aims at the trajectory of regional σ_1 during Pyrenean tectonics.

Joint clustering in the Blue Lias at Lilstock is interesting for two reasons. First, these rocks record dramatic swings in orientation of joint sets in the vicinity of normal faults, much like those reported by Rawnsley et al. (1992) for other places along the English coastline. However, J_3 joints were not observed near the normal faults. Second, this is also an example, like the Brallier, where there is a non-orthogonality between joints and bedding. Here the mechanism for tilting joints relative to bedding is believed to be a shear traction set up by regional stresses rather than local bending stresses, as was the case for the Brallier (Engelder & Peacock 2001). Nevertheless, these joints, like those in the Brallier and Eagle formations of Elk Basin, are in the strike orientation.

The strike joints of the J_3 set in the Blue Lias of the Lilstock buttress anticline display $\sigma = 4.3°$ and $\kappa = -0.15$ (Fig. 1e). The negative κ is indicative of a range of orientation data that witness the relative inhomogeneity of the stress field at the time of joint propagation during the development of the buttress anticline at Lilstock. Small samples of joints within the Blue Lias compare with those from the Brallier, Chinle and Eagle formations. Twenty-one draws of four orientation data collected at random produces a range for α_{95} between 2.9° and 9.5°, with a median of 7.4°. Here, again, is such a large α_{95} that one cannot be sure with four measurements whether J_3 is coaxial with the buttress anticline at Lilstock Beach. Larger samples demonstrate that this is true.

An application of α_{95}: a bootstrap test for two overlapping fracture sets

In many instances, the assessment of joints and other fracture sets is based on their appearance in cliff faces and other places where good exposures in plan view (i.e. pavements) are not available. With just vertical faces available, visual inspection is rarely sufficient to recognize the presence of more than one joint set when two sets have formed at a small dihedral angle. Here statistical inferences are necessary. We first develop a bootstrap test for discriminating multiple joint sets in an outcrop where pavement view confirms the presence of two closely oriented

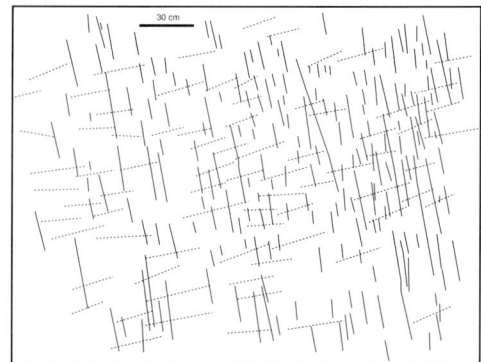

Fig. 3. Tracing of veins (solid lines) and the longest dimension of the hexagonal blocks defined by mudcracks (dashed lines) from the Dome de Barrot. This sample comes from a large block that is not in place, hence the absolute orientation of the veins and finite strain are know only from adjacent outcrops (see Siddans et al. 1984).

joint sets. Then, we apply this bootstrap test to exposures where pavement surfaces are unavailable.

Dome de Barrot, Alpes Maritimes

We start with an example from a Permian tectonite of the Dome de Barrot in the Alpes Maritimes (France) where veins cut through deformed mudcracks (Fig. 3). The Permian rocks of the Alpes Maritimes (France) consist of a sequence of continental redbeds that accumulated in discrete basins during the earliest phases of the breakup of Pangaea (Faure-Maret 1955). During the Miocene Alpine orogeny, these redbeds were coupled to basement deformation under a cover detaching along Triassic evaporites (Goguel 1936). Because of this coupling to basement, strain in the Permian redbeds is believed to be comparable to Alpine strain in the basement massifs including the Argentera and Maures-Esterel (Siddans et al. 1984). The veins of Figure 3 propagated during Alpine deformation.

We use veins as a proxy for joints in this case because the mechanics of vein propagation follow the same physical laws as those for joint propagation (Pollard & Segall 1987). We can draw some general conclusions from Figure 3. First, the veins (i.e. filled joints) seem to have propagated in the y–z-plane of this tectonite, thus raising the possibility that the veins are coaxial with finite strain in these Permian redbeds. If the veins are coaxial with flattening of the mudcracks, both deformation processes may have been driven under the same tectonic stress field and may be contemporaneous. Second, the veins are not all perfectly parallel. Third, the veins do not cross-cut nor do they abut. We then wonder whether we are looking at a natural spectrum of orientations from one set of veins or whether there are two vein sets that signal a small change in the orientation of the driving stress. If the latter is true then at least one of the vein sets is not coaxial with the total finite strain in this tectonite, an issue that we deal with later in this chapter.

The veins in the Permian section of the Dome de Barrot display $\sigma = 3.8°$ and $\kappa = 0.4$ (Fig. 1f). When $\kappa \leq 1$, there is the possibility that more than one joint set contributed to the sample. In the case of Dome de Barrot, two closely oriented vein sets appear to have formed based on the histogram for vein strike that shows two peaks separated by 5° (Fig. 1f). If, in fact, this exposure contains two vein sets, then the σ is not large relative to that for a single vein set.

The α_{95} for a random draw of four orientation data from the Permian marl is between 1.4° and 8.3°, with a median of 4.7° (Fig. 2f). Other than the clustering of joints within the Middlesex black shale, no suite of α_{95} data has smaller minimum (i.e. a tighter cluster) for four random samples. However, the unusual characteristic to the α_{95} sampling for the Dome de Barrot is that the range of α_{95} for seven random samples has a minimum value (i.e. 2.5°) that is nearly double in size from the smallest α_{95} in the four-sample draw (i.e. arrow in Fig. 2f). The minimum value for α_{95} in all other samples discussed so far has decreased when seven-sample draws are taken (i.e. arrow in Fig. 2b). When going from four-sample draws to seven-sample draws at Dome de Barrot, this increase in α_{95} is a consequence of sampling two vein sets. When taking some small samples, all data may, by random chance, come from one of the two sets and are, therefore, tightly clustered. This results in a small α_{95}. Other small samples take data from both joint sets and have a concomitant larger α_{95}. A tight cluster also reflects the fact that both of the vein sets in the Dome de Barrot are relatively homogeneous in orientation, a characteristic found only in the Middlesex black shale. Seven-sample draws all included data from both vein sets in this example.

Cretaceous basin, Bohemian Massif

The high stand of the Cretaceous seas flooded a large basin on the NE side of the Bohemian Massif in the Czech Republic. This Cretaceous basin is elliptical with its long axis trending NW–SE. On the Bohemian Massif the Turonian Jizera Formation is characterized by marls, interlayered marly siltstones and glauconitic sandstone. Steeply dipping faults are common within the Cretaceous basin. Post-Cretaceous jointing pervades the Turonian section to the extent that most exposures along the Orlice River are well jointed with more than one joint set. The most promi-

nent joint set strikes NE–SW, which is across the axis of the Cretaceous basin but the joints are in the strike orientation relative to NW-verging thrusts of the western Carpathians, which are located to the SE.

The character of jointing within marls appears quite different from the jointing within the glauconitic sandstone beds. The vertical growth of joints in the more homogeneous marls is extensive, much like the vertical joints in the Middlesex Formation of the Appalachian Plateau. A closer inspection, however, reveals that the individual joints are not as systematic as those of the Middlesex. In contrast, the joints of the glauconitic sandstone beds include a well-developed set that is contained within individual beds between 20 and 40 cm thick. More widely spaced joints may be traced vertically through several beds. Like their counterparts in the thick marls, joints within the glauconitic sandstone beds appear systematic on first glance but are also found to be less systematic upon closer inspection.

Jointing in the thicker marls was sampled near the railway station at Usti nad Orlici and about 100 m west of a high-angle normal fault. Jointing in the glauconitic sandstone beds was sampled along the River Orlice at Chocen, about 500 m from the nearest prominent normal fault. Both exposures are cliff faces so we were unable to observe joints in plan view. Both samples proved to have a larger σ than any of the previously described data sets (Fig. 1g, h). A positive κ ($= 2.0$) in the thicker marl supports the premise that we did measure just one joint set but this set is characterized by a large dispersion. The joints with the glauconitic sandstone beds show a prominent peak but have a smaller κ. The smaller κ is largely a consequence of several joints that were about 20° anticlockwise from the peak.

The α_{95} for a random sample of four orientation data from the marl is between 6.3° and 19.2°, with a median of 10.8°. This large range reflects the relatively non-systematic nature of joint development within the marls, presumably because of the exposure's proximity to the nearby steeply dipping fault. The α_{95} for a random sample of four orientation data from the glauconitic sandstone beds is between 1.9° and 16.7°, with a median of 8.5°. This reflects the more systematic organization of jointing in the bedded sandstones. Like our sample from the Dome de Barrot, random samples of seven data within the glauconitic sandstone have a larger α_{95} ($= 3.7°$) than the minimum for four random samples. In fact, 50 random samples are required before the α_{95} is reduced below the minimum for four random samples. By applying the same rational developed for orientation data from Dome de Barrot, we conclude that the glauconitic sandstone contains more than one joint set. The second set is not as common but can be seen in the histogram for the glauconitic sandstone (Fig. 1h).

An application of α_{95}: a bootstrap test for coaxial deformation

Standard techniques for strain analysis, such as the Fry (1979) centre-to-centre technique, are effective tools for measuring the orientation of the strain ellipse. However, in its original form, it offers no mechanism for assessing confidence in the orientation. We can develop a sense of confidence through a calculation of the cone of confidence for orientation data. We will apply this technique to measure the relationship between the orientation of veins in the Permian rocks of the Dome de Barrot and the strain ellipse as recorded in mudcracks (i.e. Siddans *et al.* 1984). We focus on this hypothesis because the veins of the Dome de Barrot appear to be coaxial with the flattening direction of the mudcracks and, yet the veins show no tendency to be buckled or otherwise deformed. Hence, the veins may post-date the syn-flattening strain of the mudcracks.

Coaxial deformation

Increments of strain are coaxial only if there is no component of rotational strain at any stage during the deformation (Ramsay & Huber 1983). Theoretically, coaxial strain takes place only if the driving stress field remains stationary and its principal axes correlate with the principal axes of strain throughout the deformation. In this case the finite strain is characterized by pure shear with principal axes of strain remaining in the same orientation during each increment of strain. In geological time, the tectonic stress field can change a great deal relative to a local coordinate system and consequently there has usually been some element of rotational (i.e. non-coaxial) strain when large finite strain is found in rock. Yet, tectonic stress within the Dome de Barrot appears to have remained stable from the time of the flattening of mudcracks until the time of propagation of the veins that cut the mudcracks.

There is a distinction between the strain recorded by joints and strain recorded by other structures. A joint is coaxial with one increment of strain small enough to be considered infinitesimal, whereas other strain markers record the orientation of the total-strain ellipsoid if they were present at the start of strain. A crack–seal vein may show many increments of growth and it is thus a finite-strain marker. Fibres normal to a vein wall indicate that the crack–seal vein remained coaxial with the tectonic stress field. Oblique fibres in a crack–seal vein are an indication that recracking is, in fact, guided within the vein and that the crack driving stress is no longer coaxial with the vein. A vein with oblique fibres is commonly associated with a spectrum of other veins that are misaligned, thus indicating a change in the orientation of the driving stress relative to the coordinate system of early veins.

Deformation within the Dome de Barrot: coaxial or non-coaxial?

Within the Permian section of Dome de Barrot strain indicators include mudcracks (Graham 1978), reduction spots (Siddans 1980), veins (Siddans *et al.* 1984), remanent magnetization (Henry 1973) and anisotropy of magnetic susceptibility (Siddans *et al.* 1984). In the Dome de Barrot region, in particular, veins overprint deformed mudcracks in an orientation that gives the appearance of two coaxial mesoscopic structures (Fig. 3). The problem is that with the unaided eye, we cannot tell whether these two structures are really coaxial. This is one of the reasons why we are particularly interested in understanding how a single fracture set might look from a statistical point of view. If we have a rigorous definition of a single fracture set from a simple statistical test, then we can determine the extent to which we are confident that the two mesoscopic structures are coaxial.

Using the veins and mudcracks in the Permian rocks of the Dome de Barrot, we can test for the stability of the finite-strain ellipsoid by correlating vein orientation with the finite-strain ellipse from the mudcracks. Rather than using the conventional centre-to-centre techniques for determining the orientation of the finite-strain ellipse for mudcracks (i.e. Fry 1979), we use a technique that gives some estimate of the confidence that we have in the orientation of the finite ellipse. To develop a sense of statistical confidence in the orientation of strain on the plane of bedding, we use the long axes of the blocks between the mudcracks. Each has a long dimension defined as the distance between sharp corners of the irregular hexagon that constitutes individual blocks defined by mudcracks. Because the initial shape of the hexagonal blocks is irregular, very few of the longest dimensions actually sit in the orientation of the finite-strain ellipse. However, with a large sample of the orientation of long dimensions of the many blocks, a confidence can be assigned to the calculated axis of maximum shortening (i.e. the direction perpendicular to the long axes of the hexagonal blocks). We can test for the reliability of our estimate of the stretching direction using the same α_{95} calculation that we applied to the poles to joints (Fig. 4).

In the sample coordinate system, the statistically averaged mean vector for the long axis of one section of the strain ellipsoid of the mudcracks is $346.4° \pm 1.3°$ ($n = 77$). The veins as a whole sample have a statistically averaged vector mean pole of $348.8° \pm 0.6°$ ($n = 215$). The α_{95} for the mudcracks does not encompass the vector mean pole for the veins nor do the two α_{95} cones overlap. The veins

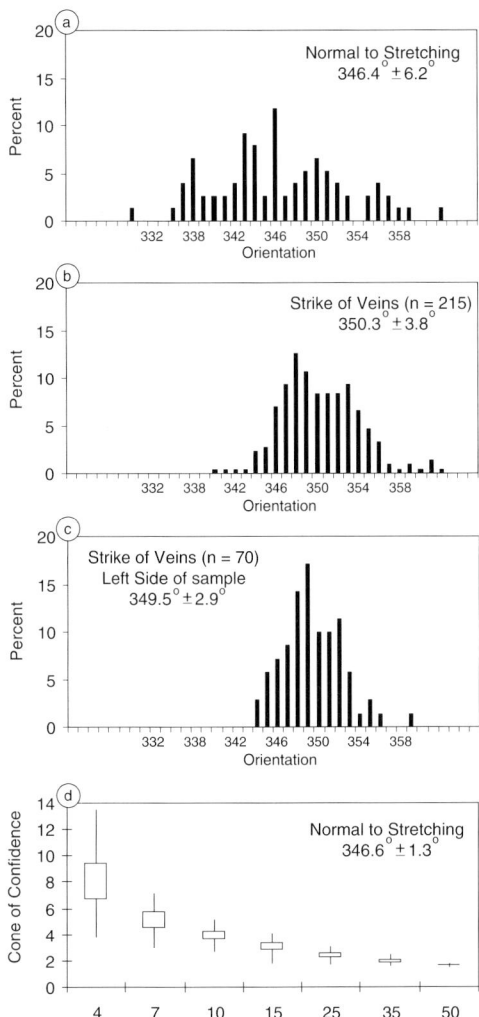

Fig. 4. Histograms of the direction of (**a**) layer-parallel shortening identified as the normal line to the long direction of each hexagonal block defined by mudcracks, (**b**) and (**c**) the strike of veins, and (**d**) box and whisker diagram for the distribution of the α_{95} cone of confidence for 21 draws of random samples ($4 \leq - \leq 50$) of the layer-parallel shortening from the Dome de Barrot. Each box defines the α_{95} for the 25th and 75th percentile.

and the mudcracks are not coaxial. This is a conclusion that is not evident with a visual inspection (Fig. 3). However, a plot of the orientation of the veins in the plane of bedding relative to the orientation of the axes of shortening in bedding show a misalignment (Fig. 4). Even if the maximum bin (i.e. 348°) were taken as the orientation of one vein set, we see that this orientation is outside the cone of confidence for the layer-parallel shortening.

In brief, most parts of the Dome de Barrot have just one vein set. Even when there are two events of vein development, neither strictly corresponds to the direction of layer-parallel shortening as indicated by the mudcracks. Therefore, we accept the hypothesis that veins and layer-parallel shortening are not coaxial. However, we are struck by the extent to which the stress state remained in approximately the same orientation (i.e. $\pm 2°$) over time.

Conclusions

Joint sets can be distinguished using line sampling of joints for which only orientation and position data are available. Systematic joint sets propagating in a non-tectonic regime show a relatively small cone of confidence (i.e. $\alpha_{95} = 0.48°$) and a peaked, postive kurtosis. Syntectonic joints or early joints that have been tilted with the growth of a fold have a larger cone of confidence (i.e. $\alpha_{95} = 1.5°–1.8°$) and less peaked but still postive kurtosis. When two closely oriented joint sets appear in the same sample set the α_{95} can increase as the selection of random samples from the entire population of joints in an exposure increases from, say, four to seven.

This work was supported by Pennsylvania's State's Seal Evaluation Consortium (SEC) and a fellowship from the French–American Foundation, Paris. P. Gilliespie, A. Whitaker, and J. Cosgrove are thanked for constructive reviews.

References

BALLY, A. W. 1983. *Seismic Expression of Structural Styles*. American Association of Petroleum Geologists: Studies in Geology Series, **15**.

DAVIS, J. C. 1986. *Statistics and Data Analysis in Geology*. John Wiley, New York.

DYER, R. 1988. Using joint interactions to estimate paleo-stress ratios. *Journal of Structural Geology*, **10**, 685–699.

EFRON, B. & GONG, G. 1983. A leisurely look at the boot-strap, the jackknife and cross-validation. *American Statistician*, **37**, 36–48.

ENGELDER, T. & PEACOCK, D. 2001. Joint development normal to regional compression during flexural-slow folding: The Lilstock buttress anticline, Somerset, England. *Journal of Structural Geology*, **23**, 259–277.

ENGELDER, T., GROSS, M. R. & PINKERTON, P. 1997. Joint development in clastic rocks of the Elk Basin anti-cline, Montana–Wyoming. *In*: HOAK, T., KLAWITTER, A. & BLOMQUIST, P. (eds) *An Analysis of Fracture Spacing Versus Bed Thickness in a Basement-involved Laramide Structure*. Rocky Mountain Association of Geologists 1997 Guidebook. Rocky Mountain Association of Geologists, Denver, CO, 1–18.

ENGELDER, T., HAITH, B. F. & YOUNES, A. 2001. Horizontal slip along Alleghanian joints of the Appalachian plateau: evidence showing that mild penetrative strain does little to change the pristine appearance of early joints. *Tectonophysics*, **336**, 31–41.

ERSLEV, E. A. 1991. Trishear fault-propagation folding. *Geology*, **19**, 617–620.

FAILL, R. T. 1973. Kink band folding, Valley and Ridge Province, Pennsylvania. *Geological Society of American Bulletin*, **84**, 1289–1314.

FAURE-MURET, A. 1955. *Etudes géologiques sure le massif de l'Argentera–Mercantour et ses envelopes sédimentaires*. Service de la Carte Géologique de France, Mémoires.

FISCHER, M.P. & WILKERSON, S. 2000. Predicting the orientation of joints from fold shape: results of pseudo-three-dimensional modeling and curvature analysis. *Geology*, **28**, 15–18.

FISCHER, M., GROSS, M. R., ENGELDER, T. & GREENFIELD, R. J. 1995. Finite element analysis of the stress distribution around a pressurized crack in a layered elastic medium: Implications for the spacing of fluid-driven joints in bedded sedimentary rock. *Tectonophysics*, **247**, 49–64.

FISHER, N. I., LEWIS, T. L. & EMBLETON, B. J. J. 1987. *Statistical Analysis of Spherical Data*. Cambridge University Press, Cambridge.

FISHER, R. W. 1953. Dispersion on a sphere. *Proceedings of the Royal Society of London*, **A217**, 295–305.

FRY, N. 1979. Random point distributions and strain measurements in rocks. *Tectonophysics*, **60**, 89–105.

GRAHAM, R. H. 1978. Quantitative deformation studies in the Permian rocks of Alpes-Maritimes. *Mémoires du Bureau de Recherches Géologiques et Minières*, **91**, 219–238.

GOGUEL, J. 1936. *Description tectonique de la bordure des Alpes de la Boéone au Var*. Service de la Carte Géologique de France, Mémoires.

HAGIN, P. N. 1997. *Joints spacing statistics in thick, homogeneous shales of the Catskill Delta complex on the Appalachian plateau: Finger lakes Region, New York*. B.S. thesis, Pennsylvania State University.

HANSEN, W. R. 1986. History of faulting in the eastern Uinta Mountains, Colorado and Utah. *In*: STONE, D. S. (ed.) *New Interpretations of Northwest Colorado Geology*. Rocky Mountain Association of Geologists, Denver, CO, 229–246.

HENRY, B. 1973. Studies of microstructures, anisotropy of magnetic susceptibility and paleomagnetism of the Permian Dome de Barrot (France): paleotectonic and paleosedimentological implications. *Tectonophysics*, **17**, 61–72.

HODGSON, R. A. 1961. Regional study of jointing in Comb Ridge–Navajo mountain area, Arizona and Utah. *AAPG Bulletin*, **45**, 1–38.

KRANZ, R. L. 1979. Crack–crack and crack–pore interactions in stressed granite. *International Journal of Rock Mechanics and Mining Science*, **16**, 37–47.

LASH, G. LOEWY, S. & ENGELDER, T. 2004. Preferential jointing of Upper Devonian black shale, Appalachian Plateau, USA: evidence supporting hydrocarbon generation as a joint-driving mechanism. *In*: COSGROVE, J. W. & ENGELDER, T. (eds) *The Initiation, Propagation, and Arrest of Joints and Other*

Fractures. Geological Society, London, Special Publications, **231**, 129–151.

MARDIA, L. 1972. *The Statistics of Orientation Data*. Academic Press, London.

MCCABE, W. S. 1948. Elk Basin Anticline, Park County, Wyoming, and Carbon County, Montana. *AAPG Bulletin*, **32**, 52–67.

MCCONAUGHY, D. T. & ENGELDER, T. 1999. Joint interaction with embedded concretions: Joint loading configurations inferred from propagation paths. *Journal of Structural Geology*, **21**, 1049–1055.

MCELHINNY, M. W. 1964. Statistical significance of the fold test in paleomagnetism. *Geophysical Journal of Astronomical Society*, **8**, 338–340.

MEANS, W. D. 1976. *Stress and Strain – Basic Concepts of Continuum Mechanics for Geologists*. Springer, New York.

NARR, W. 1991. Fracture density in the deep subsurface: Techniques with application to Point Arguello Oil Field. *AAPG Bulletin*, **75**, 1300–1323.

NEMČOK, M. BAYER, R. & MILIORIZOS, M. 1995. Structural analysis of the inverted Bristol Channel Basin: implications for the geometry and timing of fracture porosity. *In*: BUCHANAN, J. G. & BUCHANAN, P. G. (eds) *Basin Inversion*. Geological Society, London, Special Publications, **88**, 355–392.

ODE, H. 1957, Mechanical analysis of the dyke pattern of the Spanish Peaks area, Colorado. *Geological Society of America Bulletin*, **68**, 567–576.

OLSON, J. E. & POLLARD, D. D. 1989. Inferring paleostress from natural fracture patterns: A new method. *Geology*, **17**, 345–348.

POLLARD, D. D. & SEGALL, P. 1987. Theoretical displacements and stresses near fractures in rock: with applications to faults, joints, veins, dikes, and solution surfaces. *In*: ATKINSON, B. (ed.) *Fracture Mechanics of Rock*. Academic Press, Orlando, FL, 227–350.

RAMSAY, J. G. & HUBER, M. I. 1983. *The Techniques of Modern Structural Geology: Volume 1: Strain Analysis*. Academic Press, Orlando, FL.

RAWNSLEY, K. D., RIVES, T., PETIT, J.-P. HENCHER, S. R. & LUMSDEN, A. C. 1992. Joint development in perturbed stress fields near faults. *Journal of Structural Geology*, **14**, 939–951.

RUF, J. C. RUST, K. A. & ENGELDER, T. 1998. Investigating the effect of mechanical discontinuities on joint spacing. *Tectonophysics*, **295**, 245–257.

SAVALLI, L. 2003. *Mechanisms controlling rupture front geometries during joint propagation in layered clastic sediments*. MS thesis, Pennsylvania State University.

SHELDON, P. 1912. Some observations and experiments on joint planes. *Journal of Geology*, **20**, 53–70.

SIDDANS, A. W. B. 1980. Compaction, metamorphisme et structurologie des argilites permiennes dans les Alpes-Maritimes (France). *Revue de Geologie Dynamique et de Geographie Physique*, **22**, 279–292.

SIDDANS, A. W. B., HENRY, B., KLIGFIELD, R., LOWRIE, A., HIRT, A. & PERCEVAULT, M. N. 1984. Finite strain patterns and their significance in Permian rocks of the Alpes Maritimes (France). *Journal of Structural Geology*, **6**, 339–368.

SILLIPHANT, L. J., ENGELDER, T. & GROSS, M. R. 2002. The state of stress in the limb of the Split Mountain anticline, Utah: constraints placed by transected joints. *Journal of Structural Geology*, **24**, 155–172.

STONE, D. S. 1993. Basement-involved thrust-generated folds as seismically imaged in the subsurface of the central Rocky Mountain foreland. *In*: SCHMIDT, C.J., CHASE, R.B. & ERSLEV, E.A. (eds) *Laramide Basement Deformation in the Rocky Mountain Foreland of the Western United States*. Geological Society of America, Special Papers, **280**, 271–318.

SUPPE, J. 1983. Geometry and kinematics of fault-bend folding. *American Journal of Science*, **283**, 684–721.

WALPOLE, R. & MYERS, R. 1993. *Probability and Statistics for Engineers and Scientists*. Prentice-Hall, Englewood Cliffs, NJ.

WILKINS, S. J., GROSS, M. R., WACKER, M., EYAL, Y. & ENGELDER, T. 2001. Faulted joints: kinematics, displacement–length scaling relationships and criteria for their interpretation. *Journal of Structural Geology*, **23**, 315–327.

WISE, D. U. & MCCRORY, T. A. 1982. A new method of fracture analysis; azimuth versus traverse distance plots. *Geological Society of America Bulletin*, **93**, 889–897.

Stress-controlled localization of deformation and fluid flow in fractured rocks

DAVID J. SANDERSON & XING ZHANG

Department of Earth Science and Engineering, Imperial College London, Exhibition Road, London SW7 2AZ, UK

Abstract: The discrete-element method (UDEC – Universal Distinct Element Code) was used to numerically model the deformation and fluid flow in fracture networks under a range of loading conditions. A series of simulated fracture networks were generated to evaluate the effects of a range of geometrical parameters, such as fracture density, fracture length and anisotropy.

Deformation and fluid flow do not change progressively with increasing stress. Instability occurs at a critical stress and is characterized by the localization of deformation and fluid flow usually within intensively deformed zones that develop by shearing and opening along some of the fractures. The critical stress state may be described in terms of a driving stress ratio, $R = $ (fluid pressure $-$ mean stress)$/\frac{1}{2}$ (differential stress). Instability occurs where the R ratio exceeds some critical value, R_C, in the range -1 to -2.

At the critical stress state, the vertical flow rates are characterized by a large increase in both their overall magnitude and degree of localization. This localization of deformation and fluid flow develops just prior to the critical stress state and may be characterized by means of multifractals. The stress-induced criticality and localization displayed by the models is an important phenomenon, which may help in the understanding of deformation-enhanced fluid flow in fractured rock masses.

Introduction

Networks of fractures or faults in rocks may play an important role in controlling deformation and fluid flow in the crust (Barton *et al.* 1995; Zhang & Sanderson 1996). Evidence for this comes from phenomena ranging from earthquakes (Healy *et al.* 1968; Raleigh *et al.* 1976; Das & Scholz 1981; Talwani & Acree 1985) to the accumulation of hydrocarbons and hydrothermal ore deposits (Cox 1999).

The localization of deformation and fluid flow in fractures and shear zones is a common phenomenon in the upper crust (e.g. Cox 1999; Sanderson & Zhang 1999). The onset of localized deformation and fluid flow is a critical-point phenomenon, at which the mechanical and hydraulic behaviour of a previously fractured rock mass undergoes a sudden change (Zhang & Sanderson 1997, 2001; Sanderson & Zhang 1999).

In this study, we use numerical analysis to evaluate the critical stress state at which a fractured rock mass becomes unstable. A fractured rock mass is considered as a system of deformable blocks of rock separated by interfaces, usually occupied by fluid. The forces acting between all elements of this system may be in equilibrium. Changes in stress trigger instability in this system, characterized by local acceleration of movement and localization of deformation and fluid flow (Zhang & Sanderson 2001).

In this chapter we explore the relationship between the differential stress, mean stress and fluid pressure, and the critical behaviour for a range of simulated fracture networks. The approach follows that of Zhang & Sanderson (2001) and extends the earlier work of Sanderson & Zhang (1999) that examined the behaviour of three specific fracture networks. They showed that there is a critical stress state at which deformation of the rock mass leads to increased flow rate and localization of the flow. The aims of the present study are: (1) to characterize the critical state in terms of the driving stress ratio, R (see the section on 'Instability and the R ratio'); (2) to use simulated fracture networks to systematically explore the effects of some geometrical features (fracture density, fracture length and anisotropy) on the instability of fractured rock masses (see the section on 'Role of fracture-network geometry'); and (3) to examine the fractal properties of the resulting fluid flow (see the section on 'Localization of fluid flow').

Numerical modelling of critical behaviour

In this study use is made of the distinct-element method, using the UDEC code (Cundall *et al.* 1978), to simulate deformation and fluid flow in fractured rock masses in response to changes in stress. This code is suited to the modelling of the interaction of deformation and fluid flow, and allows large finite displacement and deformations, with the program recognizing new contact geometries that may arise

From: COSGROVE, J. W. & ENGELDER, T. (eds) 2004. *The Initiation, Propagation, and Arrest of Joints and Other Fractures.* Geological Society, London, Special Publications, **231**, 299–314. 0305-8719/04/$15 © The Geological Society of London 2004.

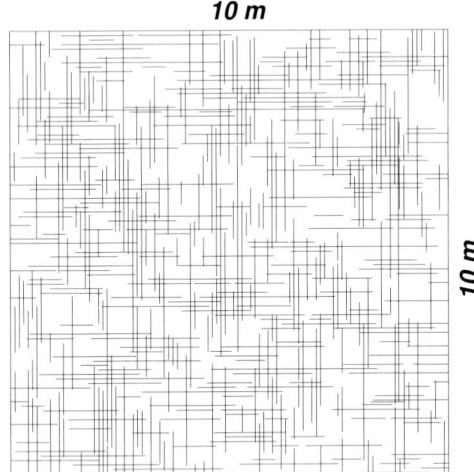

Fig. 1. A simulated fracture network consisting of two orthogonal sets of parallel fractures with a density of 7.85 m⁻¹ and an average length of 1.15 m.

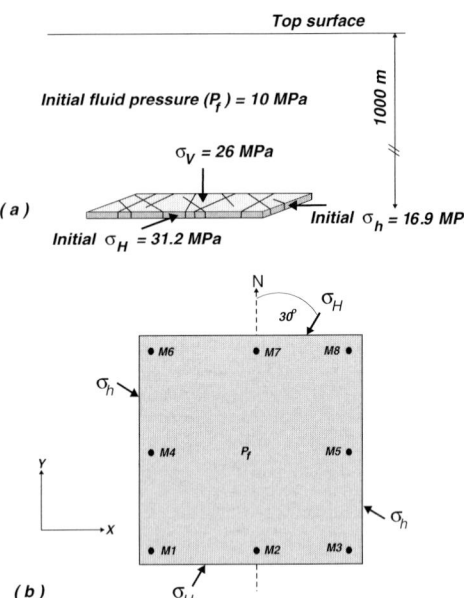

Fig. 2. (a) Initial stresses used in the modelling. **(b)** Loading directions of principal horizontal stresses and displacement monitoring system (M1–M8).

during the deformation (Zhang & Sanderson 1996, 1998, 2002). The rock blocks are modelled with a Mohr–Coulomb rheology and are considered impermeable. The fractures are treated as frictional boundaries, whose apertures are controlled by the relative movement of the blocks in response to the local stress state. Fluid flow is calculated from the aperture and pressure gradient using the cubic flow law.

The simulation is considered as a horizontal plane (10×10 m) containing a network of vertical fractures. This network consists, initially, of two sets of well-connected, orthogonal fractures with a density of 7.85 m⁻¹ and an average fracture length of 0.5 m (Fig.1). Later, the geometry of the network will be changed systematically to examine the effects of changes in fracture density, average fracture length and anisotropy.

The rock mass is considered to be at a vertical overburden stress, σ_v, of 26 MPa, corresponding to a depth of about 1000 m for a rock density of 2600 kg m⁻³. The initial major and minor principal stresses, σ_H and σ_h, are 31.2 and 16.9 MPa, respectively (Fig. 2a), with an initial hydrostatic fluid pressure of 10 MPa. The properties of the intact rock and fractures are shown in Table 1. The directions of the principal stresses are shown in Figure 2b, together with the positions of monitoring points used to measure the displacements of the sample.

Before inducing a change in stress, the fractured rock mass was allowed to reach an equilibrium state under the action of the applied boundary stresses. The stress was changed by increasing the differential stress, fluid pressure or maximum horizontal principal stress. First, successive incremental increases of

0.5 MPa in fluid pressure will be examined, with smaller changes used where necessary to more clearly define the critical state.

Movements occurred in the system after inducing a stress change. Following the rationale used by Zhang & Sanderson (2001), if the movement stopped, the rock mass was considered stable. At the early stages of deformation, the system behaved like an homogeneous medium, with no significantly localized deformation. When the stress approached a certain level, at a fluid pressure (P_f) of 13.5 MPa, the system became unstable. This was indicated by the increasing displacements and a large unbalanced force in the system. The resulting strain was concentrated along a highly deformed zone (Fig. 3a) where sliding and opening of some of the fractures produced large voids, particularly at the intersections of fractures. The displacements at the eight monitored points indicate anticlockwise rotation of blocks on either side of a right-lateral fault zone (Fig.3b).

Although the fractured rock mass deformed in a relatively homogenous manner in the early stages of loading, localized features in deformation emerge just before the critical state. This is seen clearly in Figure 4, which shows the distribution of incremental sliding along fractures during the experiment. At an early stage, a number of slip zones develop fairly homogeneously throughout the area (Fig. 4a–c). Through linkage and abandonment of more isolated fractures, more continuous slip-zones formed just

Table 1. *Parameters used in UDEC modelling*

Model parameters	Value	Units
Rock properties		
Density	2600	kg m^{-1}
Shear modulus	10	GPa
Bulk modulus	30	GPa
Tensile strength	4	MPa
Cohesion	10	MPa
Friction angle	30	°
Fracture properties		
Shear stiffness	50	GPa m^{-1}
Normal stiffness	100	GPa m^{-1}
Tensile strength	0	MPa
Cohesion	0	MPa
Friction angle	30	°
Dilation angle	5	°
Residual aperture	0.02	mm
Zero-stress aperture	0.5	mm
Fluid properties		
Density	1000	kg m^{-3}
Viscosity	0.00035	Pa s

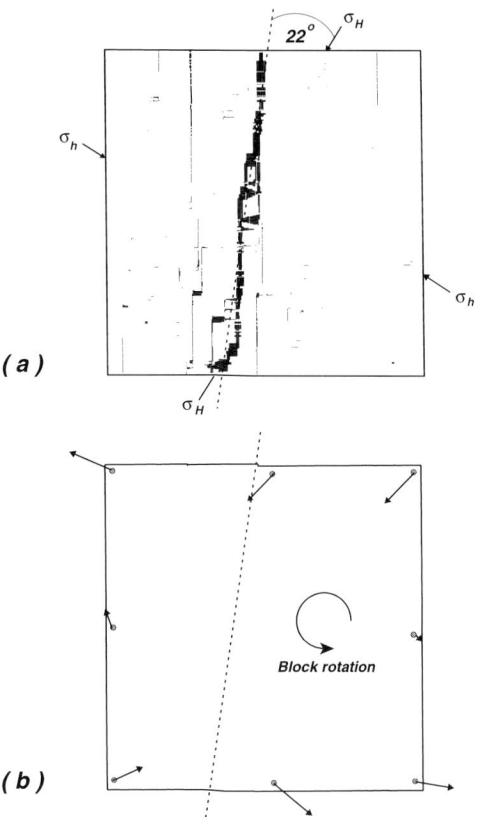

Fig. 3. (**a**) Dilational and shearing deformation of the network in Figure 1 at the critical stress state (fluid pressure = 13.5 MPa). Aperture ≥1 mm and shearing displacement ≥5 mm are shown. (**b**) Displacements at eight monitored points at the critical state, indicating slip concentration and counterclockwise block rotation.

before the critical state (Fig. 4d, e). At the critical state, a single slip-zone accommodates most of the deformation (Fig. 3a). Although this zone was present at the early stage (compare Figs 4c and e), the magnitude of incremental slip increased suddenly just before the critical state.

Detailed examination of the stress and displacements in the models indicates that mechanical heterogeneity controls the development of the deformed regions within the rock, and that such structures may be the sites of enhanced flow. What is happening is that slip on individual fracture segments creates local heterogeneity of the stress field, which in turn induces more slip. This 'feed-back' between deformation and enhanced stress leads to a 'self-organization' of the deformation in localized areas (Fig. 4) and is a common feature at the critical state.

In addition to the shear displacement, dilational deformation also develops during the movements (Fig. 5). The opening of fractures is strongly related to the direction of loading, with the most open being at a low angle to σ_H. The maximum aperture increases with increasing fluid pressure. Just before the critical stress state, a series of well-developed dilational jogs develop in the main shear zone, which consist of short opening fractures (in the *x*-direction) linked by longer sliding fractures (in the *y*-direction). These structures resemble field examples of fractures linked by dilational jogs (Sibson 1989), which develop into pull-aparts (Peacock & Sanderson 1995) as seen in Figure 3(a).

Clearly in Figures 3–5, the orientation of fractures relative to the far-field stresses is an important factor, but the numerical experiments indicate two points about this relationship that need careful consideration.

- The effective normal stress on both fracture orientations would be expected to be compressive not tensile, although stress heterogeneity may lead to small tensile stresses developing locally within the network. This emphasizes the role of slip rather than hydraulic opening of fractures in the development of the networks.
- Both fracture orientations, at 90° to one another, contribute to the fluid flow, although both cannot be critically stressed in the terms of Barton *et al.* (1995), i.e. optimally oriented for frictional sliding in response to the far-field stresses. In this case, the dilational jogs, parallel to the *x*-axis in Figure 5, would lie outside the critically stressed

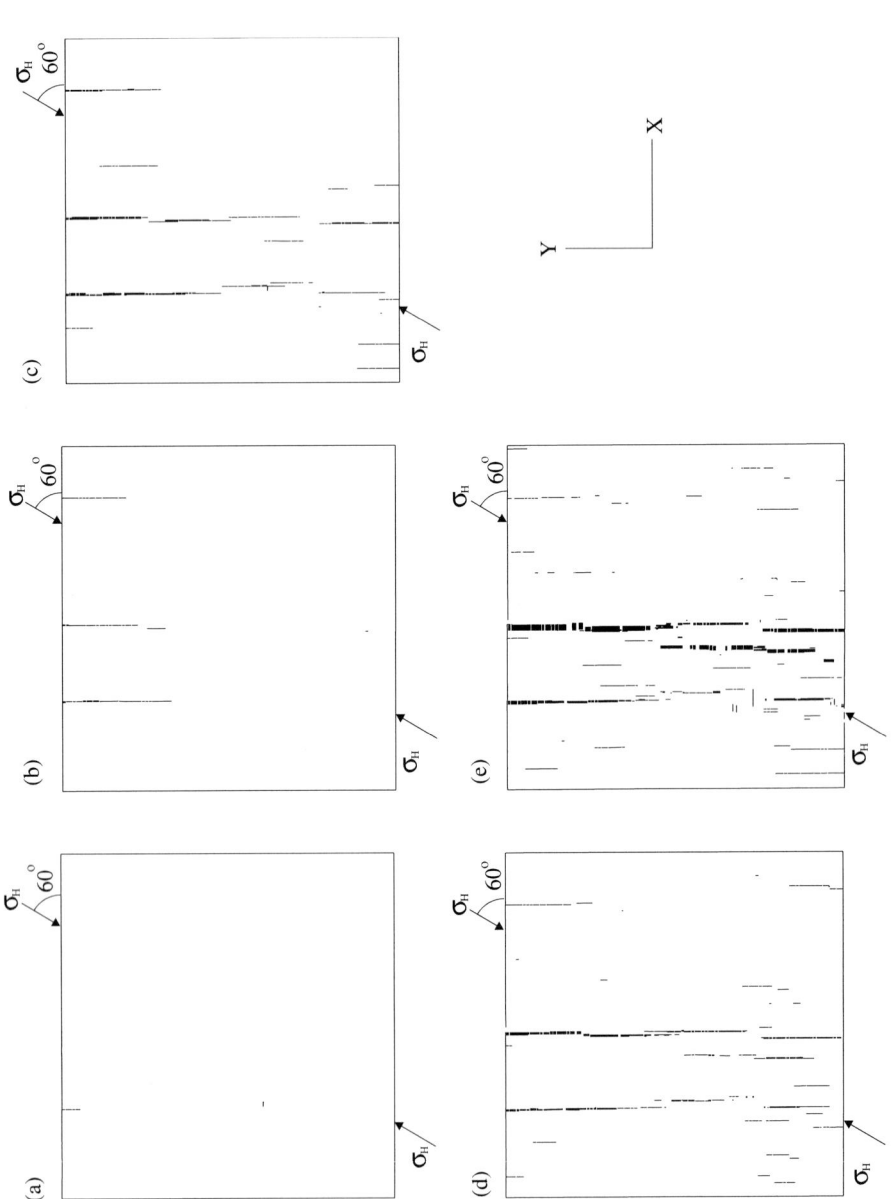

Fig. 4. Incremental shear displacements along fractures during stable movements of the network in Figure 1 caused by a series of changes in stress. (**a**) Fluid pressure = 10 MPa and maximum incremental slip = 1.23 mm; (**b**) fluid pressure = 11 MPa and maximum incremental slip = 1.8 mm; (**c**) fluid pressure = 12 MPa and maximum incremental slip = 2.8 mm; (**d**) fluid pressure = 12.5 MPa and maximum incremental slip = 3.0 mm; (**e**) fluid pressure = 13 MPa and maximum incremental slip = 9.5 mm. Slip is proportional to the width line, with the finest line representing a shear displacement of 1 mm.

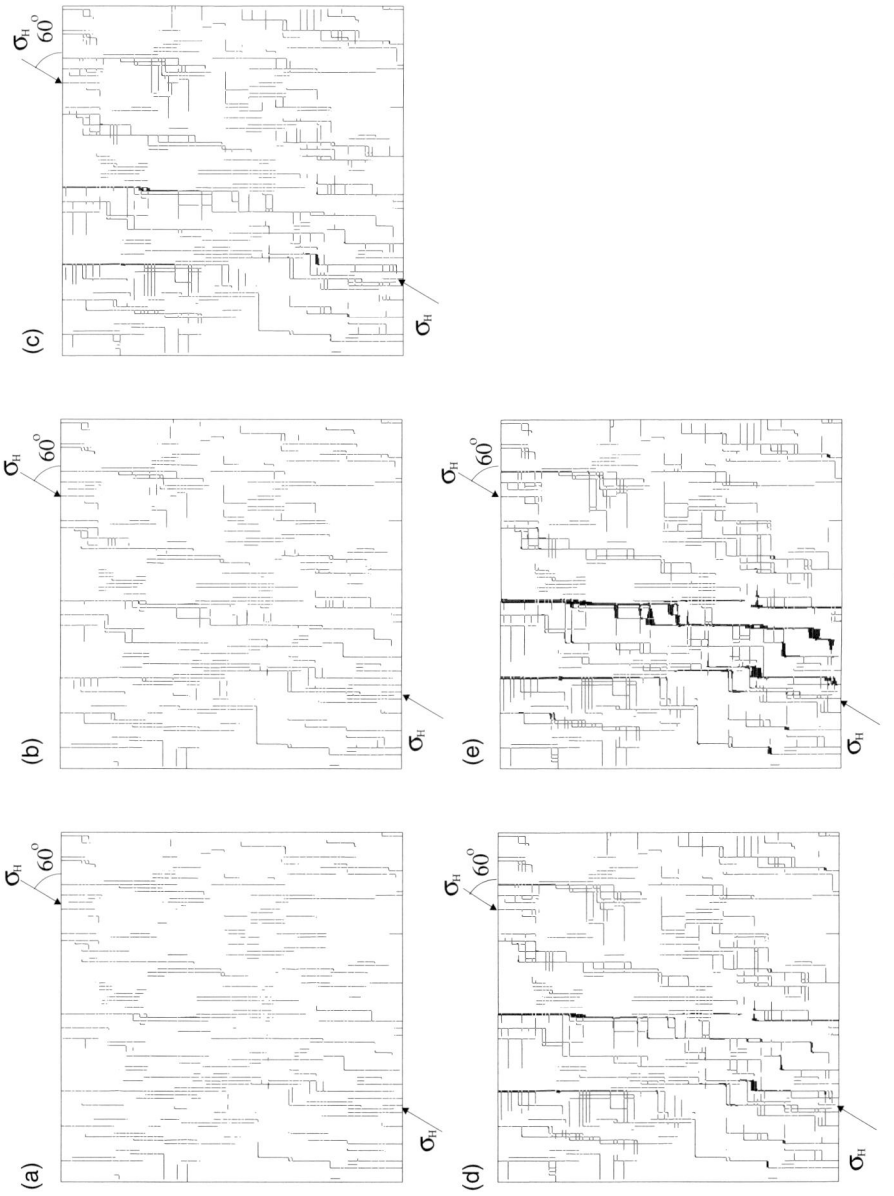

Fig. 5. Evolution of dilational deformation during stable movements of the network in Figure 1 caused by a series of changes in stress. (**a**) Fluid pressure = 10 MPa and maximum aperture = 0.6 mm; (**b**) fluid pressure = 11 MPa and maximum aperture = 0.8 mm; (**c**) fluid pressure = 12 MPa and maximum aperture = 2.5 mm; (**d**) fluid pressure = 12.5 MPa and maximum aperture = 5.1 mm; (**e**) fluid pressure = 13 MPa and maximum aperture = 12.4 mm. Aperture is proportional to the width line, with the finest line representing an aperture of 0.35 mm.

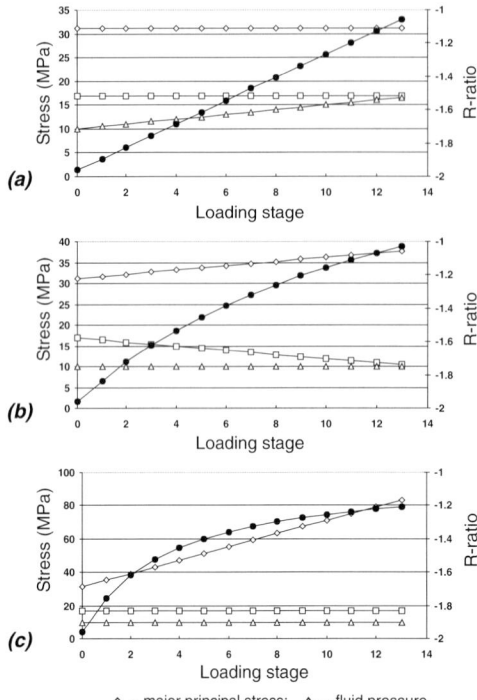

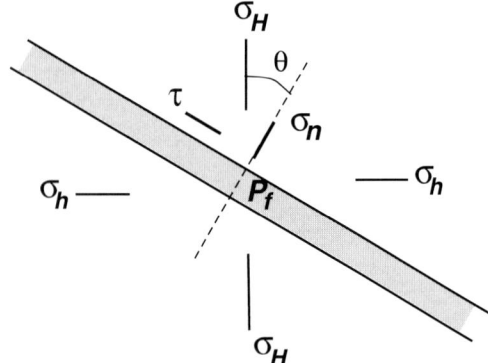

Fig. 7. Model for the opening and slip on a single fracture with normal at θ to major principal stress (σ_H).

Fig. 6. Three loading schemes and *R* ratios used in modelling. Scheme 1 increases the fluid pressure so that the differential stress remains constant and the mean stress decreases; scheme 2 increases the major principal stress and decreases the minor principal stress so that the mean stress is constant and the differential stress increases; scheme 3 increases the major principal stress only so that the differential and mean stresses increase.

region defined by Barton *et al.* (1995) and, hence, would not be expected to control the flow.

Thus, although the orientation of fractures relative to the stress field is important, it is not the only factor contributing to flow in fractures.

Instability and the *R* ratio

In the previous model the change in 'effective' stress was caused by an increase in fluid pressure (Fig. 6a), but this could also be achieved with other loading conditions. Two additional loading schemes have been used. One involved increasing the major principal stress and decreasing the minor principal stress, thus increasing the differential stress whilst maintaining a constant mean stress and fluid pressure (Fig. 6b). The other involved increasing the major principal stress with constant minor principal stress

and fluid pressure, thus increasing both the differential stress and mean stress (Fig. 6c).

Using the three loading schemes, three different specifications of the critical stress state were obtained for the fracture network in Figure 1, which are:

- Scheme 1: fluid pressure = 13.5 MPa; maximum differential stress = 14.3 MPa; mean stress = 24.05 MPa.
- Scheme 2: fluid pressure = 10 MPa; maximum differential stress = 18.3 MPa; mean stress = 24.05 MPa.
- Scheme 3: fluid pressure = 10 MPa; maximum differential stress = 24.3 MPa; mean stress = 29.05 MPa.

These are just three specific cases of a range of stress states at which the fractured rock may become unstable. Do they have any common features?

As all fractures are parallel to one of the principal stresses, σ_V, the normal (σ_n) and shear stresses τ for a single, infinitely long, straight fracture can be expressed in terms of the principal stresses and orientation of the fracture, θ (Fig. 7), as follows:

$$\sigma_n = \tfrac{1}{2}(\sigma_H + \sigma_h) + \tfrac{1}{2}(\sigma_H - \sigma_h)\cos 2\theta \quad (1a)$$

$$\tau = \tfrac{1}{2}(\sigma_H - \sigma_h)\sin 2\theta. \quad (1b)$$

The condition for opening of fractures can be expressed as:

$$P_f \geq \sigma_n. \quad (2)$$

Substituting equation (1a) in (2) and arranging the stress terms on the left-hand side gives:

$$R = [P_f - \tfrac{1}{2}(\sigma_H + \sigma_h)] \,/\, [\tfrac{1}{2}(\sigma_H - \sigma_h)] \geq \cos 2\theta. \quad (3)$$

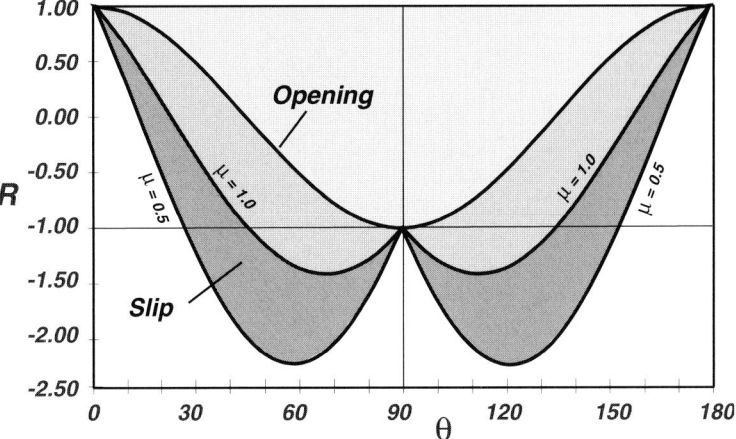

Fig. 8. Plot of driving stress ratio (R) against orientation (θ) showing conditions for opening and sliding of a single fracture. The sliding condition for coefficients of friction (μ) of 0.5 and 1 bracket likely slip conditions for rock. The light shaded area shows where hydraulic opening of fractures is possible, and intermediate shaded area where slip is possible.

This expresses the condition for opening of fractures in terms of a driving stress ratio (R) as suggested by Delaney *et al.* (1986) and subsequently used to examine the opening of igneous sheets by Jolly & Sanderson (1995, 1997).

The condition for frictional sliding can be expressed (e.g. Byerlee 1978) as:

$$|\tau| \geq \mu \, (\sigma_n - P_f) \qquad (4)$$

where μ is the coefficient of sliding friction. Substituting equations (1a) and (1b) in (4) and arranging the stress terms on the left-hand side gives:

$$R = [P_f - \tfrac{1}{2}(\sigma_H + \sigma_h)] \, / \, [\tfrac{1}{2}(\sigma_H - \sigma_h)] \geq$$
$$\cos 2\theta - |(1/\mu) \sin 2\theta|. \qquad (5)$$

Note that equations (3) and (5) describe the relationship of the critical stress components to the opening and sliding of a single fracture. These equations are plotted in Figure 8. Equation (3) defines a condition for opening, above which (light shaded area) hydraulic opening of fractures is possible. The slip condition has been plotted for $\mu = 0.5$ and 1.0, spanning the likely values for most rocks (e.g. Byerlee 1978), above which (intermediate shaded area) slip on fractures is possible. It is clear from this diagram than slip will generally occur at lower driving stress ratios than opening.

It is tempting to suggest that the R ratio may be used to link the instability of a fractured rock and the critical stress state. For the fracture network in Figure 1, three driving stress ratios (R_C) were calculated at the critical stress states mentioned above, which have

values between -1.48 and -1.57 (see Table 2). As the model mainly involves sliding on the set of fractures at $\theta = 60°$ to σ_H and setting the coefficient of sliding friction to $\mu = 0.7$ (equivalent to a friction angle of 30° and dilation angle of 5°), direct application of equation (5) produces $R_C = -1.74$. Slip on the other set of fractures, which are at $\theta = 30°$ to σ_H, would require $R_C = -0.74$. These values set lower and upper bounds to the expected value of R_C. The observed value, $R_C = -1.5$, lies within this range, being only slightly larger than that predicted for a through-going fracture at $\theta = 60°$.

Another way of looking at the relationship between the network model and equation (5) is to consider the resultant shear zone at $\theta = 68°$ (Fig. 3a), which would produce $R_C = -1.5$, for $\mu \cong 0.9$, probably not an unreasonable value for the effective behaviour of the fracture networks. In the rest of the chapter we will summarize and discuss the critical state in terms of the R ratio. The exact stress conditions for any model can be found in Tables 2 and 3.

Role of fracture-network geometry

A series of simulated fracture networks has been generated to examine the behaviour of deformation and fluid flow. This allows a systematic study of the interplay and relative importance of the two key factors – fracture-network geometry and applied stress state. The simulated fracture networks all comprise two orthogonal sets of parallel, fractures. Three groups of simulated networks are discussed, in which the fracture density, mean fracture length and the anisotropy of fracture networks are systematically

Table 2. *Loading conditions, geometrical parameters and driving stress ratios (R_C) for models with different fracture densities*

Fracture network	d1	d2	d3*	d4	d5	Loading scheme
	(average fracture length = 1.15 m, see Fig.11)					
Fracture density (m^{-1})	4.3	6.1	7.85	9.7	11.6	
Major horizontal stress, σ_H (MPa)	31.2	31.2	31.2	31.2	31.2	1
	35.7	35.2	33.2	32.2	32.2	2
	66.2	53.2	41.2	35.2	34.2	3
Minor horizontal stress, σ_h (MPa)	16.9	16.9	16.9	16.9	16.9	1
	12.4	12.9	14.9	15.9	15.9	2
	16.9	16.9	16.9	16.9	16.9	3
Fluid pressure, P_f (MPa)	16.3	15.0	13.5	12.0	12.0	1
	10.0	10.0	10.0	10.0	10.0	2
	10.0	10.0	10.0	10.0	10.0	3
Differential stress, $q = \sigma_H - \sigma_h$ (MPa)	14.3	14.3	14.3	14.3	14.3	1
	23.3	22.3	18.3	16.3	16.3	2
	49.3	36.3	24.3	18.3	17.3	3
Mean stress, $p = (\sigma_H + \sigma_h)/2$ (MPa)	24.05	24.05	24.05	24.05	24.05	1
	24.05	24.05	24.05	24.05	24.05	2
	41.55	35.05	29.05	26.05	25.55	3
Effective mean stress, $p' = p - P_f$ (MPa)	7.75	9.05	10.55	12.05	12.05	1
	14.05	14.05	14.05	14.05	14.05	2
	31.55	25.05	19.05	16.05	15.55	3
R ratio, $R_C = -2p'/q$	−1.084	−1.266	−1.476	−1.685	−1.685	1
	−1.206	−1.260	−1.536	−1.724	−1.724	2
	−1.280	−1.380	−1.568	−1.754	−1.798	3

*Used for the example described in the text.

Table 3. *Loading conditions, geometrical parameters and driving stress ratios (R_C) for models with different fracture sizes*

Fracture network	s1	s2	s3*	s4	s5	s6	Loading scheme
	(fracture density = 7.85 m^{-1}, see Fig.12)						
Average fracture length (m)	0.61	0.81	1.15	1.47	1.85	2.11	
Major horizontal stress, σ_H (MPa)	31.2	31.2	31.2	31.2	31.2	31.2	1
	34.2	33.2	33.2	32.7	32.45	32.2	2
	49.2	42.2	41.2	39.2	36.2	35.7	3
Minor horizontal stress, σ_h (MPa)	16.9	16.9	16.9	16.9	16.9	16.9	1
	13.9	14.9	14.9	15.4	15.65	15.9	2
	16.9	16.9	16.9	16.9	16.9	26.9	3
Fluid pressure, P_f (MPa)	14.5	13.5	13.5	13.0	12.8	12.5	1
	10.0	10.0	10.0	10.0	10.0	10.0	2
	10.0	10.0	10.0	10.0	10.0	10.0	3
Differential stress, $q = \sigma_H - \sigma_h$ (MPa)	14.3	14.3	14.3	14.3	14.3	14.3	1
	20.3	18.3	18.3	17.3	16.8	16.4	2
	32.3	25.3	24.3	22.3	19.3	18.8	3
Mean stress, $p = (\sigma_H + \sigma_h)/2$ (MPa)	24.05	24.05	24.05	24.05	24.05	24.05	1
	24.05	24.05	24.05	24.05	24.05	24.05	2
	33.05	29.55	29.05	28.05	26.55	26.3	3
Effective mean stress, $= p' - P_f$ (MPa)	9.55	10.55	10.55	11.05	11.25	11.55	1
	14.05	14.05	14.05	14.05	14.05	14.05	2
	23.05	29.55	29.05	28.05	26.55	26.3	3
R-ratio, $R_C = -2p'/q$	−1.336	−1.476	−1.476	−1.545	−1.573	−1.615	1
	−1.384	−1.536	−1.536	−1.624	−1.673	−1.724	2
	−1.472	−1.545	−1.568	−1.619	−1.715	−1.734	3

*Same as d3 in Table 2.

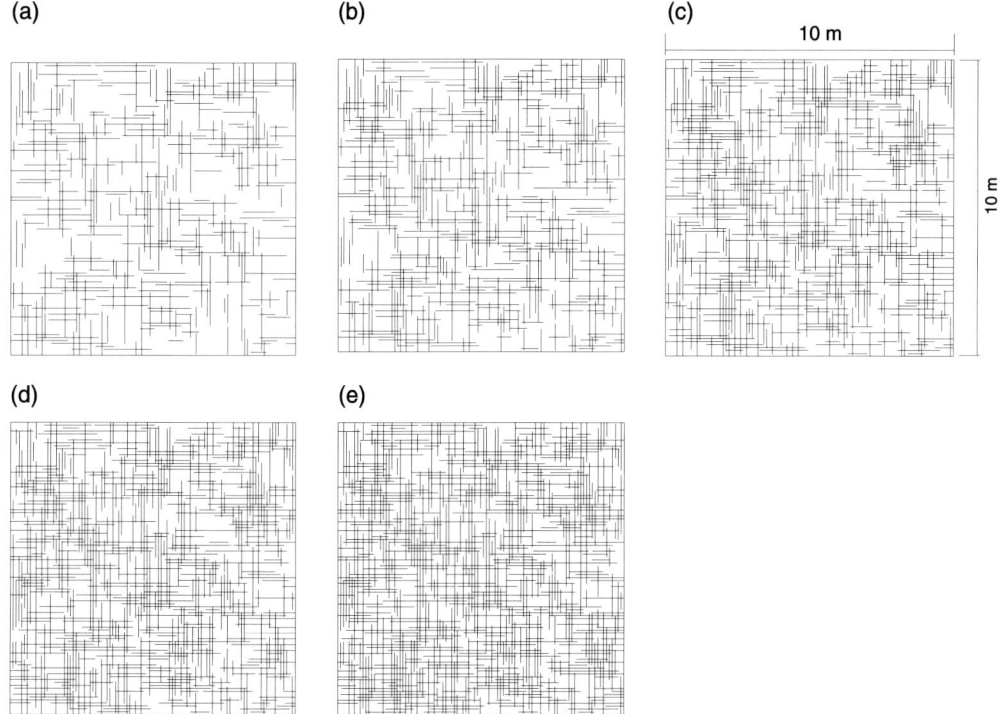

Fig. 9. Fracture networks (group A) having different fracture densities between 4.3 and 11.6 m^{-1}, and an average fracture length of 1.15 m.

varied. The modelling involves similar procedures to those described by Sanderson & Zhang (1994) and Zhang & Sanderson (2001, 2002), which are as follows.

- The simulated region was a square of 10×10 m, in which the fracture trace-length varied from a fixed upper limit of 2.5 m to a lower limit of between 0.5 and 2 m. Thus, the fractures are small in relation to the size of the simulated region, which contains no single through-going fracture.
- Trace-lengths were sampled from a power-law distribution over a narrow size range, where the number of fractures (N) of length (L) had a form $N \propto L^{-E}$ (Segall & Pollard 1983; Barton & Hsieh 1989; Heffer & Bevan 1990; Jackson & Sanderson 1992). The exponent (E) could be varied, but a value of 1.2 was used in this study. The exponent and the narrow size range make such truncated power-law distributions very similar to log-normal distributions characterized by an average length.
- The coordinates of the centre of a fracture were selected randomly from a uniform distribution within the simulated region. A procedure of self-avoiding generation was used, such that new

fractures were selected only if they were located at a minimum distance to previously generated fractures. The minimum distance can be varied, but was set at 50 mm (i.e. 0.5% of the size of the square) in this modelling.

- Fractures were generated sequentially, according to the above rules. As more fractures were added the density of the fractures increased, where fracture density is defined as the total length of the fracture-trace per unit area.

Fracture density

In this group of simulations the two fracture sets are statistically equal, having the same density and average fracture length of 1.15 m (range 0.5 and 2.5 m). The overall density was varied between 4.3 and 11.6 m^{-1} (Fig. 9), the lowest density (4.3 m^{-1}) corresponding to the percolation threshold, below which the fracture network consisted of individual fractures and locally connected clusters, with no spanning (or critical) cluster (e.g. Stauffer & Aharony 1991; Zhang & Sanderson 1994). At or above this critical density, a continuous cluster formed and connected all sides of the square (Zhang & Sanderson 1998).

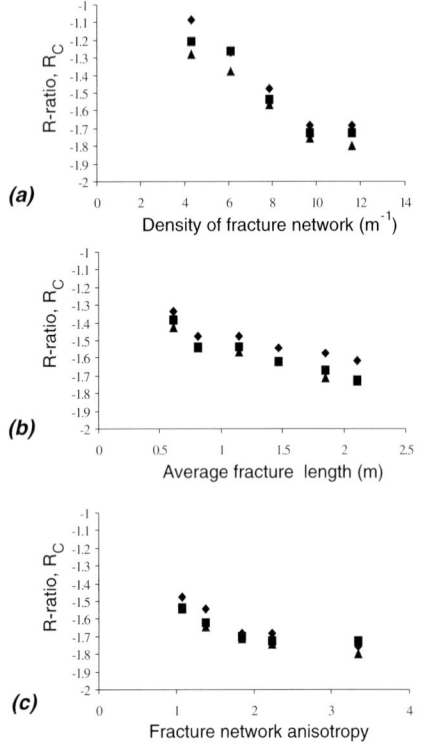

(a)

(b)

(c)

■ -- loading scheme 1; ◆ -- loading scheme 2; ▲ -- loading scheme 3.

Fig. 10. Effects of fracture network geometry on R_C ratio under different loading schemes. (**a**) Fracture density; (**b**) average fracture length; (**c**) fracture-network anisotropy.

Figure 10(a) shows the variation of R_C plotted against fracture density. The results and input parameters for these models are given in Table 2. For fracture densities from 4.3 to 11.6 m^{-1}, R_C varies from -1.08 to -1.80, respectively. There is a steady decrease in R_C with increasing fracture density indicating that stability is reduced where the rock is more highly fractured.

Fracture length

In this group of simulations the fracture density was held constant at 7.85 m^{-1}, with both sets being equally represented. The average length of fractures was varied between 0.61 and 2.11 m (Fig. 11). The upper limit was set at 2.5 m, and the lower limit selected at 0.1, 0.2, 0.5, 1, 1.5 and 2 m, to produce the variation in average fracture length.

The R_C-ratio varies between -1.34 and -1.73 (Fig. 10b), with R_C decreasing with increasing fracture length. The results and input parameters for these tests are shown in Table 3. For the same frac-

ture density, networks with fewer, but larger, fractures become unstable at lower driving stress ratios. If this trend were continued until the individual fractures extend through the square then, as argued earlier, the driving stress ratio would be given by equation (5) and is $R_C \cong -1.75$ (see the discussion at end of the previous section). As can be seen, this value is approached at average fracture lengths of over about 2 m (i.e. about 20% of the size of the square).

Anisotropy

In this group of simulations the total fracture density was kept constant at 7.85 m^{-1}, but the average fracture lengths in the x- and y-direction were varied. The lower limit was fixed at 1 m in the y-direction, and in the x-direction the lower limit was varied between 0.1 and 1 m. The variation of length in the two directions led to a variation in fracture density with direction producing the anisotropy. The anisotropy, as defined by Zhang & Sanderson (1995), is simply the ratio of fracture densities of the two sets and is between 1 and 3.35 (Fig. 12, Table 4).

The R_C-ratio decreased somewhat, from -1.5 to -1.8 as the anisotropy of the fracture networks increased from 1 to 3.35 (Fig. 10c). This change indicates that as the fracture network becomes dominated by one set of fractures the R_C-ratio appropriate to the dominant orientation applies. In this set of simulations the dominant fractures are at $\theta = 60°$ to σ_H, which from equation (5) would produce $R_C \cong -1.74$, in good agreement with the modelled results. Clearly as networks become more anisotropic, both in terms of the proportion of fractures in the dominant set and as these fractures become longer, then the orientation of the fractures in relation to the loading direction has a considerable impact on the critical stress state. Similar effects were seen in the modelling of naturally fracture networks (Zhang & Sanderson 1997; Sanderson & Zhang 1999).

Localization of fluid flow

The openings created during deformation of a fractured rock mass are likely to cause a dramatic change in the magnitude and pattern of flow rates, in particular the vertical flow due to "holes" at the intersections of fractures (Figs 3a and 5). In this chapter, the flow rate, in the vertical direction through the network, was calculated for a 16 × 16 grid. This was carried out using the cubic law (Snow 1968) and/or the pipe formula (Sabersky *et al.* 1989), depending on the shape of openings (Sanderson & Zhang 1999), using a hydraulic gradient of 1000 Pa m^{-1}.

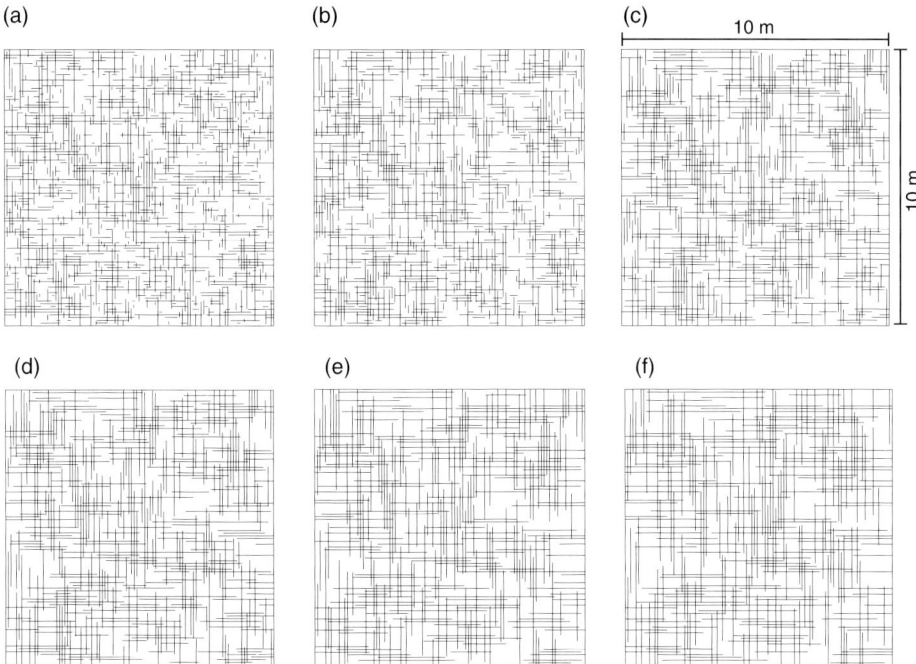

Fig. 11. Fracture networks (group B) having different average length between 0.61 and 2.11 m, and a fracture density of 7.85 m^{-1}.

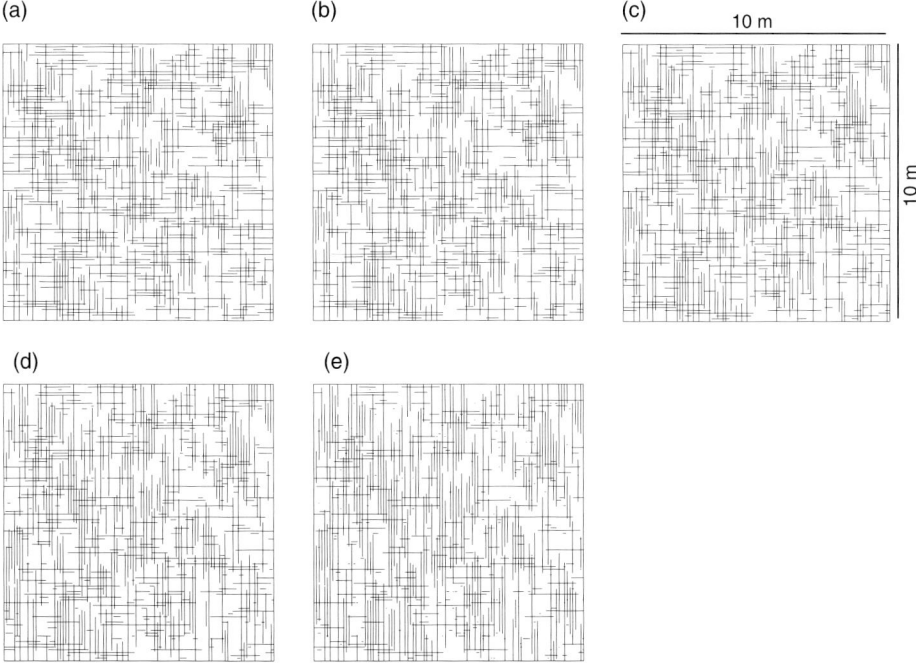

Fig. 12. Fracture networks (group C) having different anisotropy coefficient between 1.08 and 3.35, and a fracture density of 7.85 m^{-1}.

Table 4. *Loading conditions, geometrical parameters and driving stress ratios (R_C) for models with different anisotropy*

Fracture network	a1	a2	a3	a4	a5	Loading scheme
	(average fracture length = 1.15 m and fracture density = 7.85 m^{-1}, see Fig.10)					
Fracture anisotropy	1.08	1.38	1.85	2.23	3.35	
Major horizontal stress, σ_H (MPa)	31.2	31.2	31.2	31.2	31.2	1
	33.2	32.7	32.2	32.2	32.2	2
	42.2	38.2	36.2	35.4	34.2	3
Minor horizontal stress, σ_h (MPa)	16.9	16.9	16.9	16.9	16.9	1
	14.9	15.4	15.9	15.9	15.9	2
	16.9	16.9	16.9	16.9	16.9	3
Fluid pressure, P_f (MPa)	13.5	12.75	12.0	12.0	11.5	1
	10.0	10.0	10.0	10.0	10.0	2
	10.0	10.0	10.0	10.0	10.0	3
Differential stress, $q = \sigma_H - \sigma_h$ (MPa)	14.3	14.3	14.3	14.3	14.3	1
	18.3	17.3	16.3	16.3	16.3	2
	25.3	21.3	19.3	18.5	17.3	3
Mean stress, $p = (\sigma_H + \sigma_h)/2$ (MPa)	24.05	24.05	24.05	24.05	24.05	1
	24.05	24.05	24.05	24.05	24.05	2
	29.55	27.55	26.55	26.15	25.55	3
Effective mean stress, $p' = p - P_f$ (MPa)	10.55	11.3	12.05	12.05	12.55	1
	14.05	14.05	14.05	14.05	14.05	2
	19.55	17.55	16.05	16.15	15.55	3
R ratio, $R_C = -2p'/q$	−1.476	−1.584	−1.685	−1.685	−1.755	1
	−1.536	−1.624	−1.703	−1.724	−1.724	2
	−1.545	−1.648	−1.715	−1.746	−1.798	3

Figure 13 shows the evolution of the vertical flow rate prior to the critical state (see Figs 4 and 5). From fluid pressures of 10–12.5 MPa there is little change in the flow rate, but just before the critical fluid pressure the flow rate starts to increase markedly. Note that in Figure 13c the scaling of the flow-rate axis is 60 times that in Figure 13a and b. The average vertical flow rates in the 10 × 10 m square increased by about one order of magnitude just before the critical state (Fig. 14a), whereas the highest value in a single subarea increased by about three orders of magnitude. This indicates that flow rates do not just increase in overall magnitude, but also become highly localized.

Figure 14b shows a log–log plot of the cumulative frequency of flow rates in the 256 subareas; curves A–C corresponding to the flow rates shown in Figure 13. Curve A has a log-normal distribution with a median flow rate of just under 0.02×10^{-6} m^{3}s^{-1}, this value being determined by the initial aperture ascribed to the fractures and the hydraulic gradient. Curve C corresponds to the distribution just before the critical state and has a power-law distribution; two subareas, circled in Figure 14b, contribute about 50% of the total flow.

Although the rock mass was initially homogenous, subject to simple boundary conditions, and the flows are described by the cubic law or the pipe formula, the system as a whole behaves in a highly non-linear manner, showing the characteristics of a critical point phenomenon; it appears to undergo a phase change at some critical value of the R ratio. As discussed earlier, this results in self-organization of deformation that occurs spontaneously on the backbone of the fracture network, which leads to fluid-flow localization. Such a systems typically exhibit highly non-linear behaviour near the critical state, producing a power-law (fractal) scaling of features.

One way of looking at the localization of flow within the networks is to consider the proportion of flow in each subelement and to characterize the distribution of flow rates using multifractal techniques. If P_i is the proportion of the total flow in the ith subelement of size r, the general fractal dimension, D_q, at order q can be defined as (e.g. Schroeder 1991; Cowie *et al.* 1995; Sanderson & Zhang 1999):

$$D_q = \lim \frac{\log \sum_{i=1}^{N} P_i^q}{\log r} \qquad (6)$$

Figure 15 shows a plot of D_q, against q for the flow-rate distributions in Figure 13. As all fractures have a residual aperture, and hence some flow, it follows that at $q = 0$ the capacity dimension of flow is that of the fracture network, which for a random distribution is $D_0 = 2$. Berkowitz & Hadad (1997) have carried out multifractal analyses of fracture

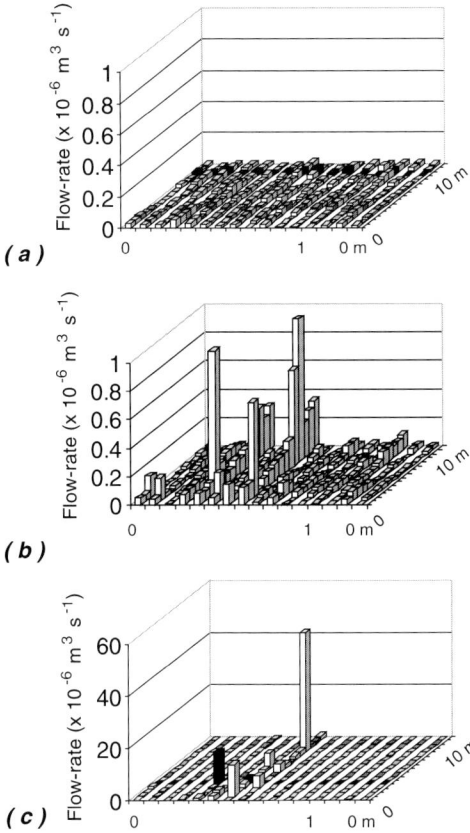

Fig. 13. Evolution of vertical flow rates, at a hydraulic gradient of 1000 Pa m^{-1}, during stable movements of the network in Figure 1 caused by a series of changes in stress (see Fig. 3). (**a**) Fluid pressure = 10 MPa; (**b**) fluid pressure = 12.5 MPa; (**c**) fluid pressure = 13 MPa.

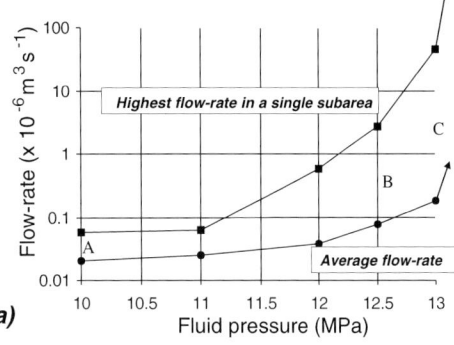

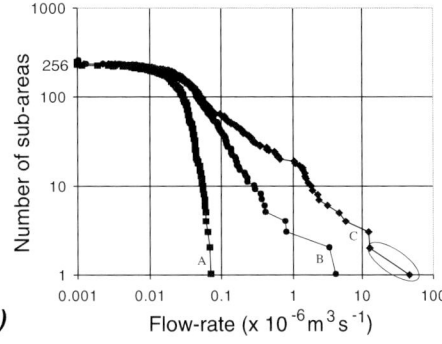

Fig. 14. (**a**) Variations of the highest value in a single subarea and the average flow rates during stable movements of the network in Figure 1 caused by a series of changes in stress. (**b**) Log–log plot of number of subareas with vertical flow rate greater than specified value for the network in Figure 1. A, B and C correspond to increasing fluid pressure as in Figure 13. Curve A is close to a log-normal distribution with a median flow rate of just under 0.02×10^{-6} m^3 s^{-1}; curve C is just before the critical state and corresponds to a power-law distribution with the two subareas (circled) contributing to 50% of the total flow.

density in natural fracture networks and conclude that connected fracture networks generally have D_0 between 1.7 and 2; similar results were obtained by Zhang & Sanderson (1994) from simulations of connected networks. Values close to 2 would also be expected from bond percolation models (e.g. Stauffer & Aharony 1991).

At higher orders ($q>0$), the fractal dimensions are dominated by the distribution of high-flow subareas and the fractal dimension decreases with increasing q. At hydrostatic pressure (10 MPa) the decrease is small, to a value of about 1.8 at $q=6$ (Fig. 15). Just before the critical fluid pressure, however, the network shows a rapid decrease in fractal dimension with increasing q to a value of about 0.5 at $q=6$ (Fig. 15). This confirms that the high-flow regions are clustered, rather than homogeneously distributed, as at lower fluid pressures. This change in dimensionality of the flow is important when interpreting well-

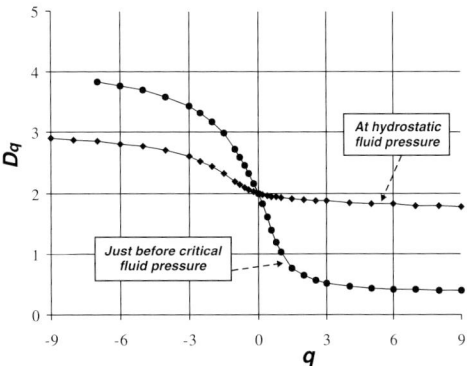

Fig. 15. Multifractal analysis for the vertical flow in the network in Figure 1, comparing the variation of D_q with power q at hydrostatic and critical fluid pressures.

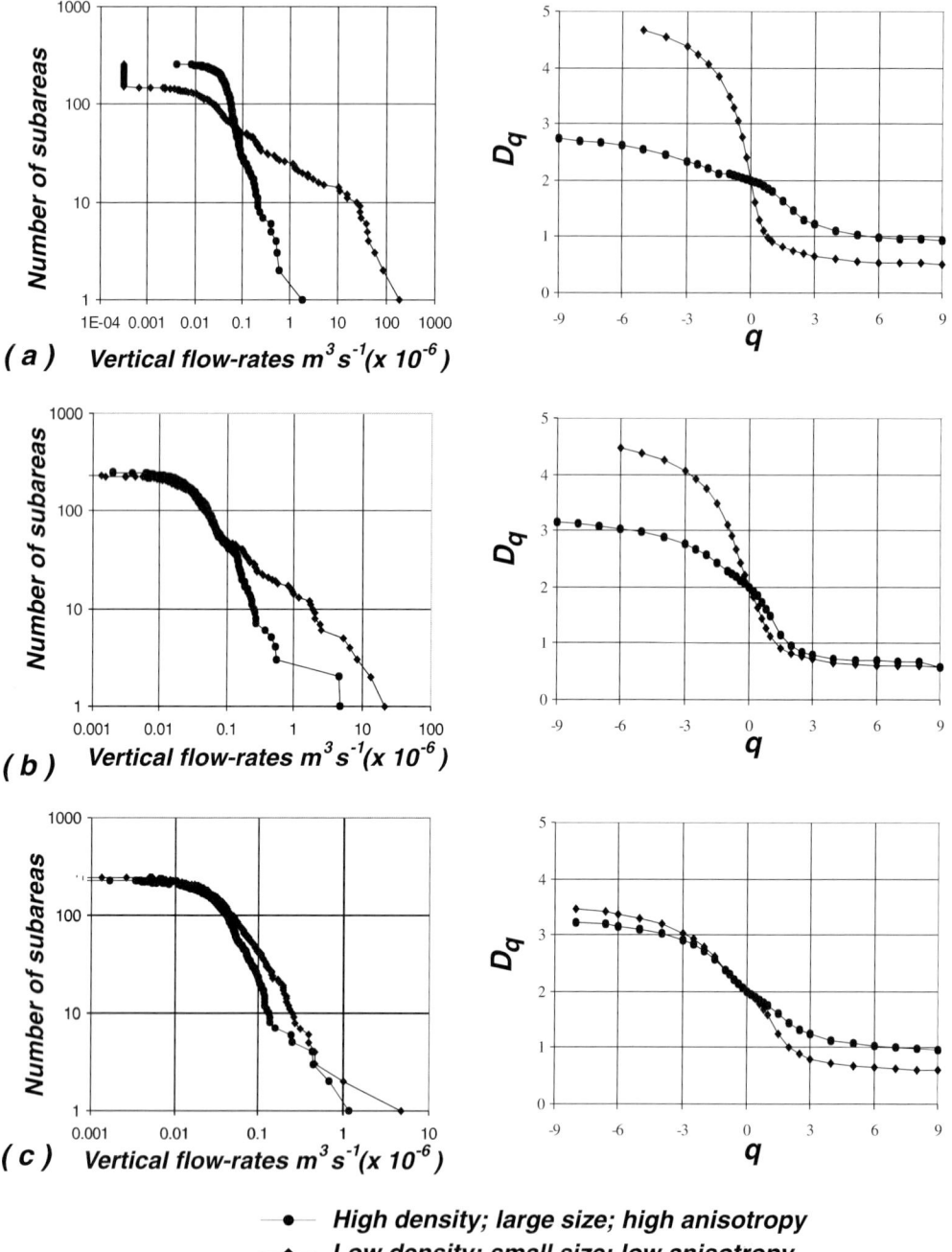

Fig. 16. Effects of fracture network geometry on the distribution of vertical flow (left) and multifractal dimensions (right) just before the critical fluid pressure. (**a**) Varying fracture density; (**b**) varying average fracture length; (**c**) varying anisotropy coefficient.

test data and may be a way of recognizing the critical state in the absence of detailed mapping of variation in flow rate or permeability.

Figure 16 shows the distribution of vertical flow rates and associated multifractal analyses for different network geometries just before the critical fluid pressures. These results show that lower fracture density, smaller fracture size and lower anisotropic coefficient promoted greater localization and lower fractal dimensions, although the effect of fracture density is greatest.

Conclusions

The critical stress state at which a fractured rock mass becomes unstable has been determined for a series of simulated networks, with the deformation and fluid flow being modelled by distinct-element methods. All of these models show that instability occurs as a critical stress state is approached. The models consider well-connected fracture networks with no further initiation or propagation of fractures. Flow is considered to be entirely through the fracture network, with no contribution from the rock matrix.

Bearing these assumptions in mind, the following conclusions can be drawn from the numerical models.

- Localization of deformation is an important characteristic of a fractured rock mass, and occurs when a critical stress state is reached, at which the rock mass becomes mechanically unstable. Prior to the critical stress state, displacement and strain within the fractured rock mass are fairly homogenously distributed.
- At the critical stress state, large displacements accompany yield of the rock mass. Near this point considerable localization of deformation and fluid flow occur, and the rock mass as a whole becomes more permeable.
- The critical stress can be expressed in terms of the driving stress ratio R_C, which links the differential stress, mean stress and fluid pressure (equation 5). For typical values of coefficients of sliding friction in rocks, R_C has a value between −1 and −2 for a wide range of fracture geometries examined.
- Fracture density has an important effect on the R_C, which decreases with increasing fracture density. In the models examined, for fracture density ranging from 4.3 and 11.6 m^{-1}, R_C decreases from −1.08 to −1.8 (Fig. 10a).
- At a given fracture density, R_C decreases with increasing fracture size. For average fracture lengths ranging from 0.61 and 2.11 m, at a density of 7.85 m^{-1}, R_C decreases from −1.34 to −1.73 (Fig. 10b).

- Approaching the critical state, multifractal analysis indicates that the high-order fractal dimensions are much lower. This is consistent with a considerable localization (or clustering) of the flow. In networks with a higher fracture density, larger fracture size and greater anisotropy this is even more pronounced.

Clearly, fractured rock masses exhibit very complex behaviour as the critical stress state is approached. This study has only focused on some of the important geometrical aspects in well-connected fracture networks. The direction of principal stresses is another important factor controlling the deformation (see Sanderson & Zhang 1999), especially for those fracture networks with a high anisotropy in geometry. The geomechanical properties of the fractures and rock blocks may also contribute significantly to the deformation and, hence, fluid flow. However, the numerical models presented here demonstrate that deformation and fluid flow in fractured rock has a highly non-linear dependence on stress, exhibiting critical-point behaviour. Fracture-network geometry plays a significant role in controlling rock-mass behaviour, hence, improved characterization of fracture-network geometry is essential for understanding deformation and fluid flow.

This work is part of a project supported by the NERC Micro-to-Macro Programme (grant GST/03/2311). The authors wish to thank R. Harkness and the Geomechanics Research Group at the University of Southampton for continued support, and A. Barker for advice at various stages of the project. Constructive reviews of an earlier draft by S. Zhang and G. Couples helped improve the paper.

References

BARTON, C. A., ZOBACK, M. D. & MOOS, D. 1995. Fluid flow along potentially active faults in crystalline rock. *Geology*, **23**, 683–686.

BARTON, C. C. & HSIEH, P. A. 1989. Physical and hydraulic-flow properties of fractures. *In*: *28th International Geological Congress*, T385. American Geophysical Union, Fieldtrip Guidebook.

BERKOWITZ, B. & HADAD, A. 1997. Fractal and multifractal measurements of natural and synthetic fracture networks. *Journal of Geophysical Research*, **101**, 12005–12218.

BYERLEE, J. D. 1978. Friction of rocks. *Pure and Applied Geophysics*, **116**, 615–626.

COX, S. F. 1999. Deformational controls on the dynamics of fluid flow in mesothermal gold systems. *In*: MCCAFFREY, K. J. W., LONERGAN, L. & WILKINSON, J.J. (eds) *Fractures, Fluid Flow and Mineralization*. Geological Society, London, Special Publications, **155**, 123–140.

COWIE, P. A., SORNETTE, D. & VANNESTE, C. 1995. Multifractal scaling properties of a growing fault population. *Journal of Geophysics International*, **122**, 457–469.

CUNDALL, P. A., MARTI, J., BERESFORD, P., LAST, N. C. & ASGAIN, M. 1978. *Computer modelling of jointed rock masses*. US Army Engineers, WEST Technical Report, **N-74–8**.

DAS, S. & SCHOLZ, C. H. 1981. Theory of time-dependent rupture of the earth. *Journal of Geophysical Research*, **86**, 6039–6051.

DELANEY, P. T., POLLARD, D. D., ZIONEY, L. I. & MCKEE, E.H. 1986. Field relations between dikes and joints: emplacement processes and paleostress analysis. *Journal of Geophysical Research*, **91**, 4920–4938.

HEALY, J. H., RUBEY, W. W., GRIGGS, D. T. & RALEIGH, C.B. 1968. The Denver earthquakes. *Science*, **161**, 1301–1310.

HEFFER, K. J. & BEVAN, T.G. 1990. Scaling relationship in natural fractures: Data, theory and application. *Society of Petroleum Engineers*, **SPE20981**, 367–379

JACKSON, P. & SANDERSON, D. J. 1992. Scaling of fault displacements from the Badajoz–Cordoba shear zone, SW Spain. *Tectonophysics*, **210**, 179–190.

JOLLY, R. J. H. & SANDERSON, D. J. 1995. Variation in the form and distribution of dykes in the Mull swarm, Scotland. *Journal of Structural Geology*, **17**, 1543–1557.

JOLLY, R. J. H. & SANDERSON, D.J. 1997. A Mohr circle construction for the opening of a pre-existing fracture. *Journal of Structural Geology*, **19**, 887–892.

PEACOCK, D. C. P. & SANDERSON, D. J. 1995. Pull-aparts, shear fractures and pressure solution. *Tectonophysics*, **241**, 1–13.

RALEIGH, C. B., HEALY, J. H. & BREDEHOEFT, J. D. 1976. An experiment in earthquake control at Rangely, Colorado. *Science*, **191**, 1230–1237.

SABERSKY, R. H., ACOSTA, A. J. & HAUPTMANN, E. G. 1989. *Fluid Flow*. Macmillan, New York.

SANDERSON, D. J. & ZHANG, X. 1999. Critical stress localisation of flow associated with deformation of well-fractured rock masses, with implications for mineral deposits. *In*: MCCAFFREY, K. J. W., LONERGAN, L. & WILKINSON, J. J. (eds) *Fractures, Fluid Flow and Mineralization*. Geological Society, London, Special Publications, **155**, 69–81.

SCHROEDER, M. 1991. *Fractal, Chaos, Power Laws*. Freeman, New York.

SEGALL, S. W. & POLLARD, D. D. 1983. Joint formation in granite rock of the Sierra Nevada. *Geological Society of America Bulletin*, **94**, 563–575.

SIBSON, R. H. 1989. Earthquake faulting as a structural process. *Journal of Structural Geology*, **11**, 1–14.

SNOW, D. T. 1968. Rock fracture spacings, openings and porosities. *Journal of Soil Mechanics Foundation. Division, ASCE*, **94** (SM1), 73–91.

STAUFFER, D. & AHARONY, A. 1991. *Introduction to Percolation Theory*. Taylor & Francis, London.

TALWANI, P. & ACREE, S. 1985. Pore pressure diffusion and the mechanism of reservoir-induced seismicity. *Pure and Applied Geophysics*, **122**, 947–965.

ZHANG, X. & SANDERSON, D. J. 1994. Fractal structure and deformation of fractured rock masses. *In*: KRUHL, J., (ed.) *Fractal and Dynamic Systems in Geoscience*. Springer, Basel, 37–52.

ZHANG, X. & SANDERSON, D. J. 1995. Anisotropic features of geometry and permeability in fractured rock masses. *Engineering Geology*, **40**, 65–75.

ZHANG, X. & SANDERSON, D. J. 1996. Numerical modelling of the effects of fault slip on fluid flow around extensional faults. *Journal of Structural Geology*, **18**, 109–119.

ZHANG, X. & SANDERSON, D. J. 1997. Effects of loading direction on localized flow in fractured rocks. *In*: YUAN, J.-X. (ed.) *Computer Methods and Advances in Geomechanics*. A. A. Balkema, Rotterdam, 1027–1032.

ZHANG, X. & SANDERSON, D. J. 1998. Numerical study of critical behaviour of deformation and permeability of fractured rock masses. *Marine and Petroleum Geology*, **15**, 535–548.

ZHANG, X. & SANDERSON, D. J. 2001. Evaluation of instability in fractured rock masses using numerical analysis methods: effects of fracture geometry and loading direction. *Journal of Geophysical Research*, **106**, 26671–26687.

ZHANG, X. & SANDERSON, D. J. 2002. *Numerical Modelling and Analysis of Fluid Flow and Deformation of Fractured Rock Masses*. Pergamon, Oxford.

Indentation pits: a product of incipient slip on joints with a mesotopography

TERRY ENGELDER[1], KAREL SCHULMANN[2] & ONDREJ LEXA[2]

[1] Department of Geosciences, Pennsylvania State University, University Park, PA 16802, USA
(e-mail: engelder@geosc.psu.edu)
[2] Institute of Petrology and Structural Geology, Universita Karlova, Alberta 6, Praha 128 43,
Czech Republic

Abstract: The mechanism for structural damage during incipient slip on joints within the Melechov Granite, Czech Republic, changes with the misalignment of the joint's mesotopography, largely a plumose surface morphology. Prior to slip, the joint surfaces are well mated so that contact area is organized on a microscopic scale. During the first phase of slip, diffusion-mass transfer is the active deformation mechanism between the sliding surfaces of the joints, as indicated by the extensive growth of crystal-fibre lineations characteristic of slickenside surfaces. After slip of the order of 1 cm or more, the mesotopography becomes mismatched and the contact area is reorganized to form indentation pits aligned on the ridges of hackle plumes. Indentation pits, that are testimony to a brittle process, are generated by the excavation of Hertzian ring cracks that propagate under contact loading of a brittle substrate. The depth of the indentation pits increases with contact width, suggesting that indentation creep is active. Following indentation along Hertzian ring cracks the slip mechanism transforms to a frictional abrasion. The distribution of indentation track lengths is consistent with laboratory wear grooves generated during earthquake-like stick–slip sliding. The elliptical shape of the indentation pits indicates a gradual decrease in contact area, a process that is consistent with a slip-weakening mechanism during a stick–slip cycle.

Joints appear as early structures within many tectonic settings, in igneous rocks of convergent margins as well as in sedimentary rocks of extensional basins (e.g. Pollard & Aydin 1988). Because joints form early in the tectonic cycle, one presumes that there should have been ample opportunity for a shear traction to drive slip and subsequent overprinting by any of a number of processes reflecting frictional wear. Yet, joints persist even when host rocks have been subject to a complex tectonic history (Nickelsen 1979; Gray & Mitra 1993). In some tectonically complex areas, slip is difficult to detect even when it is clear that the joints were subject to a shear traction as a consequence of changes in the orientation of the remote stress field (Younes & Engelder 1999). On occasion, when joints do slip, there is no evidence for frictional wear on these incipient faults (Engelder *et al.* 2001; Silliphant *et al.* 2002). In other instances, frictional wear during slip on joints is seen in its most common manifestation, the slickenside surface (Hancock 1985). When joints of several sets are decorated with slickensides, they serve as a basis for fault-slip analyses to determine the orientation the regional stress field (Angelier 1979).

Frictional wear during the formation of lineations on slickenside surfaces includes both brittle and ductile processes (Wilson & Will 1990). Lineations on slickenside surfaces may consist of scratches, commonly called wear grooves, wear tracks or tool marks

(e.g. Engelder 1974; Fleuty 1975; Hancock 1985; Doblas 1998). These, along with streaks and trails, are a manifestation of brittle wear processes (e.g. Tjia 1967). Ductile wear shows up as the removal of steps and minor elevations by pressure solution to form slickolites (Arthaud & Mattauer 1972; Davis & Reynolds 1996). In other instances, ductile wear leads to the formation of striations that more closely resemble experimentally deformed paraffin wax (Means 1987). There are also examples reported of crystal fibres growing to form lineations on the leeward side of small steps and asymmetric elevations (Durney & Ramsay 1973). The small steps and asymmetric elevations tend to cause fracture dilation upon slip and thus indicate frictional contact (Petit 1987).

Despite the large population of joints present during the tectonic cycle, instances of a combination of brittle and ductile wear processes operating either simultaneously or serially during slip on joint surfaces are either uncommon or difficult to recognize. There are several reasons for this result. First, a multiple-step tectonic process is required where the generation of an initial joint is followed by the reorientation of the stress field to subject that joint to a shear traction (Engelder *et al.* 2001). Stress field reorientation over geological time may be less common than presumed. Second, it does not take much slip before all evidence of the initial joint surface is removed by fault-related abrasion and the

From: COSGROVE, J. W. & ENGELDER, T. (eds) 2004. *The Initiation, Propagation, and Arrest of Joints and Other Fractures.* Geological Society, London, Special Publications, **231**, 315–324. 0305-8719/04/$15 © The Geological Society of London 2004.

Moldanubian Plutons

Fig. 1. Tectonic map of the Bohemian Massif showing the location of the Melechov Granite.

generation of fault gouge (Scholz 1987). Third, many faults start as en echelon cracks, hence bypassing the need for stress realignment after joint propagation (e.g. Martel *et al.* 1988). The purpose of this chapter is to describe the transition from ductile slip to abrasional wear on joints that have become incipient faults. The evolution of slickensides on these joint surfaces is of particular interest because the development of indentation pits provide clues about the evolution of rock friction during fault slip.

Rock friction and the mechanism for sliding of rock in the brittle regime is largely a reflection of the behaviour of contacts during abrasional wear. One model for frictional slip, particularly the earthquake-generating stick–slip mechanism, focuses on the locking and breaking of contacts during sliding (Byerlee 1967). Stick–slip requires that stationary contacts create a higher friction than that present during fault slip (Rabinowicz 1958). Stationary contacts develop a higher friction by a time-dependent behaviour arising from static fatigue under these contacts (Dieterich 1972; Scholz & Engelder 1976). An early explanation for stick–slip focuses on a model for which frictional contacts weaken with slip (Byerlee 1970). However, stick–slip oscillations are best explained if incipient slip during contact rupture is a velocity-weakening process (i.e. Ruina, 1983; Scholz 1998, 2002). Theories for stick–slip by velocity weakening are best tested in the laboratory using relatively clean joint surfaces where frictional abrasion is minimal (Marone 1998). Appropriate field examples demonstrating contact behaviour during stick–slip are far less common. While inferences from field observations are restricted to slip weakening, they are nevertheless instructive for understanding the evolution of friction in nature. An opportunity for a case study of natural contacts is found on joints that have slipped a small amount (<10 cm) within a granite of the Bohemian Massif, in the Czech Republic.

The geology of the Melechov Massif

The Bohemian Massif encompasses a large suite of outcrops within the Variscan (Devonian–Carboniferous) Orogen of Europe (Schulmann *et al.* 1994). Within the massif are NE-SW-trending Neo-Proterozoic blocks surrounding a high-grade orogenic root domain in the centre – the Moldanubian zone. The central part of the Moldanubian zone is intruded by a composite crustal batholith, the Central Moldanubian Pluton (Fig. 1). The Melechov Massif,

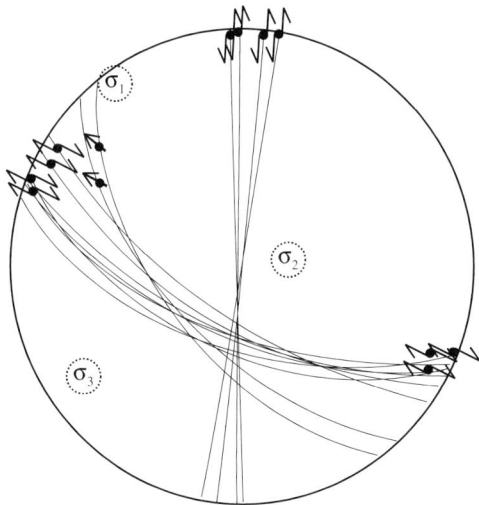

Fig. 2. Lower-hemisphere projection of joints within the Dolni Brezinka quarry of the Melechov Granite. Orientation of the principal stresses is based on the inverse method for fault-slip data.

Fig. 3. A joint surface within the Melechov Granite at the Dolni Brezinka quarry, Svetla, Czech Republic (joint surface A). Tracks of indentation pits can be seen running subvertically on this joint face (white arrows). This joint dips toward the camera (WSW) at 72°. Slip on the joint is subhorizontal with a dextral sense.

the subject of this study, is composed of a complex of granitic bodies separated from the northern edge of the Central Moldanubian Pluton by metamorphic rocks of the so-called monotonous series of the Moldanubian zone. The marginal parts of the Melechov Massif are built of fine- to medium-grained granites. The age of the granite established by the Rb–Sr method gave 303 ± 6 Ma (Scharbert & Vesela 1990). A fine- to medium-grained two-mica (biotite–muscovite) granite forms the rim of the entire massif. The central part of the massif is built of coarse-grained to porphyritic two-mica (biotite–muscovite) granite of the Melechov type. This pluton is of elliptical shape, elongated in the NNE–SSW direction.

The entire massif exhibits a concentric zoning with corresponding orientation of foliation in adjacent metamorphic rocks. Granites of the Melechov Massif to the west are enveloped by the rocks consisting of biotite and sillimanite–biotite paragneisses with cordierite, locally migmatized and containing bodies of marbles, calc-silicate rocks, amphibolites and quartzites. The northern and southern mantle of the massif is built of rocks of the Moldanubian monotonous series.

Fractures in the Melechov Granite

The Melechov Granite at the Dolni Brezinka quarry contains several fracture sets with two prominent sets, one striking at approximately 118° and dipping about 72° SSW and the other vertical set striking 5° (Fig. 2). Fibre lineations on the 118°-striking fracture set have

a small rake ($<5°$) and steps on the fibres indicate slip of the hanging wall to the WNW (i.e. top or missing half of the joint surface in Figs 3 and 4 moved to the left giving a dextral sense of slip). The 005°-striking fracture set bears fibre lineation with a small rake ($<5°$) indicating sinistral displacement. Estimated principal maximum compressive stress calculated using an inverse method is subhorizontal and oriented NW–SE (Fig. 2).

The 118°-striking fractures are planar, parallel features, a pattern that is consistent with a joint set. Surfaces of these joints are decorated with brittle indentation pits or cavities a few centimetres in diameter and generally less than 0.5 centimetres deep (Fig. 3). Rather than being uniformly scattered on the joint surface, as is common for most fault surfaces displaying tool marks (i.e. Petit 1987), these indentation pits are aligned in a series of gently curving, concentric paths. Their concentric arrangement makes a pattern much like either the plumose morphology seen on joint surfaces in sedimentary rocks (Woodworth, 1896) or rib marks seen on the surfaces of joints found in granites elsewhere within the Czech Republic (i.e. Bankwitz & Bankwitz 1984; Bahat *et al.* 2003). Topography on plumose morphology is referred to as plume barbs (Bahat 1991) or hackle plumes (i.e. Kulander & Dean 1985). Because the tracks of indentation pits radiate along irregular paths rather than forming a series of concentric rings, we favour the hackle plume (i.e. plume barb) interpretation (Fig. 3).

The indentation pits follow along a mesotopography on the joint surface (Fig. 4). We used the term

Fig. 4. Indentation pits within joint surface B in the Melechov Granite at the Dolni Brezinka quarry, Svetla, Czech Republic. The pits are elongated parallel to the slip lineation. A mesotopography (i.e. a surface roughness) in the form of a series of ridges sweeps to the upper left from the bottom of the photograph. The Czech coin is 23 mm in diameter.

mesotopography (amplitude on the order of 5 mm) to distinguish it from the microtopography (amplitude on the order of 10^{-2} mm) that may be found at the grain scale on a fresh joint surface. Microtopography may also have a grain that gives rise to a very fine plume pattern on a joint surface in sandstone (Bahat & Engelder 1984). This mesotopography has a trough to peak elevation of a few millimetres, as is common for the plumose morphology on other granites (Bankwitz & Bankwitz 1984).

Indentation pits

Indentation trails are best developed where the mesotopography is at a high angle to the slip lineations (Fig. 3). On some joints most indentation pits are approximately circular with a depth that often exceeds the amplitude of the mesotopography on the joint surfaces (Figs 5 and 6). On other joints indentation pits are elliptical with their long axis aligned parallel to crystal fibre lineations characteristic of slickensided surfaces (Fig. 7). Because the long axes of the elliptical indentations parallel the slickenside lineations, there is little doubt that the indentation pits are tool marks produced by frictional abrasion on the joint surface (Doblas 1998). Many of the indentation pits contain tension cracks that have the characteristic of chatter marks left by tools on fault surfaces or along glacially carved outcrops (Willis & Willis 1934; Tija 1967) (Figs 6 and 7). Such chatter marks are the manifestation of Hertzian ring cracks generated during asperity indentation (Lawn 1993). The deepest part of the pit is often at the back end of the pit (i.e. the right-hand side of the pits shown in Figs 6 and 7). Hence, the deepest portion of the pit is

located at the trailing edge of the asperity that is responsible for the pit and concomitant elliptical groove, if present.

Three data were collected when documenting the geometry of individual indentation pits on three 118°-striking joints surfaces: length parallel to the crystal fiber lineation, width normal to the crystal fibre lineation and depth of the pit (Fig. 5). There are no offset markers on the joint surface so we have no independent indication of the magnitude of slip along these three joints. We assume that total slip distance is proportional to the sum of the length of the crystal fibre growth and the degree of ellipticity of the indentation pits. Those joints with more elliptical indentations are presumed to reflect a greater total slip distance.

The indentation pits show a gradual progression in excavation from a more or less circular hole to an elliptical cavity with a length (i.e. the dimension in the slip direction) nearly three times the width of the indentation. Each of the joints shows enough slip to have developed crystal fibre lineations, but for one the mode of the length/width ratio for various indentation pits was close to 1 (Fig. 8a). With the joint showing a mode close to 1, there is a range of length/width ratios for the indentation pits including a number of pits with a value of less than 1. We presume that this scatter in the data indicates that the contact areas responsible for the indentation pits were irregular in shape, with some having a longer dimension in a direction normal to the slip lineation. The fact that the mode is not centred on 1 but rather shifted to a value slightly higher than 1 indicates ongoing slip with the incipient excavation of a groove in the direction of slip. Of course, as slip progresses the mode for the length/width data increases from

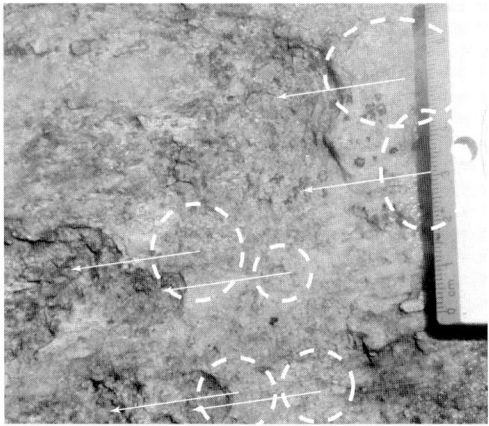

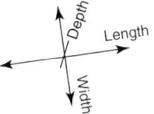

Fig. 5. Indentation pits on joint surface A. Arrows parallel the slip lineation as indicated by the orientation of crystal fibre growth. The arrows indicate the direction and distance of motion of the top (forward) surface during development of the crystal fibre growth. Hence, slip on the joint was subhorizontal with a dextral sense.

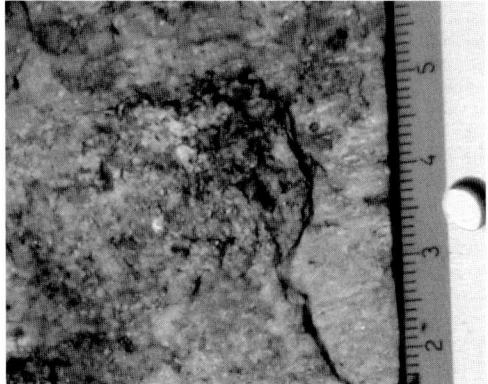

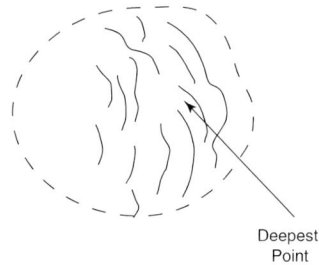

Fig. 6. Indentation pit on joint surface B. The drawing of the pit shows the location of Hertzian ring cracks in the back (bottom) block that developed as the top (forward) block moved to the left.

1.1 to 1.3 to 1.5, respectively (Figs 8a, b and c). By the time slip has generated a length/width mode of 1.5 there are no indentation pits with a length/width ratio <1. At this point excavation during slip has compensated for any irregularities in the shapes of the contact area.

Depth of excavation is a function of the size of the contact area, as indicated by the correlation between depth and width of the indentation pit (Fig. 9). Although the exact timing (i.e. syn-slip v. post-slip) for excavation of the indentation pits is unclear, the present depth of the indentation pits correlates with the depth of penetration of the initial ring cracks. Depth of excavation is, however, not a function of the amount of slip or length of the cut made by the contact, as indicated by the fact that pit depth is not a function of the length of the indentation pit but correlates very nicely with the width of the pit regardless of its ellipticity. Pit excavation can also be viewed in terms of a plot of length against width (Fig. 10): initial excavation plots with a slope of 1 (i.e. surface A). Of course, as excavation continues with slip, length increases without concomitant increase in either width or depth (i.e. surfaces C). This is somewhat contrary to the conventional view of tool marks that become progressively deeper as an asperity is dragged through a substrate (i.e. Engelder 1974; Doblas 1998).

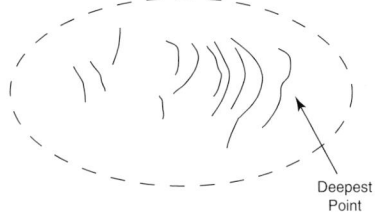

Fig. 7. Indentation pits on joint surface C. The drawing of the pit shows the location of Hertzian ring cracks in the back (bottom) block that developed as the top (forward) block moved to the left.

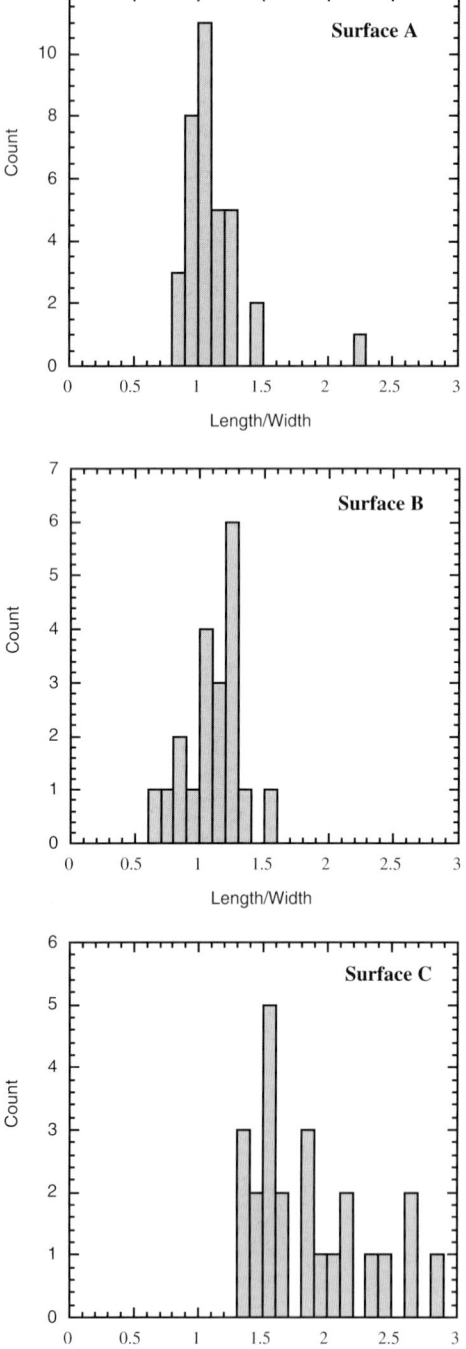

Fig. 8. Histogram for the ratio of length/width of the indentation pits in a joint-normal view of joint surfaces A, B and C, respectively. Length is defined as the dimension of the pit parallel to the slickenside lineation and width is the dimension normal to the slickenside lineation.

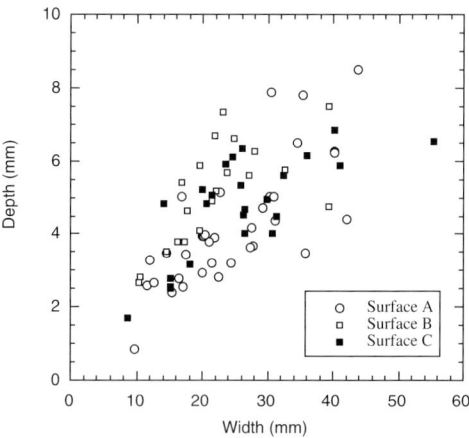

Fig. 9. Depth as a function of width of indentation pits on joint surfaces A, B and C, respectively.

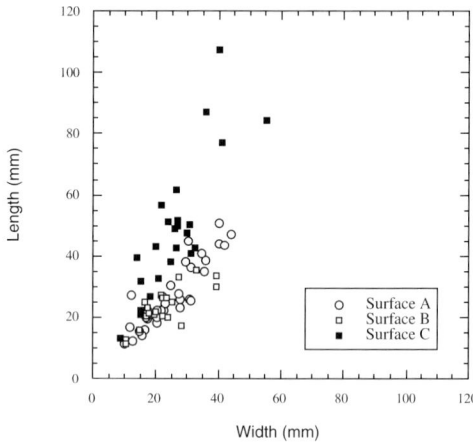

Fig. 10. Length as a function of width of indentation pits on joint surfaces A, B, and C respectively.

Discussion

Transition from ductile slip to abrasional wear

The indention pits cut off crystal fibre lineation, and crystal fibre lineation is never seen growing within indentation pits. This is direct evidence in favour of a transition in slip mechanism from ductile creep to abrasional wear. Initially, the joint surfaces are well matched so that the mesotopography associated with plume morphology is well mated. Contact points are small, probably on the grain scale or smaller, so that microtopography controls the slip mechanism. At this stage, crystal fibre growth accompanied slip.

Hence, we infer that as long as the joint surfaces were in reasonably good contact without contact stress being focused on the mesoscopic scale, creep by diffusion-mass transfer mechanism was responsible for ductile 'fault' slip. It may be that at the microscopic scale, point contact stress favoured pressure solution rather than brittle indentation. Abrasional wear by brittle indentation starts after some finite amount of 'fault' slip. The length of the crystal fibre lineation indicates that this slip may have been of the order of 1–3 cm (Fig. 5). At this stage during slip, surface contact reorganizes to support occasional mesoscopic contact areas.

Brittle indentation requires the appropriate contact area. In the literature on slickenside surfaces, such contact area comes from structures known as steps (Hancock 1985), slickenside roches moutonnées (Tjia 1967) or knobby elevations (Doblas 1998). Enough of the morphology of the original joints is visible in the Melechov Granite to infer that the tops of the plumose mesotopography on the joint surface account for the distribution of contact points. In particular, it is clear that once the joints had slipped 1–3 cm (i.e. the wavelength of the mesotopography on the joint surface) there is sufficient mismatch in the shape of the surfaces so that contact area was reorganized from the initial condition presented by the initially well-fitted joint surfaces.

In summary, crystal fibre lineation provides direct evidence that the joint surfaces slipped by a ductile mechanism prior to the generation of indentation pits. After slipping, the joints no longer fit in the nearly perfect match that would have been present just after joint propagation.

Brittle indentation

Once the mismatch between surfaces becomes sufficient, contact is localized. At this stage indentation pits seem to have originated under static contact points by Hertzian intentation (Lawn 1993). Cracks are driven into the substrate at the edge of local contact points. This is certainly consistent with the trailing end of the indentation pits being deepest following the initiation of slip. This heavy fracturing may have allowed for later excavation of the indentation pits by rapid erosion of the cracked contact area after the removal of the hanging-wall block. The indentation pits grow under an asperity indentation mechanism called indentation creep (Westbrook & Jorgensen 1968). Indention creep has the time-dependent effect that allows an increased penetration depth with time of loading (Scholz & Engelder 1976). As the surfaces close by indention creep, more points along the surface come in contact, leading to the linear distribution of indendation pit depths as a function of width (Fig. 9).

The size and shape of the contact area (i.e. the tops of the mesotopography on the plumose morphology) are reflected in the geometry of the indentation pits. The size of the contact areas was not uniform. The areas and vertical depth of the indentation pits scale with each other, and their size conforms with the mesotopography of the wall that has been removed. Finally, the mesoscopic contact points are, on average, initially circular in map view.

The indentation pits on two of the three joint surfaces (i.e. A and B) have a length/width ratio close to unity (Fig. 10). This behaviour suggests that once indentation creep was initiated following 1–3 cm of slip as indicated by the growth of fibre lineation, the joint surfaces locked before further slip. Indentation creep has the effect of raising the frictional resistance to slip. On surface C, additional slip is by an abrasional wear mechanism with indentation pits leaving a track of ring cracks (Fig. 7).

The elliptical indentation pits on joints of the Melechov Granite are similar to a group of tool marks called 'V' or crescentic markings (Doblas 1998). Commonly, the deepest part of the marking is found at the trailing edge of the excavation tool. These most closely resemble gouging/plucking markings, except that the excavation pits contain chatter marks and they tend not to be carrot-shaped features. Laboratory sliding friction experiments typically show carrot-shaped wear grooves, but the sharp end of these grooves point in the direction of motion of the surface in which the grooves lie (Engelder & Scholz 1976). This means that experimentally produced grooves get deeper at the leading edge of the excavation tool, whereas in this natural example the deepest edge of the pit is found at the trailing edge. Because of their shape, the elliptical excavation pits on joints in the Melechov Granite are thought to reflect slip weaking as frictional slip is reinitiated after a period of stationary contact.

Significance relative to stick–slip and microearthquake generation

To better understand the significance of the indentation pits on the joints of the Melechov Granite, we turn to the laboratory experiments where tool marks have been produced (e.g. Engelder 1974, 1976). Tool marks are produced on highly polished Westerly Granite when sliding takes place in compression above 30 MPa confining pressure on surfaces inclined at 35° to the cylinderical axis. Often slip is by earthquake-like stick–slip where tool marks are equal to or less than slip during individual stick–slip events. The normalized length distribution of tool marks in feldspar in the Westerly Granite experiments shows that the mode for these data is

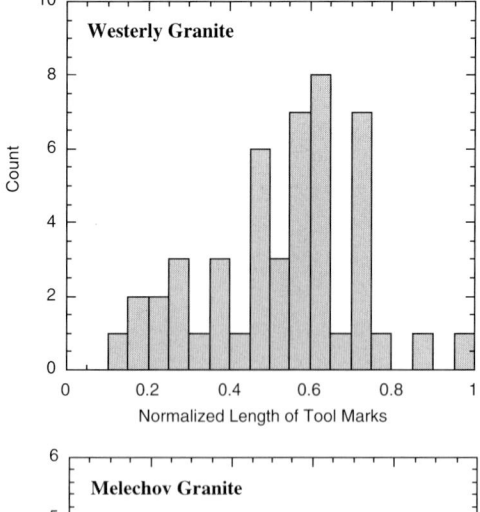

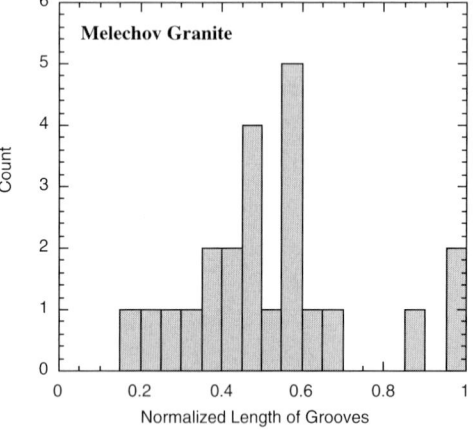

Fig. 11. Normalized length of tool marks developed during frictional wear on a polished surface of Westerly Granite (slip normalized by longest groove). Measurements taken on the large quartz grain in the upper portion of figure 5 in Engelder (1976). The normalized length of grooves developed during frictional wear on joint surface C in the Melechov Granite.

approximately the length of the slip event (Fig. 11) (see fig. 6 in Engelder 1976). This is true for both quartz cutting into feldspar and feldspar smears on quartz (see figs 7 and 8 in Engelder 1976). The length distribution data gradually increase to the mode and then fall off abruptly. In the experiments there are some tool marks that are longer than individual slip events, perhaps as an indication that some tools survive through more than one stick–slip event.

A plot of the normalized groove lengths on joint surface C of the Melechov Granite shows a similar distribution as that seen on the Westerly Granite (Fig. 11). This comparison ties the mechanism for frictional slip under brittle conditions in the field to

the mechanism for frictional slip in the laboratory. One interpretation is that tool marks on surface C in the Melechov Granite were produced during one earthquake-like slip event. If this were the case, then we are looking at a surface that slipped stably as a ductile fault to produce crystal fibre lineations during early slip. This early stage of slip produces the mismatch of the joint surface and concomitant reorganization of contact area for the initiation of indention creep. During further evolution, joints in the Melechov Granite slipped by a very different mechanism, a brittle indentation. Regardless, the parallelism between the slip lineation and the long axis of the tool marks suggests that the orientation of the critically resolved shear stress did not change during the switch from sliding by ductile creep to brittle wear. Then, further slip on surface C was accompanied by a gradual decrease in width and depth of penetration of Hertzian cracks. This means that the contact area was gradually decreasing as slip reinitiated. Although the detailed reasons for the decrease in contact area are unknown, the decreasing depth of penetration is not consistent with a concomitant higher normal stress under contact areas. Hence, the frictional force decreases with additional slip, a characteristic that is consistent with a slip-weakening model. The slip distance during slip weakening is approximately 2.5 cm, as indicated by subtracting the width of the indentation pits on surface C from their length.

The Melechov joints may be a very small-scale model for the creeping portions of larger-scale crustal faults. Portions of the San Andreas fault zone are known to produce most of its slip aseismically while generating large numbers of microearthquakes that occur in streaks (e.g. Rubin *et al.* 1999). While not aligned in the direction of slip, it is clear that brittle slip on the Melechov joints is locally concentrated, particularly if each indentation pit is considered the hypocentre of a microearthquake. The analogy with a creeping fault is further strengthened by the well-developed fibre lineation that grows during aseismically slippage.

Conclusions

Joints in the Melechov Granite, in the Czech Republic, contain clear evidence for a transformation in slip mechanism during initial rock sliding. Early in the slip history of these joints, slip was by a diffusion-mass transfer that led to the growth of a crystal fibre lineation. This initial slip caused well-mated joint surfaces to become misaligned. With the misalignment, the contact area reorganized. With a reorganized contact area, indentation creep was initiation by penetration of Hertzian ring fractures, a brittle mechanism. Indentation creep lead to the generation of indentation pits that follow the meosto-pography of a plumose morphology on the joint surfaces. Further slip is accompanied by indentation creep under the reorganized contact points. The subsequent wear grooves are elliptical indention pits indicating slip of about 2.5 cm with the deep end of the indention pit near the trailing end of the pit, as would be expected for a slip-weakening friction model for stick–slip.

We thank V. Dvorakova and C. Marone for discussing ideas in this manuscript, and D. Bahat for reviewing an early version of this paper. The field work was supported with funds from the Pennsylvania State Seal Evaluation Consortium (SEC).

References

ANGELIER, J. 1979. Determination of the mean principal stress from a given fault population. *Tectonophysics*, **56**, T17–T26.

ARTHAUD, R. & MATTAUER, M. 1972. Sur l'origine tectonique de certains joints stylolitiques parallèles a la stratification; leur relation avec une phase de destension (exemple du Languedoc). *Bulletin de la Societe Geologique de France, Serie 7*, **XIV**, 12–17.

BAHAT, D. 1991. *Tectonofractography*. Elsevier, Amsterdam.

BAHAT, D. & ENGELDER, T. 1984. Surface morphology on cross-fold joints of the Appalachian Plateau, New York and Pennsylvania. *Tectonophysics*, **104**, 299–313.

BAHAT, D., BANKWITZ, P. & BANKWITZ, E. 2003. Preuplift joints in granites: Evidence for subcritical and post-critical fracture growth. *Geological Society of America Bulletin*, **115**, 148–165.

BANKWITZ, P. & BANKWITZ, E. 1984. Die Symmetrie von Kluftoberflächen und ihre Nutzung für eine Paläospannungsanalyse. *Zeitschrift für Geologische Wissenschaften*, **12**, 305–334.

BYERLEE, J. D. 1967. Theory of friction based on brittle fracture. *Journal of Applied Physics*, **38**, 2928–2934.

BYERLEE, J. D. 1970. The mechanics of stick-slip. *Tectonophysics*, **9**, 475–486.

DAVIS, G. H. & REYNOLDS, S. J. 1996. *Structural Geology of Rocks and Regions*. Wiley, New York.

DIETERICH, J. H. 1972. Time-dependent friction in rocks. *Journal of Geophysical Research*, **77**, 3690–3697.

DOBLAS, M. 1998. Slickenside kinematic indicators. *Tectonophysics*, **295**, 187–197.

DURNEY, D. W. & RAMSAY, J. G. 1973. Incremental strains measured by syntectonic crystal growths. *In*: DEJONG, K. & SCHOLTEN, R. (eds) *Gravity and Tectonics*. Wiley, New York.

ENGELDER, J. T. 1974. Microscopic wear grooves on slickensides. Indicators of paleoseismicity. *Journal of Geophysical Research*, **79**, 4387–4392.

ENGELDER, J. T. 1976. Effect on scratch hardness on frictional wear and stick-slip on Westerly granite and Cheshire quartzite, *In*: STRENS R. (ed.) *The Physics and Chemistry of Rocks and Minerals*. Wiley, New York, 139–150.

ENGELDER, J. T. & SCHOLZ, C. H. 1976. The role of asperity indentation and ploughing in rock friction: II Influence of relative hardness and normal load. *International Journal of Rock Mechanics and Mining Science*, **13**, 155–163.

ENGELDER, T., HAITH, B. F. & YOUNES, A. 2001. Horizontal slip along Alleghanian joints of the Appalachian plateau: evidence showing that mild penetrative strain does little to change the pristine appearance of early joints. *Tectonophysics*, **336**, 31–41.

FLEUTY, M. J. 1975. Slickensides and slickenlines. *Geological Magazine*, **112**, 319–322.

GRAY, M. B. & MITRA, G. 1993. Migration of deformation fronts during progressive deformation: evidence from detailed structural studies in the Pennsylvania Anthracite region, U.S.A. *Journal of Structural Geology*, **15**, 435–450.

HANCOCK, P. L. 1985. Brittle microtectonics: principles and practice. *Journal of Structural Geology*, **7**, 437–457.

KULANDER, B. R. & DEAN, S. L. 1985. Hackle plume geometry and joint propagation dynamics. *In*: STEPHANSSON, O. (ed.) *Fundamentals of Rock Joints*. Centek, Luleå, Sweden, 85–94.

LAWN, B. 1993. *Fracture of Brittle Solids*, 2nd edn. Cambridge University Press, Cambridge.

MARONE, C. 1998. Laboratory-derived friction laws and their application to seismic faulting. *Annual Reviews of Earth and Planetery Science*, **26**, 643–696.

MARTEL, S. J., POLLARD, D. D. & SEGALL, P. 1988. Development of simple strike-slip fault zones, Mount Abbot Quadrangle, Sierra Nevada, California. *Geological Society of America Bulletin*, **100**, 1451–1465.

MEANS, W. D. 1987. A newly recognized type of slickenside striation. *Journal of Structural Geology*, **9**, 585–590.

NICKELSEN, R. P. 1979. Sequence of structural stages of the Allegheny orogeny at the Bear Valley Strip Mine, Shamokin, Pennsylvania. *American Journal of Science*, **279**, 225–271.

PETIT, J.-P. 1987. Criteria for sense of movement on fault surfaces in brittle rocks. *Journal of Structural Geology*, **9**, 597–608.

POLLARD, D. D. & AYDIN, A. 1988. Progress in understanding jointing over the past century. *Geological Society of America Bulletin*, **100**, 1181–1204.

RABINOWICZ, E. 1958. The intrinsic variables affecting the stick-slip process. *Proceedings of the Physical Society, London*, **71**, 668–675.

RUBIN, A. M., GILLARD, D. & GOT, J.-L. 1999. Streaks of microearthquakes along creeping faults. *Nature*, **400**, 635–641.

RUINA, A. 1983. Slip instability and state variable friction laws. *Journal of Geophysical Research*, **88**, 10359–10370.

SCHARBERT, S. & VESELA, M. 1990. Rb–Sr systematics of intrusive rocks from the Moldanubicum around Jihlava. *In*: MINARIKOVA, D. & LOBITZER, H. (eds), *Thirty Years of Geological Co-operation between Austria and Czechoslovakia*. Vydal Ustred. Ustrav Geol., Prague.

SCHOLZ, C. H. 1987. Wear and gouge formation in brittle faulting. *Geology*, **15**, 493–495.

SCHOLZ, C. H. 1998. Earthquakes and friction laws. *Nature*, **391**, 37–42.

SCHOLZ, C. H. 2002. *The Mechanics of Earthquakes and Faulting*, 2nd edn. Cambridge University Press, New York.

SCHOLZ, C. H. & ENGELDER, T. 1976. The role of asperity indentation and ploughing in rock friction: I Asperity creep and stick–slip. *International Journal of Rock Mechanics and Mining Science*, **13**, 149–154.

SCHULMANN, K., MELKA, R., LOBKOWICZ, M., LEDRU, P., LARDEAUX, J.-M. & AUTRAN, A. 1994. Contrasting styles of deformation during progressive nappe stacking at the southeastern margin of the Bohemian Massif (Thaya Dome). *Journal of Structural Geology*, **16**, 355–370.

SILLIPHANT, L. J., ENGELDER, T. & GROSS, M. R. 2002. The state of stress in the limb of the Split Mountain Anticline, Utah: constraints placed by transected joints. *Journal of Structural Geology*, **24**, 155–172.

TIJA, H. D. 1967. Sense of fault displacements. *Geologie en Mijnbouw*, **46**, 392–396.

WESTBROOK, J. H. & JORGENSEN, P. J. 1968. Effects of water desorption on indentation microhardness anisotropy in minerals. *American Mineralogist*, **53**, 1899.

WILLIS, R. & WILLIS, R. 1934. *Geological Structures*. McGraw-Hill, New York.

WILSON, C. J. L. & WILL, T. M. 1990. Slickenside lineations due to ductile processes. *In*: KNIPE, R. J., & RUTTER, E. H. (eds) *Deformation Mechanisms, Rheology and Tectonics*. Geological Society, London, Special Publications, **54**, 455–460.

WOODWORTH, J. E. 1896. On the fracture system of joints, with remarks on certain great fractures. *Boston Society of Natural History, Proceedings*, **27**, 163–183.

YOUNES, A. & ENGELDER, T. 1999. Fringe cracks: Key structures for the interpretation of progressive Alleghanian deformation of the Appalachian Plateau. *Geological Society of America Bulletin*, **111**, 219–239.

Index

Page numbers in italics, e.g., *153* refer to figures. Page numbers in **bold**, e.g. **173**, signify entries in tables.

330   INDEX